Aquaculture Perspective of Multi-Use Sites in the Open Ocean

Bela H. Buck · Richard Langan
Editors

Aquaculture Perspective of Multi-Use Sites in the Open Ocean

The Untapped Potential for Marine
Resources in the Anthropocene

 Springer Open

Editors
Bela H. Buck
Marine Aquaculture, Maritime Technologies
and ICZM
Alfred Wegener Institute Helmholtz Centre
for Polar and Marine Research (AWI)
Bremerhaven
Germany

Richard Langan
School of Marine Science and Ocean
Engineering
University of New Hampshire
Durham, NH
USA

ISBN 978-3-319-51157-3   ISBN 978-3-319-51159-7   (eBook)
DOI 10.1007/978-3-319-51159-7

Library of Congress Control Number: 2016960276

Cover illustration: Extractive Aquaculture in a German Wind Farm by Alexandra Haasbach.

Printed on acid-free paper

This Springer imprint is published by Springer Nature
The registered company is Springer International Publishing AG
The registered company address is: Gewerbestrasse 11, 6330 Cham, Switzerland

Preface

The Global Imperative to Develop New Models of Open Ocean Aquaculture for Accelerating Large-Scale Food and Energy Production

This edited volume "Aquaculture Perspective of Multi-Use Sites in the Open Ocean: The Untapped Potential for Marine Resources in the Anthropocene" comes at a critical time for our planet. A 2015 article in *Science* updated the long-range population projections of the United Nations in 1992. In contrast to the 1992 UN estimate, the Science paper showed no stabilization of the world's population by 2100, and that there was an 80% chance that the world's population, currently 7.2 billion, will reach 9.6 billion by 2050, and up to 12.3 billion in 2100. Much of the increases result from growth in Africa and Asia. Many of the nations on these two continents consume aquatic foods as their main sources of animal proteins. Accelerated demands for aquatic proteins in these regions will mean that in the future they will not export their products to Europe and North America anymore but use them in domestic markets.

Adding to a projected increased demand for aquatic foods is the recent call from the World Committee on Food Security that aquatic foods from fisheries and aquaculture be included in dietary planning to fight malnutrition in low income and food deficient nations (High Level Panel of Experts 2014).

Unprecedented—really shocking—ocean warming is causing the movement of temperate species north into the peri-Arctic seas and has led to the development of new fisheries, notably in the Barents Sea. For example, the Norwegian-Russian Fisheries Joint Commission allocations for cod and haddock for 2015 is 1,072,000 MT (Barents Observer 2015), nearly equal to the entire aquaculture production of the continent of Africa (FAO 2016). Alarmists of the late 2000s did overly exaggerate the global collapse of nearly all marine fisheries, and the downward trends of overfished stocks seem to be stabilizing. The recent models of Barange et al. (2014) show that by 2050 global capture fisheries will likely remain within +/- 10% of present global yields. But it is clear that the world's oceans and large lakes cannot take any more fishing pressure, and they cannot (and will not) produce any more aquatic foods for humanity, and that seawater aquaculture (mariculture) is the only solution to the world's food supply but also to avoid a major biodiversity crisis (Costa-Pierce 2016).

Most of the planet's population has decided that its primary habitat in the Anthropocene is on the world's coasts, there are fewer and fewer coastal options to develop coastal marine aquaculture. A *New York Times* series of articles on polluted seas a few years ago, see especially their article "In China, Farming Fish in Toxic Waters" (Barboza 2007) brings the challenge clearly to the world. In the world's "seafood nations" where aquatic foods are the most important to people, coastal oceans and large lakes are damaged from land-based pollutants and toxicants, making these most nutritious, nutrient-dense protein foods…toxic!

Global food security, human health and overall human welfare are in serious jeopardy as the production of living marine/aquatic resources can no longer be sustained by aquatic ecosystems and natural fisheries production. Expansion of agriculture cannot meet future these protein needs without massive impacts on forests, wildlife, nature reserves and parks (Zabel et al. 2014; Costa-Pierce 2016). The only way we can proceed as an educated species is to develop mariculture that does not harm capture fisheries, promote unsustainable agriculture, or damage the integrity of aquatic ecosystems.

The drivers for open ocean aquaculture and offshore energy production are not only food, trade, electricity, and technology. There are powerful social and ethical concerns afloat. In some sort of weird "food insanity", many Western nations import most of the seafood they eat, and export most of what they catch or produce. These nations are far too dependent on imports from aquaculture systems in nations where aquaculture is threatened by coastal urbanization, industrialization, water pollution, and overall environmental degradation. Such "food insane nations" also have a moral and ethical responsibility to develop large-scale open ocean aquaculture to feed their own people and not take these valuable foods from undernourished, food scarce nations. Similarly, a continued reliance on fossil fuels by developed and developing nations is causing massive global climate change, a path that is neither environmentally sound nor economically sustainable.

There are valuable examples in this book of the development of robust submerged technologies; but there remains an urgent need to accelerate the education and training of open ocean aquaculture engineers; and to open wide the "aquaculture toolbox" to greater amounts of innovation. Norway leads the way globally in this regard; and New Zealand too, with that country now having about 10,000 ha of permitted areas for open ocean shellfish farming.

The use of offshore wind farms as multi-use platforms appears to be especially promising. A chapter in this volume reports that a 10% reduction in of O&M costs is possible for wind farms if aquaculture was combined. As shown in Germany's pioneering efforts, such combined food-energy systems in the offshore requires a high level of research and development funding as not only the development of innovative technologies is required, but also the need for marine spatial planning, and transparent, adaptive management processes to ensure economies of scope and scale, for spatial efficiency and conflict resolution, and for the development of innovative policies and financial instruments.

There are few interdisciplinary departments or learned academic R&D centers focused on the multiple disciplines that intersect with open ocean aquaculture, and even fewer looking into multiple uses of offshore structures. Education and training networks are needed to provide the required multidisciplinary and interdisciplinary expertise for the safe and professional operations of multi-use systems. Aquaculture and energy production should align with planned ocean monitoring networks since these systems have significant potential to serve as oceanic environmental quality monitoring stations. Ecological design, engineering, and ecosystem based management approaches to develop open ocean aquaculture research and education innovation centers would produce design and performance optimizations that could potentially benefit all stakeholders plus increase research and development funding and boost the regional innovation economy. The Bremerhaven Declaration on Offshore Aquaculture clearly states that these priorities can only be met by the example of Germany who is funding an internationally important offshore experimental platform (Rosenthal et al. 2012).

Offshore aquaculture developments, alone or in conjunction with other uses will require much higher inputs of capital and new levels of cooperation from a wide range of social, technological, economic, and natural resource users, and a greater degree of cooperation and collaboration with other industry sectors. Transparent strategies need to be developed with the strong participation of all affected stakeholders interested in the social-ecological design and engineering of innovative offshore multi-use systems. In this context, multi-use systems if intelligently designed can be incorporated into programs for cooperative fisheries restoration and aquatic ecosystem management strategies. In this regard, multi-use systems that incorporate aquaculture development need to be guided by international policy instruments such as the FAO Code of Conduct for Responsible Fisheries (FAO 1995) and the FAO Guidelines for an Ecosystems Approach to Aquaculture (FAO 2010).

The Editors and authors of this volume are among a select group of pioneers who over the past 20+ years have studied, spoke about, and practiced open ocean aquaculture, and have recognized the many opportunities for combining uses of the ocean such as energy and food production. They have garnered wisdom from direct experiences, from many a day operating aquaculture systems on the rough waters of the Northwest Atlantic Ocean, the North Sea, and elsewhere. They have gathered in this volume a diverse group of interdisciplinary aquaculture scientists and engineers to discuss what they consider are the most important recent science and policy developments. Their views as thought leaders are globally important for the future of protein and nutrient-rich foods on Earth, not only for the future of open ocean aquaculture in combination with other offshore uses. It is our desire that this volume stimulate further research and development into the combined production of energy and seafood in open ocean environments. Success in this endeavor is paramount to the sustainability of systems critical to maintaining healthy oceans, meeting the nutritional needs of a growing world population, and to the very survival of the Earth's ecosystems good and services.

Barry A. Costa-Pierce
Department of Marine Sciences
University of New England (UNE)
Biddeford, ME, USA

Richard Langan
Judd Gregg Marine Research Complex
School of Marine Science and Ocean Engineering
University of New Hampshire, New Castle, NH, USA

References

Barange, M., Merino, G. Blanchard, J. L., Scholtens, J., Harle, J., Ellison, E. H. et al. (2014). Impacts of climate change on marine ecosystem production in fisheries-dependent societies. *Nature Climate Change, 4*, 211–216.

Barents Observer. (2015). *Norway and Russia agree on Barents Sea quotas for 2015*. Kirkenes, Norway. Retrieved October 20, 2016, from http://barentsobserver.com/en/nature/2014/10/norway-and-russia-agree-barents-sea-quotas-2015-10-10.

Costa-Pierce, B. A. (2016). Ocean foods ecosystems for planetary survival in the Anthropocene. In E. M. Binder (Ed.), *World nutrition forum: Driving the protein economy* (pp. 301–320). Austria: Erber, AG.

FAO. (1995). *Code of conduct for responsible fisheries*. Rome: FAO. Retrieved October 21, 2016, from http://www.fao.org/docrep/005/v9878e/v9878e00.HTM.

FAO. (2010). Aquaculture development. 4. Ecosystem approach to aquaculture. *FAO Technical Guidelines for Responsible Fisheries*, (no. 5, suppl. 4). Rome: FAO.

FAO. (2016). *The state of world fisheries and aquaculture 2016: Contributing to food security and nutrition for all*. Rome: FAO.

High Level Panel of Experts. (2014). Sustainable fisheries and aquaculture for food security and nutrition. In *A report by the high level panel of experts on food security and nutrition of the committee on world food security*. Rome: FAO.

Barboza, D. (2007). *In China, farming fish in Toxic Waters*. New York Times. Retrieved October 1, 2016, from http://www.nytimes.com/2007/12/15/world/asia/15fish.html?pagewanted=all&_r=0.

Rosenthal, H., Costa-Pierce, B. A., Krause, G., & Buck, B. H. (2012). Bremerhaven declaration on the future of global open Ocean Aquaculture. *Part I: Preamble and recommendations. Part II. Recommendations on subject areas and justifications*. Bremerhaven, Germany: BiS. Retrieved October 19, 2016, from http://www.bis-bremerhaven.de/sixcms/media.php/631/BremerhavenDeclaration3-Part1_05-2013_L02.pdf.

Zabel, F., Putzenlechner, B., & Mauser, W. (2014). *Global agricultural land resources—A high resolution suitability evaluation and its perspectives until 2100 under climate change conditions*. PLOS doi:10.1371/journal.pone.0107522

Contents

Editors and Contributors

About the Editors

Prof. Dr. Bela H. Buck studied neurophysiology and marine biology at the University of Bremen, at the Institute for Marine Research in Kiel and at the Leibniz Center for Tropical Marine Ecology (ZMT) in Bremen (all in Germany). In the years 1999/2000 he was involved in research projects concerning the aquaculture of giant clams at the Great Barrier Reef Marine Park Authority (GBRMPA), the James Cook University (JCU) and the Australian Institute for Marine Science (AIMS) in Townsville (all in Australia), in which he got is graduation as a marine biologist.

Since 2001 he has been engaged in projects regarding offshore aquaculture (especially as multifunctional use of offshore wind farms) at the Alfred Wegener Institute Helmholtz Centre for Polar and Marine Research (AWI) in Bremerhaven/Germany. He conducted his Ph.D. in 2001–2004 in various aspects of offshore aquaculture related to technology, biology, legislation and ICZM issues within the German Bight (grade of excellent/highest distinction). From 2005 to 2007 Dr. Buck was postdoc at the AWI and is the head of the working group "Marine Aquaculture, Maritime Technologies and ICZM".

He was responsible to establish the Institute for Marine Resources (IMARE), in which he was the head of the section "Marine Aquaculture" as well as the Vice Director.

In July 2007 he was given a professorship for "Applied Marine Biology" from the University of Applied Sciences in Bremerhaven including the right to award doctorates at the University of Bremen. From 2012 to 2015 he was the President of the German Aquaculture Association and is since then nominated as honorary president. Bela H. Buck is a member of the steering committee of the German Agricultural Research Alliance (DAFA).

Today, Bela H. Buck is involved in various projects concerning the cultivation of marine plants/animals, the development of technological design and the realization of pilot projects to commercial enterprises. He is in cooperation with various national/international institutions. He is the founder for a new RAS plant (2.4 Mio €) for aquaculture research, which was inaugurated in March 2011. Bela H. Buck won three prices during his scientific career (e.g. The Price for interdisciplinary research from the Chamber of Commerce).

Prof. Dr. Richard Langan is a Research Professor in the School of Marine Science and Ocean Engineering and the Director of Coastal and Ocean Technology Programs at the University of New Hampshire, USA. He received a Bachelors Degree in Biology from Lehigh University, and both Masters and Ph.D. from the University of New Hampshire. Dr. Langan's Masters thesis topic was the ecology of anadromous fish in New England tidal rivers and his Ph. D. research focused on by-catch and discards in the commercial otter trawl fisheries in New England waters. Dr. Langan has been involved in fisheries and aquaculture research and development as both a scientific researcher and commercial entrepreneur for nearly four decades. He has more than fifty peer-reviewed publications and his work has received hundreds of citations.

At the University of New Hampshire, Dr. Langan has held positions as a Research Scientist and Director of the Jackson Estuarine Laboratory; Director of the Cooperative Institute for Coastal and Estuarine Environmental Technology which focused on development and application of innovative technologies to address coastal water quality and habitat restoration; Director of the National Estuarine Research Reserve Science Collaborative, which supported collaborative environmental research at the 43 Research Reserve sites across the USA; and the Director of the Atlantic Marine Aquaculture Center, which conducted research, development, and demonstration on sustainable open ocean aquaculture of finfish and shellfish. Prior to his tenure at the University, Dr. Langan spent 5 years as first mate on the commercial fishing trawlers *F/V Scotsman* and *F/V Captain Gould*; 4 years as owner of a seafood retail, wholesale, and restaurant business; and 5 years as the owner and operator of a commercial oyster farm.

In recent years, Dr. Langan has continued to work with the commercial fishing community and interested entrepreneurs to develop molluscan shellfish aquaculture businesses, focusing on oysters in estuarine waters and mussels and scallops in offshore waters. He has been successful in transferring technology developed the University to commercial practitioners. He is also working with the US government Protected Species Program to address the potential for entanglement of marine mammals and turtles in open ocean aquaculture gear.

Contributors

Bela H. Buck Marine Aquaculture, Maritime Technologies and ICZM, Alfred Wegener Institute Helmholtz Centre for Polar and Marine Research (AWI), Bremerhaven, Germany; Applied Marine Biology, University of Applied Sciences Bremerhaven, Bremerhaven, Germany

Michael D. Chambers School of Marine Science and Ocean Engineering, University of New Hampshire, Durham, NH, USA

Thierry Chopin Canadian Integrated Multi-Trophic Aquaculture Network, University of New Brunswick, Saint John, NB, Canada

John S. Corbin Aquaculture Planning and Advocacy LLC, Kaneohe, HI, USA

Detlef Czybulka University of Rostock, Rostock, Germany

Arne Fredheim Department of Aquaculture Technology, SINTEF Fisheries and Aquaculture, Sluppen, Trondheim, Norway

David Fredriksson Department of Naval Architecture and Ocean Engineering, United States Naval Academy, Annapolis, MD, USA

K. Gee Helmholtz Zentrum Geesthacht, Geesthacht, Germany

A. Gimpel Thünen-Institute of Sea Fisheries, Hamburg, Germany

M. Gopnik Independent Consultant, Washington DC, USA

Nils Goseberg Ludwig-Franzius-Institute for Hydraulic, Estuarine and Coastal Engineering, Leibniz Universität Hannover, Hannover, Germany

Britta Grote Alfred Wegener Institute Helmholtz Centre for Polar and Marine Research (AWI), Marine Aquaculture, Maritime Technologies and ICZM, ZMFE Zentrum Für Maritime Forschung und Entwicklung, Bremerhaven, Germany

Kevin Heasman Cawthron Institute, Nelson, New Zealand

José Joaquín Hernández-Brito PLOCAN (Consorcio Para La Construcción, Equipamiento Y Explotación de La Plataforma Oceánica de Canarias), Telde, Spain

Poul Holm School of Histories and Humanities, Trinity College Dublin, Dublin 2, Ireland

John Holmyard Offshore Shellfish Ltd., Brixham, Devon, UK

Jeffrey B. Kaiser Marine Science Institute's Fisheries and Mariculture Laboratory, The University of Texas at Austin, Port Aransas, TX, USA

Hauke L. Kite-Powell Marine Policy Center, Woods Hole Oceanographic Institution, MS 41, Woods Hole, MA, USA

Job Klijnstra Endures BV, Den Helder, The Netherlands

Gesche Krause Alfred Wegener Institute Helmholtz Centre for Polar and Marine Research (AWI), Earth System Knowledge Platform (ESKP), Bremerhaven, Germany; SeaKult—Sustainable Futures in the Marine Realm, Bremerhaven, Germany

Sander Lagerveld Wageningen University and Research—Wageningen Marine Research, Den Helder, The Netherlands

Richard Langan School of Marine Science and Ocean Engineering, University of New Hampshire, Durham, NH, USA

Scott Lindell Biology Department, Woods Hole Oceanographic Institution, Woods Hole, MA, USA

Pedro Mayorga EnerOcean S.L, Málaga, Spain

Eirik Mikkelsen Departement of Social Science, Norut Northern Research Institute AS, Tromsø, Norway

Katja Mintenbeck Alfred Wegener Institute Helmholtz Centre for Polar and Marine Research (AWI), Integrative Ecophysiology/Marine Aquaculture, Bremerhaven, Germany

Arkadiusz Mochtak University of Rostock, Rostock, Germany

Nancy Nevejan Laboratory of Aquaculture & Artemia Reference Center, Department of Animal Production, Faculty of Bioscience Engineering, Ghent University, Ghent, Belgium

Nikos Papandroulakis Hellenic Centre for Marine Research, Institute of Marine Biology Biotechnology and Aquaculture, Heraklion, Crete, Greece

Bernadette Pogoda Alfred Wegener Institute Helmholtz Centre for Polar and Marine Research (AWI), Marine Aquaculture, Maritime Technologies and ICZM, ZMFE Zentrum Für Maritime Forschung und Entwicklung, Bremerhaven, Germany; Faculty of Applied Marine Biology, University of Applied Sciences Bremerhaven, Bremerhaven, Germany

Christine Röckmann Wageningen University and Research—Wageningen Marine Research, Den Helder, The Netherlands

Torsten Schlurmann Ludwig-Franzius-Institute for Hydraulic, Estuarine and Coastal Engineering, Leibniz Universität Hannover, Hannover, Germany

Maximilian F. Schupp Alfred Wegener Institute Helmholtz Centre for Polar and Marine Research (AWI), Marine Aquaculture, Maritime Technologies and ICZM, ZMFE Zentrum Für Maritime Forschung und Entwicklung, Bremerhaven, Germany

John Stavenuiter Asset Management Control Centre, Den Helder, The Netherlands

Selina M. Stead School of Marine Science and Technology, Newcastle University, Newcastle, UK

Vanessa Stelzenmüller Thünen-Institute of Sea Fisheries, Hamburg, Germany

Claudia Thomsen Phytolutions GmbH, Bremen, Germany

Sjoerd van der Putten TNO Structural Dynamics, Van Mourik Broekmanweg 6, Delft, The Netherlands

Lara Wever Forschungszentrum Jülich GmbH, Jülich, Germany

Mathieu Wille Laboratory of Aquaculture & Artemia Reference Center, Department of Animal Production, Faculty of Bioscience Engineering, Ghent University, Ghent, Belgium

Xiaolong Zhang Endures BV, Den Helder, The Netherlands

Abbreviations

A	Exposed surface area
ABNJ	Area Beyond National Jurisdiction
ABRC	Aquatic Biomass Research Center
ACOE	Armey Corps of Engineers
ADV	Acoustic Doppler Velocimetry
AHP	Analytical Hierarchy Process
AMC	Asset Management Control
AOGHS	American Oil and Gas Historical Society
APA	Aquaculture Planning and Advocacy
AQU	Aquamats®
ARIES	Artificial Intelligence for Ecosystem Services
Art.	Article
ASAIM	American Soybean Association International Marketing
ASW	Artificial Seaweed Collector
AWI	Alfred Wegener Institute Helmholtz Centre for Polar and Marine Research
ß	Angle between cable and horizontal axis
BAH	Biologische Anstalt Helgoland (Biological Institute Helgoland)
BFN	Bundesamt für Naturschutz (Federal Agency for Nature Conservation (BfN)
BMP	Best Management Practices
BNatSchG	Bundesnaturschutzgesetz (German Federal Nature Conservation Act)
BRE	Bronx River Estuary
BSH	Bundesamt für Seeschifffahrt und Hydrographie (Federal Maritime and Hydrographic Agency)
BWB	Bundesamt für Wehrtechnik und Beschaffung (Federal Office of Defence and Precurement)
C	Carbon
CAPEX	Capital expenditures
C_D	Drag coefficient

CEFAS	Centre for Environment, Fisheries, and Aquaculture Science
CI	Condition index
CIMTAN	Canadian Integrated Multi-Trophic Aquaculture Network
CLME	Caribbean Large Marine Ecosystem
C_M	Dynamic drag coefficient
CO_2	Carbon dioxide
COC	Coconut Rope
d	Day
DA	Department of the Army
DHA	Docosahexaenoic acid
DIN	Dissolved inorganic nitrogen
DOWES	Dutch Offshore Wind Energy Services
DRG	Drag
DSP	Diarrheic shellfish poisoning
dw	Dry weight
EATiP	European Aquaculture Technology and Innovation Platform
EC	European Commission
EEZ	Economic Exclusive Zone
e.g.	Exempli gratia (for example)
EIA	Environmental Impact Assessment
ELG	Effluent Limitation Guidelines
EM	Electron Microscope
EPA	Environmental Protection Agency, EicosaPentaenoic Acid
EU	European Union
EuRG	Ernte- und Reinigungsgerät (Harvest and cleaning device)
F	Force
FAO	Food and Agriculture Organisation of the United Nations
F_b	Buoyancy forces
FDA	Food and Drug Administration
F_{DRG}	Drag force
FE	Finite element
FHI	Fish Health Inspectorate
FINO	Forschungsplattformen in Nord- und Ostsee (Research Platforms in the North Sea and the Baltic Sea)
F_m	Mass forces
FP7	7th Framework Programmes for Research and Technological Development
FPN	Forschungsplattform Nordsee (Research Platform "North Sea")
F_s	Force in the cable
FSA	Food Services Agency
g	Gravity
GAO	General Accounting Office
GAR	Galician Rope
GCFMC	Gulf Coast Fishery Management Council
GG	Grundgesetz (Basic Law for the Federal Republic of Germany)

GIS	Geographic Information System
GIS-MCE	GIS-based Multi-Criteria Evaluation
GPS	Global Positioning System
GRT	Gross register tonnage
GW	Gigawatt
GWEC	Global Wind Energy Council
HLS	Harvard Law School
H_{max}	Maximum wave height
HOARP	Hawaiian Offshore Aquaculture Research Project
HSVA	Hamburg Ship Model Basin
HSWRI	Hubbs Sea World Research Institute
ICES	International Commission for the Exploration of the Seas
ICT	Information and communications technology
ICZM	Integrated Coastal Zone Management
i.e.	Id est (that is to say, that is)
ILVO	Instituut voor Landbouw- en Visserijonderzoek (Institute for Agricultural and Fisheries Research)
IMTA	Integrated Multi-Trophic Aquaculture
InVEST	Integrated Valuation of Ecosystem Services and Trade-offs
IQF	Individually Quick Frozen
IRR	Internal rate of return
IWGA	Interagency Working Group on Aquaculture
Km	Kilometers
kN	Kilo Newton
KW	Kilowatt
LAD	Ladder Collector
LCM	Life Cycle Management
LIS	Long Island Sound
LEC	Looped Christmas Tree
LL	Longline
LOC	Leaded Christmas Tree
M	Material
m	Mass, meter
m_a	Additional mass
MAP	Modified Atmosphere Packaging
MaRS	Marine Resource System
MaxEnt	Maximum Entropy modelling
m_c	Collector mass
MEAB	Millennium ecosystem assessment
MFA	Marine Fisheries Agency
MIC	Microbial corrosion
MIMES	Multi-scale Integrated Models of Ecosystem Services
m_l	Line mass
MMC	Multipurpose Marine Cadastre
MMO	Marine Management Organization

MMS	Minerals Management Service
Mmt	Millions of metric tons
MPA	Marine protected area, Magnuson-Stevens Act
MSAC	Marine Special Area of Conservation
MSFD	Marine Strategy Framework Directive
MSP	Marine Spatial Planning
mt	Metric tonnes
MTBF	Mean Time Between Failure
MTTR	Mean Time To Repair
MUPs	Multi-Use Platforms
MW	Megawatt
MWh	Megawatt hour
MYM	Multi-Year Maintenance
N	Nitrogen
NASA	National Aeronautics and Space Administration
NE	Natural England
NFL	Naue® Fleece
NGO	Non-governmental organisation
NIMBY	"Not In My Back Yard"
nm	Nautical miles
NMFS	National Marine Fisheries Service
no.	Number
NOAA	National Oceanic and Atmospheric Administration
NOC	National Ocean Council
NODV	Northeast Ocean Data Viewer
NPDES	National Pollution Discharge Elimination System
NPV	Net present values
NSGLC	National Sea Grant Law Center
NTC	Nutrient trading credit
O	Origin
OCS	Outer Continental Shelf
ODC	Ocean Discharge Criteria
ODAS	Ocean Data Acquisition System
OIE	World Organization for Animal Health
OFW	Offshore wind farms
O&M	Operation & Maintenance
OOA	Open Ocean Aquaculture
OPEX	Operational expenditures
OSPAR	Oslo-Paris Commission
OTEC	Ocean Thermal Energy Conversion
OWA	Ordered Weighted Average approach
OWEC	Offshore wind energy converters
OWF	Offshore wind farm
OWMF	Offshore wind-mussel farm
ρ	Density of water

P	Phosphors
PAR	Photosynthetically active radiation
PE	Polyethylene
PEST	Model-independent Parameter Estimation
PIV	Particle Image Velocimetry
PL	Phospholipids
PLOCAN	PLataforma Oceánica de las CANarias
POM	Particulate Organic Matter
PP	Polypropylene
PSP	Paralytic Shellfish Poison
PRD	Protected Resources Division
PUFA	PolyUnsaturated Fatty Acids
PV	PhotoVoltaic
Q	Displaced water volume
RA	Regional Administrator
RAS	Recirculating Aquaculture System
RB	Research Boat
REF	Reference Collector
R&D	Research & Development
R,D&D	Research, Development and Demonstration
ROGF	Regional Ocean Governance Framework
ROI	Return Of Investment
RV	Research Vessel
S	Size
SAMS	Scottish Association of Marine Science
S&F	"Set and Forget"
SDVO	Stichting voor Duurzame Visserijontwikkeling Visserijontwikkeling (Foundation for sustainable development of fisheries)
SES	Social-Ecological System, Seaweed Energy Solutions AS
SeeAnlV	Seeanlagenverordnung (Marine Facilities Ordinance)
SFA	State Fisheries Agency
SGR	Specific Growth Rate
Sintef	Stiftelsen for industriell og teknisk forskning (The Foundation for Scientific and Industrial Research)
SL	Service load
SOSSEC	Submersible Offshore Shellfish and Seaweed Cage
SRB	Sulfate reducing bacteria
SSC	Self-Sinking Collector
SWOT	Strengths, Weaknesses, Opportunities, and Threats
TAG	Triacylglycerols
TEAL	Transport, Energy, Aquaculture, and Leisure
TLP	Tension-Leg Platform
TNO	Toegepast Natuurwetenschappelijk Onderzoek (Dutch Organization for Applied Scientific Research)
UK	United Kingdom

UKL	United Kingdom Legislation
UN	United Nations
UNCLOS	United Nations Convention on the Law of the Sea
UNFAO	United Nations Food and Agriculture Organization
UNH	University of New Hampshire
USA	United States of America
USDA	United States Department of Agriculture
USDOC	United States Department of Commerce
UV	Ultraviolet
V	Velocity, Volume
VMS	Vessel Monitoring System
VOWTAP	Wind Technology Advancement Project
WAS	World Aquaculture Society
WQS	Water Quality Standards
WSA	Water and Shipping Agency
WTG	Wind turbine generator
ZAF	Zentrum für Aquakulturforschung (Center for Aquaculture Research)

Chapter 1
Introduction: New Approaches to Sustainable Offshore Food Production and the Development of Offshore Platforms

Poul Holm, Bela H. Buck and Richard Langan

Abstract As we exhaust traditional natural resources upon which we have relied for decades to support economic growth, alternatives that are compatible with a resource conservation ethic, are consistent with efforts to limit greenhouse emissions to combat global climate change, and that support principles of integrated coastal management must be identified. Examples of sectors that are prime candidates for reinvention are electrical generation and seafood production. Once a major force in global economies and a symbol of its culture and character, the fishing industry has experienced major setbacks in the past half-decade. Once bountiful fisheries were decimated by overfishing and destructive fisheries practices that resulted in tremendous biomass of discarded by-catch. Severe restrictions on landings and effort that have been implemented to allow stocks to recover have had tremendous impact on the economy of coastal communities. During the period of decline and stagnation in capture fisheries, global production from aquaculture grew dramatically, and now accounts for 50% of the world's edible seafood supply. With the convergence of environmental and aesthetic concerns, aquaculture, which was already competing for space with other more established and accepted uses, is having an increasingly difficult time expanding in nearshore waters. Given the

P. Holm (✉)
School of Histories and Humanities, Trinity College Dublin,
2 College Green, Dublin 2, Ireland
e-mail: holmp@tcd.ie

B.H. Buck
Marine Aquaculture, Maritime Technologies and ICZM, Alfred Wegener
Institute Helmholtz Centre for Polar and Marine Research (AWI),
Bussestrasse 27, 27570 Bremerhaven, Germany

B.H. Buck
Applied Marine Biology, University of Applied Sciences Bremerhaven,
An der Karlstadt 8, 27568 Bremerhaven, Germany

R. Langan
School of Marine Science and Ocean Engineering,
University of New Hampshire, Durham, NH 03854, USA

© The Author(s) 2017
B.H. Buck and R. Langan (eds.), *Aquaculture Perspective of Multi-Use Sites in the Open Ocean*, DOI 10.1007/978-3-319-51159-7_1

1

constraints on expansion of current methods of production, it is clear that alternative approaches are needed in order for the marine aquaculture sector to make a meaningful contribution to global seafood supply. Farming in offshore marine waters has been identified as one potential option for increasing seafood production and has been a focus of international attention for more than a decade. Though there are technical challenges for farming in the frequently hostile open ocean environment, there is sufficient rationale for pursuing the development of offshore farming. Favorable features of open ocean waters include ample space for expansion, tremendous carrying and assimilative capacity, reduced conflict with many user groups, lower exposure to human sources of pollution, the potential to reduce some of the negative environmental impacts of coastal fish farming (Ryan 2004; Buck 2004; Helsley and Kim 2005; Ward et al. 2006; Langan 2007), and optimal environmental conditions for a wide variety of marine species (Ostrowski and Helsley 2003; Ryan 2004; Howell et al. 2006; Benetti et al. 2006; Langan and Horton 2003). Those features, coupled with advances in farming technology (Fredheim and Langan 2009) would seem to present an excellent opportunity for growth, however, development in offshore waters has been measured. This has been due in large part to the spill over from the opposition to nearshore marine farming and the lack of a regulatory framework for permitting, siting and managing industry development. Without legal access to favorable sites and a "social license" to operate without undue regulatory hardship, it will be difficult for open ocean aquaculture to realize its true potential. Some parallels can be drawn between ocean aquaculture and electricity generation. Continued reliance on traditional methods of production, which for electricity means fossil fuels, is environmentally and economically unsustainable. There is appropriate technology available to both sectors, and most would agree that securing our energy and seafood futures are in the collective national interest. The most advanced and proven renewable sector for ocean power generation is wind turbines, and with substantial offshore wind resources in the, one would think there would be tremendous potential for development of this sector and public support for development. The casual observer might view the ocean as a vast and barren place, with lots of space to put wind turbines and fish farms. However, if we start to map out existing human uses such as shipping lanes, pipelines, cables, LNG terminals, and fishing grounds, and add to that ecological resource areas that require some degree of protection such as whale and turtle migration routes, migratory bird flyways, spawning grounds, and sensitive habitats such as corals, the ocean begins to look like a crowed place. Therefore, when trying to locate new ocean uses, it may be worthwhile to explore possibilities for co-location of facilities, in this case wind turbines and fish and shellfish farms. While some might argue that trying to co-locate two activities that are individually controversial would be a permitting nightmare, general agreement can probably be reached that there are benefits to be gained by reducing the overall footprint of human uses in the ocean. Meeting the challenges of multi-use facilities in the open ocean will require careful analysis and planning; however, the opportunity to co-locate sustainable seafood and renewable energy production facilities is intriguing, the concept is consistent with the goals of Marine Spatial Planning and ecosystem based management, and therefore worthy of pursuit.

1.1 Aquaculture—A Historical Overview

The transition on land from hunting to agriculture took thousands of years. In the oceans, the transition from capture fishing to modern aquaculture production happened in just two human generations. As late as 1965, a major review did not pay much attention to the potential of aquaculture. Christy and Scott (1965) considered marine farming only in passing for oysters and other mollusks and predicted that increasing use of fresh and brackish ponds in low-income countries might be a means to increase local protein output. By 2009, however, the world's human population consumed more cultured fish than was caught in the wild. Indeed, technological progress has been so rapid that the number of species domesticated for aquaculture now exceeds the number of species domesticated on land (Duarte et al. 2007). Monoculture currently dominates production and a few species, carps, shrimp, prawns, salmon and trout, make up half of total production (Asche et al. 2008). We have long since lost sight of the implications and consequences of culturing the land—but a similar process is now taking place at sea and we hardly notice it. As this is a change similar to the agricultural revolution on land which by archaeologists was identified as the "Neolithic Revolution", in all of these senses we are living through a 'Neolithic' revolution of the oceans.

The fact that aquaculture, for all practical concerns, is a very recent phenomenon may explain many of its characteristics. The industry has experienced almost exponential growth while it has suffered heavily from the spread of diseases in monoculture farms and has been severely impacted and restructured as a result of boom-and-bust growth, particularly around 1990. What was to the first generation of aquaculture entrepreneurs a business of trial-and-error and reliance on the family and local work force is now a globalized corporate enterprise. Science and public management largely saw their roles in the early period as ones of support and encouragement but have now developed agendas of inquiry and management. Major problems such as feed and access to marine space loom large as future threats to the industry. All of these characteristics identify aquaculture as an industry in an early rather than mature phase of growth.

Aquaculture comprises both fresh and brackish water production as well as marine aquaculture (sometimes called mariculture). There is no clear-cut dividing line between working in the different environments, partly because some of the major cultured finfish such as salmon are anadromous, and partly because of learning and innovation across the sectors. Fresh and brackish water aquaculture are characterised by relatively small-scale operations while marine aquaculture is now dominated by larger scale operations. Since the 1980s, marine aquaculture has contributed between 50 and 60% of global traded production volume with a

Fig. 1.1 Historical image of an aquaculture farming enclosure (CC 2016)

decreasing trend, while traded value is down from 40 to 32% in 2013 relative to fresh and brackish species (FAO 2016).

The origins of fish farming may be found far back in time. Freshwater aquaculture emerged at least 3000 years ago in East Asia and the Middle East. Carp in particular became a very important local food source for China and developed significantly in the eighteenth and nineteenth centuries (Li 1997). Similarly, the origins of inshore saltwater aquaculture go far back in time and seem to have originated independently in several regions. Roman towns relied on closed inshore lagoons for the provision of saltwater fish and shellfish at least two millennia ago (McCann 2003). The Polynesian culture of the Hawaiian Islands had developed extensive inshore lagoon fish farming centuries before the arrival of Europeans (Ziegler 2002). Little historical research has been undertaken but it is clear that in Europe, fishponds throughout the pre-modern period were prestigious undertakings, which catered to the needs of affluent consumers who added diversity and freshness to their meals at very high costs (Serjeantson and Woolgar 2006) (Fig. 1.1). The system often depended on collecting and transferring young fish or shellfish to an artificial environment. In Asia, fish farming was organically linked to local aquatic ecosystems such as rice paddies and sewage systems. This extensive fish farming practice was relatively inexpensive and therefore catered to the needs of farmers as well as the elite.

The bottleneck to European fish farming was the development of artificial hatcheries. In Germany, Jacobi (1768) cracked the code to enable external fertilization of brown trout and salmon by extracting and mixing eggs and sperm from mature trout and successfully cultivated the offspring. News of his discovery spread slowly and reached the United Kingdom in the 1830s and Norway by 1850 when it led to a short burst of mostly failed experiments (Hovland et al. 2014). This was a time of great expansion of cities and markets. However, this was also a time when oceanic fisheries made great strides forward in the development of trawling and propulsion (Cushing 1988). Steam trawlers soon provided such abundant landings that interest in aquaculture subsided in Europe except for a minor trout industry.

Only in the 1960s did a new and sustained push for aquaculture begin. It came about as a result of independent developments in many regions of the world including Japan, China, the United States, and Europe. The origin of these developments is so far poorly researched with the notable exception of the important Norwegian case (Hovland et al. 2014). Our lack of knowledge is compounded by the fact that the statistical evidence for aquaculture only becomes solid in the 1980s. Nevertheless, it is safe to say that the first major increase took place in Southeast Asia. The expansion built on traditional techniques, but was supported by state policies. The widespread use of explosives and poison in local fisheries had led to a rapid depletion of marine resources. The governments of the Philippines, Indonesia and Thailand saw aquaculture as a means to feed a growing population and as a source of employment for fishermen who were losing their jobs. This local development prepared fish farmers to seize the opportunity when—as a consequence of overfishing—the market for penaeid shrimp soared in Japanese and Western markets around 1970. Over the next twenty years, coastal mangrove forests were cut down to give way for fish and shrimp ponds that were used for extensive practices relying on the natural productivity of the environment. In the mid to late 1980s a new intensive form of cultivation in Southeast Asia gained ground. Intensive farming involves controlling the environment by means of pumps, aerators and generators as well as access to quality feed. Globalisation of markets enabled the injection of capital into an industry that rapidly became dominated by a few large business groups. The giant tiger shrimp (*Penaeus monodon*) was the most profitable commodity and quickly dominated production (Butcher 2004). With concentration came, however, increased threats from diseases that caused heavy losses to the industry, which responded by injections of antibiotics, salinization of lands and expanding production to new fertile mangrove forests (Zink 2013; Hall 2003). In short, the environmental consequences of this early success of the industry were dire while profits were high.

The rise of aquaculture depended on a concomitant dramatic increase in the availability of feed. In the early 1970s, Peruvian (Glantz 1986) and Danish (Holm et al. 1998) fish reduction operations provided millions of tonnes of fishmeal and fish oil as cheap and efficient feed for agriculture as well as aquaculture. As these pelagic resources proved volatile and vulnerable to overexploitation, the aquaculture industry faced a potentially limiting factor to growth. Despite massive improvements in feed technology and the introduction of vegetable (but omega 3-deficient) protein, the feed problem remains a major challenge both to future growth and public perception (Natale et al. 2013).

Science and technological innovation pointed to a way forward. Aquaculture science developed in the United States in particular, and by 1969 the World

Mariculture Society was established, renamed in 1986 the World Aquaculture Society or WAS (Avault and Guthrie 1986). Today the WAS has more than 3000 members in about 100 countries. Although the match of science and industry seemed straightforward, the differing aims and measures of success of business and academia have caused friction at times. Perhaps the most staggering example of the success and problems of the marriage is that of Norwegian salmon farming. Local fishermen and craftsmen began experimenting with salmon farming in Norwegian fjords in the late 1960s. When an entrepreneur developed a cheap open net cage in 1970, the industry took off on a staggering growth trajectory. By 1990 the processes of globalisation and capitalisation had created a Norwegian multi-billion dollar industry. The ripple effects of foreign emulation and Norwegian investments created similar large-scale operations in Chile and smaller-size industries in Canada, Scotland and elsewhere. Almost from the beginning, the Norwegian state and industry joined forces to ensure that science matched industry needs. The state provided lucrative land and water licenses for industry and invested heavily in research. Population genetics in particular perfected the cultured salmon species, while epidemiological research was crucial in combating diseases. Scientists became divided, however, in their belief in the sustainability of the industry. The concern became obvious in 2010 when one director of research predicted a tenfold increase of production while the Director of the Department of the Environment advocated a halving of the industry out of concern for wild salmon and the natural environment (Hovland et al. 2014). Such disagreement, based on differing scientific measures and methods, indicates the degree to which aquaculture had rapidly moved from being a productive force to also being a risk.

The rise of aquaculture depended on the willingness of buyers to substitute cultured products for wild fish. While environmental concerns have been voiced by some consumer organisations, the overall picture is one of market acceptance (Natale et al. 2013). The rise in Western per capita demand for fish was associated with the recognition of fish as a health food, including farmed fish, and has enabled a doubling of fish consumption in developed countries since 1950. Local communities have also largely embraced aquaculture as a source of income despite environmental and land issues. A comparative study of the introduction of forestry and aquaculture industries in South East Asia showed that public perception of aquaculture entrepreneurs—including big companies—was positive while forestry companies were resented as intrusions by big capital (Hall 2003). Similarly, the public perception of aquaculture in Chile was a warm welcome because of job creation. The demands on land and water access are now being identified in Norway as a major challenge to the public perception of the industry (Hovland et al. 2014).

While most of the world's aquaculture is still conducted in semi-intensive operations there is no doubt that the direction is towards increased control of production (Asche et al. 2008). Fish farmers have leap-frogged the technological innovation that took millennia on land and use advanced technologies in a marine environment—of which, paradoxically, we have not yet developed a scientific understanding. The expansion of aquaculture has brought about lower fish prices and introduced jobs and income to developing countries as well as to regions in the developed world that might otherwise have been depopulated. The industry is, however, highly volatile and subject to major threats such as disease and marine space access. Aquaculture has often been branded as a *Blue Revolution* akin to the 1970s *Green Revolution* of agriculture. The comparison is apt in terms of the contribution by the industry to increased food security as production has vastly exceeded population growth and offset the stagnation of capture fish landings. It is also apt in terms of the increased use of medicine, toxins and other technology for the production of natural resources and in terms of species manipulation such as by turning a species like salmon into a semi-vegetarian. Aquaculture may be seen as a true harbinger of the human condition in the Anthropocene—an epoch in which humans have become the main geo-biological agent.

Despite the tremendous growth in aquaculture over the past five decades, seafood demand is projected to outpace supply by 40 million metric tons by 2030 (FAO 2006). With capture fisheries stagnant, and space constraints on continued expansion of nearshore aquaculture, it is clear that alternative means of production are needed. There are two potential means by which marine aquaculture production can expand either land-based recirculating aquaculture systems or development of the technological capacity for farming in exposed oceanic locations. A Canadian study indicated that land based systems are not yet profitable for full grow out of larger fish while they are highly efficient and provide better environmental controls for the production of juvenile fish (Boulet et al. 2010). Recent investments in land-based systems in Europe indicate the potential of cutting transportation costs by locating the industry close to market when energy and water resources are available. Land-based systems may also help alleviate the problem of coastal land access by restricting the use of marine ponds for mature fish. Thus, technological advances in land-based systems have the potential to change the parameters of aquaculture in the future.

1.2 Moving Aquaculture Operations Offshore

Farming in offshore marine waters is the other potential option for increasing seafood production and has been a focus of international attention for more than two decades. Though there are technical challenges for farming in the frequently hostile open ocean environment, there is sufficient rationale for pursuing the development of offshore farming. Favorable features of open ocean waters include ample space for expansion, tremendous carrying and assimilative capacity, reduced conflict with many user groups, lower exposure to human sources of pollution, the potential to reduce some of the negative environmental impacts of coastal fish farming, and optimal environmental conditions for a wide variety of marine species, to name a few (Buck 2002, 2004; Ryan 2004; Langan 2007; Langan and Horton 2003; Ostrowski and Helsley 2003; Buck et al. 2004; Helsley and Kim 2005; Benetti et al. 2006; Howell et al. 2006; Ward et al. 2006). A recent study conducted in New Zealand indicated an additional benefit that open ocean locations may be subjected to less biofouling (Atalah et al. 2016), a costly maintenance operation for coastal aquaculture. Those features, coupled with advances in farming technology (Fredheim and Langan 2009) would seem to present an excellent opportunity for growth, however, development in offshore waters has been limited. The reasons for this vary depending on location, but include risk aversion, lack of access to capital, clear identification of ownership, and unresolved technological issues (Langan 2012). In some countries, like the USA and some EU member states, lack of a regulatory structure for permitting offshore farms has been an impediment (e.g. Cicin-Sain et al. 2001; Buck et al. 2003), as well as opposition form environmental NGOs (Langan 2012).

Similar to the recent developments in aquaculture, the energy sector has undergone significant changes. Over the course of the last decade, the establishment of offshore wind farms as a sustainable and economically viable form of energy production has generated interest in the potential for optimizing use of offshore sites to include other activities. Thus, consideration of multiple uses of offshore renewable energy systems in the design phase so that the economic benefits from a unit area of sea can be maximized in a sustainable way has been a central research topic since the year 2000 (Buck 2001).

1.3   The Multi-use Concept

One particular area of interest is combining energy and aquaculture-based seafood production within the same ocean footprint (Buck and Krause 2012, 2013). Interest in marine aquaculture in exposed oceanic locations has been explored as a stand-alone activity, however, commercial development has been thus far very limited. The stability of offshore energy production structures (e.g. wind turbine and oil drilling platforms) is an attractive feature for a suite of requirements for aquaculture production, including attachment points for mooring cages and longlines, and for mounting feeding, hatchery and nursery systems. Though desirable attributes for energy and seafood production may not exist at all offshore sites, there is likely a subset of locations that are suitable, acceptable and economically viable. Thus, the slogan "Maximizing the benefit of a piece of land" (Buck 2009a) is a potential solution to foster offshore multi-use concepts of renewable energy systems, but also from any other offshore installation type, such as other renewable energy installations (e.g. tidal energy) or oil and gas (Kaiser et al. 2011).

1.3.1   Pilot Projects in Russia

The first synergy of offshore platforms with aquaculture was initiated in the Caspian Sea (27 km off the Turkmenian shore) in 1987 (Fig. 1.2a–d), where a fish farm was moored next to an oil rig (Bugrov 2016). Unfortunately, high operating costs led to

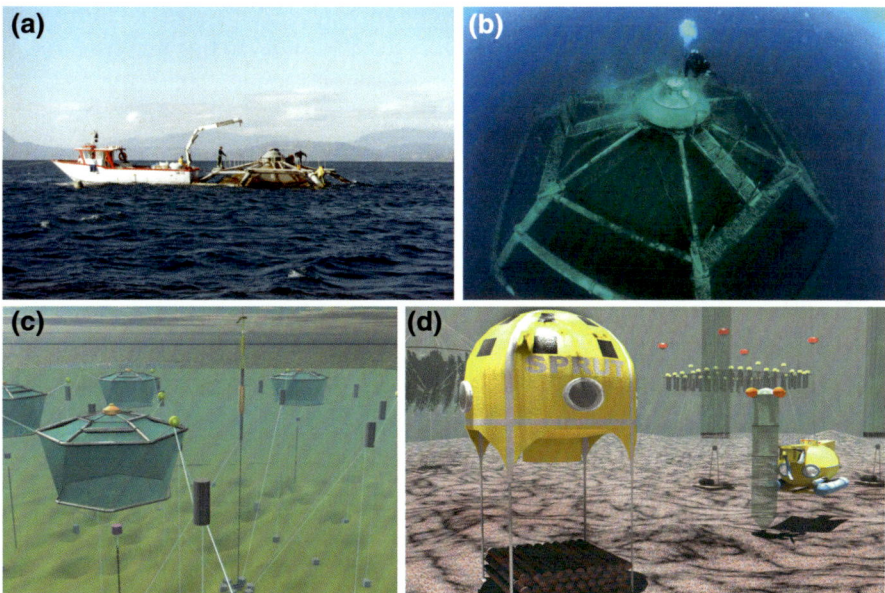

Fig. 1.2 Submersible cage complex "Sadco-Kitezh" (consisting of 6 individual cage modules) disposed next to an offshore oil-rig in the Caspian Sea in 1988, floating (**a**) and submerged (**b**); **c** displays a concept for a series of submerged cages and **d** shows a collection of submergible devices for fish and bivalves and floating seaweed a test site following an IMTA-concept (Bugrov 2016 following Buck et al. 2004)

a shutdown of this enterprise at a very early stage (Bugrov 1992, 1996). Over the past 25 years more than 1000 oil and gas structures were installed in the same area and more than 300 in the Black Sea. The amount of time to decommission these platforms takes on average of one year and international experience in disassembling those platforms showed that the average cost of disassembly works is several million Euros (Bugrov 1991). Resigning the dismantling of the platforms and therefore saving costs could support a cross-subsidasation of aquaculture. This would have had an influence on the commercial potential of these multi-use concept, however, that was not taken into account at that time.

1.3.2   Pilot Projects in the USA

In the Gulf of Mexico the cumulative costs of a total removal of oil rigs had reached an estimated $1 billion by the year 2000 (Dauterive 2000). In this respect the search for a way of conversion of such structures became more important and initiated the search for alternatives. Operators have recognized that during a rig's productive years, significant marine life aggregates on and around its structures. This is also caused by the fact that marine areas occupied by offshore platforms are off limits for commercial fishing vessels due to safety reasons (Berkenhagen et al. 2010). This results in an increase in biomass of fish or other species and/or a greater number of species in this area aggregating at the artificial reefs. These areas then can be considered as more or less a marine protected area (MPA). Marine scientists have therefore suggested preserving much of this marine life and encouraging further natural productivity (Jensen et al. 2000). While the operator benefits by avoiding the substantial cost of removal, populations of marine species benefit from the refuge the structures provide. These findings encouraged recreational fisherman, divers, offshore oil and gas operators, aquaculturists and others who could benefit from the increased density to establish the "Rigs-to-Reefs" program in American and European Seas (Reggio 1987), where decommissioned offshore oil and gas rigs were turned into artificial reefs. Since then many scientists have reported that these artificial reefs increase the number and diversity of marine organisms adjacent to these sites (e.g. Bohnsack et al. 1994; Zalmon et al. 2002) including many commercially important fish, shellfish and crustacean species (Bohnsack et al. 1991; Jensen et al. 2000).

To this point, some efforts have been carried out to successfully install offshore aquaculture constructions as pilot systems even in the open Pacific but none have so far reached a continuous commercial operation. In particular, projects carried out in the USA were of prime importance for the successful installation of various offshore systems (e.g. Loverich 1997, 1998; Loverich and Gace 1997; Braginton-Smith and Messier 1998; Loverich and Forster 2000). These efforts led to the idea to include various disused oil platforms in the Gulf of Mexico in a multi-use concept (Miget 1994; Wilson and Stanley 1998) (Fig. 1.3a–b).

The *National Sea Grant College Program* funded such research projects to explore offshore sites for stand-alone mariculture purposes. The *Open Ocean Aquaculture Program* at the University of New Hampshire is one of the few attempts made so far (Ward et al. 2001) as well as the *Hawaiian Offshore Aquaculture Research Project* (HOARP) (Ostrowski and Helsley 2003). Due to the

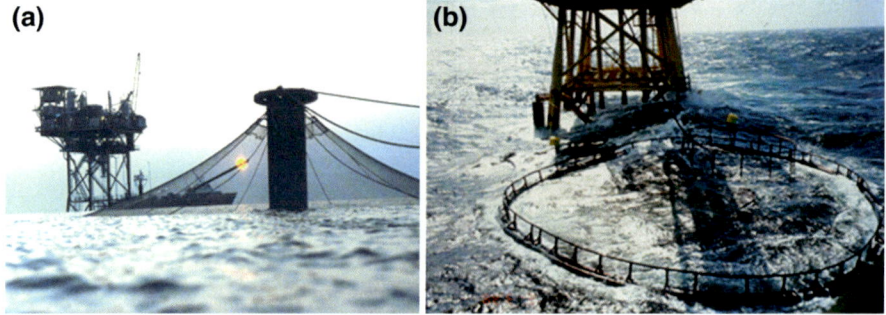

Fig. 1.3 **a** Ocean Spar Cage deployed next to an offshore oil rig in federal waters 22 miles off Mississippi in the Gulf of Mexico (Bridger 2004); **b** typical 4–5 m winter seas moving through the cage and platform site

technological capacity of the US and their extended marine areas, the movement of aquaculture activities into offshore areas gained momentum for a period of time (Dalton 2004) and has encouraged other western countries to follow.

Several studies have estimated that tons to tens of tons of wild fish congregate in the immediate area around fish farms in both warm and cold-temperate environments (Dempster et al. 2004, 2009; Leonard et al. 2011). For some species, artificial reefs can increase the availability of critical habitat (Polovina and Sakai 1989) for feeding, spawning, and juvenile refugia (Jensen et al. 2000) in addition to reducing the detrimental impacts on existing habitats by mobile gear exclusion (Claudet and Pelletier 2004). Additionally, these constructions can be helpful in developing cost effective fishing practices by reducing displacement cost for the inshore fleets and reducing competition for territory between fishermen. The question whether artificial reefs close to aquaculture sites would decrease the impact of cultured fish waste on the surrounding ecosystem has been suggested as a topic of research (Buschmann et al. 2008).

1.3.3 Pilot Projects in Germany

In Germany, the plans for the massive expansion of wind farms in offshore areas of the North Sea triggered the idea of a combination of wind turbines with other uses. Various multi-use concepts were followed led by tourism, marine protected areas (MPAs), passive fishery actions as well as desalination and research, just to name a few (see Fig. 1.4). Another concept is to co-use wind farm installations with extensive aquaculture of native bivalves and macroalgae (e.g. Buck 2002, 2004; Buck and Buchholz 2004; Buck et al. 2008, 2012; Lacroix and Pioch 2011). Due to the fact that offshore wind farms provide an appropriately sized area free of commercial shipping traffic (as most offshore wind farms are designed as restricted-access areas due to hazard mitigation concerns), projects on open ocean aquaculture have been carried out since 2000 in the German Bight (Buck 2001). Further expansion towards finfish culture has since then been proposed and carried out in land-based facilities with regard to system design and coupling technologies for submersible fish cages as well as Integrated Multi-trophic Aquaculture (IMTA) and site-selection.

The combination of wind energy and aquaculture enterprises was already proven in China in the early 1990s (Chunrong et al. 1994), however, these wind turbines were land-based and used to enhance dissolved oxygen in the water column as well as to increase fishpond temperature. Today, many other research institutes have adopted this concept and have conducted feasibility studies within their coastal and EEZ waters, in Denmark, The Netherlands, Belgium, the UK, USA and others (Figs. 1.4 and 1.5; Wever et al. 2015).

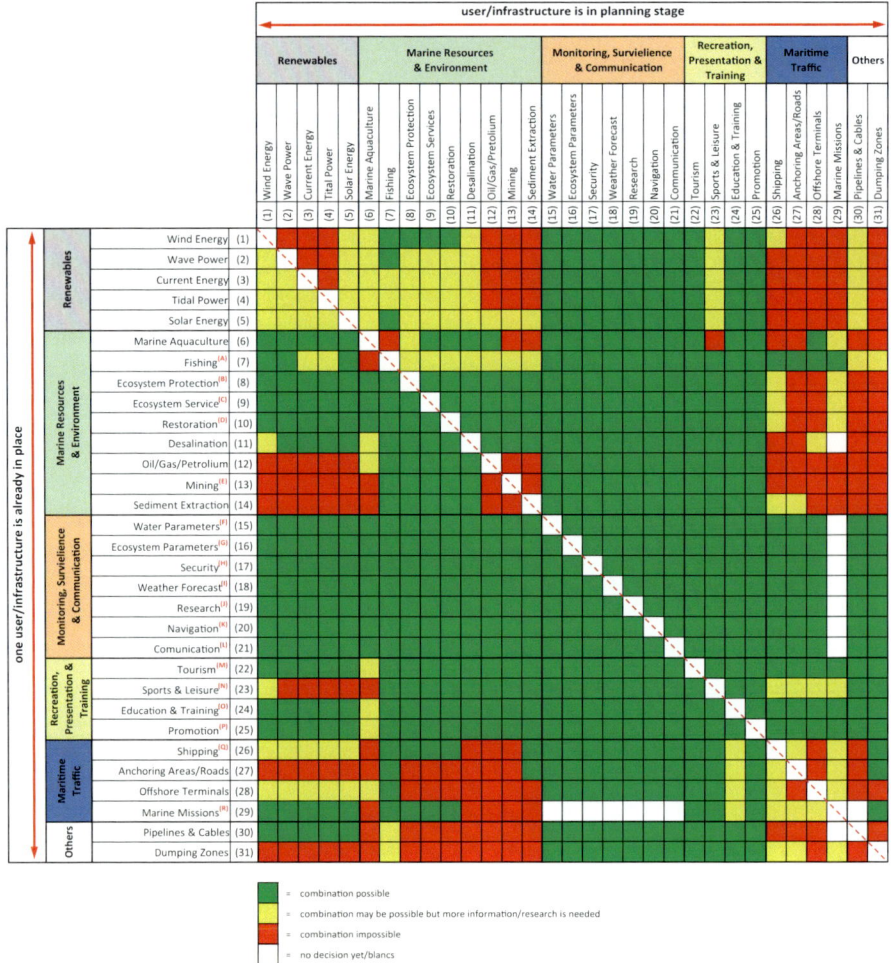

Fig. 1.4 Assessment of potential uses and achievable multi-use options for users/installations, which are already at sea, and for those, which are currently in planning phase. *A* includes traditional fishing techniques without seabed connection (due to avoid any contact to ground cables/pipes) as well as sustainable/passive and recreational fishing; *B* includes MPA's, nature conservation, compensatory measures; *C* including e.g. buffer zones, nutrient cycling, primary production, etc.; *D* includes e.g. shellfish or seaweed restoration and rehabilitation; *E* includes e.g. manganese/copper/cobalt and others; *F* includes the monitoring of oceanographic parameters (salinity, pH, temperature, O_2, etc.), chemical parameters (nitrite, nitrate, phosphate, etc.), as well as harmful substances (toxins, heavy metals, etc.); *G* includes mapping of flora and fauna, other habitat parameters; *H* includes the surveillance of the national/EEZ territory (traffic of drugs or other illegal goods, illegal passage of persons and equipment, etc.), as well as security on the entire traffic (commercial and recreational); *I* includes also tsunami watch; *J* includes marine/coastal research on moving platforms (vessels, buoys, etc.) and fixed platforms (research stations); *K* includes e.g. radar; *L* includes telephone and network cables as well as wireless systems; *M* includes sport fishing, diving, daily visiting tourists with interacting interests, etc.; *N* includes e.g. sailing regatta, races, etc.; *O* includes security training for work at platforms/vessels for the offshore industry or to teach students; *P* includes preparation of advertising films, movies; *Q* includes commercial as well as recreational shipping; *R* includes marine practice areas, firing and torpedo areas as well as submarine areas (Modified after Buck (2013), images by AWI/Prof. Dr. Bela H. Buck)

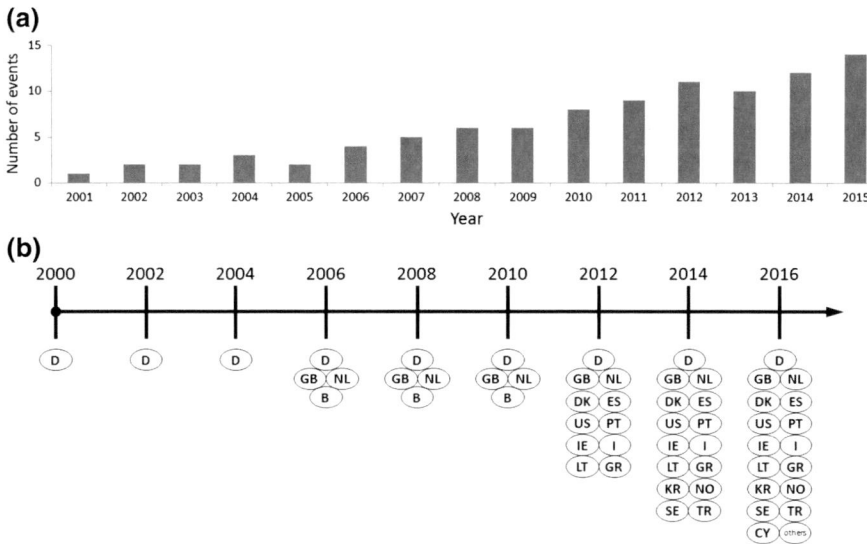

Fig. 1.5 a Graph shows a time scale with the number of events worldwide in which the combination of aquaculture within offshore wind farms was discussed; **b** countries involved in aquaculture wind farm combinations on a time line. Both images modified after Wever et al. 2015

1.4 Initiation to the Topic

For this rather risky and expensive development to happen in practical terms, an understanding of basic needs, such as design requirements, data acquisition, site specifications, operation and maintenance issues, etc. is required. Offshore structures will need to be modified or adapted to accommodate other uses without compromising functionality and safety. Indeed, this move further from shore and into higher energy open ocean environments has created demand for new vessels for installation, operation, maintenance and decommissioning. While it is clear that multi-uses will require multiple types of service vessels, there will be areas of overlap where economies of scale can be achieved, for example in the transport of technicians. Technologies for aquaculture in exposed environments are still in the early stages of development, and combined use at energy production sites will require some rethinking of engineering design.

Other combinations of offshore uses are possible, thus supporting the trend to combine expensive infrastructure and collocate it in offshore areas (Buck 2009b). In this respect a great deal of discussion has begun on moving various kinds of uses to regions where more space is available, focusing specifically on resources, which could become scarce in the near future (e.g. production of food). However, one has to keep in mind that plausibility and profitability are incontrovertible constraints to any enterprise offshore, especially when combining them into a multi-use concept. Some concepts to move industrial interests off the coast did not fulfil these

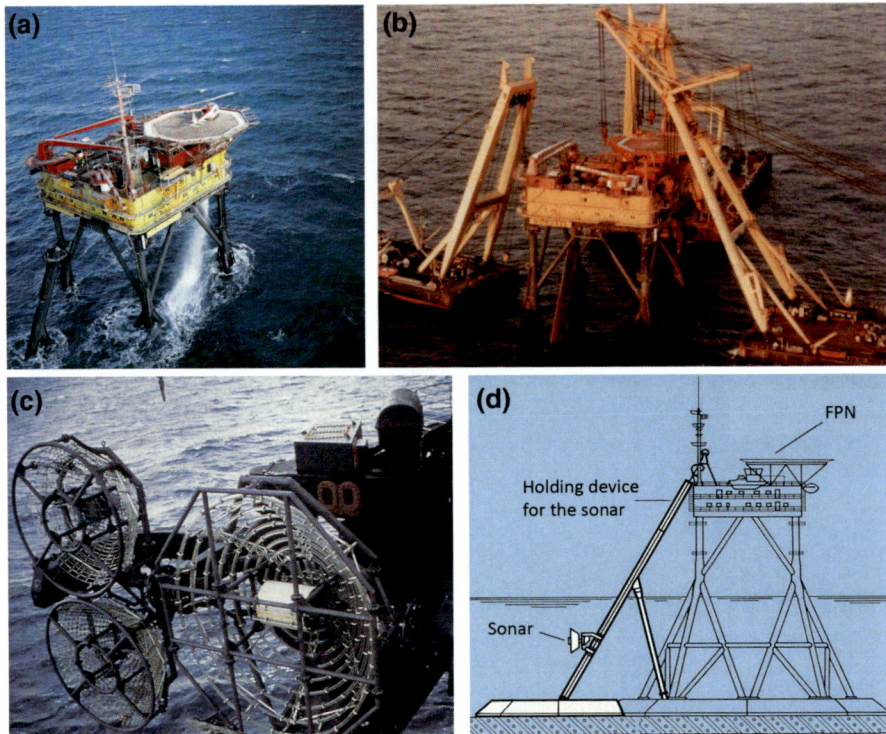

Fig. 1.6 a Research platform "Nordsee" (FPN) about 75 km off the German mainland;
b dismantling of the platform 20 years later; **c** sonar research device from the Federal Office of
Defense and Procurement (BWB) before positioning at the basement of the foundation; **d** drawing
of the sonar research device. All images modified after IMS 2016

requirements. For instance, the ChevronTexaco Corp plan to construct a US\$ 650
million offshore liquefied natural gas receiving and re-gasification terminal with
accommodation for personnel (to be located 13 km off the coast of Baja California,
Mexico) (ChevronTexaco 2003) could not be realized as originally conceived due
to escalating costs. The *Forschungsplattform Nordsee* (FPN, Research Platform
"North Sea"), which was constructed for 35 million DM[1] in 1974 about 75 km NW
off the Island of Helgoland (Germany) housed 14–25 people, a helicopter landing
site as well as a jetty (Fig. 1.6a–d), and were equipped for a number of different
functions, including marine ecology, oceanography, and climate research by natural
scientists, underwater technology and sensors by engineers as well as defence
technology by the former Federal Office of Defence and Procurement (BWB).
However, even this met, over the course of time, a similar fate. The platform was
dismantled in 1993 due to high maintenance and operational costs (Dolezalek 1992).

---

[1]DM = Former German currency, 1 DM ≈ 0.5 €.

Hundreds of offshore future visions, such as the concepts for space, land and sea of *Agence Jacques Rougerie Architecte* (Rougerie 2012) or the carbon-neutral self-sufficient offshore farming platform, called *Equinox* (FDG 2011), exist on paper, but are yet far away from practical realization. Other uses that could have an economic potential but have not been realized so far are passive fishing in combination with other uses in the open ocean. Furthermore, there is strong interest in the production of freshwater off the coast in areas with a significant lack of freshwater supply (He et al. 2010). Although there has been plenty of research into the use of renewable energy to power the desalination process (Carta et al. 2003; Forstmeier et al. 2007; Heijman et al. 2010) no offshore demonstration has been carried out so far.

This book pulls the different strands of investigations in this new emerging field together and provides an overview of the current state-of-the-art of the research fields involved. Out of an array of different possible offshore renewable energy systems, offshore wind farms are most advanced in practical terms. Thus, the expertise focuses strong on these systems and its potential link with offshore aquaculture. The suitability of aquaculture production together with or at offshore wind energy sites will be discussed in detail.

References

Asche, F., Roll, K. H., & Tveterås, S. (2008). Future trends in aquaculture: Productivity growth and increased production. In: M. Holmer, K. Black, C. M. Duarte, N. Marba & I. Karakassis (Eds.), *Aquaculture in the ecosystem* (pp. 271–92). Berlin: Springer.

Atalah, J., Fletcher, L. M., Hopkins, G. A., Heasman, K., Woods, C. M. C., & Forrest, B. M. (2016). Preliminary assessment of biofouling on offshore mussel farms. Journal of the World Aquaculture Society.

Avault, J. W., & Guthrie, P. W. (1986). The formation and history of the world mariculture society, 1969–1986. *Journal of the World Aquaculture Society, 17,* 64–71.

Benetti, D., O'Hanlon, B., Brand, L., Orhun, R., Zink, I., Doulliet, P., Collins, J., Maxey, C., Danylchuk, A., Alston, D., & Cabarcas, A. (2006). Hatchery, on growing technology and environmental monitoring of open ocean aquaculture of cobia (*Rachycentron canadum*) in the Caribbean. World Aquaculture Society, Abstract. In *Proceedings of Aquaculture 2006*, Florence Italy.

Berkenhagen, J., Doring, R., Fock, H., Kloppmann, M., Pedersen, S. A., & Schulze, T. (2010). Conflicts about spatial use between wind farms and fisheries—What is not implemented in marine spatial planning. *Inf. Fischereiforsch., 57,* 23–26.

Bohnsack, D. E., McCellan, D. B., & Hulsbeck, M. (1994). Effects of reef size on colonization and assemblage structure of fishes at artificial reefs off Southeastern Florida. *Mar. Sci., 55,* 796–823.

Bohnsack, J. A., Johnson, D. L., & Ambrose, R. E. (1991). Ecology of artificial reef habitats and fishes. In W. Seaman & L. M. Sprague (Eds.), *Artificial habitats for marine and freshwater fisheries* (pp. 61–108). Cambridge, MA, USA: Academic Press.

Boulet, D., Struthers, A., & Gilbert, E. (2010). Feasibility study of closed-containment options for the British Columbia aquaculture industry, Retrieved June 12, 2016, from http://publications.gc.ca/pub?id=9.694792&sl=0

Braginton-Smith, B., & Messier, R. H. (1998). Design concepts for integration of open ocean aquaculture and Osprey™ Technology. In W. H. Howell, B. J. Keller, P. K. Park, J. P. McVey, K. Takayanagi & Y. Uekita (Eds.), *Nutrition and Technical Development of Aquaculture, Proceedings of the Twenty-Sixth U.S.-Japan Aquaculture Symposium*, Durham/New Hampshire/USA, September 16–18, 1997 (pp. 239–245). UJNR Technical Report No. 26, Durham, University of New Hampshire Sea Grant Program.

Bridger, C. J. (Ed.) (2004). *Efforts to develop a responsible offshore aquaculture industry in the Gulf of Mexico: A compendium of offshore aquaculture consortium research*. MASGP-04-029. Mississippi-Alabama Sea Grant Consortium, Ocean Springs, MS.

Buck, B. H. (2001). *Combined utilization of wind farming and mariculture in the North Sea*. Alfred-Wegener-Institute for Polar and Marine Research, 2000–2001 Report: 33–39.

Buck, B. H. (2002). *Open Ocean Aquaculture und Offshore-Windparks: Eine Machbarkeitsstudie über die multifunktionale Nutzung von Offshore-Windparks und Offshore-Marikultur im Raum Nordsee*. Reports on Polar and marine research. Bremerhaven: Alfred Wegener Institute for Polar and Marine Research, pp. 412–252.

Buck, B. H. (2004). *Farming in a high energy environment: potentials and constraints of sustainable offshore aquaculture in the German Bight (North Sea)*. Germany: Dissertation. University of Bremen.

Buck, B. H. (2009a). Meeting the quest for spatial efficiency: Progress and prospects of extensive aquaculture within offshore wind farms in Europe. The Ecology of Marine Wind Farms: Perspectives on Impact Mitigation, Siting, and Future Uses. Keynote Speaker, 8th Annual Ronald C. Baird Sea Grant Science Symposium, November 2–4, 2009, Newport, Rhode Island (USA).

Buck, B. H. (2009b). Seaweed aquaculture in the open coastal waters, aquaculture and food. *Special edition, 4*, 100–103.

Buck, B. H. (2013). Upscaling aquaculture operations in offshore environments—challenges and possibilities in Europe. Submariner-Conference - Innovative uses of Baltic marine resources in the light of the EU Blue Growth initiative. 5–6, September 2013, Gdansk (Poland).

Buck, B. H., & Buchholz, C. M. (2004). The offshore-ring: A new system design for the open ocean aquaculture of macroalgae. *Journal of Applied Phycology, 16*, 355–368.

Buck, B. H., & Krause, G. (2013). *Short expertise on the potential combination of aquaculture with marine-based renewable energy systems*. SeaKult-Sustainable Futures in the Marine Realm. Expertise für das WBGU-Hauptgutachten "Welt im Wandel: Menschheitserbe Meer". Wissenschaftlicher Beirat der Bundesregierung Globale Umweltveränderungen (WBGU).

Buck, B. H., & Krause, G. (2012). Integration of aquaculture and renewable energy systems, In R. A. Meyers (Ed.), *Encyclopaedia of sustainability science and technology* (Vol. 1). Springer Science+Business Media LLC. Chapter No. 180, pp. 511–533.

Buck, B. H., Krause, G., Rosenthal, H., & Smetacek, V. (2003). Aquaculture and environmental regulations: The German situation within the North Sea. In: A. Kirchner (Ed.), *International marine environmental law: institutions, implementation and innovation*, vol. 64. The Hague: International Environmental. Law and Policies Series of Kluwer Law International.

Buck, B. H., Krause, G., & Rosenthal, H. (2004). Extensive open ocean aquaculture development within wind farms in Germany: The prospect of offshore co-management and legal constraints. *Ocean and Coastal Management, 47*, 95–122.

Buck, B. H., Krause, G., Michler-Cieluch, T., Brenner, M., Buchholz, C. M., Busch, J. A., et al. (2008). Meeting the quest for spatial efficiency: Progress and prospects of extensive aquaculture within offshore wind farms. *Helgoland Marine Research, 62*, 269–281.

Buck, B. H., Dubois, J., Ebeling, M. W., Franz, B., Goseberg, N., Hundt, M., et al. (2012). *Mulitple Nutzung und Co-Management von Offshore-Strukturen: Marine Aquakultur und Offshore-Windparks*. Open Ocean Multi-Use (OOMU)-Projektbericht, gefördert durch das Bundesministerium für Umwelt, Naturschutz und Reaktorsicherheit.

Bugrov, L. (1991). Alternative using of petroleum-gas structures in the caspian and the black seas for fish-farming and fishing. In *5th International Conference on Aquatic Habitat Enhancement* (Artificial Habitats for Fisheries), Long Beach, California, USA.

Bugrov, L. (1992). Rainbow trout culture in submersible cages near offshore oil platforms. *Aquaculture, 100,* 169.

Bugrov, L. (1996). Underwater fish-farming technology for open sea areas: Review of a 10-year experience. In *Open Ocean Aquaculture* (pp. 269–296). Portland, Maine: UNHMP-CP-SG-96-9. Sea Grant College Program.

Bugrov, L. (2016). *Personal communication and provision of images.* Saint Petersburg, Russia: State Research Institute on Lake and River Fisheries.

Buschmann, A. H., Hernández-González, M. C., Aranda, C., Chopin, T., Neori, N., Halling, C., & Troell, M. (2008). Mariculture waste management. In S. E. Jørgensen & B. D. Fath (Ed.), *Ecological engineering. Vol. 3 of encyclopedia of ecology* (pp. 2211–2217). Oxford: Elsevier.

Butcher, D. (2004). *The closing of the frontier. A history of the marine fisheries of Southeast Asia c. 1850–2000.* Singapore: ISEAS/KITLV.

Carta, J. A., Gonzalez, J., & Subiela, V. (2003). Operational analysis of an innovative wind powered reverse osmosis system installed in the Canary Islands. *Solar Energy, 75,* 153–168.

CC. (2016). Custodian Consultancy Ltd., 1B Damastown Way, Mulhuddart, Dublin 15, D15 NNOF.

ChevronTexaco. (2003). ChevronTexaco plans offshore LNG terminal. *Filtration Industry Analyst, 12,* 3.

Chunrong, W., Liangjie, D., Guoxi, L., & Yuxian, X. (1994). Utilizing wind energy to develop aquaculture industry. In L. Nan, G. Best, C. C. de Carvalho Neto, (Eds.), *Integrated energy systems in China—The cold Northeastern region experience.* Rome: Food and Agriculture Organisation of the United Nations.

Cicin-Sain, B., Bunsick, S. M., DeVoe, R., Eichenberg, T., Ewart, J., Halvorson, H., et al. (2001). *Development of a policy framework for offshore marine aquaculture in the 3–200 Mile U.S. Ocean Zone.* Report of the Centre for the Study of Marine Policy, University of Delaware.

Claudet, J., & Pelletier, D. (2004). Marine protected areas and artificial reefs: A review of the interactions between management and scientific studies. *Aquatic Living Resource, 17,* 129–138.

Cushing, D. H. (1988). *The provident sea.* Camebridge: Camebridge University Press.

Christy, F. T., & Scott, A. (1965). *The common wealth in ocean fisheries. Some problems of growth and economic allocation.* Baltimore: Johns Hopkins U.P.

Dalton, R. (2004). Fishing for trouble. *Nature, 431,* 502–504.

Dauterive, L. (2000). *Rigs-to-reef policy, progress, and perspective.* OCS Report (U.S. Department of the Interior), pp. 1–12.

Dempster, T., Sanchez-Jerez, P., Bayle-Sempere, J. T., & Kingsford, M. J. (2004). Extensive aggregations of wild fish at coastal sea-cage fish farms. *Hydrobiologia, 525,* 245–248.

Dempster, T., Uglem, I., Sanchez-Peres, P., Fernandez-Jover, D., Bayle-Sempere, J., Neilsen, R., et al. (2009). An ecosystem effect of salmonid farming: Extensive and persistent aggregations of wild fish. *Marine Ecology Progress Series, 385,* 1–14.

Dolezalek, H. (1992). *Oceanographic research towers in European waters.* ONR Europe Reports. Office of Naval Research European Office 92-7-R.

Duarte, C. M., Marbà, N., & Holmer, M. (2007). Rapid domestication of marine species. *Science, 316,* 382–383.

FAO. (2016). *Food and agriculture statistics service, fisheries and aquaculture department.* Rome: Food and Agriculture Organization of the United Nations (Accessed June 24, 2016).

FAO. (2006). *FAO—fisheries technical paper. No. 500. State of world aquaculture 2006.* Rome: Food and Agriculture Organization of the United Nations.

FDG. (2011). Equinox—A carbon-neutral self-sufficient offshore farming platform. Retrieved June 15, 2016 from http://www.formationdesign.com

Forstmeier, M., Mannerheim, F., D'Amato, F., Shah, M., Liu, Y., Baldea, M., et al. (2007). Feasibility study on wind-powered desalination. *Desalination, 203,* 463–470.

Fredheim, A., & Langan, R. (2009). Advances in technology for offshore and open ocean aquaculture. In G. Burnell & G. Allen (Eds.), *New technologies in aquaculture: Improving production efficiency, quality and environmental management* (pp. 914–942). Cambridge, UK: Woodhead Publications.

Glantz, M. H. (1986). Man, state, and fisheries: An inquiry into some societal constraints that affect fisheries management. *Ocean Development and International Law, 17*, 191–270.

Hall, D. (2003). The international political ecology of industrial shrimp aquaculture and industrial plantation forestry in Southeast Asia. *Journal of Southeast Asian Studies, 34*, 251–264.

He, X., Li, C., & Gu, W. (2010). Research on an innovative large-scale offshore wind power seawater desalination system. In *World Non-Grid-Connected Wind Power and Energy Conference (WNWEC)*.

Helsley, C. E., & Kim, J. K. (2005). Mixing downstream of a submerged fish cage: A numerical study. *IEEE Journal of Oceanic Engineering, 30*, 12–19.

Heijman, S. G., Rabinovitch, E., Bos, F., Olthoff, N., & van Dijk, J. C. (2010). Sustainable seawater desalination: Stand-alone small scale windmill and reverse osmosis system. *Desalination, 248*, 114–117.

Holm, P., Byskov, S., & Toft-Hansen, S. (1998). *Proteiner fra havet: Fiskemelsindustrien i Esbjerg, 1948–1998*. Esbjerg: Fiskeri- og Søfartsmuseet.

Howell, W. H., Watson, W. H., & Chambers, M. D. (2006). Offshore production of cod, haddock and halibut. CINEMar/Open Ocean Aquaculture Annual Progress Report for the Period from January 01–31, 2005. Final Report for NOAA Grant No. NA16RP1718, interim Progress Report for NOAA Grant No. NA04OAR4600155.

Hovland, E., et al. (2014). *Norges Fiskeri- og Kysthistorie V*. Fagbokforlaget: Bergen.

IMS. (2016). IMS ingenieurgesellschaft mbH, Hamburg.

Jacobi, S. L. (1768). Von der künstlichen Erzeugung der Forellen und Lachse. Lippische Intelligenzblätter, pp. 697–704; 709–720. Retreived May 12, 2016, from http://s2w.hbz-nrw.de/llb/periodical/pageview/124064

Jensen, A., Collins, K. J., & Lockwood, A. P. (2000). *Artificial reefs in European seas*. Dordrecht: Kluwer Academic Publishers.

Kaiser, M. J., Snyder, B., & Yu, Y. (2011). A review of the feasibility, costs, and benefits of platform-based open ocean aquaculture in the Gulf of Mexico. *Ocean and Coastal Management, 54*, 721–730.

Lacroix, D., & Pioch, S. (2011). The multi-use in wind farm projects: more conflicts or a win-win opportunity? *Aquatic Living Resources, 24*, 129–135.

Langan, R., & Horton, C. F. (2003). Design, operation and economics of submerged longline mussel culture in the open ocean. *Bulletin of the Aquaculture Association of Canada, 103*, 11–20.

Langan, R. (2007). Results of environmental monitoring at an experimental offshore farm in the Gulf of Maine: Environmental conditions after seven years of multi-species farming. In: C. S. Lee & P. J. O'Bryen (Eds.), *Open ocean aquaculture—Moving forward* (pp. 57–60). Oceanic Institute.

Langan, R. (2012). Ocean cage culture. In J. H. Tidwell (Ed.), *Aquaculture production systems*. Oxford, UK: Wiley-Blackwell.

Leonhard, S. B., Stenberg, C., & Støttrup, J. (2011). *Effect of the horns rev 1 offshore wind farm on fish communities. Follow-up seven years after construction*. DTU Aqua, Orbicon, DHI, NaturFocus. Report commissioned by The Environmental Group through contract with Vattenfall Vindkraft A/S.

Li, B. (1997). *Agricultural development in Jiangnan, 1620–1850, studies on the Chinese Economy*. London: St. Martin's Press.

Loverich, G. F. (1997). *A summary of the case against the use of gravity cages in the sea farming industry*. Nr. 091997.

Loverich, G. F. (1998). Recent practical experiences with ocean Spar^R offshore sea cages. In R. R. Stickney (Ed.), *Joining forces with industry—open ocean aquaculture* (pp. 78–79).

Proceedings of the 3rd Annual International Conference, May 10–15, Corpus Christi, Texas. TAMU-SG-99-103, Corpus Christi, Texas Sea Grant College Program.

Loverich, G. F., & Gace, L. (1997). The effect of currents and waves on several classes of offshore sea cages. In C. Hesley (Ed.), *Open ocean aquaculture: chartering the future of ocean farming* (pp. 131–144). Proceedings of an International Conference, April 23–25, 1997, Maui, Hawaii. UNIHI-Seagrant-CP-98-08, Maui, University of Hawaii Sea Grant College Program.

Loverich, G. F., & Forster, J. (2000). Advances in offshore cage design using spar buoys. *Marine Technology Society Journal, 34,* 18–28.

McCann, M. (2003). *The Roman port and fishery of Cosa.* Rome: American Academy. in Rome.

Miget, R. J. (1994). The development of marine fish cage culture in association with offshore oil rigs. In K. L. Main & C. Rosenfeld (Eds.), *Culture of high value marine fishes in Asia and the United States* (pp. 241–248). Proceedings of a Workshop in Honolulu, Hawaii, August 8–12, 1994. The Oceanic Institute.

Natale, F., Hofherr, J., Fiore, G., & Virtanen, J. (2013). Interactions between aquaculture and fisheries. *Marine Policy, 38,* 205–213.

Ostrowski, A. C., & Helsley, C. E. (2003). The Hawaii offshore aquaculture research project: Critical research and development issues for commercialization. In: C. J. Bridger & B. A. Costa-Pierce (Eds.), *Open ocean aquaculture: From research to commercial reality* (pp. 119–128). Baton Rouge, Louisiana, USA: The World Aquaculture Society.

Polovina, J. J., & Sakai, I. (1989). Impacts of artificial reefs on fishery production in Shimamaki. *Japan Marine Science, 44,* 997–1003.

Reggio, V. C. (1987). *Rigs-to-reefs: The use of obsolete petroleum structures as artificial reefs.* OCS Report MMS 87-0015. New Orleans, U.S. Department of the Interior, Minerals Management Service, Gulf of Mexico OCS Region.

Rougerie (2012) Agence Jacques Rougerie Architecte, Péniche Saint Paul, Port des Champs-Elysées 75008 Paris. Retrieved June 2, 2016, from http://www.rougerie.com/12.html

Ryan, J. (2004). *Farming the deep blue.* Bord Iascaigh Mhara Technical Report. Dublin, Ireland: BIM.

Serjeantson, D., & Woolgar, C. M. (2006). Fish consumption in medieval England. In C. M. Woolgar, D. Serjeantson, & T. Waldron (Eds.), *Food in medieval England.* Oxford: Oxford University Press.

Ward, L. G., Grizzle, R. E., Bub, F. L., Langan, R., Schnaittacher, G., & Dijkstra, J. (2001). *New Hampshire open ocean aquaculture demonstration project, site description and environmental monitoring report on activities from fall 1997 to winter 2000.* Durham, New Hampshire: University of New Hampshire, Jackson Estuarine Laboratory.

Ward, L. G., Grizzle, R. E., & Irish, J. D. (2006). *UNH OOA environmental monitoring program, 2005.* CINEMar/Open Ocean Aquaculture Annual Progress Report for the Period from January 01, 2005 to December 31, 2005. Final Report for NOAA Grant No. NA16RP1718, Interim Progress Report for NOAA Grant No. NA04OAR4600155.

Wever, L., Krause, G., & Buck, B. H. (2015). Lessons from stakeholder dialogues on marine aquaculture in offshore wind farms: Perceived potentials, constraints and research gaps. *Marine Policy, 51,* 251–259.

Wilson, C. A., & Stanley, D. R. (1998). Constraints of operating on petroleum platforms as it relates to mariculture: Lessons from research. In R. R. Stickney (Ed.), *Joining forces with industry—Open ocean aquaculture* (p. 60). Proceedings of the 3rd Annual International Conference, May 10–15, Corpus Christi, Texas. TAMU-SG-99-103, Corpus Christi, Texas Sea Grant College Program.

Zalmon, I. R., Novelli, R., Gomes, M. P., & Faria, V. V. (2002). Experimental results of an artificial reef programme on the Brazilian coast north Rio de Janeiro. *ICES Journal of Marine Science, 59,* 83–87.

Ziegler, A. C. (2002). *Hawaiian natural history, ecology and evolution.* Honolulu: University of Hawaii Press.

Zink, E. (2013). *Hot science, high water. Assembling nature, society and environmental policy in contemporary Vietnam.* Copenhagen: Nordic Institute of Asian Studies.

Part I
Species, Techniques and System Design

Chapter 2
Offshore and Multi-Use Aquaculture with Extractive Species: Seaweeds and Bivalves

Bela H. Buck, Nancy Nevejan, Mathieu Wille, Michael D. Chambers and Thierry Chopin

Abstract Aquaculture of extractive species, such as bivalves and macroalgae, already supplies a large amount of the production consumed worldwide, and further production is steadily increasing. Moving aquaculture operations off the coast as well as combining various uses at one site, commonly called multi-use aquaculture, is still in its infancy. Various projects worldwide, pioneered in Germany and later accompanied by other European projects, such as in Belgium, The Netherlands, Norway, as well as other international projects in the Republic of Korea and the USA, to name a few, started to invest in robust technologies and to investigate in system design needed that species can be farmed to market size in high energy environments. There are a few running enterprises with extractive species offshore, however, multi-use scenarios as well as offshore IMTA concepts are still on project scale. This will change soon as the demand is dramatically increasing and space is limited.

B.H. Buck (✉)
Marine Aquaculture, Maritime Technologies and ICZM, Alfred Wegener Institute
Helmholtz Centre for Polar and Marine Research (AWI),
Bussestrasse 27, 27570 Bremerhaven, Germany
e-mail: Bela.H.Buck@awi.de

B.H. Buck
Applied Marine Biology, University of Applied Sciences Bremerhaven,
An der Karlstadt 8, 27568 Bremerhaven, Germany

N. Nevejan · M. Wille
Laboratory of Aquaculture & Artemia Reference Center,
Department of Animal Production, Faculty of Bioscience Engineering,
Ghent University, Campus Coupure F, Coupure Links 653, 9000 Ghent, Belgium

M.D. Chambers
School of Marine Science and Ocean Engineering, University of New Hampshire,
Morse Hall, Room 164, Durham, NH, USA

T. Chopin
Canadian Integrated Multi-Trophic Aquaculture Network, University of New Brunswick,
100 Tucker Park Road, Saint John, NB E2L 4L5, Canada

© The Author(s) 2017
B.H. Buck and R. Langan (eds.), *Aquaculture Perspective of Multi-Use Sites in the Open Ocean*, DOI 10.1007/978-3-319-51159-7_2

2.1   Sustainable Aquaculture

The development of sustainable aquaculture is aimed at insuring that commercial aquaculture has minimal adverse effects on the environment. One way to achieve this goal is through the development of improved methods of waste management for land based, coastal and offshore aquaculture by combining extractive and fed aquaculture, also referred to as integrated multi-trophic aquaculture (IMTA) systems.

Fish excrete nitrogen (N), phosphorus (P) and carbon (C) (Beveridge 1987; Mugg et al. 2000; Neori et al. 2004, 2007; Corey et al. 2014). Nearly 50 kg N and 7 kg P can be released per ton of finfish produced per year (Chopin et al. 1999; Kautsky et al. 1999; Troell et al. 2003; Kim et al. 2013). In coastal waters, high levels of these nutrients can trigger harmful microalgal blooms (red tides) and contribute to excessive growth of nuisance or opportunistic macroalgae (green and brown tides), which in turn have negative consequences on coastal ecosystems and economies. These nutrients could instead be used to support the growth of economically important seaweeds, which would compete for nutrients with nuisance species, especially in nearshore coastal environments, hence mitigating these potentially adverse environmental impacts (Neori et al. 2004, 2007; Chopin et al. 2008; Pereira and Yarish 2008; Abreu et al. 2009, 2011b; Buschamnn et al. 2008; Corey et al. 2012, 2014; Kim et al. 2013, 2014a, 2015a). Seaweeds take up N, P and C, which they use for growth and production of proteins and energy storage products (mostly carbohydrates). When seaweeds are harvested from IMTA or nutrient bio-extraction systems, the nutrients are also removed from the environment. Seaweeds can then be used on for bio-based, high-valued compounds for human consumption, protein sources in finfish aquaculture diets, sources of phycocolloids, cosmeceuticals, nutraceuticals and other biochemicals, and for low-value commodity energy compounds such as biofuels, biodiesels, biogases and bioalcohols (Horn et al. 2000; Smit 2004; Chopin et al. 2011; Cornish and Garbary 2010; Gellenbeck 2012; Kim 2011).

Integration of shellfish with cage culture of fish can also help to reduce the risk of eutrophication since the particulate organic matter (POM) produced by fish (wasted feed and faeces) and the increased plankton production serve as excellent feed and are filtered out by these organisms. Faster growth (between 30 and 40% greater) of bivalves near fish cages has been reported with contributions of fish feed and fish faeces varying between 5–28% and 4–35%, respectively (Wallace 1980; Jones and Iwama 1991; Stirling and Okomus 1995; Buschmann et al. 2000; Lefebvre et al. 2000; Lander et al. 2004; Peharda et al. 2007; Chopin et al. 2008; Sara et al. 2009; Handå et al. 2012; Jiang et al. 2013; Dong et al. 2013). Other studies, however, did not observe a difference in growth near fish cages (Mazzola and Sara 2001; Navarrete-Mier et al. 2010). Several explanations for these contradicting results have been given by e.g. Troell and Norberg (1998), Troell et al. (2011), Handå (2012) and Reid et al. (2013): (1) the POM generated by the fish culture doesn't increase the seston concentration significantly due to dilution;

(2) the shellfish culture infrastructures were not adequately positioned to intercept the POM plume; (3) the POM produced by fish sinks very quickly to the bottom where it is no longer accessible for the filter feeders; (4) shellfish don't adapt fast enough to the pulse feeding system used for feeding the fish; (5) they only filter particulate waste product when the natural plankton production is low; (6) the seston concentration and size reach the limits where pseudo-faeces are produced.

There are limited interactions between seaweed and bivalve cultures as seaweeds feed on inorganic nutrients and bivalves on organic nutrients from the water column and, therefore, can be regarded as co-cultures with separate nutrient models, although the excretion of metabolic waste products by bivalves enhances the availability of inorganic nutrients (Jansen 2012).

2.2 Introduction to Extractive Species

Marine extractive species include a large variety of species, which can be subdivided into three main groups among animals and algae: (1) filter feeders, such as oysters and mussels, (2) deposit feeders, such as polychaetes, sea urchins and sea cucumbers, as well as (3) dissolved nutrient absorbers, such as microalgae and macroalgae. These species act as living filters and can be raised without supplemental feed as they take up nutrients for nourishment from the surrounding water column. While filter and deposit feeders preferentially use small and large POM for their nutrition, algae extract dissolved inorganic nutrients (DIN) from the water column. The POM mainly consists of naturally occurring seston and uneaten fish feed, faeces and bacterial matts in aquaculture operations. The dissolved fraction consists of inorganic nitrogen (N), phosphorus (P) and carbon (C) available from nature and released from fed aquaculture operations. As deposit feeders are not yet often used in offshore environments or in multi-use platforms, they are not discussed further within this chapter.

2.3 IMTA on Offshore Applications

There have been technology exchanges between Asian and western countries. For example, traditional seaweed cultivation technologies have been exported from Asia to the West, while the concept of ecosystem services (Costanza et al. 1997) has been transferred to Asia. Scientists in Asia and the West are integrating these technologies to increase production in an environmentally friendly manner. In turn, bivalve cultivation was mainly developed in Europe (and to some extend in New Zealand) and expertise and technology were transferred to North and South American countries, as well as Asia, Australia and countries in the Pacific Ocean.

One conceptual approach driven by the various stakeholders using coastal waters is to transfer aquaculture operations away from nearshore areas. Moving the production of seaweeds and bivalves off the coast in offshore environments could solve spatial and environmental problems currently existing in coastal seas. The production of extractive species in offshore areas could solve the global issue of providing large amounts of biomass to the end user. There are several reports available examining the feasibility of moving seaweed and bivalve aquaculture towards the more hostile environment of the open ocean, in combination or not with offshore structures. Nevertheless, plans to connect such culture devices to offshore foundations of wind turbines are still in their infancy (Buck and Krause 2012). A non-exhaustive list is presented below:

- 2000: Open Ocean Aquaculture within Offshore Wind Farms. A Feasibility Study Concerning the Multifunctional Use of Offshore Wind Farms and Offshore Aquaculture within the North Sea (Buck 2002)
- 2004: Farming in a High Energy Environment: Potentials and Constraints of Sustainable Offshore Aquaculture in the German Bight (North Sea) (Buck 2004, 2007)
- 2012: Integration of Aquaculture and Renewable Energy Systems (Buck and Krause 2012)
- 2013: Short Expertise on the Potential Combination of Aquaculture with Marine-Based Renewable Energy Systems (Buck and Krause 2013)
- 2013: Aquaculture in Welsh Offshore Wind Farms: A Feasibility Study into Potential Cultivation in Offshore Wind Farm Sites (Syvret et al. 2013).
- 2013: Triple P Review of the Feasibility of Sustainable Offshore Seaweed Production in the North Sea (van den Burg et al. 2013)
- 2014: Combining Offshore Wind Energy and Large-Scale Mussel Farming: Background & Technical, Ecological and Economic Considerations (Lagerveld et al. 2014)
- 2015: Go Offshore—Combining Food and Energy Production (Carlberg and Christensen 2015)
- 2015: Aquaculture Pilot Scale Report (TROPOS 2015).

2.4 Extractive Species Aquaculture

2.4.1 Seaweeds

Seaweeds are part of the cultural heritage of Asian countries, much more so that in Western countries. Nevertheless, according to a recent archaeological study, cooked and partially eaten seaweed were found at a 14,000-year-old site in southern Chile, suggesting seaweeds have been part of human diet for a very long time and in other parts of the world as well (Dillehay et al. 2008). Global seaweed aquaculture production represents 49.1% of total world mariculture production by weight, with an annual value of US $6.4 billion (FAO 2014; Chopin 2014). From an estimated

10,500 species of seaweeds, only six genera provide 98.9% of the production and 98.8% of the value: *Saccharina, Undaria, Porphyra, Gracilaria, Kappaphycus* and *Sargassum*. Unfortunately, not much of that is produced in the Western World as 96.3% of seaweed aquaculture is concentrated in six Asian countries: China (with over 54.0% of production), Indonesia, the Philippines, the Republic of Korea, Japan and Malaysia (Chopin 2014). Currently a total of 54,000 t of seaweed have been cultivated in the Americas and Europe with an annual value of US \$51 million in 2013 (FAO 2016a, b), which is even less than the value that Korea exported to the U.S. in the same period (US \$67 million; Meekyiung Kim, Korea Agro-Trade Center pers. comm.). Seaweed production in the Americas and Europe is still in its early stages compared to the vast production in Asia. Although seaweed aquaculture is a fairly new industry in the Americas and Europe, the market demand is expected to increase rapidly due to an increasing consumer demand for new protein sources and healthy food supplements and the food industry's interest in sustainable textural additives (Buchholz et al. 2012).

There are some characteristics that will foster the production of seaweed worldwide. They provide ecosystem services, which need to be recognized and valued appropriately (Chopin 2014). One often forgotten function of seaweeds is that they are excellent nutrient scrubbers and can be used for nutrient biomitigation of fed aquaculture or other sources of nutrification. Seaweeds can be cultivated without the addition of fertilizers and agrochemicals, especially in an IMTA setting, where the fed aquaculture component provides the nutrients. Seaweed cultivation does not require more arable soil and transformation of land for agricultural activities with accompanying loss of some ecosystem services. If appropriately designed, it can be seen as engineering new habitats and harboring thriving communities, and can be used for habitat restoration. Moreover, it does not need irrigation, on a planet where access to water of appropriate quality is becoming more and more an issue. As photosynthetic organisms, seaweeds are the only aquaculture component with a net production of oxygen. All other fed and organic extractive components are oxygen consumers. Hence, seaweeds contribute to the avoidance of coastal hypoxia. While performing photosynthesis, seaweeds also absorb carbon dioxide and hence participate in carbon sequestration, even if in a transitory manner. Consequently, they could be a significant player in the evolution of climate change, slowing down global warming, especially if their cultivation is increased and spread throughout the world. By sequestering carbon dioxide and increasing pH in seawater, seaweeds could also play a significant role in reducing ocean acidification at the coastal level (Clements and Chopin 2016).

The IMTA concept (Chopin et al. 2001; Troell et al. 2003; Neori et al. 2004; Chopin 2006) is a good illustration of how to take advantages of the ecosystem services provided by extractive species and fits very well within the concept of circular economy (Pearce and Turner 1989). Moreover, seaweed production doesn't compete with other food productions while delivering new biomass flows for animal feed, food, and non-food products (Buck 2004). For instance, sugars, proteins and fatty acids (principally omega 3s) from seaweeds can form an alternative to soya and fishmeal. Further, the use of seaweeds for the production of chemicals and biofuels could be a climate-friendly alternative to fossil raw materials.

Seaweed aquaculture technologies for land based, as well as sheltered nearshore areas, have developed dramatically in Asia and to a lesser extend in South America. In contrast, technologies to move offshore were developed in Americas and Europe over the last 15 years. There are still challenges to overcome, which vary depending on species, technologies and countries. In Europe, mainly kelps were tested in first offshore trials during the early 1990s (Lüning and Buchholtz 1996). France and Germany were the first investing countries for offshore seaweed cultivation followed by UK and Ireland, 10 years later. Today, the Netherlands, Denmark, Spain, Portugal, as well as Norway, are carrying out seaweed farming trials off the coast. Most of the investigations are still carried out on a pilot scale. Only a few farms off the coast are in early commercial operations, such as enterprises in Germany (e.g. CRM 2001), Norway (e.g. SES 2015a, b), and The Netherlands (Hortimare 2016).

2.4.2 Bivalves

In 2014, the annual production of bivalves, including mussels and oysters, reached approximately 7 million t, which is equivalent to one quarter of the annual seaweed production (approximately 27 million t) (FAO 2016a, b). In Europe, the blue mussel (*Mytilus edulis*) and the Mediterranean mussel (*Mytilus galloprovincialis*) are the main cultured species reaching approximately 500,000 t annually. The main producers are Spain (220,000 t), Italy (80,000 t), France (75,000 t), the Netherlands (55,000 t), the UK (23,000 t) and Germany (6000 t) (FAO 2016a, b). Other important countries outside of Europe producing mussels are China (800,000 t), Chile (240,000 t), Thailand (125,000 t), South Korea (52,000 t), New Zealand (38,000 t), Canada (26,000 t), Brazil (20,000 t) and the Philippines (19,000 t). In 2014, approximately 5.2 million t of oysters were produced worldwide, dominated by the Pacific oyster (*Crassostrea gigas*). In Europe, the introduced Pacific oyster reaches ≈90,000 t headed by France with ≈75,000 t, followed by Ireland with 9000 t, the Netherlands with ≈2500 t and Norway with ≈2000 t. Other countries outside of Europe are China with ≈4.3 million t, South Korea with ≈285,000 t, Japan with ≈185,000 t, the USA with ≈125,000 t, the Philippines, Taiwan and Thailand with ≈22,000 t, and Australia and Canada with ≈12,000 t. The European flat oyster (*Ostrea edulis*) is cultured to a minor extent only with approximately 2500 t in 2014 led by France and followed by Ireland, Spain and the Netherlands (FAO 2016a, b).

2.5   Cultivation Technologies, Challenges and Future Directions in Major Cultured Extractive Species

Below we provide some basic information on the cultivation of red and brown macroalgae to allow a better understanding of the entire production cycle. Mussel and oyster cultivation methods are explained as well. However, as this book is mainly focusing on the future development of extractive species off the coast we do not go in detail with the complete reproduction of these species and will describe further the cultivation only with regard to the transfer offshore and its relevance to modern technical design. Further information into the biology and cultivation can be gained by reading the literature on fundamental seaweed and bivalve biology and ecology, such as Kim (2012), Wiencke and Bischof (2012) and Hurd et al. (2014) for seaweeds, as well as Matthiessen (2001), Gosling (2003) and Shumway (2011) for bivalves.

2.5.1   Red Seaweeds

2.5.1.1   *Pyropia* and *Porphyra* ('Gim' in Korean or 'Nori' in Japanese)

Although a total of 138 species of *Pyropia* and *Porphyr*a are currently accepted taxonomically (Guiry and Guiry 2015), only 4 major species (*Py. yezoensis, Py. tenera, Py. haitanensis* and *Po. umbilicalis*) have been cultivated, mostly in China, the Republic of Korea and Japan (99.99% of total production) and to some extend in the USA and Europe (FAO 2016a, b). The culture methods of *Pyropia/Porphyra* in these three Asian countries are basically similar, with minor modifications (Sahoo and Yarish 2005; Pereira and Yarish 2008, 2010). For example, some farmers use free-living conchocelis for seeding while others use conchocelis on oyster shells (He and Yarish 2006). Seedlings may be outplanted in the open water farms using one of three cultivation methods: fixed pole, semifloating raft, or floating raft (Sahoo and Yarish 2005). The epiphyte control techniques are also different, based on the cultivation techniques. Most Chinese farms, and some Korean and Japanese farms use desiccation control methods by exposing the *Pyropia/Porphyra* nets to the air to kill epiphytes and competing organisms (e.g. *Ulva* spp.), while most Korean and Japanese farmers use the more efficient pH control method by spraying organic acids onto the nets (Pereira and Yarish 2010; Kim et al. 2014b) which is, however, more costly. The uncontrolled fouling organisms reduce the quality of the product. Recent reports suggest that the world's largest macroalgal blooms of *Ulva prolifera* are trapped on rafts of *Pyropia* farms in the Southern Yellow Sea of China (Liu et al. 2009; Zhang et al. 2014, 2016) before moving towards the Qingdao region. The initial propagules may, however, originate further south on the large intertidal radial mudflat in the Jiangsu Province where they find favorable conditions, enriched by nutrients discharges from coastal land-based animal aquaculture, before developing into green tides (Liu et al. 2013). In mid-1990s, Coastal Plantations International attempted to cultivate *Pyropia* (*Porphyra*) in the open waters of Maine, USA (Chopin et al. 1999; McVey et al. 2002) but was unsuccessful for economic and managerial reasons. More recently, *Porphyra umbilicalis* cultivation in Maine has been taken up by Brawley and her colleagues (Blouin et al. 2011). A seaweed manual has been published for seed-stock production of *Pyropia/Porphyra* in the USA (Redmond et al. 2014) with accompanying videos in English. However, *Pyropia/Porphyra* cultivation is still in its nascent stage in North America and Europe commercial enterprises. Therefore, it is critical to develop local cultivars and cultivation techniques suitable for the local environments and boutique markets. Selective breeding (intra-specific and inter-specific) of cultivated *Pyropia/Porphyra* has been intensively studied in Asia (Miura 1984; Shin 1999, 2003; Niwa et al. 2009). Genetic improvement has resulted in superior strains with higher growth rates, better flavor, darker color and higher tolerance to diseases (Chen et al. 2014).

Since 1880, Earth's average surface temperature has risen by about 0.8 °C and the majority of that warming has occurred in the past three decades (NASA 2015). Recently, NASA and NOAA reported that 2014 ranks as Earth's warmest year since 1880. The 10 warmest years since records were registered, have occurred since 2000, with the exception of 1998. Therefore, development of new strains with high thermo-tolerances will be necessary for the development of a sustainable seaweed aquaculture industry. With current technology, six to eight harvests can be obtained from the same nets during one growing season. Expansion of the growing season by using new strains will, therefore, result in higher production for growers. These current and developing cultivation technologies for *Porphyra/Pyropia* can be used for offshore cultivations. Simulation programs such as the hydrodynamic simulation program ANSYS-AQWA are useful tools to analyze the dynamic characteristics of a seaweed culture system (Lee et al. 2013).

2.5.2   Brown Seaweeds

2.5.2.1   *Saccharina* and *Undaria*

Kelps have been utilized mostly for human consumption, but recently it's been increasingly utilized as abalone feed thanks to relatively low production cost (Hwang et al. 2013). For example, over 60% of the total production of *Saccharina* and *Undaria* in the Republic of Korea was used in the abalone industry in 2012.

In western countries, kelp species (primarily the sugar kelp, *Saccharina latissima*) have been cultivated during the last two decades in the Atlantic Ocean (e.g. USA,

Canada, Norway, Scotland, Ireland, Sweden and Germany; Buck and Buchholz 2004a, 2005; Chopin et al. 2004; Barrington et al. 2009; Broch et al. 2013; Marinho et al. 2015; Kraemer et al. 2014; Kim et al. 2015a) and *Macrocystis* in the eastern Pacific Ocean (e.g. Chile; Buschmann et al. 2008). The kelp aquaculture industry in western countries has become one of the fastest growing industries.

For both *Saccharina* and *Undaria*, cultivation begins with zoospores (meiospores) for seeding. The seeding methods are a bit different between Asia (use of seed frames) and the West (use of seedspools), mainly due to the nursery capacities and the scale of operations of the open water farms (Redmond et al. 2014). However, the open water cultivation techniques use very similar longlines. Once the seedstring is outplanted in the open water farms, the kelp thalli will grow up to 2–5 m in length, even to 10 m sometimes (Pereira and Yarish 2008; Redmond et al. 2014).

Facing climate change, many efforts are deployed to develop kelp strains that grow fast, are resistant to disease and tolerate higher temperatures. Selective breeding and intensive selection of kelp strains in Asia, however, have reduced the genetic diversity and narrowed the germplasm base of the varieties in cultivation. This has led to decreased adaptability of these varieties to environmental changes and jeopardizes the industry expansion in Asia (Kawashima and Tokuda 1993; Li et al. 2009; Robinson et al. 2013). In North America and Europe, strain development will be a challenge. The collection of zoospore "seeds" has relied mostly on wild sources. Development of a seedbank for kelp species will provide a sustainable and reliable source of seedstock without impacting the natural beds of seaweeds. Having seaweeds with desirable morphological and physiological traits will also enhance production capacity of the seaweed industry. Other challenges in these parts of the world are the legal aspects and the social syndrome known as NIMBY ("Not In My Back Yard"). For example, in the USA at least 120 federal laws were identified that affect aquaculture either directly (50 laws) or indirectly (70 laws) and more than 1200 state statutes regulate aquaculture in 32 states (Getchis et al. 2008). Regulatory complexity is further increased when towns or counties are given jurisdiction over local waters. Social resistance has also been a major factor limiting the growth of aquaculture in the USA. The nearshore waters of the USA are heavily used, having both recreational (boating, fishing, swimming) and aesthetic (ocean and bay views from waterfront homes) values. For that reason, offshore cultivation has been suggested as an alternative to avoid stakeholder conflicts (Langan and Horton 2005; Rensel et al. 2011; Buchholz et al. 2012).

Since nutrients may be limiting in offshore environments, an IMTA systems may be more appropriate than monoculture (Troell et al. 2009; Buchholz et al. 2012). Considering current cultivation techniques, the kelp species should be one of the most appropriate groups to cultivate offshore due to their low maintenance requirement and ease of harvest in comparison to other cultured species. The IMTA practices of offshore farms are unlikely to experience user conflicts, which may result in fewer restrictions on farm size and greater economies of scale.

2.5.2.2 *Sargassum*

Sargassum is the most common brown macroalgae found in temperate, tropical, and subtropical waters worldwide with more than 346 species (Guiry and Guiry 2015). *Sargassum* species have traditionally been utilized for food and medicine in Asia. They continue to be wild harvested and cultivated in Japan, China and the Republic of Korea, for human consumption as sea vegetables and for use as a medicinal "seaweed herbs". Locally known as the "black vegetable" in China, *Sargassum* is valued for its high nutritional content and nutty flavor. It is added to salads, soups or vegetable dishes (Xie et al. 2013). *Sargassum* is utilized in Chinese medicine as an expectorant for bronchitis, and to treat laryngitis, hypertension, infections, fever, and goiter (Hou and Jin 2005). *Sargassum fusiforme* (formally "*Hizikia fusiformis*") cultivation was initiated in the early 1980s. Thus, the production and economic value is still low, approximately 150,000 t of production with an annual value of US $70 million in 2013 (FAO 2014). Nearly all *Sargassum* production takes place in China, including several species such as *S. thunbergii*, *S. fulvellum*, *S. muticum*, and *S. horneri* (Xie et al. 2013). In the Republic of Korea, *S. fusiforme* and *S. fulvellum* are being cultured (Hwang et al. 2006).

Traditional culture methods initially relied on the use of wild seedlings collected from natural beds. Groups of 3–4 seedlings, 5–10 cm in length, were inserted onto seeding ropes at intervals of 5–10 cm. These smaller seeding lines were then attached to a main longline placed at depths of 2–3 m, and cultivated from November to May (Sohn 1998; Hwang et al. 2006; Redmond et al. 2014). This dependency on wild seedlings resulted in the overharvesting of natural beds, so new culture methods were developed. Holdfast-derived seeding was the first step towards developing more efficient culture techniques for *Sargassum*. This type of culture takes advantage of the perennial nature of the holdfast, allowing farmers to reuse the holdfasts from the previous year's crops (Hwang et al. 1998). While plants may still be sourced from wild beds, the attached holdfasts can be re-used for the next season's crop by over-summering in the sea after harvest until the next growing season. While this allows for reuse of existing cultured plants, the resulting harvestable biomass tends to diminish after each year. Today, *Sargassum* lines are seeded with juvenile plants obtained from reproductive adults. Obtaining seedlings through sexual reproduction allows for mass production of new plants for seeding, and results in higher biomass yields (Hwang et al. 2007; Peng et al. 2013; Redmond et al. 2014). Fertilized eggs are gathered from mature fronds and "seeded" onto seed string by allowing juveniles to attach to seed lines with newly forming rhizoids. Once attached, seedlings are cultured in a nursery until ready for out-planting at sea, where they are transferred to submerged long lines until harvest. The attached holdfasts can also be re-used for multiple years without any further initiation of culture ropes. This is an economically reasonable cultivation method, but there are problems with fouling organisms. New technologies to reduce fouling problems are in urgent need for the sustainability and growth of the *Sargassum* aquaculture industry.

Sargassum is also an appropriate species to cultivate offshore. It even has certain advantages over kelps, such as a higher market value ($500 per t) and the unique cultivation technology of using a perrenating holdfast for multiple years.

2.5.3   Bivalves

2.5.3.1   Mussels

Mytilids present a high fecundity and the larval phase is mobile and free-living, a fact that facilitates a widespread distribution. The culture technique used is greatly influenced by the seed availability. Mussel culture worldwide is mainly relying on wild seed, either by capturing competent (=ready to settle) larvae by means of mussel seed collectors or by fishing young mussel seed (2 cm) from natural mussel banks, which is predominantly applied in The Netherlands, Germany and Denmark.

Mussels can be farmed on-bottom or off-bottom. On-bottom culture is based on the principle of transferring wild seed from areas where they have settled naturally to areas where they can be placed in lower densities to increase growth rates, and that facilitate harvest and predation control (Seed and Suchanek 1992; Seaman and Ruth 1997). Off-bottom cultivation methods includes (1) the bouchot method, where piles with seeded mussel ropes are planted in the intertidal zone and mussel spat grows to market size, (2) rafts, where ropes are seeded with young mussels and suspended vertically from a floating raft construction, and (3) longline culture, where mussel spat is attached to vertical ropes, called seed collectors, or on continuous ropes (e.g. New Zealand), which hang on a horizontal backbone including floating devises to support buoyancy (Buck et al. 2006b). Seed collectors may consist of coconut fiber ropes placed in the intertidal zone (France), short polyethylene droppers from rafts (Spain), specially designed collector droppers with increased surface (e.g. "christmas tree"), fixed on longlines or collector nets (e.g. Smartfarm®). Production time from settlement to market size depends very much on the species and production location (temperature and phytoplankton concentration) but varies between 18 and 36 months.

Longlines and raft systems are the two commonly and most successfully used mussel-farming systems. The longline system consists of one to two suspended horizontal headlines made of 12–15 cm polypropylene rope about 100–220 m in length and anchored at both ends. Single headlines are sometimes preferred as the gear required to lift and harvest a single line is far more economical and less labor intensive. The headlines are connected to large floats (usually of plastic, each with about 300 kg displacement), which together support a large number of 5 m culture droppers (ropes). The number of droppers used varies with hydrological and biological conditions but is normally in the order of 400–500 per (double headline) longline. Secure anchoring is necessary to prevent the longlines shifting in heavy seas or strong currents. In soft bottom sediments, traditional concrete or steel anchors of 250–500 kg are used. Depending upon the amount of movement encountered, one or two anchors are placed at the end of headlines. At exposed sites, a mid-anchor may be needed to control movement and stretch; screw anchors may also be used to maintain the position of the longline. Continuous seeding of entire loops of culture rope over 200 m long—rather than 5 m lengths—is now common and was invented in New Zealand. Harvesting has also become mechanized: one machine lifts and strips the ropes, washes, separates and packs the mussels into bags. Longlines may be sunk to greater depths to reduce aesthetic objections. The aquaculture area can then be shared with recreational fishing and pleasure boats. Raft culture is similar in many respects to longline culture, except that the raft commonly swings on a single mooring. As with longline culture, ropes are suspended from the raft. Most harvesting is done by hand since this technique has far less scope for mechanization, but over the last decade, mechanization also started. Raft culture in Europe is typically found in Spain. The operations generally are smaller in scale compared to longlining ones. However, the raft does provide a valuable depot structure.

Mussel spat is rarely produced in hatcheries because the production cost relative to the market value of the commercial product doesn't justify the effort. There are exception though along the West coast of USA and Canada (British Columbia) where the introduced blue mussels *Mytilus edulis* and *M. galloprovincialis* are being produced in hatcheries. Currently, the largest larval production takes place in Nelson (New Zealand) for New Zealand green-lipped mussels, *Perna canaliculus*.

2.5.3.2   Oysters

Oysters can also be cultured on- and off-bottom. Off-bottom cultivation strategies include (1) lantern nets, where oysters are placed in little bags suspended in the water column, or (2) *poche* culture, where the oysters are stuffed in bags, which are put on metal frames along the beach at low tide, and finally (3) in trays (Buck et al. 2006a). These systems are places on the seafloor or are deployed in a suspended manner below longline or other floating carrying devices as mentioned in the previous paragraph.

2.6   Current Status of Offshore Seaweed and Bivalve Production and Their Potential for Multi-Use

When looking at the offshore production of extractive species in a multi-use concept, it can be concluded that most projects were conducted in Europe and some others in Asia and the USA. In this section we provide a review on the potential and constraints of offshore seaweed and bivalve research and production in terms of biology, technology, and system design, as reported for different countries.

2.6.1   Germany

Germany is one of the leading countries in offshore seaweed science with first trials of moving seaweed operations offshore in the early 1990s. Therefore, the developments conducted in Germany are discussed in detail in another chapter (Chap. 11 "The German Case Study" within this book) while here, we only provide a short summary.

Seaweed aquaculture research at offshore sites in Germany started in 1992 with longline, ladder, grid and ring constructions (Lüning and Buchholz 1996) (Fig. 2.1a–d). Next to designing systems, research focused on reproduction, seeding strategies, growth performance and site selection (Buck and Bucholz 2004a, 2005; Lüning and Buchholz 1996).

Laminarian species seem to be the most robust species with offshore aquaculture potential in the North Sea, followed by *Palmaria* and *Ulva*. First experiments on laminarian species showed that they adapt to strong currents after being transferred at sea as young individuals. The offshore plants grew well at these exposed sites (Buck and Buchholz 2004a, b, 2005).

Fig. 2.1 (**a–d**): **a** shows the first offshore ladder construction (*double longline*) for hostile environments (Lüning and Buchholz 1996); **b** displays a grid construction off the offshore island of Helgoland; **c** shows a ring construction at harvest; **d** shows the ring construction in a multi-modular mode. *Images* (**a–d**) AWI/Cornelia Buchholz

The technology (Fig. 11.32 and 11.33 in Chap. 11 "German Case Study") used at that time was modified in terms of easy handling (deployment, maintenance, harvesting), robustness and connectedness to the foundations of offshore wind farms. Unfortunately, the longline and grid systems installed in harsh offshore conditions were not robust enough as there was a considerable stress on support material and algae during floatation mode (Lüning and Buchholz 1996; Buck 2004; Buck and Buchholz 2004a, b). As the idea of utilizing the foundations of offshore wind generators for the fixation of aquaculture systems is intriguing, these co-use concepts were the main driver to keep on working with seaweeds offshore infrastructures (e.g., Buck 2002; Krause et al. 2003; Buck et al. 2004). As a consequence, the ring construction was modified several times until it reached its final design and resisted any kind of possible sea condition in the North Sea. This offshore ring device is the first modern structure worldwide that enables mass seaweed cultivation on an industrial scale in the world's oceans. Another seaweed project led by German scientists plans to integrate *Saccharina latissima* cultures within a projected wind farm in Nantucket Sound (Massachusetts, USA) (Buck et al. 2011).

Mytilid larval appearance and settlement of *Mytilus* post-larvae at offshore test collectors, which were placed in the vicinity of offshore wind farms in the German Bight, were suitable for spat collection as well as for grow-out (Buck 2017). However, at some offshore test sites grow-out has to be economically calculated, as settlement might not be sufficiently dense. Mussels (*Mytilus edulis*) and oysters (*Ostrea edulis*) were successfully cultivated in and in the vicinity of the wind farms *Nordergründe* (17 nautical miles off Bremerhaven) and *Butendieck* (ODAS, 14 nautical miles off the Island of Sylt), as well as in North-South and West-East transects crossing all potential wind farms in the German Bight (Brenner et al. 2007; Buck 2007, 2017; Buck et al. 2006b). Health conditions with regard to infestations of macro-parasites, fitness and growth performance for both species were excellent (Pogoda et al. 2011, 2012, 2013; Brenner et al. 2007, 2012, 2014). Various technical solutions to connect submerged infrastructures to a windmill foundation or to deploy it centrally into a wind farm area were worked out (Buck 2007; Buck et al. 2006b, see also Chap. 11 "German Case Study"). To measure the forces impinging on the entire backbone, wave and current load cells were integrated in the system and artificial test bodies were used (Fig. 2.2a–f). Economic feasibilities studies were carried out as well as a protocol for a one-step mussel cultivation method that doesn't require any thinning procedure till reaching market size (Buck et al. 2010).

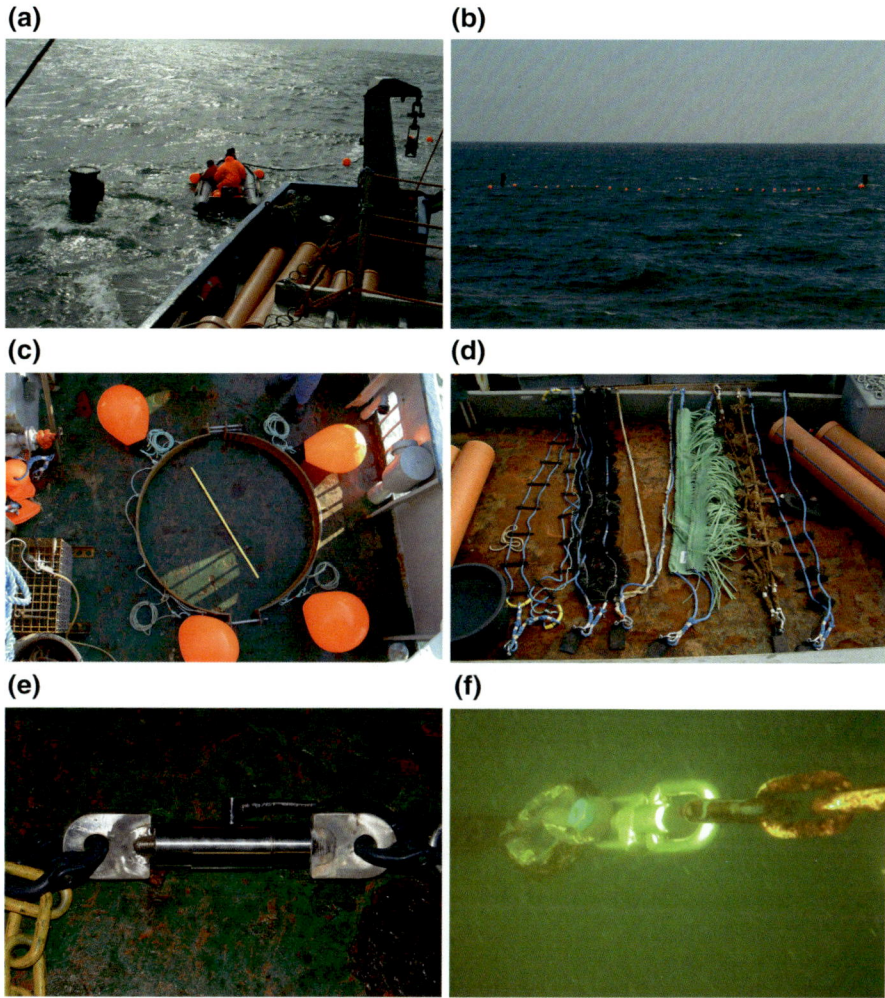

Fig. 2.2 (**a–f**): ODAS test site 14 nautical miles off the coast of the island of Sylt. **a** Displays the set-up of parts of the longline to an offshore test pole (symbolizing the foundation of an offshore wind turbine); **b** shows the floating buoys marking the submerged longline; **c** demonstrates the connection piece from the foundation to the holding device at one end of the submerged longline; **d** shows different spat collectors as well as artificial test bodies; and underwater load cells **e** on board of the vessel as well as **f** in operation under water

2.6.2 Belgium

The Belgian part of the North Sea is more or less 3454 km^2, or represents only 0.5% of the total surface of the North Sea. The coastline is only 65 km. Depth reaches a maximum of 45 m.

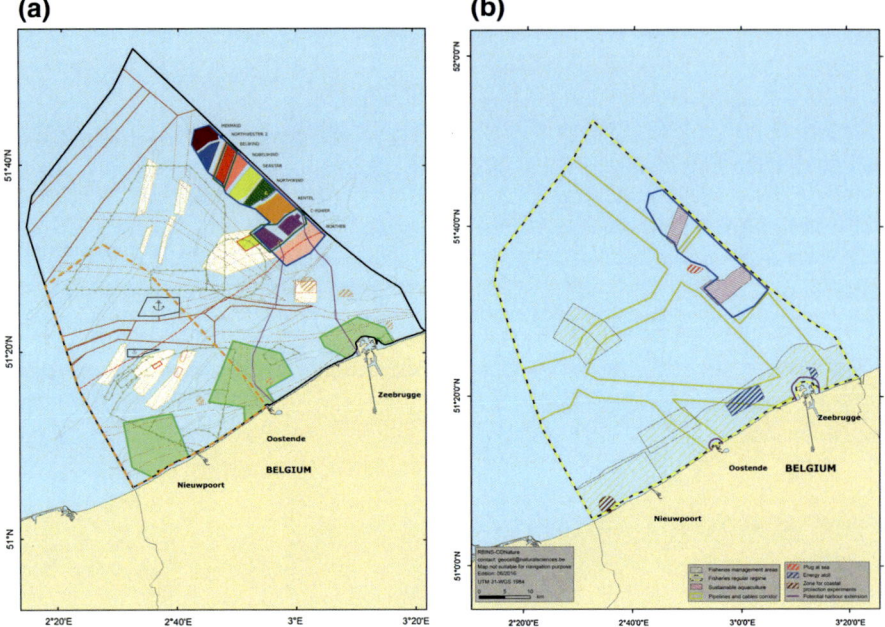

Fig. 2.3 (**a–b**): location of wind energy concessions (**a**) and aquaculture zones (**b**) in the Belgian part of the North Sea (Vigin et al. 2016)

In early 2014, a new Marine Spatial Plan (MSP) was introduced that outlines which activities can take place where and under which conditions. The plan is valid for a period of 6 years until 2020, after which modifications can be introduced. According to the MSP, commercial aquaculture is only possible in two zones which fall within the concessions for wind energy: the concessions of C-power (zone 1) and Belwind (zone 2), which are located on the Thorntonbank and the Bligh Bank, respectively (Fig. 2.3). Aquaculture in these zones needs to fulfill two requirements: (1) the owner of the wind energy concession has to approve and (2) the aquaculture activity contributes to the reduction of the eutrophication levels of the concession zones.

Between 1997 and 2011, there were several initiatives to culture the blue mussel *M. edulis* in the Belgian part of the North Sea. The summary of events given below is based on information collected from personal communication with Willy Versluys and the interim report of The Institute for Agricultural and Fisheries Research (ILVO) (Delbare 2001). Starting in 1998, several production systems and mussel seed capture devices were tested. One of the first designs consisted of a longline of 200 m with anchors of 1 t at each end. Every 10 m there was a float to carry the V-shaped, 5 m long mussel ropes. Mussels were allowed to settle on the ropes and stayed there until market size. Pegs prevented the biomass of sliding (Fig. 2.4a). Six of these structures with a total length of 1200 m were hung at sea

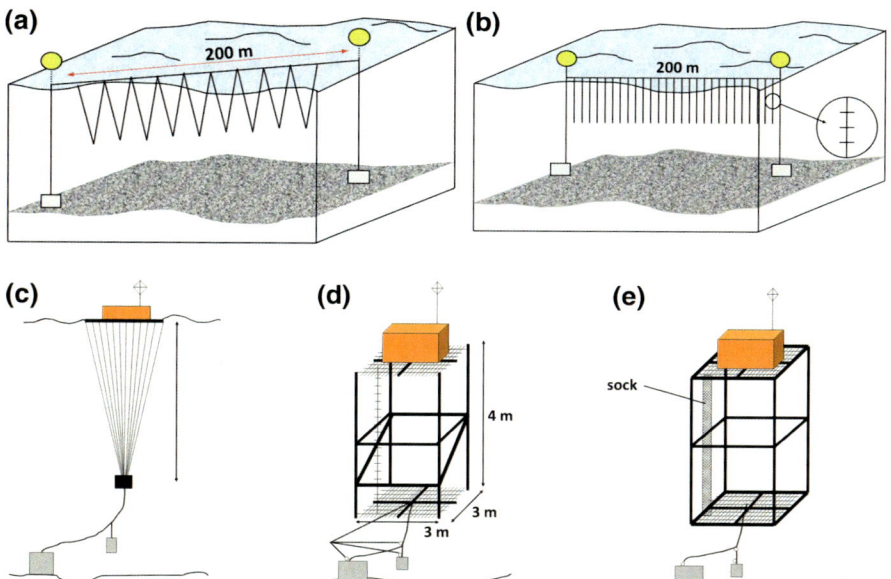

Fig. 2.4 (**a–e**): **a** construction of a mussel hang culture (used in 5b project 1998); **b** construction for spat collection (used by José Reynaert in 2001); **c–d** Hang culture system with and without closed frame (used in Pesca-Project 2000) **e** construction for grow-out. *Graphics*: Delbare (2001)

(location: top of Buitenratel) but were destroyed by boat traffic. Mussel growth results were, however, promising.

In 2000, a more robust construction was developed for the zone D1 (zone of 0.09 km², north of Noordpas, near buoy D1, extension of sandbank Smal bank, 13 km of the coast at Nieuwpoort) where boat traffic is not allowed. A first design consisted of the floating frame with 45 mussel ropes of 10 m whose ends were attached at one point. A weight prevented the ropes of friction against each other and to get entangled (Fig. 2.4c–d). The whole construction was anchored with a heavy weight and foreseen of a contra-weight to compensate for the tugging. Later, it was decided to include a closed frame so the whole construction could be lifted at once into the boat. By the end of 2000, however, the cages were lost due to insufficient anchoring and floating devices.

In 2001, two different systems were designed for mussel seed collection and grow-out. To capture mussel seeds, a longline of 200 m with floats was anchored by means of two anchors of 1 t at each side, in the zone D1. A total of 20 polypropylene ropes of 5 m length and 12 mm thickness were attached to the longline (Fig. 2.4b). The grow-out system consisted of floating cages with socks hung inside (Fig. 2.4e). Despite the prohibition for boats to enter the D1 zone (quite shallow), more than half of the longlines were destroyed and the grow-out cages dislodged by boat traffic. Mussels grew very well on the sunken cages, reaching a size of 6–7 cm in 10 months, despite the high densities. From 2002 to 2006, José Reynaert continued his endeavor with private money and with scientific support

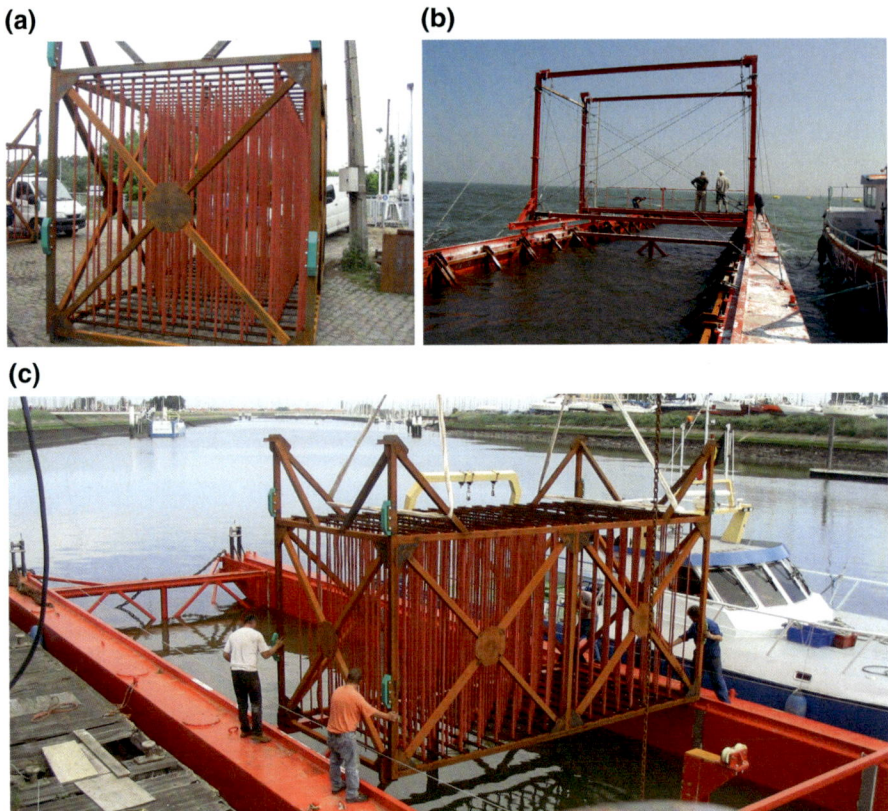

Fig. 2.5 (**a–c**): culture system for the "Belgica Mussel" in 2009: here, the aquaculturist worked with floating pontoons because of the shallow depth and strong currents. **a** Metal frame as holding device for the mussel collectors, **b** view from the front panel, **c** transfer at sea. *Photographs* (**a**) and (**c**) Willy Versluys; (**b**) Kris Van Nieuwenhove

from Ilvo. He developed a floating cage that contained several frames with mussel ropes that could slide in and out of the cage. Results have not been communicated.

This system was further improved in 2009 by Reynaert and Versluys. It became a floating pontoon with 8 cages (Fig. 2.5a–c). Each cage contained 160 hollow plastic tubes of 3 m which served not only as pole around which the mussel socks were wrapped but they also increased the buoyancy of the construction (Fig. 2.5c). The cages measured 34 m by 8 m and remained 3.4 m above the water and 5 m under the water. However, the pontoon soon twisted under the increasing weight of the growing mussels, making it impossible to lift the cages out of the frame for harvesting or maintenance. Mussels were harvested in June–July and the quality was high. The mussels were commercialized as the "Belgica" mussels. Production did not meet the expectations because of technical problems and in 2008 the whole crop had to be destroyed because of a suspicion of diarrhetic shellfish poisoning (DSP).

Fig. 2.6 (a–c): Buoy-collector-system used by SDVO (2005–2010); **a** device on board of the deployment vessel, **b–c** collection mussel buoy at deployment. *Photograph* Kris Van Nieuwenhove

Between 2005 and 2010, another subsidized project started independently in the Belgian North Sea and was carried out by Stichting voor Duurzame Visserijontwikkeling (SDVO). They developed round floating buoys with mussel ropes wrapped around them (Fig. 2.6a–c). More than one hundred of these buoys were put in the sea but data on production are not available. A large number of these devices got lost or were swept onto the beach. The mussels were commercialized as the "Flandres Queen Mussel". All efforts were stopped in 2010.

In conclusion, there is quite some experience in mussel culture in open sea in the Belgian part of the North Sea. Despite the fact that all the constructions proved themselves to be inadequate to deal with the rough North Sea conditions, the experiments demonstrated that the biological and chemical conditions in open sea along the Belgian North Sea coast are very suitable for growing mussels. The harvested mussels were of excellent quality and could be harvested earlier than the mussels of the Netherlands, the major supplier of mussels to Belgium. Obtaining natural spatfall was also never a problem.

Designing a right infrastructure and anchoring system that guarantees a reliable mussel production, that can cope with the very rough conditions of the North Sea and that offers enough protection against boat traffic is the big challenge for the future. In 2017, a new project ("Edulis") will explore the possibilties to grow blue mussels and collect blue mussel seed (both *Mytilus edulis*) in the windmill parcs

C-Power and Belwind, making use of semi-submerged longlines (backbone 58 m) that will be held in place by weight anchors at a depth of 5 m under the surface. The project is mainly funded by private partners and coordinated by Ghent University, Laboratory of Aquaculture & Artemia Reference Center with the support of the Alfred-Wegener-Institute (AWI) in Germany. At the same time, another initiative, Value@sea will start growing seaweeds (*Undaria pinnatifida*, *Saccharina latissima* and *Porphyra sp.*) in combination with different species of bivalves closer to the Belgian coast (in front of Koksijde). It will use submerged longlines (backbone 100 m, 1.5 m under the surface) which will be secured with screw anchors. Again, this project is largely supported by the private sector and coordinated by Ilvo.

2.6.3   Norway

Regarding the cultivation of seaweeds, the report "A new Norwegian bioeconomy based on cultivation and processing of seaweeds: Opportunities and R&D needs" (Skjermo et al. 2014) discusses the use of seaweed cultivation in offshore environments as one of the opportunities. When moving off the coast, the authors evaluate the best opportunity for the seaweed industry to co-use existing offshore structures, such as wind farms. One driver for the preparation of this study was the existing problems associated with vast amounts of nutrients originating from salmonid farms. In 2011, the Norwegian Research Council funded a project called SWEEDTECH, which looked into the development of a cost efficient system in order to start with large scale offshore seaweed cultivation. This project also included seaweed seeding strategies, development of carrier material, design and development of a structural rig as well as the development of alternative deployment and harvesting methods (SES 2015a, b). As an outcome of SWEEDTECH and as a result of the current development of seaweed cultivation off the coast of Norway, AquaCulture Engineering AS (ACE) and SINTEF Fisheries and Aquaculture will establish a new site of 3 ha with an expected yearly production of 1500 t of laminarian macroalgae (ACE 2015).

The company Seaweed Energy Solutions AS (SES) was involved in concepts to upscale seaweed cultivation off the coast (Bakken 2013). As a consequence, SES patented (SES 2015a) the first modern structure to enable mass seaweed cultivation on an industrial scale in Norway. This structure, called *Seaweed Carrier* (Fig. 2.7a–e), is described as being a sheet-like structure that basically copies a very large seaweed blade, moving freely back and forth through the sea from a single mooring on the seabed. The carrier can withstand rough water, has few moving parts, has low cost and allows for easy harvesting. The way the carrier has been put into practice to date has been limited to two major tangible prototypes (SES 2015b): (1) A semi-rigid truss prototype (90 m long) with 8 carrier nets of 50 m width and 5–10 m depth, which had been developed for energy-scale, fully mechanized operations. This undertaking is, however, presently suspended. (2) A flexible hybrid long line system for semi-exposed water, as used in the 100 t pilot project in Frøya. The carriers were

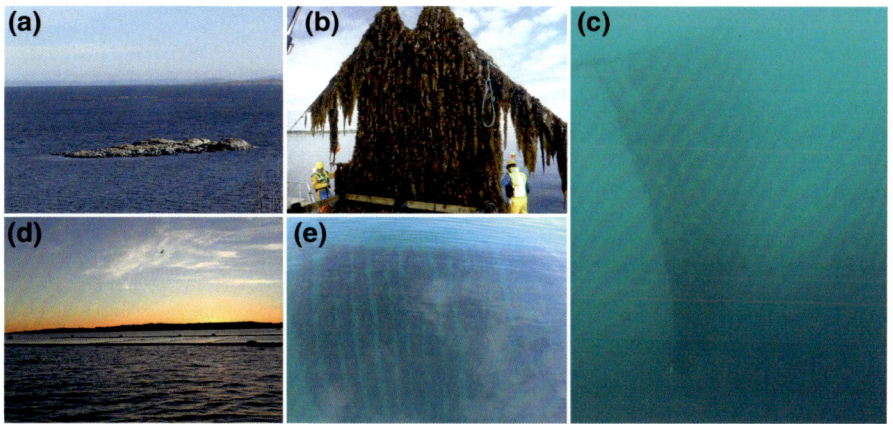

Fig. 2.7 (**a–e**): **a** and **d** Site of the installation off the coast of Frøya; **b** harvest of seaweed growing on the *Carrier*; **c** and **e** underwater image of the kelp growing on the carrier devices; all images provided by SES (2015a, b)

attached as 6.5 m long and 5 m deep two-dimensional flexible units and showed very good growth. One single longline operated as a backbone for 20 of these carriers. Until today, a mechanization of the process has not been implemented due to lack of funds. SES focus is now to pursue a gradual development of this aspect.

SES' vision with this system is to allow seaweed cultivation in deeper and more exposed waters, opening the way for large scale production which is necessary to make seaweed a viable source of energy. Furthermore, they can be placed in a co-use concept with other marine operations, such as oil and gas, offshore wind and offshore wave ventures.

Regarding the production of bivalves, the collaboration between Sintef and Statoil as part of the European-funded research project Mermaid (Innovative multipurpose offshore platforms: planning, design and operation) has led to the design of 2 a possible exploitation models where salmon, mussels and seaweeds are grown together in a windmill park in the North Sea and Atlantic Ocean, respectively.

The North Sea model (southern North Sea at 40 m water depth and 100 km off from the coast line) estimates the electricity annual production to be 3300 GW h^{-1}, at an annual average wind speed of 9.5 m s^{-1}, based on 10 MW WTS power production characteristics (Fig. 2.8). The annual salmon production is estimated at 60,000–70,000 t based on a fish production of 20 kg m^{-3} (maximum 25 kg m^{-3}) and a fish survival of 88–95%. In financial terms, the salmon production would yield a total of 240–280 million € at 4 € kg^{-1}, which accounts for 50–60% of the annual electricity yield. In addition, the production of blue mussels and seaweed (e.g. sugar kelp) is estimated to reach 20,000–30,000 and 160,000–180,000 t respectively, representing roughly 20–30 and 160–210 million €, respectively (at 1 € kg^{-1} for both mussels and seaweed) (He et al. 2015).

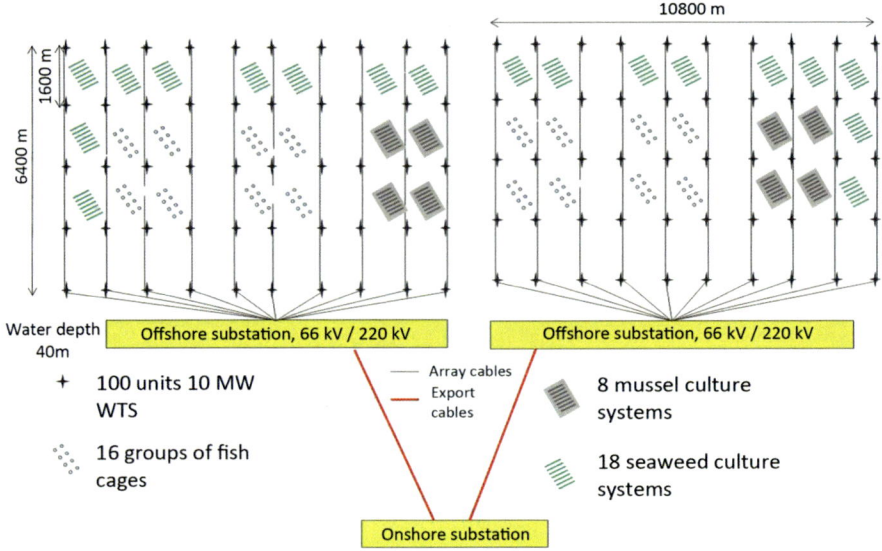

Fig. 2.8 Sintef-Statoil IMTA design for the North Sea (He et al. 2015)

2.6.4 Denmark

In Denmark, a feasibility study was carried out to look into mussel production on longlines in the windmill park Nysted. The park contains 7 windmills and measures 24 km (Christensen et al. 2009 in Verhaeghe et al. 2011). To our knowledge, no mussel production is taking place at the moment.

2.6.5 The Netherlands

For the production of fish and animal feed, as well as biofuels and energy, one of the first trials to farm seaweeds off the coast within a wind farm site was conducted by Ecofys and Hortimare in 2012. The system (Fig. 2.9a–h) that covers 400 m^2 (20 m × 20 m) consists of a set of steel cables, submerged 2 m below the sea surface and held by anchors and floating buoys. In between, horizontal nets (10 m × 10 m), seeded with laminarian sporophytes, were suspended between those cables. In the plans was to test if the design is suitable for seaweed cultivation in offshore wind farms in the North Sea.

Bartelings et al. (2014) carried out a desk study to look into the possibilities to combine offshore aquaculture in windmill parks on the Dutch Continental Plateau. The study concluded that mussel culture and mussel seed culture on longlines are the most promising options within the Dutch context. It could potentially lead to a reduction of 40% of the total cost per MWh and the Dutch mussel sector would be

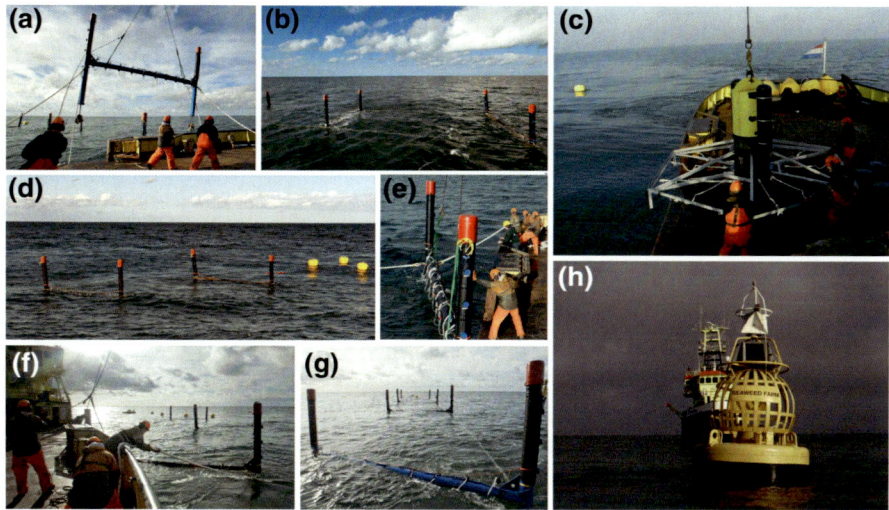

Fig. 2.9 Installation of the innovative H-frame seaweed farm offshore off Texel (Patent No. P94437EP00) as well as a seaweed test module. **a** Launching of the installation in October 2012 with RV Terschelling; **b** H-frame after launching, total length 120, 10 m wide; **c** launches of a test module 15 km out from coast Texel; **d** mooring by 2 anchors at both ends, 4 yellow buoys kept the mooring points floating; **e** work at the mooring in connection with the H-frame during deployment; **f** installation work; **g** side view on the entire installation; **h** work offshore completed at the test farm of Noordzeeboerderij off Texel. *Images* (**a**, **b**, **d**–**g**) modified after Hortimare (2016), *images* **c** and **h** by Schuttelaar and Partners, *other images* are provided by Hortimare (2016)

offered a new market opportunities of 100,000 t. That is twice the current production. Fish culture in cages was considered impossible by the authors because of the shallow depth and no opinion was formulated towards the production of seaweeds. Lagerveld et al. (2014) also simulated the feasibility of combining an offshore wind farm with an offshore mussel farm in the Dutch North Sea, assuming certain synergies between the two activities.

As a result, Stichting Noordzeeboerderij (the North Sea Farm) is committed to develop a mass cultivation of seaweeds in the Dutch North Sea. In June 2015, Hortimare successfully executed a small scale test on the Dutch North Sea, 10 km North-East from the Island of Texel. This test site includes brown macroalgae, such as *Saccharina latissima* as well as *Laminaria digitata*. The test farm for *S. latissima* off Texel has a size of one km², the *L. digitata* farm at the location called Proefboerderij is slightly smaller. Another test site is planned a few kilometers off the coast from Scheveningen (The Hague).

The planned construction of the new windmill park in Borsele (capacity of 1050–2100 MW, 22 km offshore of the Province of Zeeland, with a depth between 10 and 40 m), that will start in the second half of 2016 has led to strong discussions about offshore aquaculture possibilites in the park. Van den Burg et al. (2016) report that aquaculture activities are not included in the tender and no specific advantage will be granted to the candidates who will include that option in their proposal. Four major advantages of culturing seaweed in the windmill park of Borsele have been identified: uptake of nutrients and reduction of eutrophication, attenuation of waves and hence less erosion of the windmill foundations, contribution to fish resources, and increased regional support through job creation.

2.6.6 France

In France, longline systems are used offshore along the Mediterranean and the Atlantic coasts. The Mediterranean farms have been in use since the late 1970s and are all fully submerged and produce up to 10,000 t/year, although this has decreased in recent years due to predation and water quality. The techniques and equipment are fully described by Danioux et al. (2000). Offshore long lines have been used since the mid-nineties on the Atlantic coast and the coast of Charentais, Pertuis Breton and Pertuis Antioch. It is believed that there are several hundred lines in this region. Production is around 5000 t from these offshore long lines. Expansion will depend on access to new space and improvement in techniques (Ögmundarson et al. 2011).

The project "WINSEAFUEL" was funded by the French National Research Agency and was one of the first French projects with an offshore multi-use aspect (Lasserre and Delgenès 2012). The aim of the project was to produce biomethane and bioproducts from macroalgae cultivated in the open sea next to offshore wind farm facilities (Langlois et al. 2012). A life cycle assessment for the production of biogas from offshore-cultivated macroalgal feedstock in a European framework was carried out, as well as production optimization from hatchery phase to offshore (Marfaing 2014).

In the North of France, at the border with Belgium, mussels are being produced in Dunkerque (North Sea) using the longline system. Ropes and anchors are made extra strong to cope with the local environmental conditions. The project started in 2006 and obtained its first good results in 2009. An annual production of 1000 t is obtained nowadays. They are produced and commercialized by the fishermen's cooperative "Cooperative maritime de Dunkerque" as "Les Moules de Dunkerque" or "Les Moules du Banc de Flandre".

2.6.7 United Kingdom

In 2014, Kames Fish Farming Ltd. installed two offshore seaweed longline trials for research commissioned by the Bangor University's School of Ocean Science and the Scottish Association of Marine Science (SAMS), respectively (KFM 2014). The design consists of a small submerged rope grid supported on foam filled cushion buoys and oceanic mussel headline floats. The location of the first longline at

Conwy Bay north of the Menai Straits is exposed to a high wave regime combined with a high tidal level accompanied with strong current velocities. The second longline was moored north of Oban (Argyll) Lynn of Lorne at Port a Bhuiltin.

Mussel bottom culture (Deepdock Ltd.) was tested in the windmill park of North Hoyle OWF (Wales) in 2010. The windmill park contains 30 monopiles and the depth was 10 m at low tide. The mussels grew well, but unexplainable mortality occurred at harvest. To our knowledge, no offshore mytiliculture is going on at the moment.

The Kentish Flats offshore wind farm about 10 km off the coast of Kent provides a permit to the fishing community to access the wind farm area for trawling of bivalves as well as for passive fishing strategies (Brown and May 2016). Studies had been undertaking with regard to various potential impacts on commercial fishing including the impacts of loss of fishing area resulting from offshore wind farm installations for environmental impact assessments supporting planning applications, compensation modelling and preparation of safety strategies. Mussels were caught at spat size and used for on-bottom cultivation. Other wind farms, such as Gwynt y Môr, North Hoyle, Lynn and Inner Dowsing wind farms were used for natural stock relay and ranching projects (CEFAS 2014). No commercial enterprise is running to this day.

2.6.8   Italy

Offshore long line mussel production has been in place in the Adriatic for some years and plays an increasingly significant part of production (Danioux et al. (2000). Experiments were carried out in the central part of the Adriatic sea (Conero Promotory, Ancona province) in Italy to assess the potential to produce mussels (*M. galloprovincialis*) and oysters (*Crassostrea gigas*) in submerged structures in combination with artificial reefs in locations where traditional shellfish farming in not possible (Bombace et al. 2000; Fabi and Spagnolo 2001) (Fig. 2.10). The base units of the reef consist of concrete blocks of 8 m^3 with holes of different diameters. The blocks are piled up into pyramids of 2 or 3 layers and layed down at depths of 10–20 m. The distance between the pyramids is 15–50 m. The units at the top are connected to each other with cables that support nets to collect the mussel seed or longlines systems. Concrete shellfish cages with dropper lines are placed between the pyramids.

The structures are seeded naturally with wild mussel seed and an average of 20–55 kg/m^2/year is produced by the cages and the blocks. A reef of 116 pyramids and 48 cages produces 260 t of mussels yearly on average. Based on this, the costs of a reef are paid back after 5 years. Thanks to the dynamic environment, the environmental impact is reduced to a minimum (Fabi et al. 2009).

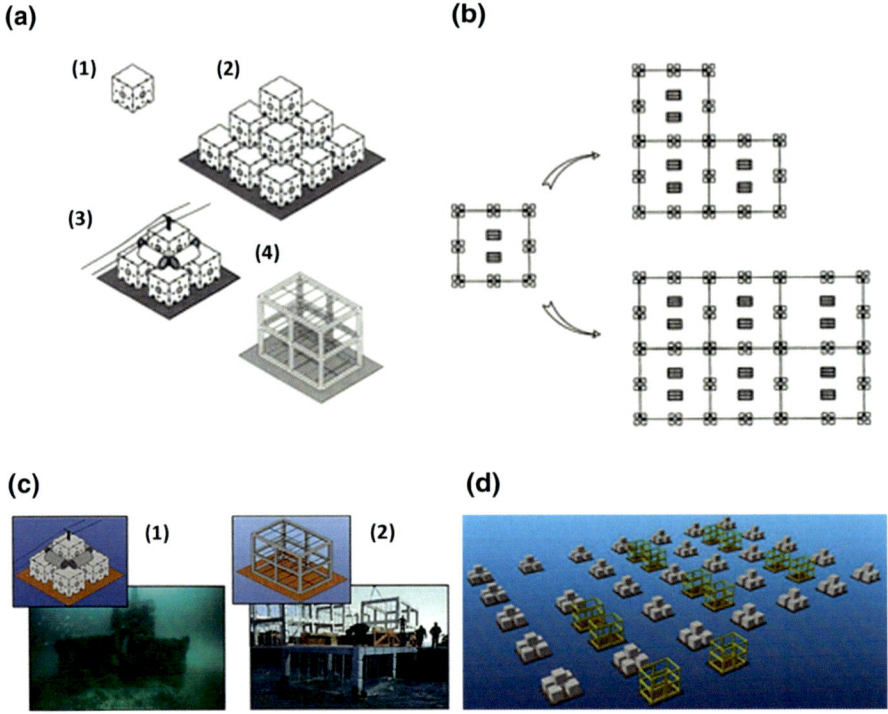

Fig. 2.10 (**a–d**): **a** examples of artificial reefs in the Adriatic sea (*1*), (*2*) pyramids, (*3*) connection for longlines, and (*4*) cages for shellfish (Bombace et al. 2000; Tassetti et al. 2015); **b** and **d**: layout of artifical reefs, in groups of 8 pyramids with 2 shellfish cages in the center (Bombace et al. 2000; Tassetti et al. 2015); **c** two-layer pyramids of concrete blocks (*1*) and concrete cages for mussel culture (*2*) (Tassetti et al. 2015)

2.6.9 *The United States*

In the USA, there have been discussions in the last years in co-locating aquaculture production together with offshore wind energy projects. One example to combine extractive species with offshore wind turbines was planned for the wind farm *Cape Wind* in Massachusetts (USA) (Lapointe 2013). Buck et al. (2011) prepared a first economic feasibility study on the production of seaweeds within this wind farm. Another attempt at multi-use of an offshore wind installation with mussels was conducted at Block Island Wind Farm, off the coast of Rhode Island (Voskamp 2010). Further, the Wind Technology Advancement Project (VOWTAP) conducted initial engineering, designing, and permitting for offshore wind energy installations off the coast of Virginia while taking aquaculture into account (Moura et al. 2011). They planned to collaborate with other local users by using the wind farm infrastructure to enhance the profitability of fisheries or aquaculture operations. However, the current expansion of natural gas production, with a concomitant decrease in gas

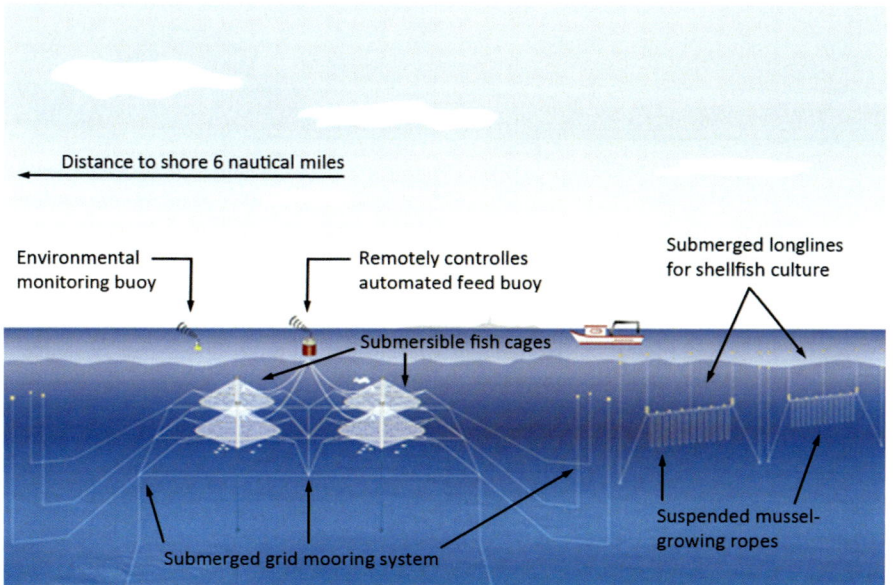

Fig. 2.11 Diagram of the University of New Hampshire's offshore installation depicting submerged fish cages, submerged shellfish longlines, automated feeding buoy and oceanographic instrumentation

prices, will likely have a significant dampening effect on the development of operational offshore wind energy projects or sites, which in turn would decrease the chance for multi-use aspects.

The University of New Hampshire (UNH) Open Ocean Aquaculture Project (http://amac.unh.edu) had a demonstration unit of 12 ha at 10 km off the coast in the Gulf of Maine that included finfish culture in submerged cages and shellfish (mussels and scallops on submerged longlines (Langan 2009) (Fig. 2.11).

Water depth is 52 m on average and the area is completely exposed (waves > 9 m high). The submerged longline system consists of a longline (130 m long, 28 mm polysteel rope) with 3 floats (2 for the anchoring lines and one in the middle). Each anchoring line (28 mm polysteel rope) is kept in place with a weight of 4 t and is connected directly to the anchor at an angle of 45°. The longline is kept at a depth of 10–12 m and carries mussel ropes that form loops of 7–12 m. In 1999, the first two submerged longlines were installed. They are still in place and tended by commercial operators though the anchors and the floats have been replaced several times. Three km to the south (Boars Head), there is a commercial production with 10 longlines (Verhaeghe et al. 2011).

More recently, UNH and New Hampshire Sea Grant have been developing IMTA platforms for marine fish, shellfish and seaweeds (Fig. 2.12a–b). IMTA is where the culture of a fed product (i.e. fish) is combined with the culture of extractive species that bio-mitigate nutrients from the farm and surrounding waters.

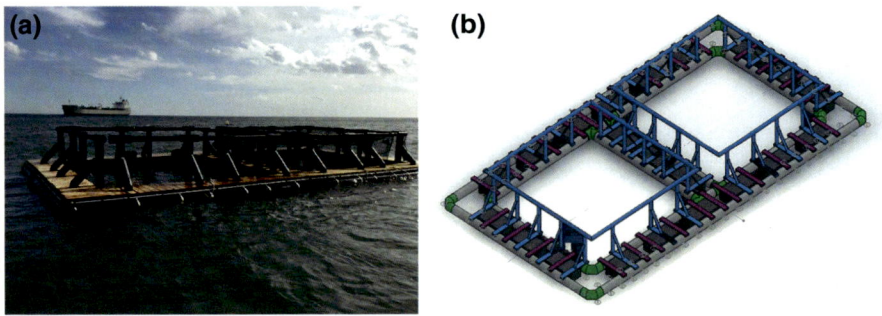

Fig. 2.12 (**a–b**): **a–b** open ocean, integrated multi-trophic aquaculture raft being evaluated at the University of New Hampshire (**a**), first pilot scale set-up built after the model (**b**)

This creates product diversification, improves farm output, helps maintain a clean environment and is more socially acceptable. IMTA promotes economic and environmental sustainability by converting byproducts and uneaten feed from fed organisms into harvestable crops, thereby reducing eutrophication, and increasing economic diversification (Neori et al. 2004; Troell et al. 2003). IMTA has been investigated extensively in Atlantic Canada with salmon, blue mussels and kelps (Chopin et al. 2004, 2013; Ridler et al. 2006, 2007; Robinson et al. 2007).

The University has been engaging commercial fishermen on small-scale, IMTA as a means to diversify their income. By integrating the production of steelhead trout (*Oncorhynchus mykiss*), blue mussels (*Mytilus edulis*) and sugar kelp (*Saccharina latissima*) on a floating platform, fishermen have been able to grow and sell new sources of seafood (Fig. 2.12a–b). The IMTA model was initially adopted due to concerns by the United States Environmental Protection Agency of nitrogen inputs from fish feed in coastal areas that were already nitrogen impaired. Working with state and federal agencies, a nitrogen mass balance model was derived and a IMTA demonstration project was launched in 2012 (Chambers 2013).

The project sourced rainbow trout (known as steelhead in the ocean) from a local hatchery and stocked them diretly into seawater cages at 250 g. New Zealand fuzzy rope was suspended around the cage perimeter to collect mussel spat. The fuzzy rope, made from loops of polyester line, provides abundant surface area for mussel settlement. Mussels typically spawn twice a year in the Gulf of Maine and adhere to bottom substrate and materials in the water column (Langan and Horton 2005). Sugar kelp is endemic to New England waters and naturally settles on subsurface substrate. Kelp sorus tissue (sporophyte) was collected from mature kelp blades growing on the seacages and spawned in captivity. With the help of Ocean Approved in Portland, ME (http://www.oceanapproved.com/) and Maine Sea Grant, kelp spores were successfully spawned onto twines that were later seeded onto vertical and horizontal growout lines at the cage site.

The UNH IMTA project demonstrated that nitrogen extraction by mussels (2.0% tissue and shell), kelp (2.4%), and trout (40% at harvest) can exceed nitrogen input from trout production. With a 4:1 ratio, 4 t of mussels/kelp to 1 t of trout, more nitrogen can be extracted from the environment then inputs from feed thus having a net ecosystem benefit (Chambers 2013). Per these results, UNH was funded to design, construct and evaluate a robust, open ocean raft that is currently in the field.

This information has aided regulatory agencies in their decision making for permitting aquaculture. In addition, IMTA has created new sources of sustainable, local seafood and employment, helping fishermen diversify into seafood production while continuing to fish.

2.6.10   The Republic of Korea

Following the South-west offshore wind farm development plan, the construction of the first large scale offshore wind farm in the Republic of Korea in combination with seaweeds and bivalves started in 2016 (Figs. 2.13a–d and 2.14a–b). The co-location concept was suggested to the local communities in 2013 and was recently accepted. The major issue of co-using was the local acceptance by stakeholders, such as fisheries, local officials and the fisheries cooperative union. The multi-use concept was inspired by the shellfish and seaweed cultivation trials in Germany (North Sea) and Wales (North Hoyle and Gwynt y Môr). The objective of this project is, on one hand, the development of technology for co-locating fisheries and aquaculture with offshore wind farm, and, on the other hand, the multi-use concept which should improve the social acceptance of offshore wind farms. The first trials include IMTA systems that combine fish, sea cucumber, oysters and seaweeds (KEPCO and KIOST 2016).

Provincial governments have been developing offshore seaweed cultivation technologies. The target species for these offshore systems included, besides *Porphyra*, the species *Saccharina* and *Undaria*. The Aquatic Biomass Research Center (ABRC) carried out an offshore seaweed aquaculture project to produce seaweed biomass for biofuels (Chung et al. 2015). The objectives of this study were (1) to select appropriate seaweed species and to develop seed planting techniques for high density mass production, (2) to develop a Tension-Leg Platform (TLP) type seaweed cultivation system, and (3) to develop automatic out planting and harvesting systems. The TLP system provides a stable platform for seaweed cultivation even in the offshore environment. During phase I (2010–2013), a brown seaweed, *Saccharina japonica*, was successfully cultured on the TLP system near Geumil-do Island, Wando, Jeollanamdo. This TLP system is now installed near Cheongsan Island, between Wando and Jeju Island, growing several species, including *Saccharina japonica, Ecklonia cava, E. stolonifera, Sargassum horneri, Myagropsis myagroides,* and some others. The cost for a TLP system is estimated at $500,000 for a 1 ha seaweed farm. Although this study showed potential to grow seaweeds in the offshore environment using the TLP system, limiting high production costs are the most challenging part.

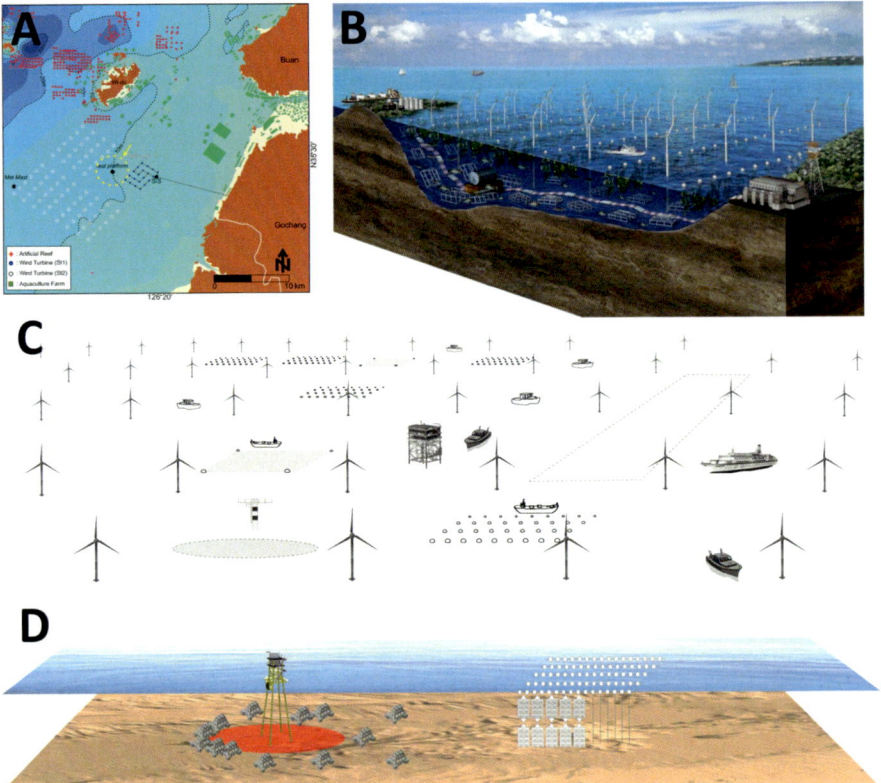

Fig. 2.13 (**a–d**): **a** co-location area for aquaculture within offshore wind farms off the coast of southern-east South-Korea; **b** drawing of the multi-use concept at the wind farm site shown in a profile view; **c** complete wind farm-aquaculture site for the first development stage; **d** wind farm foundation with artificial reefs and aquaculture installations in its vicinity. Modified after KEPCO and KIOST (2016)

2.6.11   China

Sea ranching as a commercial activity is mainly developed in China and Japan and can also be deployed in offshore areas. In China, there is the example of Zhangzidao Fishery Group Co. Ltd. The farm has a size of 40,000 ha of which 26,500 ha are used for the culture of pectinidae, sea urchin, abalone, and the sea cucumber *Apostichopus japonicus*. In 2005, the site produced 28,000 t of product with a value of US$ 60 million. It is located near the Zhangzidao islands, north of the Yellow Sea, 40 miles away from the continent of Liaoning Province. Shellfish, macroalgae, crustaceae and echiniderms are grown at depths of 1–40 m in an area that is characterized by strong currents (max. 100 cm s^{-1}).

On the other hand, suspended culture in open sea in China is practiced in Sungo (Sanggou) Bay (13,000 ha in total), east of Shandong Peninsula and is one of the

Fig. 2.14 (**a–b**): Deployment of the first wind farm foundation at the co-location wind farm project in the south-west of South Korea. **a** Foundation during mooring procedure; **b** preparation of the foundation for potential multi-uses. Modified after KEPCO and KIOST (2016)

most important mariculture regions for the scallop *Chlamys farreri* and the kelp *Saccharina japonica* in northern China (Fig. 2.15). The abalone *H. discus hannai* is also cultured there, and to a lesser degree the blue mussel *M. edulis*. The bay stretches 8 km from the coast, has a depth of 20–30 m and has currents up to 60 cm s^{-1}. It has been estimated that dissolved nitrogen excreted by scallops in the Bay (2 billion individuals) amounts to 284 t during a kelp culturing period. Similarly, the inorganic nitrogen excretion by mussels in the Bay (0.27 billion individuals) amounts to more than 11 t. Together with the excretion of other fouling animals such as sea squirt and oyster, the total inorganic nitrogen excretion of cultivated and fouling animals in the Bay amounts to more than 300 t. Twenty thousand tons of dried kelps can be produced annually through uptake of inorganic nitrogen from the Bay (Troell et al. 2009) (Fig. 2.15).

(a)   **(b)**

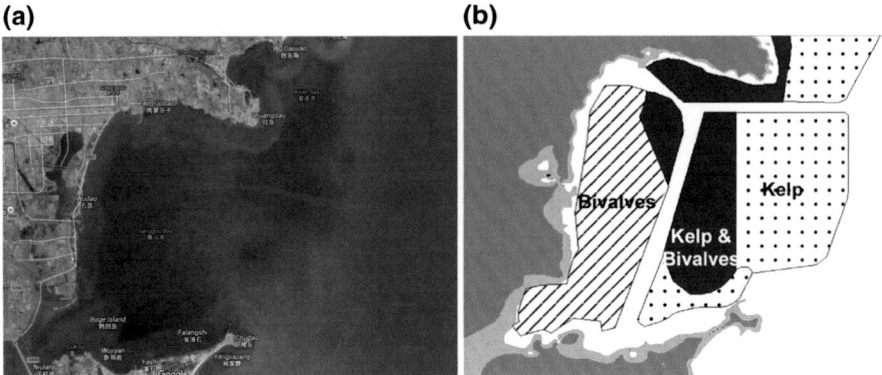

Fig. 2.15 a–b: a map of Sanggou Bay including the aquaculture site (Google Earth 2016);
b drawing of the aquaculture area after site selection (Nunes et al. 2003)

The aquaculture cages shown in Fig. 2.16a–b have been proposed to co-use the space within offshore wind farms under open sea conditions. These cages are planned to be used in an IMTA mode with seaweed and bivalves as well as sea cucumbers. The co-location of aquaculture (small scale) will provide insights into the real barriers and risks encountered. The operational experience gained will encourage the stakeholders' interests in the development of commercial co-location.

Using offshore wind technology has the potential to help accelerate the movement of aquaculture to open water sites in China where the water quality is better. Combining both offshore wind farm with aquaculture meets the challenge of both the production of clean energy and high quality seafood while maintaining minimum environmental impact.

More than 50 experts from energy, vessels and fishery research groups and industries gathered at Jiangsu University of Science and Technology, China, to discuss: Going offshore: Combining offshore renewable energy and high-quality seafood production. The plan was to set-up a center for offshore aquaculture and renewable energy. The work plan for the center for combining offshore renewable energy and aquaculture have been proposed and discussed.

Chinese energy companies presented the energy technology development map including 44 offshore wind farms in China. The fishery research institute and university presented the Chinese aquaculture farms and the challenges.

2.7   Ecosystem Services

Most studies on macroalgae from temperate regions used as biofilters in IMTA focus on *Saccharina latissima*, *Alaria esculenta, Ulva* spp., *Gracilaria* spp. and *Pyropia/Porphyra*, which are well-established aquaculture species and whose nutrient uptake abilities are high compared to most other seaweeds (e.g. Chopin

(a)

(b)

Fig. 2.16 a–b: proposed aquaculture cages for open sea conditions in combination with wind farm foundadtions. These are planned to be used in an IMTA mode

et al. 2004; Kang et al. 2014; Martinez-Aragon et al. 2002; Neori et al. 2000, 2004; Msuya and Neori 2008; Abreu et al. 2013; Sutherland et al. 2011). However, in the case of eco-intensification of offshore aquaculture operations, the macroalgae species additionally needs to be robust enough to withstand a high energy environment (Buck and Buchholz 2005). First attempts to use *Saccharina latissima* as well as *Palmaria palmata* within offshore IMTA systems was investigated by Grote (2016). First attempts to calculate nutrient budgets of Saccharina latissima and Palmaria palmata for the use as offshore IMTA candidate was investigated by Grote (2016) as well as Grote and Buck (2017).

Recently, Kim et al. (2014a, 2015a) cultivated *Gracilaria tikvahiae* and *Saccharina latissima* in open water nutrient bioextraction farms in Long Island Sound (LIS) and the Bronx River Estuary (BRE), and evaluated the nutrient bioextraction performance in urbanized estuaries. They estimated that the biomass yields of *G. tikvahiae* and *S. latissima* could be up to 21 and 62 metric tons fresh weight per hectare, respectively. The potential nitrogen removal could exceed 320 kg N ha^{-1} year^{-1} from LIS (280 kg ha^{-1} from *Saccharina* and 40 kg ha^{-1} from *Gracilaria*) and 430 kg N ha^{-1} year^{-1} from the BRE site (280 kg ha^{-1} from *Saccharina* and 150 kg ha^{-1} from *Gracilaria*).

Seaweed is also an important CO_2 sink and the duration of net CO_2 removal can be extended if the biomass is used in environmentally friendly ways (Chung et al. 2013). Kim et al. (2014a, 2015a) estimated that *Gracilaria* and *Saccharina* could sequester up to 300 (LIS) and 727 kg C ha^{-1} (BRE), and 1800 (LIS) and 1350 kg C ha^{-1} (BRE), respectively. Together, over 2000 kg C ha^{-1} could be removed by alternating these two species at the seaweed farm sites. An economic value of nutrient bioextraction was estimated. Considering the most recent nutrient credit values in the USA for these two elements (US \$12.37 kg^{-1} N, US \$6.00–US \$60.00 mt^{-1} C (as CO_2), Stephenson and Shabman 2011; CDP 2013; CT DEEP 2014; Tedesco et al. 2014), the potential economic values of C and N removal could exceed \$3000 ha^{-1}, which could be additional income for seaweed growers beyond the value of seaweed products, if seaweed aquaculture is incorporated in the Connecticut Nitrogen Credit Trading Program and a carbon-pricing scheme (CDP 2013).

Considering the global seaweed production and tissue carbon and nitrogen contents for each species, total extractive nitrogen and carbon by seaweed aquaculture can be estimated. Considering a 10:1 DW:FW ratio and average values of nitrogen (*Pyropia/Porphyra*: 5.5%, *Gracilaria*: 3.0%, *Kappaphycus/Eucheuma*: 1.7%, kelp: 2.0% and *Sargassum*: 4.1%) and carbon (*Pyropia/Porphyra*: 38%, *Gracilaria*: 28%, *Kappaphycus/Eucheuma*: 29%, kelp: 30% and *Sargassum*: 34%), the total nitrogen and carbon removal by these five major aquaculture groups is approximately 54 million t of nitrogen per year and 700 million t of carbon per year (equivalent to 2600 million t of CO_2), respectively (Asare and Harlin 1983; Gerard 1997; Schaffelke and Klumpp 1998; Gevaert et al. 2001; Schaffelke 2001; Chung et al. 2002; Rawson et al. 2002; Sahoo and Ohno 2003; Dean and Hurd 2007; Kim et al. 2007, 2014a, 2015a; Buschmann et al. 2008; Abreu et al. 2009; Robertson-Andersson et al. 2009; Levine and Sahoo 2010; Broch et al. 2013). This is, in fact, a significant amount of carbon and nitrogen removal. In 2013, global nitrogen discharge to coastal waters and the open ocean via leaching and riverine transport was estimated to be up to 70 billion t (Fowler et al. 2013). During the same period, carbon emissions due to fossil fuel use (and cement production) was 9.9 billion t. Seaweed aquaculture during the same period has removed approximately 0.13% of nitrogen discharge from leaching and riverine transport and 6.6% of carbon emission via fossil fuel use.

The extractive sequestration of nutrients by seaweeds provides ecosystem services that need to be recognized and valued appropriately (Chopin 2014). Much has been said about carbon sequestration and the development of carbon trading taxes. In coastal environments, mechanisms for the recovering of nitrogen and phosphorus should also be highlighted and accounted for in the form of nutrient trading credits (NTCs), a much more positive approach than taxing. If the composition of sea-weeds can be averaged at around 0.35% nitrogen, 0.04% phosphorus and 3.00% carbon, and nutrient trading credits are valued at US $10–30/kg, $4/kg and $30/mt for the respective compounds, the ecosystem services for nutrient biomitigation provided by the 23.8 million t of worldwide annual seaweed aquaculture can be valued at between $892.5 and $2.6 billion—as much as 40% of its present commercial value. This significant value is, however, never noted in any budget sheet or business plan, as seaweeds are typically valued only for their biomass and food trading values. It is interesting to note that the above value for carbon is per tonne, whereas those for nitrogen and phosphorus are per kilogramme, a factor of 1000. Nobody seems to have picked up on that when looking at the sequestration of elements other than carbon. Moreover, having organisms able to accumulate phosphorus is becoming increasingly attractive, as this element will soon be in short supply.

The recognition of the ecosystem services provided by extractive species and the implementation of NTCs would give a fair price to extractive aquaculture (Chopin et al. 2010, 2012). They could be used as financial and regulatory incentive tools to encourage single-species aquaculturists to contemplate IMTA as a viable option to their current practices. Unfortunately, the Western world's animal-biased aqua-culture sector and coastal managers and regulators often do not appreciate the ecosystemic roles of seaweeds and fail to take advantage of the environmental, economic and societal benefits provided by seaweeds and other extractive species, which have their full justification to be included in multi-use offshore platforms. Education and dissemination of the concept of ecosystem services will be key for their recognition by the general public, proper valuation by economic markets (Barbier 2013; Costanza et al. 2014), and appropriate use for the emergence of innovative aquaculture practices.

2.8   Concluding Remarks and Outlook

There is increasing concern about the negative consequences of intensive fish aquaculture and the need to remediate its consequences. The two strategies to meet the requirements for more space allotted to aquaculture have been and will continue to be the following: (1) offshore aquaculture, that to date seems very expensive and technically demanding, but due to ample space available, will allow considerable mass production and (2) the very promising but likewise complex IMTA approach.

A combined design of fish cages in the foundation of the turbines in addition to the extractive components of IMTA systems was first discussed at the World Aquaculture Conference in Korea in 2008 (McVey and Buck 2008). This led to a new project in the German Bight (Offshore Site-Selection, Buck and Krause 2012), where for the first time kelp species were tested in an IMTA approach offshore with partners from the offshore wind industry. Future multi-use development will require a great deal of coordination between offshore energy industries, the commercial offshore aquaculture sector, research institutions and international government agencies in order to be successful.

Both seed recruitment and growing and fattening phases are subject to changes in environmental conditions: mussel seed recruitment requires in most of the cases a natural settlement of the larvae on the collection systems, unlike, for instance, in the fish farming industry, where juveniles are bred in hatcheries. As for the feeding of the mussels, it also relies on naturally occurring phytoplankton present in the water column, as opposed to fish that are fed with manufactured fishmeal. In this sense mussel aquaculture depends highly on environmental determinants and their temporal variability. Chile that produces roughly 240,000 t of *Mytilus chilensis* per year experienced a crisis in 2009 when the phytoplankton concentrations decreased to such an extent that mussel growth came to a standstill (Serramalera 2015). Projected climate change poses multiple risks to mussel farming because of the increased frequency of extreme events that might lie outside the realm of present day experience (Serramalera 2015; Adger et al. 2005; Searle and Rovira 2008).

Natural variation in seed collection is inevitable. Environmental temperature for example has an impact on the egg quality and quantity produced by the female broodstock (Bayne et al. 1978) but also on the primary productivity which can have a major impact on larval development and survival. Predation on larvae also can vary significantly between years and is unpredictable. For example, in 2012 Chile found itself in a critical situation when spatfall in the mussel production areas was very poor for unexplained reasons (Carrasco et al. 2014).

Production of bivalves in hatcheries is a valid alternative and technically very feasible but it is not economically viable to produce hatchery seed of mussels in Europe currently because of the low market prices of the end-product. Production of triploid mussels may offer an interesting alternative, as is currently the case for the cupped oyster *Crassostrea gigas*, because this technique may also induce faster growth and lead to year-round supply (since the animals do not reproduce).

The IMTA multi-crop diversification approach (fish, seaweeds and invertebrates) could be an economic risk mitigation and management option to address pending climate change and coastal acidification impacts (Chopin 2015; Clements and Chopin 2016). Multi-species and multi-use systems could not only bring increased profitability per operation/cultivation unit, but also environmental sustainability and societal acceptability.

References

Abreu, M. H., Pereira, R., Yarish, C., Buschmann, A. H., & Sousa-Pinto, I. (2011b). IMTA with *Gracilaria vermiculophylla*: Productivity and nutrient removal performance of the seaweed in a land-based pilot scale system. *Aquaculture, 312,* 77–87.

Abreu, M. H., Pereira, R., Yarish, C., Buschmann, A. H., & Sousa-Pinto, I. (2013). IMTA with *Gracilaria vermiculophylla*: Productivity and nutrient removal performance of the seaweed in a land-based pilot scale system. *Aquaculture, 312,* 77–87.

Abreu, M. H., Varela, D. A., Henriquez, L., Villarroel, A., Yarish, C., Sousa-Pinto, I., et al. (2009). Traditional vs. integrated multi-trophic aquaculture of *Gracilaria chilensis* C. J. Bird, J. McLachlan & E. C. Oliveira: Productivity and physiological performance. *Aquaculture, 293,* 211–220.

ACE. (2015). ACE expands with seaweed site. Accessed November 6, 2015, at http://aceaqua.no/en/2015/09/ace-expands-with-seaweed-site

Adger, W. N., Hughes, T. P., Folke, C., Carpenter, S. R., & Rockström, J. (2005). Social-ecological resilience to coastal disasters. *Science, 309,* 1036–1039.

Asare, S. O., & Harlin, M. M. (1983). Seasonal fluctuations in tissue nitrogen for five species of perennial macroalgae in Rhode Island Sound. *Journal of Phycology, 19,* 254–257.

Bakken, P. (2013). Seaweed to bio-fuels future direction. In *International Seaweed Symposium*, April 23, 2013, Bali (Indonesia).

Barbier, E. B. (2013). Valuing ecosystem services for coastal wetland protection and restoration: Progress and challenges. *Resources, 2,* 213–230.

Barrington, K., Chopin, T., & Robinson, S. (2009). Integrated multi-trophic aquaculture (IMTA) in marine temperate waters. In D. Soto (Ed.), *Integrated mariculture: A global review* (pp 7–46). FAO fisheries and aquaculture technical paper, No. 529. Rome: FAO.

Bartelings, H., Burg van den, S., Jak, R., Jansen, H., Klijnstra, J., Lagerveld, S., et al. (2014). Combining offshore wind energy and large-scale mussel farming: Background and technical, ecological and economic considerations. In S. Lagerveld, C. Rockmann & M. Scholl (Eds.), *Imares Wageningen UR* (117 pp). Report C056/14.

Bayne, B. L., Holland, D. L., Moore, M. N., Lowe, D. M., & Widdows, J. (1978). Further studies on the effects of stress in the adult on the eggs of *Mytilus edulis*. *Journal of the Marine Biological Association of the United Kingdom, 58,* 825–841.

Beveridge, M. (1987). *Cage aquaculture* (p. 351). Farnham, England: Fishing News Ltd.

Blouin, N. A., Brodie, J. A., Grossman, A. C., Xu, P., & Brawley, S. H. (2011). Porphyra: A marine crop shaped by stress. *Trends in Plant Science, 16,* 29–37.

Bombace, G., Fabi, G., & Fiorentini, L. (2000). Artificial reefs in the Adriatic Sea. In A. C. Jensen, K. J. Collins & A. P. M. Lockwood (Eds.), *Artificial reefs in European seas* (pp. 31–63). Kluwer Academic Publishers.

Brenner, M., Broeg, K., Frickenhaus, S., Buck, B. H., & Koehler, A. (2014). Multi-biomarker approach using the blue mussel (*Mytilus edulis L.*) to assess the quality of marine environments: Season and habitat-related impacts. *Marine Environmental Research, 95,* 13–27.

Brenner, M., Buchholz, C., Heemken, O., Buck, B. H., & Koehler, A. (2012). Health and growth performance of blue mussels (*Mytilus edulis* L.) from two different hanging cultivation sites in the German Bight: A nearshore—Offshore comparison. *Aquaculture International, 20,* 751–778.

Brenner, M., Buck, B. H., & Köhler, A. (2007). New concepts for the multi-use of offshore wind farms and high quality mussel cultivation. *Global Aquaculture Advocate, 10,* 79–81.

Broch, O. J., Ellingsen, I. H., Forbord, S., Wang, X., Volent, Z., Alver, M. O., et al. (2013). Modelling the cultivation and bioremediation potential of the kelp *Saccharina latissima* in close proximity to an exposed salmon farm in Norway. *Aquaculture Environment Interactions, 4,* 186–206.

Brown & May (2016). Brown & May Marine Ltd. Accessed May 18, 2016, at http://www.brownandmaymarine.com

Buchholz, C. M., Krause, G., & Buck, B. H. (2012). Seaweed and man. In C. Wiencke & K. Bischof (Eds.), *Seaweed biology, ecological studies* (pp. 471–493). Springer, Berlin.

Buck, B. H. (2002). Open Ocean Aquaculture und Offshore Windparks Eine Machbarkeitsstudie über die multifunktionale Nutzung von Offshore-Windparks und Offshore-Marikultur im Raum

Nordsee. Reports on Polar and Marine Research, Alfred Wegener Institute for Polar and Marine Research, Bremerhaven, 412, 252 pp.

Buck, B. H. (2004). *Farming in a high energy environment: Potentials and constraints of sustainable offshore aquaculture in the German Bight (North Sea)*, PhD Thesis, Alfred Wegener Institute for Polar and Marine Research Bremerhaven, 214 pp., University of Bremen.

Buck, B. H. (2007). Experimental trials on the feasibility of offshore seed production of the mussel *Mytilus edulis* in the German Bight: Installation, technical requirements and environmental conditions. *Helgoland Marine Research, 61,* 87–101.

Buck, B. H. (2017). Mytilid larval appearance and settlement at offshore wind farm sites in the German Bight - a trial to estimate multi-use potentials for bivalve seed collection and cultivation. *Helgoland Mar Research.*

Buck, B. H., & Buchholz, C. M. (2004a). The offshore-ring: A new system design for the open ocean aquaculture of macroalgae. Journal of Applied Phycology, *16,* 355–368.

Buck, B. H., & Buchholz, C. M. (2004b). Support device for the cultivation of macro organisms in marine waters. Patent: PCT/DE2005/000234.

Buck, B. H., & Buchholz, C. M. (2005). Response of offshore cultivated *Laminaria saccharina* to hydrodynamic forcing in the North Sea. *Aquaculture, 250,* 674–691.

Buck, B. H., & Krause, G. (2012). Integration of aquaculture and renewable energy systems. In R. A. Meyers (Ed.), *Encyclopaedia of sustainability science and technology* (Vol. 1, pp. 511–533). Springer Science+Business Media LLC (Chapter No. 180).

Buck, B. H., & Krause, G. (2013). Short expertise on the potential combination of aquaculture with marine-based renewable energy systems. SeaKult-Sustainable Futures in the Marine Realm. Expertise für das WBGU-Hauptgutachten „Welt im Wandel: Menschheitserbe Meer". Wissenschaftlicher Beirat der Bundesregierung Globale Umweltveränderungen (WBGU). 58 pp.

Buck, B. H., Ebeling, M., & Griffin, R. (2011). Seaweed aquaculture as a co-use in an offshore wind farm—Feasibilities and constraints of economic viability for cape wind (Massachusetts). AWI-Report—Project No. USA 11/A01, 12 p.

Buck, B. H., Ebeling, M., & Michler-Cieluch, T. (2010). Mussel cultivation as a co-use in offshore wind farms: Potentials and economic feasibility. *Aquaculture Economics and Management, 14,* 255–281.

Buck, B. H., Krause, G., & Rosenthal, H. (2004). Extensive open ocean aquaculture development within wind farms in Germany: The prospect of offshore co-management and legal constraints. *Ocean & Coastal Management, 47*(3–4), 95–122.

Buck, B. H., Walter, U., Rosenthal, H., & Neudecker, T. (2006a). The development of Mollusc farming in Germany: Past, present and future. *World Aquaculture Magazine, 6–11,* 66–69.

Buck, B. H., Berg-Pollack, A., Assheuer, J., Zielinski, O., & Kassen, D. (2006b). Technical realization of extensive aquaculture constructions in offshore wind farms: Consideration of the mechanical loads. In *Proceedings of the 25th international conference on offshore mechanics and arctic engineering, OMAE 2006*: Presented at the 25th International conference on offshore mechanics and arctic engineering, 4–9 June 2006, Hamburg, Germany/sponsored by Ocean, offshore, and arctic engineering, ASME. New York, NY: American Society of Mechanical Engineers, pp 1–7.

Buschmann, A. H., López, D., & González, M. (2000). Cultivo integrado de moluscos y macroalgas en líneas flotantes y en estanques. In F. M. Faranda, R. Albertini, J.A. Correa (Eds.), Manejo Sustentable de los Recursos Marinos Bentónicos en Chile Centro-Sur: Segundo Informe de Avance. Pontificia Universidad Católica de Chile, Santiago, 7–16.

Buschmann, A. H., Varela, D. A., Hernández-González, M. C., & Huovinen, P. (2008). Opportunities and challenges for the development of an integrated seaweed-based aquaculture activity in Chile: Determining the physiological capabilities of Macrocystis and Gracilaria as biofilters. *Journal of Applied Phycology, 20,* 571–577.

Carlberg, L. K., & Christensen, E. D. (2015). *Go offshore—Combining food and energy production.* DTU Mechanical Engineering, Technical University of Denmark. 48 pp.

Carrasco, A., Astorga, M., Cisterna, A., Arias, A., Espinoza, V., & Uriarte, I. (2014). Pre-feasibility study for the installation of a Chilean mussel *Mytilus chilensis* (Hupé, 1854) seed hatchery in the lakes region, Chiles. *Fisheries and Aquaculture Journal, 5,* 102.

CDP. (2013). Use of internal carbon price by companies as incentive and strategic planning tool: A review of findings from CDP 2013 disclosure. CDP North America. New York. Retrieved October 23, 2015, from https://www.cdp.net/CDPResults/companies-carbon-pricing-2013.pdf

CEFAS. (2014). Centre for environment, fisheries and aquaculture science. Project partner meeting for the EFF project: Shellfish aquaculture in welsh offshore wind farms. Fishmongers' Hall, London, May 22, 2013. Access via Aquafishsolutions, Accessed May 18, 2016 at www.aquafishsolutions.com

Chambers, M. D. (2013). *Integrated multi-trophic aquaculture of steelhead, Blue Mussels and Sugar Kelp*, Ph.D. dissertation. Department of Biological Sciences, University of New Hampshire, Durham, NH.

Chen, T. T., Lin, C. M., Chen, M. J., Lo, J. H., Chiou, P. P., Gong, H. Y., et al. (2014). Principles and application of transgenic technology in marine organisms. In S. K. Kim (Ed.), *Handbook of marine biotechnology*. Berlin: Springer.

Chopin, T. (2006). Integrated multi-trophic aquaculture. What it is and why you should care … and don't confuse it with polyculture. *Northern Aquaculture, 12,* 4.

Chopin, T. (2014). Seaweeds: Top mariculture crop, ecosystem service provider. *Global Aquaculture Advocate, 17,* 54–56.

Chopin, T. (2015). Marine aquaculture in Canada: well-established monocultures of finfish and shellfish and an emerging Integrated Multi-Trophic Aquaculture (IMTA) approach including seaweeds, other invertebrates, and microbial com-munities. *Fisheries, 40,* 28–31.

Chopin, T., Buschmann, A. H., Halling, C., Troell, M., Kautsky, N., Neori, A., et al. (2001). Integrating seaweeds into marine aquaculture systems: A key toward sustainability. *Journal of Phycology, 37,* 975–986.

Chopin, T., Cooper, A., Reid, G., Cross, S., & Moore, C. (2012). Open-water integrated multi-trophic aquaculture: Environmental biomitigation and economic diversification of fed aquaculture by extractive aquaculture. *Reviews in Aquaculture, 4,* 209–220.

Chopin, T., Neori, A., Buschmann, A., Pang, S., & Sawhney, M. (2011). Diversification of the aquaculture sector. Seaweed cultivation, integrated multi-trophic aquaculture, integrated sequential biorefineries. *Global Aquaculture Advocate, 14,* 58–60.

Chopin, T., Robinson, S., Sawhney, M., Bastarache, S., Belyea, E., Shea, R., et al. (2004). The AquaNet integrated multi-trophic aquaculture project: Rationale of the project and development of kelp cultivation as the inorganic extractive component of the system. *Bulletin-Aquaculture Association of Canada, 104,* 11–18.

Chopin, T., Robinson, S. M. C., Troell, M., Neori, A., Buschmann, A. H., & Fang, J. (2008). Multitrophic integration for sustainable marine aquaculture. In S. E. Jørgensen & B. D. Fath (Eds.), *Ecological engineering. Vol. 3 of Encyclopedia of Ecology* (Vol. 5, pp. 2463–2475). Oxford: Elsevier

Chopin, T., Robinson, S., Reid, G., & Ridle, N. (2013). Prospects for Integrated Multi-Trophic Aquaculture (IMTA) in the open ocean. *Bulletin of the Aquaculture Association of Canada,111* (2), 28–35.

Chopin, T., Troell, M., Reid, G. K., Knowler, D., Robinson, S. M. C., Neori, A., et al. (2010). Integrated multi-trophic aquaculture (IMTA)—A responsible practice providing diversified seafood products while rendering biomitigating services through its extractive components. In N. Franz & C. C. Schmidt (Eds.), *Proceedings of the Organisation for economic co-operation and development (OECD) workshop "Advancing the Aquaculture Agenda: Policies to Ensure a Sustainable Aquaculture Sector", Paris, France, 15–16 April 2010: 195–217* (p. 426). Paris: Organisation for Economic Co-operation and Development.

Chopin, T., Yarish, C., Wilkes, R., Belyea, E., Lu, S., & Mathieson, A. (1999). Developing *Porphyra*/salmon integrated aquaculture for bioremediation and diversification of the aquaculture industry. *Journal of Applied Phycology, 11,* 463–472.

Christensen, H. T., Christoffersen, M., Doler, P., Stenberg, C., & Kristensen, P. S. (2009). Assessment of possibilities for line cultivation of mussels in Nysted Sea Wind Farm. Project report DTU Aqua.

Chung, I. K., Kang, Y. H., Yarish, C., Kraemer, G. P., & Lee, J. (2002). Application of seaweed cultivation to the bioremediation of nutrient-rich effluent. *Algae, 17,* 187–194.

Chung, H., Kim, N. G., Choi, K. J., & Woo, H. C. (2015). Novel ocean system for high density mass production of seaweed biomass in Korea. In *2015 World Congress on Advances in Structural Engineering and Mechanics (ASEM15).* Retrieved November 15, 2015, from http://www.i-asem.org/publication_conf/asem15/7.ICOSE15/1w/W1H.HChung.OSE.F1.pdf

Chung, I. K., Oak, J. H., Lee, J. A., Shin, J. A., Kim, J. G., & Park, K.-S. (2013). Installing kelp forests/seaweed beds for mitigation and adaptation against global warming: Korean project overview. *ICES Journal of Marine Science.* doi:10.1093/icesjms/fss206

Clements, J., & Chopin, T. (2016) Ocean acidification and marine aquaculture in North America: Potential impacts and mitigation strategies. *Reviews in Aquaculture, 0,* 1–16. doi:10.1111/raq.12140

Connecticut Department of Energy and Environmental Protection (CT DEEP). (2014). Report of the Nitrogen credit advisory board for calendar year 2013 to the joint standing environment committee of the general assembly.

Corey, P., Kim, J. K., Duston, J., & Garbary, D. J. (2014). Growth and nutrient uptake by *Palmaria palmata* (Palmariales, Rhodophyta) integrated with Atlantic halibut recirculating aquaculture. *Algae, 29,* 35–45.

Corey, P., Kim, J. K., Garbary, D. J., Prithiviraj, B., & Duston, J. (2012). Bioremediation potential of *Chondrus crispus* (Basin Head) and *Palmaria palmata*: Effect of temperature and high nitrate on nutrient removal. *Journal of Applied Phycology, 24,* 441–448.

Cornish, M. L., & Garbary, D. J. (2010). Antioxidants from macroalgae: Potential applications in human health and nutrition. *Algae, 25,* 155–171.

Costanza, R., d'Arge, R., de Groot, R., Farber, S., Grasso, M., Hannon, B., et al. (1997). The value of the world's ecosystem services and natural capital. *Nature, 387*:253–260.

Costanza, R., de Groot, R., Sutton, P., van der Ploeg, S., Anderson, S. J., Kubiszewski, I., et al. (2014). Changes in the global value of ecosystem services. *Global Environmental Change, 26,* 152–158.

CRM. (2001). *Wirtschaftliches und ökologisches Potential einer Laminarien-Farm in Deutschland (Economic and ecologic potentials of algae farming in Germany).* Kiel (Germany): Coastal Research and Management.

Danioux, C., Bompais, X., Paquotte, P., & Loste, C. (2000). Offshore mollusc production in the Mediterranean Basin. In J. Muir & B. Basurco (Eds.), *Mediterranean offshore mariculture* (pp. 115–140). Zaragoza: CIHEAM (Options Méditerranéennes: Série B. Etudes et Recherches, no 30).

Dean, P. R., & Hurd, C. L. (2007). Seasonal growth, erosion rates, and nitrogen and photosynthetic ecophysiology of Undaria pinnatifida (Heterokontophyta) in southern New Zealand. *Journal of Phycology, 43,* 1138–1148.

Delbare, D. (2001). Pesca-project: Hangmosselcultuur in Belgische kustwateren. Centrum voor landbouwkundig onderzoek-Gent, Departement Zeevisserij - afdeling Aquacultuur en Restocking. pp. 14

Dillehay, T. D., Ramrez, C., Pino, M., Collins, M. B., Rossen, J., & Pino-Navarro, J. D. (2008). Monte Verde: Seaweed, food, medicine, and the peopling of South America. *Science, 320,* 784–786.

Dong, S., Fang, J., Jansen Henrice, M., & Verreth, J. (2013). Review on integrated mariculture in China, including case studies on successful polyculture in coastal Chinese waters. Report, Asem Aquaculture Plaform, 7th framework programme.

Fabi, G., & Spagnola, A. (2001). Artificial reefs and mariculture. In J. Coimbra (Ed.), *Modern aquaculture in the coastal zone* (pp. 91–98). IOS Press.

Fabi, G., Manoukian, S., & Spagnolo, A. (2009). Impact of an open-sea suspended mussel culture on macrobenthic community (Western Adriatic Sea). *Aquaculture, 289,* 54–63.

FAO. (2014). *The state of world fisheries and aquaculture 2014.* Rome: Food and Agriculture Organization. 223 pp.

FAO. (2016a). *The state of world fisheries and aquaculture.* Rome: Food and Agriculture Organization, FAO.

FAO. (2016b) Fisheries and aquaculture information and statistics branch. Food and Agriculture Organization: Accessed April 17, 2016.

Fowler, D., Coyle, M., Skiba, U., Sutton, M. A., Cape, J. N., Reis, S., et al. (2013). The global nitrogen cycle in the twenty-first century. *Philosophical Transaction of the Royal Society B, 368,* 0130164.

Gellenbeck, K. (2012). Utilization of algal materials for nutraceutical and cosmeceutical applications—What do manufacturers need to know? *Journal of Applied Phycology, 24,* 309–313.

Gerard, V. A. (1997). The role of nitrogen nutrition in high-temperature tolerance of kelp, *Laminaria saccharina. Journal of Phycology, 33,* 800–810.

Getchis, T. S., Rose, C. M., Carey, D., Kelly, S., Bellantuono, K., & Francis, P. (2008). A guide to marine aquaculture permitting in Connecticut. Connecticut Sea Grant College Program. 140p.

Gevaert, F., Davoult, D., Creach, A., Kling, R., Janquin, M. A., Seuront, L., et al. (2001). Carbon and nitrogen content of *Laminaria saccharina* in the eastern English Channel: Biometrics and seasonal variations. *Journal of the Marine Biological Association of the UK, 81,* 727–734.

Google Earth. (2016). *Map data©2016 imagery ©2016 CNES/Astrium.* DigitalGlobe, Landsat: CNES/Spot Image.

Gosling, E. (2003). *Bivalve Molluscs: Biology, ecology and culture* (456 p). Fishing News Books Ltd.

Grote, B. (2016). Bioremediation of aquaculture wastewater: Evaluating the prospects of the red alga *Palmaria palmata* (Rhodophyta) for nitrogen uptake. *Journal of Applied Phycology, 28,* 3075–3082.

Grote, B., & Buck, B. H. (2017). The IMTA-approach for nutrient balanced aquaculture: Evaluating the potential of turbot (Scophthalmus maximus) and dulse (Palmaria palmata) from onshore RAS to offshore wind farm environment. *Aquaculture.*

Guiry, M. D., & Guiry, G. M. (2015). *AlgaeBase.* Galway: World-wide electronic publication, National University of Ireland. Accessed May 21, 2015 at http://www.algaebase.org

Handå, A. (2012). Cultivaton of mussels (Mytilus edulis): Feed requirements, storage and integration with salmon farming. *Doctoral thesis Norwegian University of Science and Technology NTNU.* Norway, 200 pp.

Handå, A., Ranheim, A., Olsen, A. J., Altin, D., Reitan, K. I., Olsen, Y., et al. (2012). Incorporation of salmon fish feed and feces components in mussels (*Mytilus edulis*): Implications for integrated multi-trophic aquaculture in cool-temperate North Atlantic waters. *Aquaculture, 370–371,* 40–53.

He, P., & Yarish, C. (2006). The developmental regulation of mass cultures of free-living conchocelis for commercial net seeding of *Porphyra leucosticta* from Northeast America. *Aquaculture, 257,* 373–381.

He, W., Yttervik, R., Olsen, G. P., Ostvik, I., Jimenez, C., Impelluso, T., & Schouten, J. (2015). A case study of multi-use platform: Aquaculture in offshore wind farms, Poster no 54, EWEA Offshore 2015 Copenhagen, March 10–12, 2015.

Horn, S. J., Aasen, I. M., & Østgaard, K. (2000). Ethanol production from seaweed extract. *Journal of Industrial Microbiology and Biotechnology, 25,* 249–254.

Hortimare. (2016). Propagating seaweed for a sustainable future. Accessed online October 22, 2016, at http://www.hortimare.com/

Hou, J., & Jin, Y. (2005). *The healing power of chinese herbs and medicinal recipes.* New York: Haworth Press Inc.

Hurd, C. L., Harrison, P. J., Bischof, K., & Lobban, C. S. (2014). In *Seaweed ecology and physiology* (2nd ed., 552 pp). Cambridge University Press.

Hwang, E. K., Cho, Y. C., & Sohn, C. H. (1998). Reuse of holdfasts in Hizikia cultivation. *Korean Journal of Fisheries and Aquatic Sciences, 32,* 112–116.

Hwang, E. K., Gong, Y. G., Hwang, I. L., Park, E. J., & Park, C. S. (2013). Cultivation of the two perennial brown algae *Ecklonia cava* and *E. stolonifera* for abalone feeds in Korea. *Journal of Applied Phycology, 25,* 825–829.

Hwang, E. K., Ha, D. S., Baek, J. M., Wee, M. Y., & Park, C. S. (2006). Effects of pH and salinity on the cultivated brown alga *Sargassum fulvellum* and associated animals. *Algae, 21,* 317–321.

Hwang, E. K., Park, C. S., & Baek, J. M. (2007). Artificial seed production and cultivation of the edible brown alga, Sargassum fulvellum (Turner) C. Agardh: developing a new species for seaweed cultivation in Korea. *Journal of Applied Phycology, 18,* 251–257.

Jansen, H. M. (2012). *Bivalve nutrient cycling-nutrient turnover by suspended mussel communities in oligotrophic fjords*, PhD-thesis. Wageningen University, The Netherlands.

Jiang, Z., Wang, G., Fang, J., & Mao, Y. (2013). Growth and food sources of Pacific oyster *Crassostrea gigas* ntegrated culture with sea bass *Lateolabrax japonicas* in Ailian Bay, China. *Aquaculture International, 21,* 45–52.

Jones, T. O., & Iwama, G. K. (1991). Polyculture of the Pacific oyster, *Crassostrea gigas* (Thunberg) with chinook salmon, *Oncorhynchus tschawytscha. Aquaculture, 92,* 313–322.

Kang, J. H., Kim, S., Joon-Baek, L., Chung, I. K., & Park, S. R. (2014). Nitrogen biofiltration capacities and photosynthetic activity of Pyropia yezoensis Ueda (Bangiales, Rhodophyta): groundwork to validate its potential in integrated multi-trophic aquaculture (IMTA). *Journal of Applied Phycology, 26,* 947–955.

Kautsky, N., Troell, M., & Folke, C. (1999). Ecological engineering for increased production and environmental improvement in open sea aquaculture. *Ecological Engineering for Wastewater Treatment, 11,* 89–97.

Kawashima, Y., & Tokuda, H. (1993). Regeneration from the callus of Undaria pinnatifida (Harvey) Suringar (Laminariales, Phaeo-phyta). *Hydrobiologia, 260*(261), 385–389.

KEPCO, & KIOST. (2016). Co-location of fisheries with offshore wind farm: An overview of research carried out in KEPCO Research Institute & KIOST. Korea Electric Power Cooperation—Research Institure (KEPCO) and Korean Institute of Ocean Science and Technology (KIOST).

KFM. (2014). Kames Fish Farming Ltd., Trial offshore seaweed longlines, Kilmelford–Argyll (Scotland/UK). Accessed June 2, 2016.

Kim, S. K. (2011). *Marine cosmeceuticals. Trends and prospects.* Boca Raton: CRC.

Kim, S.-K. (2012). In *Handbook of marine macroalgae: Biotechnology and applied phycology* (592 p). Wiley-Blackwell.

Kim, J. K., Duston, J., Corey, P., & Garbary, D. J. (2013). Marine finfish effluent bioremediation: Effects of stocking density and temperature on nitrogen removal capacity of *Chondrus crispus* and *Palmaria palmata* (Rhodophyta). *Aquaculture, 414–415,* 210–216.

Kim, J. K., Kraemer, G. P., Neefus, C. D., Chung, I. K., & Yarish, C. (2007). Effects of temperature and ammonium on growth, pigment production and nitrogen uptake by four species of *Porphyra* (Bangiales, Rhodophyta) native to the New England coast. *Journal of Applied Phycology, 19,* 431–440.

Kim, J. K., Kraemer, G. P., & Yarish, C. (2014a). Field scale evaluation of seaweed aquaculture as a nutrient bioextraction strategy in Long Island Sound and the Bronx River Estuary. *Aquaculture, 433,* 148–156.

Kim, J. K., Kraemer, G. P., & Yarish, C. (2015a). Sugar Kelp aquaculture in long island sound and the Bronx River Estuary for nutrient bioextraction associated with biomass production. Marine Ecology progress series (in press).

Kim, J. K., Mao, Y., Kraemer, G. P., & Yarish, C. (2015b). Growth and pigment content of *Gracilaria tikvahiae* under fluorescent and LED lighting. *Aquaculture, 436,* 52–57.

Kim, G. H., Moon, K.-H., Kim, J.-Y., Shim, J., & Klochkova, T. A. (2014b). A revaluation of algal diseases in Korean Pyropia (Porphyra) sea farms and their economic impact. *Algae, 29,* 249–265.

Kraemer, G. P., Kim, J. K., & Yarish, C. (2014). Seaweed aquaculture: Bioextraction of nutrients to reduce eutrophication. *Association of Massachusetts Wetland Scientists Newsletter, 89,* 16–17.

Krause, G., Buck, B. H., & Rosenthal, H. (2003). Multifunctional use and environmental regulations: Potentials in the offshore aquaculture development in Germany, rights and duties in the coastal zone, multidisciplinary scientific conference on sustainable coastal zone management, 12–14 June 2003, Stockholm (Sweden).

Lagerveld, S., Röckmann, C., Scholl, M. (2014). Combining offshore wind energy and large-scale mussel farming: Background and technical, ecological and economic considerations. IMARES Wageningen UR, Report C056/14. 117 pp.

Lander, T., Barrington, K., Robinson, S., MacDonald, B., & Martin, J. (2004). Dynamics of the blue mussel as an extractive organism in an integrated multi-trophic aquaculture system. *Bulletin-Aquaculture Association of Canada, 104,* 19–28.

Langan, R. (2009). Opportunities and challenges for offshore farming (Chapter 29). In G. Burnell & G. Allen (Eds.), *New technologies in aquaculture: Improving production efficiency, quality and environmental management.* Cambridge, UK: Woodhead Publishing Limited.

Langan, R., & Horton, F. (2005). Design, operation and economics of submerged longline mussel culture in the open ocean. In *Proceedings of the Special Sessions on Offshore Aquaculture and Progress in Commercialization of New Species Held at Aquaculture Canada™ in Victoria, BC*, October 2003. No. 103–3, pp. 11–20.

Langlois, J., Sassi, J.-F., Jard, G., Steyer, J.-P., Delgenes, J.-P., & Hélias, A. (2012). Life cycle assessment of biomethane from offshore cultivated seaweed. *Biofuels, Bioproducts and Biorefining, 6,* 387–404.

Lapointe, G. (2013). NROC white paper: Overview of the aquaculture sector in New England. 25 p.

Lasserre, T., & Delgenès, J. P. (2012). WindSeaFuel: Production de macro-algues pour une valorisation en méthane et autres bioproduits. ANR Bioénergies 2009 Poster, ANR-09-BIOE-05, Label pôle, DERBI-TRIMATEC.

Lee, S., & Hwang, H. (2013). Numerical simulation on dynamic characteristics of offshore seaweed culture facility. *Journal of Ocean Engineering and Technology, 27,* 7–15.

Lefebvre, S., Barille, L., & Clerc, M. (2000). Pacific oyster (*Crassostrea gigas*) feeding responses to a fish-farm effluent. *Aquaculture, 187,* 185–198.

Levine, I. A., & Sahoo, D. (2010). *Porphyra: Harvesting gold from the sea.* Ltd: I.K. International Publihing House Pvt. 92p.

Li, X. J., Cong, Y. Z., Yang, G. P., Qu, S. C., Li, Z. L., Wang, G. W., et al. (2009). Trait evaluation and trial cultivation of Dongfang No. 2, the hybrid of a male gametophyte clone of *Laminaria longissima* (Laminariales, Phaeophyta) and a female one of *L. japonica. JOurnal of Applied Phycology, 19,* 303–311.

Liu, D. Y., Keesing, J. K., Xing, Q. G., & Ping, S. (2009). World's largest macroalgal bloom caused by expansion of seaweed aquaculture in China. *Marine Pollution Bulletin, 58,* 888–895.

Liu, F., Pang, S. J., Chopin, T., Gao, S., Shan, T., Zhao, X., et al. (2013). Understanding the recurrent large-scale green tide in the Yellow Sea: Temporal and spatial correlations between multiple geographical, aquacultural and biological factors. *Marine Environmental Research, 83,* 38–47.

Lüning, K., Buchholz, C. (1996). Massenkultur Mariner Makroalgen bei Helgoland zur Gewinnung von Phycokolloiden und zur Verwendung als Biosorptionsmittel. BAH Report Helgoland. 75 pp.

Marfaing, H. (2014). WinSeaFuel project: Biomethane and other bio product developments from offshore seaweed culture. In *21st International Seaweed Symposium of the International Seaweed Association*, April 21–26, 2013, Bali (Indonesia) (abstract).

Marinho, G. S., Holdt, S. L., Birkeland, M. J., & Angelidaki, I. (2015). Commercial cultivation and bioremediation potential of sugar kelp, *Saccharina latissima*, in Danish waters. *Journal of Applied Phycology.* doi:10.1007/s10811-014-0519-8

Martinez-Aragon, J. F., Hernandez, I., Perez-Llorens, J. L., Vazquez, R., & Vergara, J. J. (2002). Biofiltering efficiency in removal of dissolved nutrients by three species of estuarine macroalgae cultivated with sea bass (Dicenntrarchus labrax) waste waters: 1 Phosphate. *Journal of Applied Phycology, 14,* 365–374.

Matthiessen, G. C. (2001). *Oyster culture. Fishing News Books.* 172 p.

Mazzola, A., & Sara, G. (2001). The effect of fish farming organic waste on food availability for bivalve molluscs (Gaeta Gulf, Central Tyrrhenian, MED): Stable carbon isotopic analysis. *Aquaculture, 192*, 361–379.

McVey, J., & Buck, B. H. (2008). IMTA-design within an offshore wind farm. Aquaculture for human wellbeing—The Asian perspective. The annual meeting of the world aquaculture society, 23rd May, 2008, Busan, Korea.

McVey, J. P., Stickney, R. R., Yarish, C., & Chopin, T. (2002). Aquatic polyculture balanced ecosystem management: New paradigms for seafood production. In R. R. Stickney & J. P. McVey (Eds.), *Responsible marine aquaculture* (pp. 91–104). World Aquaculture Society.

Miura, A. (1984). A new variety and a new form of Porphyra (Bangiales, Rhodophyta) from Japan: Porphyra tenera Kjellman var. tamatsuensis Miura, var. nov. and P. yezoensis Ueda form. narawaensis Miura, form. nov. *Journal of the Tokyo University of Fisheries, 71*, 1–3.

Moura, S., Lipsky, A., & Morse, M. (2011). Options for cooperation between commercial fishing and offshore wind energy industries: A review of relevant tools and best practices. *SeaPlan*, 41 p.

Msuya, F. E., & Neori, A. (2008). Effect of water aeration and nutrient load level on biomass yield, N uptake and protein content of the seaweed Ulva lactuca cultured in seawater tanks. *Journal of Applied Phycology, 20*, 1021–1031.

Mugg, J., Serrano, A., Liberti, A., & Rice, M. A. (2000). Aquaculture effluent: A guide for water quality regulators and aquaculturalists. Northeast Regional Aquaculture Center. Publication No. 00–003.

NASA. (2015). NASA, NOAA find 2014 warmest year in modern record. Accessed May 2, 2016, at http://www.giss.nasa.gov/research/news/20150116/

Navarrete-Mier, F., Sanz-Lázaro, C., & Marín, A. (2010). Does bivalve mollusc polyculture reduce marine fin fish farming environmental impact? *Aquaculture, 306*, 101–107.

Neori, A., Chopin, T., Troell, M., Buschmann, A. H., Kraemer, G., Halling, C., et al. (2004). Integrated aquaculture: Rationale, evolution and state of the art emphasizing seaweed biofiltration in modern aquaculture. *Aquaculture, 231*, 361–391.

Neori, A., Shpigel, M., & Ben-Ezra, D. (2000). A sustainable integrated system for culture of fish, seaweed and abalone. *Aquaculture, 186*, 279–291.

Neori, A., Troell, M., Chopin, T., Yarish, C., Critchley, A., & Buschmann, A. H. (2007). The need for a balanced ecosystem approach: Blue revolution aquaculture. *Environment, 49*, 36–43.

Niwa, K., Iida, S., Kato, A., Kawai, H., Kikuchi, N., Kobiyama, A., et al. (2009). Genetic diversity and introgression in two cultivated species (Porphyra yezoensis and Porphyra tenera) and closely related wild species. *Journal of Phycology, 45*, 493–502.

Nunes, J. P., Ferreira, J. G., Gazeau, F., Lenceart-Silva, J., Zhang, X. L., Zhu, M. Y., et al. (2003). A model for sustainable management of shellfish polyculture in coastal bays. *Aquaculture, 219*, 257–277.

Ögmundarson, Ó., Holmyard, J., Þórðarson, G., Sigurðsson, F., & Gunnlaugsdóttir, H. (2011). Offshore aquaculture farming—Report from the initial feasibility study and market requirements for the innovations from the project. Matís ohf/Matís Food research, Innovation and Safety.

Pearce, D. W., & Turner, R. K. (1989). *Economics of natural resources and the environment*. Johns Hopkins University Press.

Peharda, M., Zupan, I., Bavcevic, L., Frankic, A., & Klanjscek, T. (2007). Growth and condition index of mussel Mytilus galloprovincialis in experimental integrated aquaculture. *Aquaculture Research, 38*, 1714–1720.

Peng, Y., Xie, E., Zheng, K., Fredimoses, M., Yang, X., & Zhou, X. (2013). Nutritional and chemical composition and antiviral activity of cultivated seaweed Sargassum naozhouense Tseng et Lu. *Marine Drugs, 11*, 20–32.

Pereira, R., & Yarish, C. (2008). Mass production of marine macroalgae. In S. E. Jørgensen & B. D. Fath (Eds.), *Ecological engineering. Vol. [3] of Encyclopedia of Ecology* (5 Vols., pp. 2236–2247) Oxford: Elsevier.

Pereira, R., & Yarish, C. (2010). The role of *Porphyra* in sustainable culture systems: Physiology and applications. In A. Israel & R. Einav (Eds.), *Role of seaweeds in a globally changing environment* (pp. 339–354). Springer Publishers.

Pogoda, B., Buck, B. H., & Hagen, W. (2011). Growth performance and condition of oysters (*Crassostrea gigas* and *Ostrea edulis*) farmed in an offshore environment (North Sea, Germany). *Aquaculture, 319,* 484–492.

Pogoda, P., Buck, B. H., Saborowski, R., & Hagen, W. (2013). Biochemical and elemental composition of the offshore cultivated oysters *Ostrea edulis* and *Crassostrea gigas*. *Aquaculture, 319,* 53–60.

Pogoda, B., Jungblut, S., Buck, B. H., & Hagen, W. (2012). Parasitic infestations of copepods in oysters and mussels: Differences between nearshore wild banks and an offshore cultivation site in the German Bight. *Journal of Applied Ichthyology, 28,* 756–765.

Rawson, M. V., Jr., Chen, C., Ji, R., Zhu, M., Wang, D., Wang, L., et al. (2002). Understanding the interaction of extractive and fed aquaculture using ecosystem modeling. In R. R. Stickney & J. P. McVey (Eds.), *Responsible marine aquaculture* (pp. 263–2961). Oxon: CABI Publishing.

Redmond, S., Kim, J. K., Yarish, C., Pietrak, M., & Bricknell, I. (2014). Culture of *Sargassum* in Korea: Techniques and potential for culture in the U.S. Orono, ME: Maine Sea Grant College Program. Retreived April 2, 2016, from seagrant.umaine.edu/extension/korea-aquaculture

Reid, G. K., Robinson, S. M. C., Chopin, T., & MacDonald, B. A. (2013). Dietary Proportion of Fish Culture Solids Required by Shellfish to Reduce the Net Organic Load in Open-Water Integrated Multi-Trophic Aquaculture: A Scoping Exercise with Cocultured Atlantic Salmon (Salmo salar) and Blue Mussel (Mytilus edulis). *Journal of Shellfish Research* 32(2): 509–517.

Rensel J., Bright K., King, G., & Siegrist Z. (2011). Integrated fish-shellfish mariculture in Puget Sound. NOAA Final report. 3-31-2011. NA08OAR4170860.

Ridler, N., Robinson, B., Chopin, T., Robinson, S., & Page, F. (2006). Development of integrated multi-trophic aquaculture in the Bay of Fundy, Canada: A socio-economic case study. *Journal of the World Aquaculture Society, 37,* 43–48.

Ridler, N., Wowchuk, M., Robinson, B., Barrington, K., Chopin, T., Robinson, S., et al. (2007). Integrated multi-trophic aquaculture (IMTA): A potential strategic choice for farmers. *Aquaculture Economics and Management, 11,* 99–110.

Robertson-Anderssonm, D. V., Wilson, D. T., Bolton, J. J., Anderson, R. J., & Maneveldt, G. W. (2009). Rapid assessment of tissue nitrogen in cultivated *Gracilaria gracilis* (Rhodophyta) and *Ulva lactuca* (Chlorophyta). *African Journal of Aquatic Science, 34,* 169–172.

Robinson, S. M. C., Lander, T., Martin, J. D., Bennett, A., Barrington, K., Reid, et al. (2007). An interdisciplinary approach to the development of integrated multi-trophic aquaculture (IMTA): The organic extractive component. In *Proceedings of Conference on Aquaculture*. World Aquaculture Society, 786 pp.

Robinson, N., Winberg, P., & Kirkendale, L. (2013). Genetic improvement of macroalgae: Status to date and needs for the future. *Journal of Applied Phycology, 25,* 703–716.

Sahoo, D., & Ohno, M. (2003). Culture of *Kappaphycus alvarezii* in deep water and nitrogen enriched medium. *Bulletin of Marine Sciences and Fisheries, 22,* 89–96.

Sahoo, D., & Yarish, C. (2005). Mariculture of seaweeds. In R. Andersen (Ed.), *Phycological methods: Algal culturing techniques* (pp. 219–237). New York: Academic Press.

Sara, G., Zenone, A., & Tomasello, A. (2009). Growth of *Mytilus galloprovincialis* (mollusca, bivalvia) close to fish farms: A case of integrated multi-trophic aquaculture within theTyrrhenian Sea. *Hydrobiologia, 636,* 129–136.

Schaffelke, B. (2001). Surface alkaline phosphatase activities of macroalgae on coral reefs of the central Great Barrier Reef Australia. *Coral Reefs, 19,* 310–317.

Schaffelke, B., & Klumpp, D. W. (1998). Nutrient-limited growth of the coral reef macroalga *Sargassum baccularia* and experimental growth enhancement by nutrient addition in continuous flow culture. *Marine Ecology Progress Series, 164,* 199–211.

Seaman, M. N. L., & Ruth, M. (1997). The molluscan fisheries of Germany. *NOAA Technical Report NMFS, 129,* 57–84.

Searle, J. P., & Rovira, J. (2008). Cambio climático y efectos en la biodiversidad: el caso chileno. *Biodiversidad de Chile: patrimonio y desafíos* (3rd ed, pp. 502–503). Santiago: CONAMA.

Seed, R., & Suchanek, T. H. (1992). Population and community ecology of *Mytilus*. In E. Gosling (Ed.), *The mussel Mytilus: Ecology, physiology, genetics and culture* (87–169 pp). Development in aquaculture and fisheries science No. 25. Amsterdam, London, New York, Tokyo: Elsevier.

Serramalera, L. (2015). *Adaptive capacity of aquaculture: Insights from the Chilean mussel industry*. Master thesis, Erasmus mundus Master Programme EMBC.

SES. (2015a). Patented seaweed carrier system by SES, Patent No. EP 09 836 439.1. SES—Seaweed Energy Solutions, Trondheim (Norway).

SES. (2015b). Seaweed carrier system for the cultivation of macroalgae in offshore environments patented by SES—Seaweed Energy Solutions, Trondheim (Norway).

Shin, J.-A. (1999). Crossing between *Porphyra yezoensis* and *P. tenera*. *Algae, 14*, 73–77.

Shin, J.-A. (2003). Inheritance mode of some characters of *Porphyra yezoensis* (Bangiales, Rhodophyta) II. Yield, photosynthetic pigment content, red rot disease-resistance, color, luster and volatile sulfur compounds concentration. *Algae, 18*, 83–88.

Shumway, S. E. (2011). *Shellfish Aquaculture and the Environment* (528 p.). Wiley-Blackwell.

Skjermo, J., Aasen, I. M., Arff, J., Broch, O. J., Carvajal, A., Christie, H., et al. (2014). A new Norwegian bioeconomy based on cultivation and processing of seaweed: Opportunities and R&D needs. SINTEF Fisheries and Aquaculture 2014-06-03. 46 pp.

Smit, A. J. (2004). Medicinal and pharmaceutical uses of seaweed natural products: A review. *Journal of Applied Phycology, 16*, 245–262.

Sohn, C. H. (1998). The seaweed resources of Korea. In A. T. Critchley & M. Ohno (Eds.), *Seaweed resources of the world* (pp. 15–33). Yokosuka: Japan International Cooperation Agency.

Stephenson, K., & Shabman, L. (2011). Rhetoric and reality of water quality trading and the potential for market-like reform. *Journal of the American Water Resources Association, 47*, 15–28.

Stirling, H. P., & Okumus, I. (1995). Growth and production of mussels (*Mytilus edulis*) suspended at salmon cages and shellfish farms in two Scottish sea lochs. *Aquaculture, 134*, 193–210.

Sutherland, J. E., Lindstrom, S. C., Nelson, W. A., Brodie, J., Lynch, M. D. J., Hwang, M. S., et al. (2011). A new look at an ancient order: Generic revision of the Bangiales (Rhodophyta). *Journal of Phycology, 47*, 1131–1151.

Syvret, M., Fitzgerald, A., Gray, M., Wilson, J., Ashley, M., & Jones, C. E. (2013). Aquaculture in Welsh offshore wind farms: A feasibility study into potential cultivation in offshore windfarm sites. Report for the Shellfish Association of Great Britain.

Tassetti, A., Malaspina, S., & Fabi, G. (2015). Using a multibeam echosounder to monitor an artificial reef. *The International Archives of Photogrammetry, Remote Sensing and Spatial Information Sciences, XL-5/W5* (2015 Underwater 3D recording and modeling, April 16–17, 2015, Piano di Sorrento, Italy).

Tedesco, M. A., Swanson, R. L., Stacey, P. E., Latimer, J. S., Yarish, C., & Garza, C. (2014). Synthesis for management. In J. S. Latimer, M. A. Tedesco, R. L. Swanson, C. Yarish, P. E. Stacey, & C. Garza (Eds.), *Long island sound: Prospects for the urban sea* (pp. 481–539). New York: Springer.

Troell, M., Chopin, T., Reid, G., Robinson, S., & Sarà, G. (2011). Letter to the editor: Finfish-shellfish integrated multi-trophic aquaculture: Variable efficiency or experimental design issues? *Aquaculture, 313*, 171–172.

Troell, M., Hailing, C., Neori, A., Buschmann, A. H., Chopin, T., Yarish, C., et al. (2003). Integrated mariculture: Asking the right questions. *Aquaculture, 226*, 69–90.

Troell, M., Joyce, A., Chopin, T., Neori, A., Buschmann, A. H., & Fang, J. G. (2009). Ecological engineering in aquaculture: Potential for integrated multi-trophic aquaculture (IMTA) in marine offshore systems. *Aquaculture, 297*, 1–9.

Troell, M., & Norberg, J. (1998). Modelling output and retention of suspended solids in an integrated salmon–mussel culture. *Ecological Modelling, 110,* 65–77.

TROPOS. (2015). Aquaculture pilot scale report. *Hellenic Center for Marine Research,* Modular multi-use deep water offshore platform harnessing and servicing Mediterranean, subtropical and tropical marine and maritime resources—TROPOS Project, Grant Agreement Number 288192. 40 pp.

van den Burg, S., Jak, R., Smits M.-J., de Blaeij, A., Rood, T., Blanken H., et al. (2016). Zeewier en natuurlijk kapitaal; Kansen voor een biobased economy. Wageningen, LEI Wageningen UR (University & Research centre), LEI Rapport 2016-049. 36 blz.

van den Burg, S. W. K., Stuiver, M., Veenstra, F. A., Bikker, P., Lopez-Contreras, A. M., Palstra, A. P., et al. (2013). Triple P review of the feasibility of sustainable offshore seaweed production in the North Sea. Lei Wageningen UR. Lei Reports 13, 77, 105 pp.

Verhaeghe, D., Delbare, D., & Polet, H. (2011). Haalbaarheidsstudie passieve visserij en maricultuur binnen de Vlaamse windolenparken? ILVO-mededeling Nr. 99, 136 pp.

Vigin, L., Devolder, M., & Scory, S. (2016). *Kaart van het gebruik van de Belgische zeegebieden —Carte de l'usage des espaces marins belges.* Brügge: Royal Belgian Institute for Natural Sciences.

Voskamp, P. (2010). *Mussel experiment bears fruit in deep water southeast of the island.* Times: Block Island. 1 p.

Wallace, J. F. (1980). Growth rates of different populations of the edible mussel, *Mytilus edulis,* in north Norway. *Aquaculture, 19,* 303–311.

Wiencke, C., & Bischof, K. (2012). *Seaweed Biology—novel insights into ecophysiology, ecology and utilization.* Berlin and Heidelberg: Springer. 514 p.

Xie, E., Liu, D., Jia, C., Chen, X. L., & Yang, B. (2013). Artificial seed production and cultivation of the edible brown alga *Sargassum naozhouense* Tseng et Lu. *Journal of Applied Phycology, 25,* 513–522.

Zhang, J., Huo, Y., Wu, H., Yu, K., Kim, J. K., Yarish, C., et al. (2014). The origin of the *Ulva* macroalgal blooms in the Yellow Sea in 2013. *Marine Pollution Bulletin, 89,* 276–283.

Zhang, J., Kim, J. K., Yarish, C., & He, P. (2016). The expansion of *Ulva prolifera* O.F. Müller macroalgal blooms in the Yellow Sea, PR China, through asexual reproduction. *Marine Pollution Bulletin, 104,* 101–106.

Chapter 3
Technological Approaches to Longline- and Cage-Based Aquaculture in Open Ocean Environments

Nils Goseberg, Michael D. Chambers, Kevin Heasman, David Fredriksson, Arne Fredheim and Torsten Schlurmann

Abstract As the worldwide exploitation rate of capture fisheries continues, the development of sustainable aquaculture practices is increasing to meet the seafood needs of the growing world population. The demand for aquatic products was historically satisfied firstly by an effort to expand wild catch and secondly by increasing land-based and near-shore aquaculture. However, stagnation in wild catch as well as environmental and societal challenges of land-based and near-shore aquaculture have greatly promoted efforts to development farming offshore technologies for harsh, high energetic environments. This contribution thus highlights recent technological approaches based on three sample sites which reach out from sheltered near-shore aquaculture sites to sites with harsh wave/current conditions. It compares and evaluates existing technological approaches based on a broad literature review; on this basis, we then strongly advocate for presently available aquaculture technologies to merge with future offshore structures and platforms and to unveil its added value through synergetic multi-use concepts. The first example

N. Goseberg (✉)
Ludwig-Franzius-Institute for Hydraulic, Estuarine and Coastal Engineering,
Leibniz Universität Hannover, Nienburger Str. 5, 30167 Hannover, Germany
e-mail: goseberg@fi.uni-hannover.de

M.D. Chambers
School of Marine Science and Ocean Engineering, University of New Hampshire,
Morse Hall, Room 164, Durham, NH, USA

K. Heasman
Cawthron Institute, Private Bag 2, Nelson 7010, New Zealand

D. Fredriksson
Department of Naval Architecture and Ocean Engineering,
United States Naval Academy, Annapolis, MD, USA

A. Fredheim
Department of Aquaculture Technology, SINTEF Fisheries and Aquaculture,
P.O. Box 4762, 7465 Sluppen, Trondheim, Norway

T. Schlurmann
Ludwig-Franzius-Institute for Hydraulic, Estuarine and Coastal Engineering,
Leibniz Universität Hannover, Nienburger Str. 4, 30167 Hannover, Germany

© The Author(s) 2017
B.H. Buck and R. Langan (eds.), *Aquaculture Perspective of Multi-Use Sites in the Open Ocean*, DOI 10.1007/978-3-319-51159-7_3

describes the recent development of longline farming in offshore waters of New Zealand. New Zealand has designated over 10,000 ha of permitted open ocean water space for shellfish farming. The farms range from 8 to 20 km out to sea and a depth of 35–80 m of water. Research has been ongoing for the last 10 years and the first commercial efforts are now developing in the Bay of Plenty. New methods are being developed which should increase efficiency and reduce maintenance with a particular focus on Greenshell mussel (*Perna canaliculus*) and the Pacific Oyster (*Crassostrea gigas*), Flat Oyster (*Tiostrea chilensis*) and various seaweeds. The second case study involves a long-term, open ocean aquaculture (OOA) research project conducted by the University of New Hampshire. During the course of approximately 10 years, the technological aspects of OOA farming were conducted with submersible cages and longlines, surface feeding systems and real time environmental telemetry. The grow-out potential of multiple marine species such as cod (*Gadus morhua*), haddock (*Melanogrammus aeglefinus*), halibut (*Hippoglossus hippoglossus*), blue mussel (*Mytilus edulis*), sea scallop (*Placopecten magellanicus*) and steelhead trout (*Oncorhynchus mykiss*) were investigated at a site 12 km from shore. The last study presents a multi-use aspect of aquaculture for an open ocean site with fish cages attached to existing offshore wind energy foundations. Technological components such as mounting forces and scour tendencies of two different cage structures (cylindrical and spherical) were investigated by means of hydraulic scale modeling. The cages were pre-designed on the basis of linear theory and existing standards and subsequently exposed to some realistic offshore wave conditions. The wind farm "Veja Mate" in German waters with 80 planned 5 MW turbines anchored to the ground by tripiles is taken as the basis for the tested wave conditions. Based on findings stemming from the three example approaches conclusions are drawn and future research demand is reported.

3.1   Introduction

As the worldwide exploitation rate of capture fisheries continues, the development of sustainable aquaculture practices is increasing to meet the seafood needs of the growing world population. The demand for aquatic products was historically satisfied firstly by an effort to expand wild catch and secondly by increasing land-based and near-shore aquaculture. However, stagnation in wild catch as well as environmental and societal challenges of land-based and near-shore aquaculture have greatly promoted efforts to development farming offshore technologies for harsh, high energetic environments. As a consequence, ocean domestication is of key importance to maintain the ocean as a sustainable source of food, both economically and ecologically (Marra 2005).

While the annual growth rate of aquaculture production has been 6.3%, total aquaculture production grew from 34.6 million tons in 2001 to 59.9 million tons in 2010; thus it depicts the second important sector to supply the continued global demand for marine proteins (FAO 2012). In 2010, world marine farming production was estimated at 36.1 million tons with a value of US$37.9 billion (FAO 2012). Nearly all ocean farming is conducted inshore, in contrast to offshore aquaculture that is still in its infancy. Offshore aquaculture may be defined as taking place in areas of the open ocean exposed to significant wind and wave action, and where there is a requirement for equipment and servicing vessels to survive and operate in severe sea conditions from time to time (Drumm 2010).

There is an obvious, demand-driven need to develop offshore aquaculture throughout the world. However, one of the most difficult obstacles to overcome is finding locations for new aquaculture farms. Because of the difficulties associated with inshore locations, it is assumed that most new aquaculture activities will be developed offshore in the Exclusive Economic Zone (3–200 miles) where there are fewer conflicts with existing user groups, and less risk of pollution. The high energy (winds and waves) of such exposed locations, however, present significant technical challenges in the design, testing and construction of aquaculture systems that are capable of surviving in these areas. In addition to these technical challenges, there are many biological, regulatory, social and economic problems to be solved.

Despite, drivers at local and global levels provide impetus for aquaculture to move to these unprotected waters of the open sea. There are issues of competition for space with other users, problems with water quality, and oftentimes there is a negative public perception of aquaculture's environmental and aesthetic impacts (Kapetsky et al. 2013). Some of these conflicting issues are also common to the offshore wind energy and oil industry. The oil/gas industry has a long history of installed facilities in offshore locations. The recent pressure to reduce carbon dioxide emissions has additionally leveraged the worldwide planning and installation of offshore wind energy converters. Countries bordering the North and Baltic Sea with limited accessibility of inland building sites such as Denmark, Great Britain or Germany started exploring offshore wind energy feasibility in the end of the eighties (Hau and von Renouard 2006). High energy cost and natural disasters facilitated this development. It is thus natural to assess if foundation structures (piles, tripods, jackets) of existing or licensed wind energy sites could be co-used for additional economic activities such as offshore aquaculture.

Aquaculture migration towards offshore sites has undergone substantial progress in the last two decades. This progress was leveraged by advances in numerical and physical modelling of various fish cages or long line arrangements as well as by technological developments and prototype sites. In order to design an open ocean aquaculture test site near the south of the Isles of Shoals, New Hampshire a numerical model was used by Tsukrov et al. (2000) to optimize the structure. Finite element analysis was applied to study the performance of surface and submerged constructions and a simplified formulation for the simulation of the nets is presented. An improved formulation was presented by Tsukrov et al. (2003) offering a consistent net element which is generally capable to model fluid action and net inertia of fishing nets and allows to simulate environmental loadings originating from currents, waves or other mechanical impacts. Motivated by the impact of storm-waves to aquaculture facilities, a lump-mass method was used to study to the effects of currents to net-cage systems and good agreement was found between the numerical results and prior experimental findings (Huang et al. 2006).

Experimental studies most often provide the basis for numerical modelling attempts, help calibrating numerical models and verify achieved simulation results. Experimental research offers down-scaled, controllable and repeatable conditions for reliable analysis which becomes increasingly important as study sites move from more sheltered near-coast regions to high-energetic offshore sites. The effects of current-only conditions on down-weighed flexible circular nets were investigated in greater detail experimentally by Lader and Enerhaug (2005). They emphasized that forces on and deformations of flexible nets are mutually highly dependent on each other. This in particular applies to aquaculture sites residing in the open ocean with harsh environments. The suitability of a modified gravity-type fish cage was similarly investigated by experimental and numerical means under exposed environmental conditions with regular waves (DeCew et al. 2005). These controlled wave conditions allowed for the derivation of motion response in heave, surge and pitch as well as load response in the anchor and bridle lines. The findings were then extended to the irregular regime by the application a stochastic approach. Based on a combination of numerical and experimental approaches Huang et al. (2008) conjectures aquaculture activities should not be situated waters with current velocities above 1 m/s unless volume-reducing effects can be safely restricted by technological means. At the same time it is recommended to carefully consider the combined effects of currents and obliged waves before facilities are installed at new sites.

Although the tendency to aquaculture site located further off the shore is well traceable in literature there are generally much fewer sources which explicitly focus on the combination of different aquaculture applications at one location or even inclusion of non-aquaculture elements such as support structures from other industries in the open ocean. Buck and Buchholz (2005) discussed the co-use of existing or planned offshore structures for the growth of *Laminaria saccharina* by comparing drag forces measured in a current flume with those resulting from conditions in the North Sea. The authors also reported on the development of a devices to culture macroalgae in a range of offshore conditions (Buck and Buchholz 2004). Based on the analysis of the current development inside the continental shelf around the world Lacroix and Pioch (2011) plead innovative efforts to move towards eco-engineered structures such as artificial fishing reefs or the incorporation of secondary purposes within wind farms. The mutual benefits of wave energy converter foundations as artificial reefs were examined for a building site about 100 km north of Gothenburg at the Swedish coast (Langhamer and Wilhelmsson 2009). It was found that basically the amount of fish and crab on the ground of the foundations was considerably increase whereas the abundance of fish in the water column above was not substantially altered. Another example of the multi-use idea is the development of artificial reefs which oftentimes serve not only the purpose to provide valuable habitat to marine species but also serve as one additional form of coastal protection (Liu and Su 2013).

This contribution presents recent technological advances and innovative approaches towards aquaculture production in harsher oceanic conditions on the basis of three case studies involving the idea of multi-use. The recent advancements of tools and methods to study the response of aquaculture technology to environmental loading such as winds, waves, currents and other mechanical impacts describe above serve as the basis for the presented studies. Technological challenges and existing obstacles due to the high-energetic environment are discussed at the end. Within the first two case studies, currently available potential and challenges of aquaculture is discussed in light of its applicability towards harsher environments and high-energy impacts from winds and waves. Then, we extend our analysis and the conclusions drawn from those first two case studies and present a unique third case study. Therein, existing aquaculture cage technology is thoroughly tested as it is innovatively attached to a commonly used offshore wind energy converter foundation, called a tripile foundation. For the first time, such a multi-use concept was tested under laboratory conditions with valuable conclusions and inter-comparisons between existing and anticipated offshore technologies being reported in the discussion and conclusion chapter.

3.2 Case Study on Long Lines

3.2.1 Mussel Farming Development in NZ

Mussels (the Greenshell™ mussel, *Perna canaliculus*) are the main shellfish in both mass and value to be farmed in New Zealand. They have been farmed in sheltered and semi-sheltered areas since the industry started in 1979. The current production has fluctuated between 86,000 and 101,000 ton per annum between 2011 and 2014. Mussels are farmed using the New Zealand longline system, the general structure of which is discussed below. There are variations to this theme depending on location and farmer preference.

In 2003 consents for the offshore aquaculture space were lodged and the process of obtaining this space started. There were numerous reasons for the departure from the sheltered waters including user conflict and a desire to increase farm size. The proposed farms were in waters ranging from 6 to 20 km off the coast in water depths ranging from 35 to 80 m.

In 2005, the first experimental ropes were installed into the open ocean waters of Hawkes Bay. The first open ocean structure was based on the traditional mussel backbone but it was influenced by the systems used in the Coromandel in the North Eastern corner of New Zealand North Island. The Coromandel generally has higher energy than that of other culture sites such as the Marlborough Sounds.

A traditional inshore New Zealand mussel longline consists of: a mooring; chain; warp; bridal; backbone; bridal; warp; chain and mooring (Fig. 3.1). The early moorings were Danforth type anchors. These were overtaken in preference by

Cross-section of a surface longline marine farm

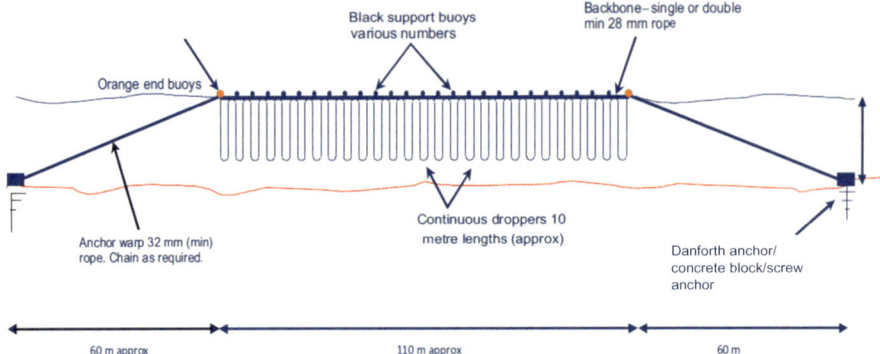

Fig. 3.1 Cross section of a New Zealand surface longline *Credit* New Zealand Marine Farmers Association

concrete shaped moorings of up to 10 metric tons (22,200 lb.) in mass. Where possible concrete mooring are being superseded by screw anchors when new ropes are being installed and the substrate is suitable.

The chain is a heavy duty chain of 6 m (20 ft.) to 15 m (50 ft.) in length. A synthetic rope of 27 mm (>1 in.) to 36 mm (1½ in.) called a warp is attached to the chain and rises to the bridal. The warp is generally three times the depth of the water. The bridal splits into two, each split attaching to a header rope on the opposite edges of the float line, respectively. This section of the mussel longline holding the buoys or floats is called the backbone. Since this backbone has a header rope on each side of the float to support the production line it is referred to as a "double backbone". The backbone extends to the opposite bridal and so on. The buoys are 1.3 m across (~4 ft. 3 in.) and spaced appropriately to support the mass of growing mussels on the backbone. Typically the backbone ranges from 100 m (328 ft.) to 200 m (656 ft.) long. A mussel longline (production rope) is hung from the backbone. It is attached on one side, descends down to 10 or 15 m, loops up to the opposite side of the backbone and crosses the gap between the buoys and the descends down again. Each loop being approximately 50 cm (20 in.) to 70 cm (28 in.) apart along the backbone. In this way, between 3000 m (9842 ft.) and 4000 m (13,123 ft.) of mussel long line is hung from a 100 m backbone.

The longline hanging from the backbone can produce between 6.5 kg (14 lb. 5 oz.) to 13 kg (28 lb. 10 oz.) per meter (or 4 lb. 12 oz. per ft. to 9 lb. 8 oz. per ft.) depending on site and situation. The time period to produce this would range from 12 to 20 months, again depending on site and location.

As the industry changed from inshore to more exposed sites the backbones were cut down to a single header rope (single backbone) i.e. there is no bridal and the warp joined directly to the backbone (Fig. 3.2) and the production longline is draped in loops along the single header rope or backbone. The backbone is submerged and has approximately 66% of the floatation attached directly to it.

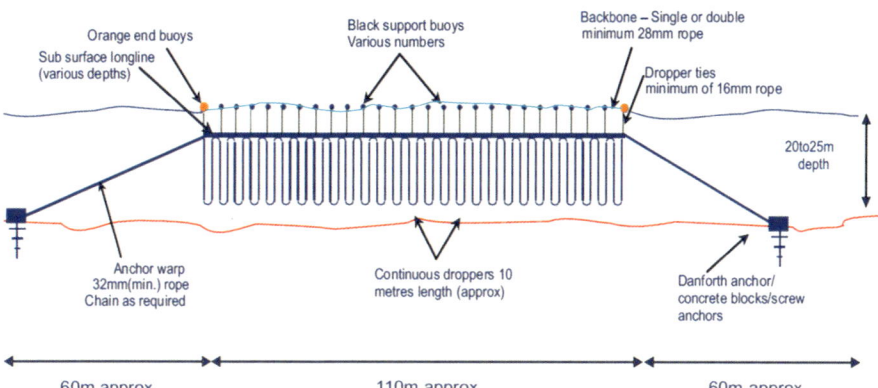

Fig. 3.2 Cross section of a New Zealand subsurface longline. *Note* More recent systems have more floats on the submerged part of the longline *Credit* New Zealand Marine Farmers Association

Additional floatation is positioned on the surface with strops extending from the surface floats to the backbone. The length of the strop dictates the depth of the backbone.

The surface floats will provide 30–40% of the floatation and also give some indication of the load the backbone is bearing and when additional floatation will be required. The backbone is installed loose enough so that a hook can be lowered from a vessel to snag the backbone and the resultant apex of the snagged backbone brought to the surface. The spacing between the droppers may be increased when compared to the inshore systems and may be as much as 1 m (3 ft. 3 in.) apart. By submerging the ropes, this system provides some protection from the wave energy experienced in open ocean situations, however, the surface floats still transfer energy to the backbone which can result in production losses and increased maintenance. In addition, there is an issue arising from the additional requirement in flotation as a result of mussel growth, i.e. once the farm has a large number of backbones on it the management of floatation will increase significantly.

There is no doubt that new systems will be developed in the future. The next generation system that is currently being developed and tested is the "Set and Forget" (S&F) system. This system, developed by the Cawthron Institute (www. cawthron.org.nz) in conjunction with an open ocean farming operation and Government funding (MBIE and Kiwinet), is a fully submersible double backbone system which will be deployed and recovered from the surface.

The S&F system (Fig. 3.3) has a similar configuration to the surface double backbone but where the bridal meets the backbone, there is a mooring directly below it (screw anchor). There are additional intermediate anchors spaced every 35 m along the backbone. These intermediate moorings are threaded through a mechanism in the S&F buoy. There are also single S&F attachment mechanisms on

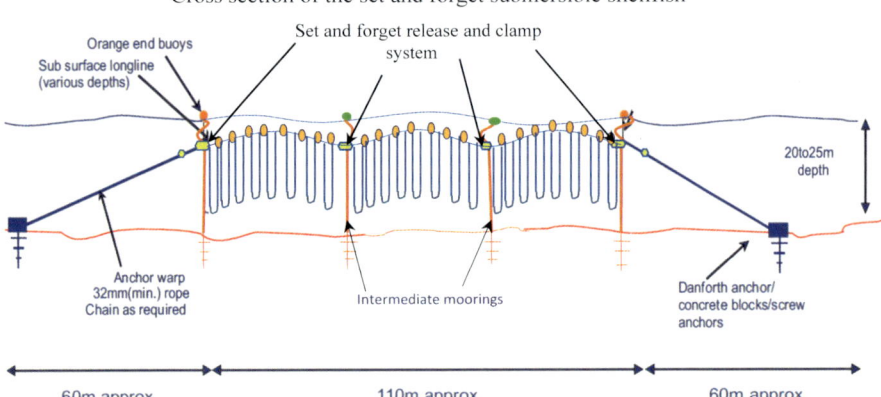

Fig. 3.3 Set and Forget system—next generation submersible backbone under testing (www. cawthron.org.nz)

each warp with a surface float. The idea is to fully seed the longline on the backbone. The backbone is then floated with sufficient buoys to support the intended harvest mass. Once seeded the backbone is pushed below the surface to the desired depth and the S&F mechanism engaged. The mechanisms on the warps are tightened to ensure the backbone does not collapse towards the center. The mussels can then be left until they are due to be harvested. No intermediate floatation is required. At harvest the mechanisms are released using a surface driven unit (physically not electronically) and the backbone rises to the surface to be harvested.

3.2.2 Oyster Farming in the Open Ocean

Oysters have also been tested on the open ocean sites. Although there are several methods used, inshore only bags have been tested in the offshore situation. Pacific oysters (*Crassostrea gigas*) have been held in purses or oyster bags (Fig. 3.4). The bags are configured one below the other in a "ladder" configuration. There are 20 bags in a ladder spaced approximately 50 cm (20 in.) apart with 50–100 oysters in each bag depending on bag size and oyster's size. Some work is required in the design of the bags to reduce the maintenance of the present ladder system. Baffles have been introduced into the bags/purses to avoid the oysters being clumped into one corner of the unit. Oysters have to be at a minimum depth below the surface to avoid being "rumbled" by the wave energy which restricts shell growth. The level of floatation has to be managed to reduce excessive energy transfer to the culture units. Oysters have shown growth rates comparable with inshore waters in North Island. The Flat oyster *Tiostrea chilensis* will be tested in the same ladder system in

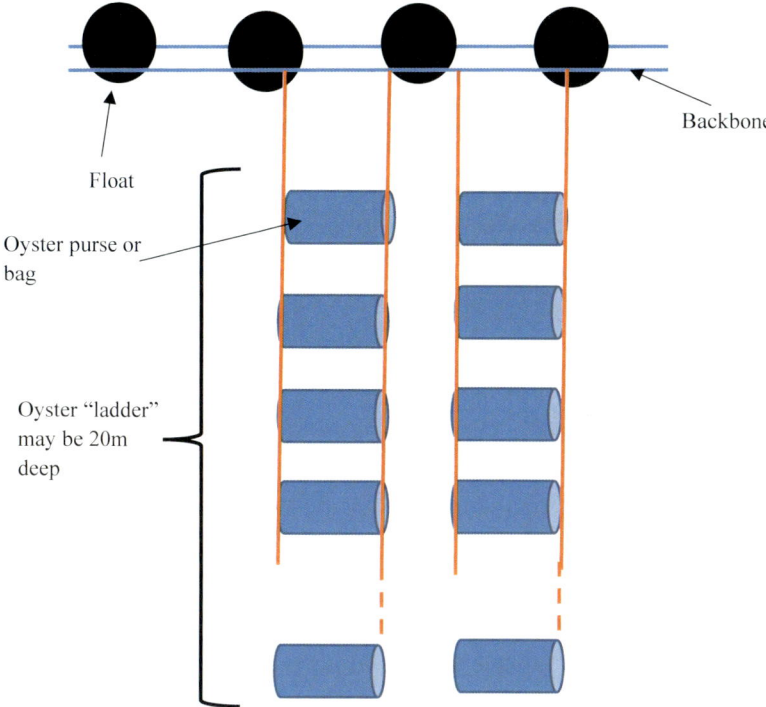

Fig. 3.4 Oyster purses being hung in a ladder configuration on the backbone

the near future on the open ocean farm. Early indicators are that flat oysters will grow in this system if they are away from direct wave energy.

3.3 Case Study on Submerged Aquaculture

3.3.1 *Open Ocean Aquaculture in New Hampshire, USA*

The University of New Hampshire (UNH) established the Open Ocean Aquaculture (OOA) research farm in 1999 (http://ooa.unh.edu/). The overall goal of the project was to stimulate the further development of commercial offshore aquaculture in New England. Also important was to work closely with commercial fishermen, coastal communities, private industry, and fellow marine research scientists to develop technologies for the aquaculture of native, cold-water finfish and shellfish species in exposed oceanic environments. The site was located 12 km offshore and in 52 m water depth, away from traditional fishing activities, recreational vessels and commercial traffic.

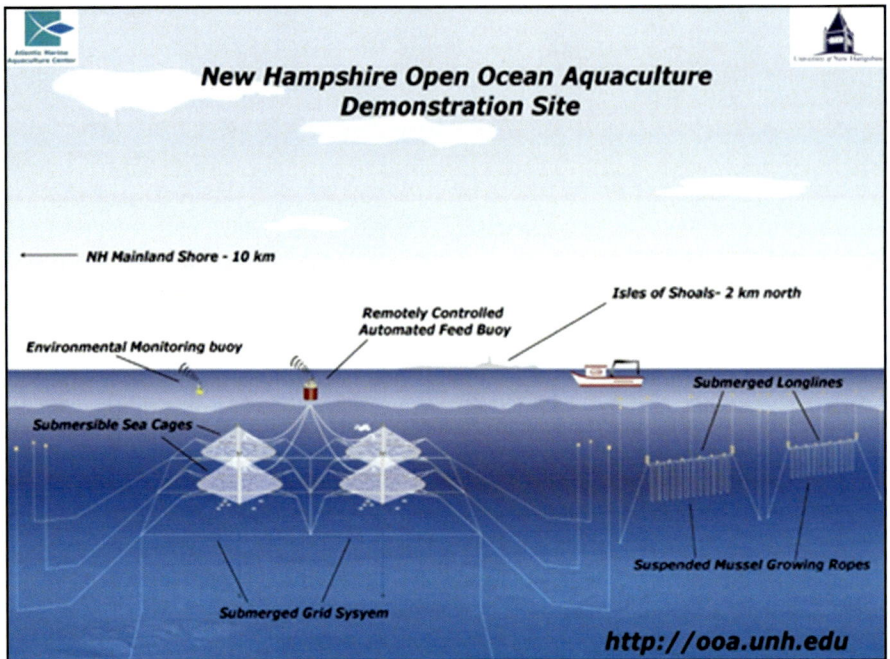

Fig. 3.5 Schematic of the University of New Hampshire Open Ocean Aquaculture site located 12 km offshore New Hampshire, USA

Prior to aquaculture systems going offshore, they were analyzed through numerical and physical scale modeling at the Jere Chase Ocean Engineering lab on campus. These tools help identify strengths and weaknesses of components, simulate failure scenarios and help determine safety factors before they deployed and field tested (Fredriksson et al. 2004; Swift et al. 1998; Tsukrov et al. 2000). The backbone of the OOA research farm was a 12 Ha, submerged grid for holding surface and sub-surface systems (Fig. 3.5). The four bay mooring had a scope of 3:1 and was held in place by 12, 1 ton embedment anchors (Fig. 3.6). The mooring complex was made from 5 cm dia. Polysteel lines that were tensioned and held in place by 1.43 m composite subsurface buoys in the corners and center of the grid (DeCew et al. 2010a, 2012; Fredriksson et al. 2004). The robust grid provided the necessary infrastructure to evaluate submersible cage systems (Chambers et al. 2011; DeCew et al. 2010b). One such system extensively tested was the Sea Station^{TM} 600 and 3000 m^3 cages (Fig. 3.7a). Using a central spar with pennant weight, the cage could be submerged to a prescribed depth or lifted by compressed air to the surface. It utilized a Spectra net that shackled to the top and bottom of the central galvanized spar and used a middle ring that gave its bi-conical shape. Also evaluated offshore in the grid was a 600 m^3 geodesic Aquapod^{TM} made with triangle panels and hard wire (Fig. 3.7b). This system was neutrally buoyant and be could set at various depths based upon the mooring configuration. Ocean Farm Technologies produces the

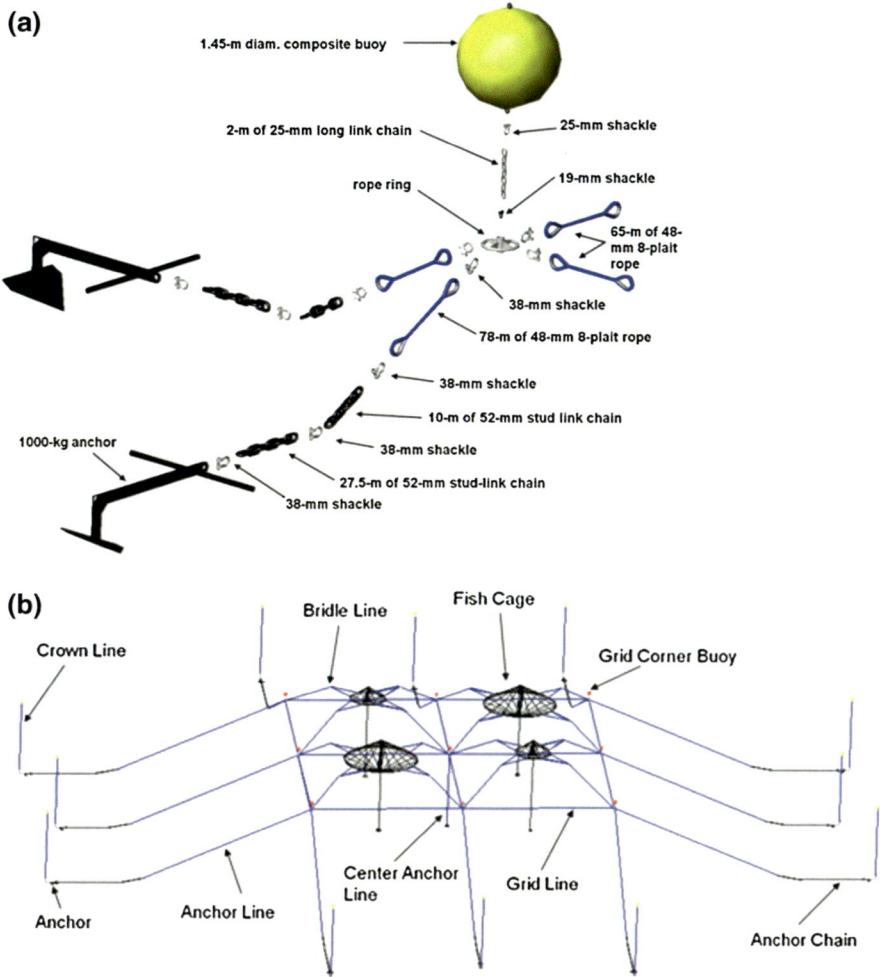

(a)

1.45-m diam. composite buoy

2-m of 25-mm long link chain

rope ring

25-mm shackle

19-mm shackle

65-m of 48-mm 8-plait rope

38-mm shackle

78-m of 48-mm 8-plait rope

38-mm shackle

10-m of 52-mm stud link chain

38-mm shackle

27.5-m of 52-mm stud-link chain

38-mm shackle

1000-kg anchor

(b)

Crown Line

Bridle Line

Fish Cage

Grid Corner Buoy

Center Anchor Line

Grid Line

Anchor

Anchor Line

Anchor Chain

Fig. 3.6 Submerged grid mooring (**a**) and corner grid assembly (**b**). The side anchor assembly was similar except for the use of only one anchor leg and a 0.95 m steel float

Aquapod™ and since their initial sea trial in NH, have made many advances to their containment system. The American Soybean Association International Marketing (ASAIM) came forth with the Ocean Cage Aquaculture Technology (OCAT) system (Fig. 3.7c). The 100 m³ cage OCAT (2 m × 4.5 m × 7 m) was constructed of HDPE pipe with galvanized steel fittings. Chain ballast hangs below the lower cage rim providing a restoring force. The net chamber is formed by attaching net panels to the cage framework. Engineers at UNH designed and evaluated components in the cage frame to make it a submersible system (DeCew et al. 2010b). Information gathered at the UNH OOA site helped these three companies improve their containment systems and increase global sales.

(a) Sea Station™   (b) Aquapod™

(c) OCAT

Fig. 3.7 Submersible cage systems (**a–c**) evaluated offshore New Hampshire. They include the Sea Station™, the Aquapod™ and the Ocean Cage Aquaculture Technology (OCAT)

Marine finfish species including summer flounder (*Paralichthys dentate*), cod (*Gadus morhua*), haddock (*Melanogrammus aeglefinus*), halibut (*Hippoglossus hippoglossus*), and steelhead trout (*Oncorhynchus mykiss*) were cultured and harvested from the research farm (Chambers and Howell 2006; Chambers et al. 2007; Howell and Chambers 2005; Rillahan et al. 2009, 2011). Hatchery production for cod, haddock and halibut was challenging with growout time in the sea cages ranging from 2.5 years (cod) to 3.5 years (halibut). Initially, fish were fed daily from a vessel through a feed hose that attached to the top of from the surface. An onboard water pump was used to mix food pellet and seawater in a funnel chamber before it was pushed down into a cage 12 m below. Autonomous feeders were not yet developed for the open ocean. Hence the design course that UNH Engineers embarked on to create three new generations of feeding systems (Rice et al. 2003). Prototype buoys evolved and scaling up to a final 20 ton version (Fullerton et al. 2004; Turmelle et al. 2006) that was fabricated by AEG in New Brunswick, CA (Fig. 3.8). The buoy hull was 8.6 × 6.9 m and had a draft of 3 m fully loaded with feed and fuel. To keep the buoy upright, 22,135 kg of concrete was poured into bottom. Fish food was pneumatically blown from a vessel into one of the four,

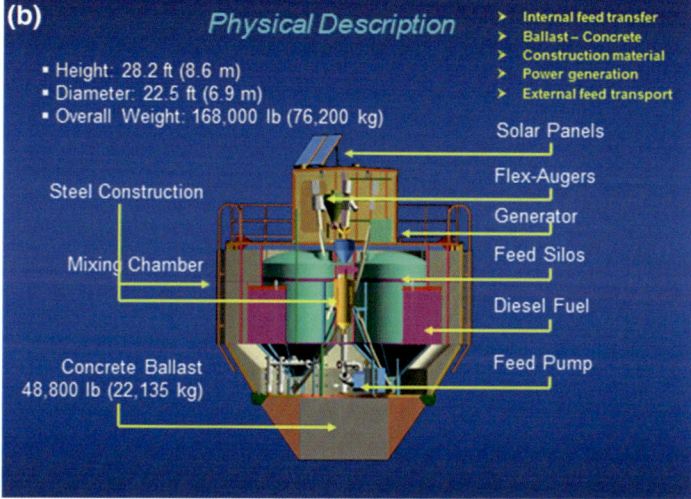

Fig. 3.8 Remote 20 ton feed buoy used to hydraulically feed four submersible sea cages up to 300 m away

5 ton silos. The house on top contained the majority of the electrical equipment as well as the generator for powering computers, pumps, valves, and augers used in daily feeding. During feed time, a flex auger would deliver feed pellets (5 mm) from the bottom of a silo to a mixing chamber in the top house. Feed rates were based upon fish species, size, biomass and temperature. In the mixing chamber, seawater was added before the feed was pumped hydraulically to individual cages. The buoy was moored adjacent to the cage grid by three, 750 ton Jeyco anchors. Feed distribution lines (4), consisting of 10 cm dia. High Density Polyethylene

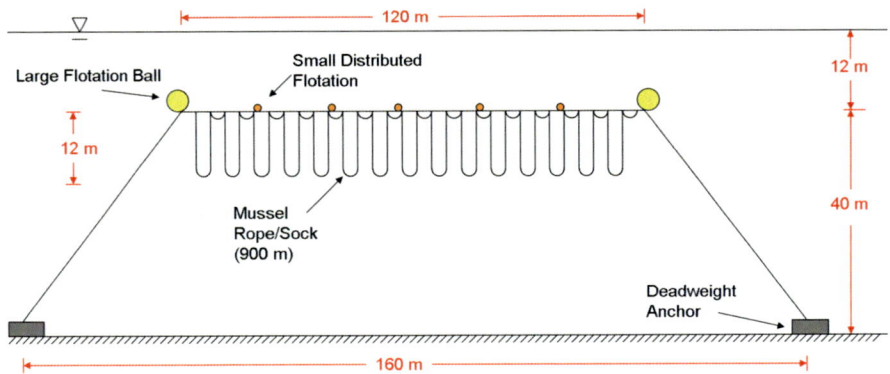

Fig. 3.9 Typical long-lines configuration for mussel grow-out. Dimensions are those used at the UNH OOA site and are representative

pipe, bolted to a manifold on the bottom of the buoy. They extended from the buoy to the submerged grid approximately 200 m away. Individual feed lines then integrated into the grid system and on to the top of each cage. Video cameras inside the cages were cabled through the feed lines back to the buoy.

Submerged shellfish longlines were designed and deployed next to the mooring grid (Figs. 3.5 and 3.9). Each 40 m long line was moored between two, 3000 kg dead weights (granite blocks) and set 12 m below surface to escape wave energies and predators (diving ducks). The backbone had surface lines so that it could be hauled up for seeding, cleaning or harvest. Approximately 900 m of mussel rope or sock could be deployed/line, able to produce between 8000 and 12,000 kg of blue mussel (*Edulis Mytilus*) per year (Langan and Horton 2003). The sea scallop (*Placopecten magellanicus*) was investigated on the submerged lines in stacked pearl nets. Problems with bio-fouling and stress from cleaning lead to low survival rates of this species.

Information important to scientists and farm managers was environmental conditions at the OOA site. This data was collected and streamed live from a single point, wave rider monitoring buoy (Irish et al. 2004, 2011). Figure 3.10 illustrates the instrumentation used at varying depths on the elastic buoy mooring. Oceanic parameters measured were wave height, current speed and direction (surface), temperature, salinity, oxygen, turbidity and fluorescence (at 1, 25 and 50 m depths).

Operations offshore were difficult to perform under winter conditions, heavy seas and with SCUBA divers. During the winter months, cages could not be accessed for several weeks at time thus minimizing days at sea for maintenance. Submerged cage systems were preferred in the North Atlantic to protect cage infrastructure and livestock. During Northeast storms, seas >10 m were recorded with currents speeds reaching 0.55 m/s. Underwater video cameras inside the cages provided insight to the fish health and feeding. This information was transmitted real time from the cages to the feed buoy and then back to shore through cellular modem.

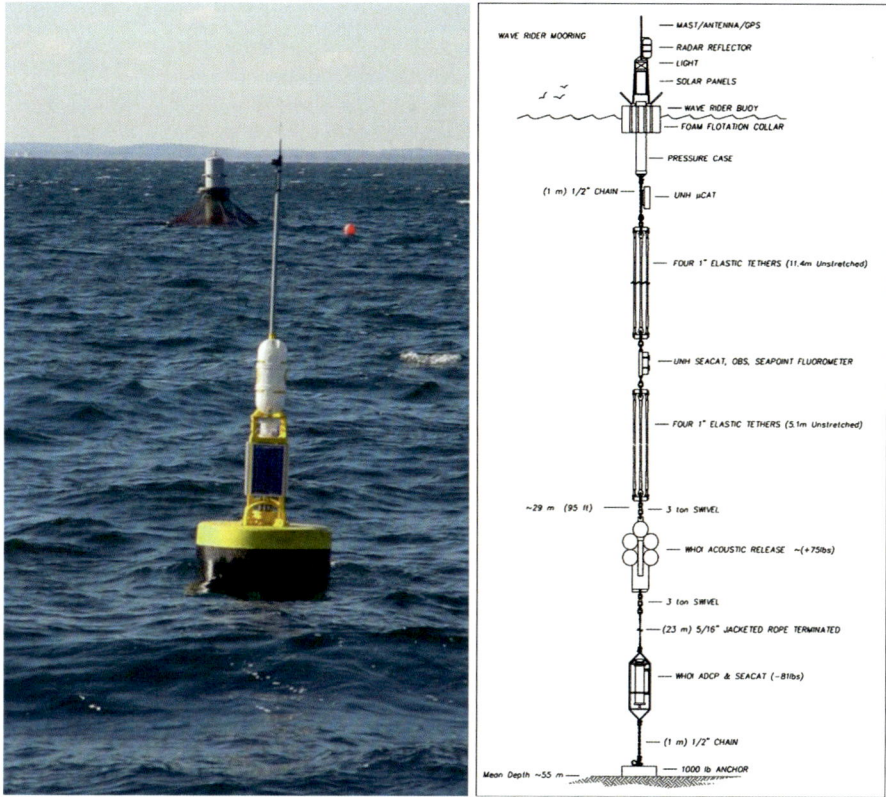

Fig. 3.10 Environmental monitoring buoy moored at the Open Ocean Aquaculture site. Oceanographic parameters measured included wave height, current speed and direction, temperature, salinity, oxygen, turbidity and fluorescence

Although the project succeeded in generating significant amounts of data and new information, the high maintenance costs, exposed nature of the site, and slow growth of marine fish species (cod, haddock and halibut) created operational and economic challenges.

3.4 Case Study on Multi-use on Open Ocean Environment

3.4.1 Methodology

This third case study picks up from the two previous case studies and details how the existing aquaculture technologies might be effectively utilized in a multi-use arrangement. In here, the multi-use of offshore wind energy and open ocean aquaculture is investigated experimentally. The location of interest for this feasibility study resides in the German bight at Veja Mate, which is reportedly 114 km off the shore in the North Sea. The reported results purely base on an experimental feasibility study since the stakeholder conflicts and the financial demand for in situ testing would have been too extreme at the point of development. The case study aimed to shed light on the general behavior of two different fish cage geometries being mounted to a tripile foundation for a 5 MW offshore wind energy converter (OWEC) at different submergence levels. An aspect of the analysis comprises the change of wave induced particle velocity and its distribution around the support structure due to the two fish cages. A second focus has been laid on the forces exposed to the tripile as a result of the additional fish cage bearing by means of force measurements under monochromatic waves. Thirdly the additional effects to potential scour around the tripile are also investigated in order to substantiate earlier findings (Goseberg et al. 2012).

The mechanical system of the fish cage and the OWEC-structure is investigated under regular wave attack with respect to (a) the deviations in the velocity field resulting from the fish cage underneath the OWEC, (b) the additional forces introduced by the fish cage and (c) the potential for scour evolution at the sea bottom of the OWEC location. Therefore a 110 m long and 2.2 m wide wave flume at the Franzius-Institute for Hydraulic, Estuarine and Coastal Engineering, Hannover, Germany is applied. The experiments are scaled at Froude similitude at a length scale of 1:40. The water depth at the wave maker (piston type) equals 1.0 m, yet to investigate scour evolution at the OWEC-structure a 0.25 m deep sand pit (fine sands of $d_{50} = 0.148$ mm) is installed at a distance of 42 m of the wave maker. At the end of the wave flume a gravel slope acts as passive wave absorption. An overview of the wave flume, its installations and a schematic of the mechanic system under investigation is given in Fig. 3.11. The two fish cage models were made of solid brass. Variations with three different surface properties (net solidity) were accomplished under the attack of monochromatic waves.

Based on a potential building site in the German bight at Veja Mate, a frequent (mean) and an extreme wave with a 50-yearly reoccurrence interval have been chosen. In model scale investigated wave heights are H = 0.04 m with T = 0.95 s for the frequent case and H = 0.28 m with T = 2.2 s for the extreme wave condition respectively. Surface elevations are measured with capacity-type wave gauges at various positions. Deviations in the velocity field around the structure and the fish cage are recorded by either ADV probe inside the fish cage or by stereo PIV measurements around the fish cage. Wave-induced forces at the constructional conjunctions between fish cage and tripile legs are taken by force transducers measuring normal forces. Force measurements are designed to consequently separate horizontal and vertical forces. Therefore all conjunctions between the fish cage and the tripile legs are designed as pendulum rods whereas the vertical forces are assumed to be gathered by a single tension rod connected to the upper tripile intersection. Compression-tension type force transducers (HBM GmbH) with a capacity of 500 N and an accuracy of 2% of its capacity were used. Scour depth evolution was manually measured with sediment gauges of 2 mm PVC sticks which were observed by underwater cameras (Abus GmbH). This optical method

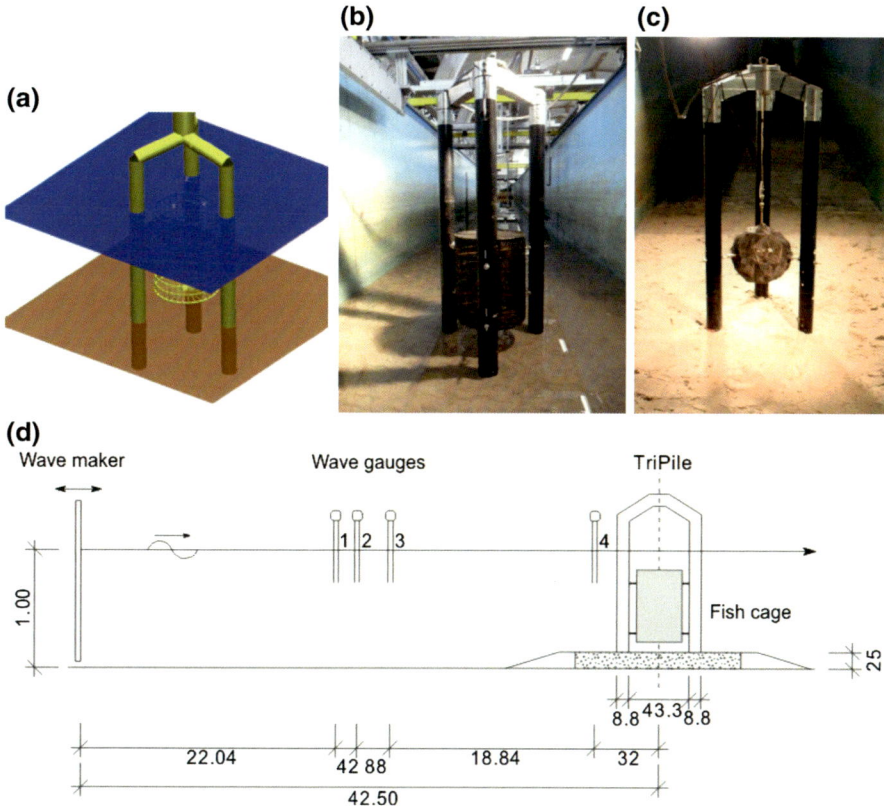

Fig. 3.11 a Schematic of the OWEC and cage arrangement. **b** Tripile model with cylindrical fish cage, **c** tripile model with spherical fish cage, and **d** longitudinal section of the wave flume with wave maker, position of wave gauges, sand pit and tripile model

where the difference between the initial and actual bed level is read from a centimeter scale on the PVC sticks based on the camera recordings gave reasonable accuracy (±0.5 cm).

3.4.2 Velocity, Force and Scour Regimes

Velocity measurements are accomplished to investigate the velocity regime in the vicinity of the OWEC-structure and its deviations by means of a fish cage installed inside the tripile legs. Foremostly, this part of the experimental program was intended to learn how velocities range inside the fish cage under wave attack in order to allow marine biologists to evaluate if fish production is feasible at all and how strong velocities influence the natural behavior of potential marine species

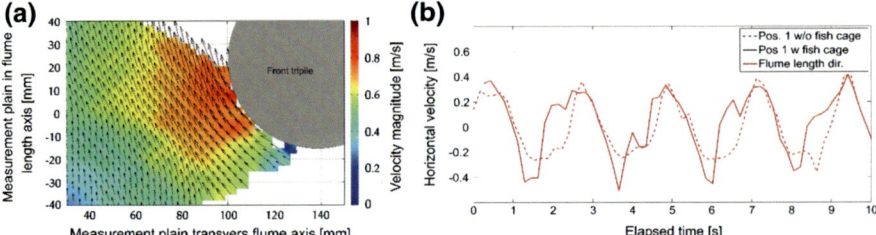

Fig. 3.12 **a** Example vector field of horizontal velocities near the front tripile for extreme wave conditions, upper measurement plain during wave crest propagation along the tripile. **b** Horizontal velocity from PIV-measurements comparing setup with and without fish cages assembled. All velocities reported are in laboratory scale. For prototype conditions multiply by $\sqrt{40}$

inside the enclosure. By means of the particle image velocimetry (PIV) technique (Raffel et al. 1998; Sveen and Cowen 2004) measurements of the velocity field very near to the tripile structure are feasible. Figure 3.12a shows an example of a velocity vector field during a propagation of a wave crest modeling extreme conditions for the cylindrical fish cage design.

Furthermore, the PIV-measurements which are recorded in stereo mode also allow for the extraction of 3D time series of velocities at a discrete position. For the evaluation of additional direct wave forces to the tripile legs the knowledge of velocity distribution is needed. Figure 3.12b hence shows the time series of horizontal velocities taken from PIV-measurements directly in front of the wave facing tripile leg at the height of the fish cage cover. Though maximum positive velocities are not significantly altered, it is obvious that negative velocities during wave trough are increased. Additionally it is apparent that phase duration of positive velocities is extended. With respect to fish cultivated in such a high energy environment a result could be that potential candidates have to be able to withstand such velocity magnitudes unharmed. Besides horizontal velocity components, vertical velocities are similar to their horizontal counterparts increased during the wave trough phase of wave passage. ADV-measurements inside the cylindrical fish cage reveal more moderate velocity changes which could be contributed to the damping effect of the modeled net material. While horizontal velocity deviations between experiments with and without fish cages are not so pronounced, it is apparent that vertical fluctuations are in a range of approx. 0.2 m/s (laboratory scale). This fact could especially influence health and behavior of flatfish which is one of the investigated candidates for the fish cages.

Force transducers in case of the cylindrical fish cage were arranged in two height levels. Force transducers FT_4 to FT_6 were mounted in the upper measurement plain connecting to the tripiles whereas the remainders (FT_1 to FT_3) were measuring forces near the bottom of the fish cages. Vertical forces were monitored with a single vertically arranged force transducer at the top cover of the cylinder. Compression forces were defined positive while tension forces were negative. Force transducers were pre-stressed and then zeroed before each experimental test for

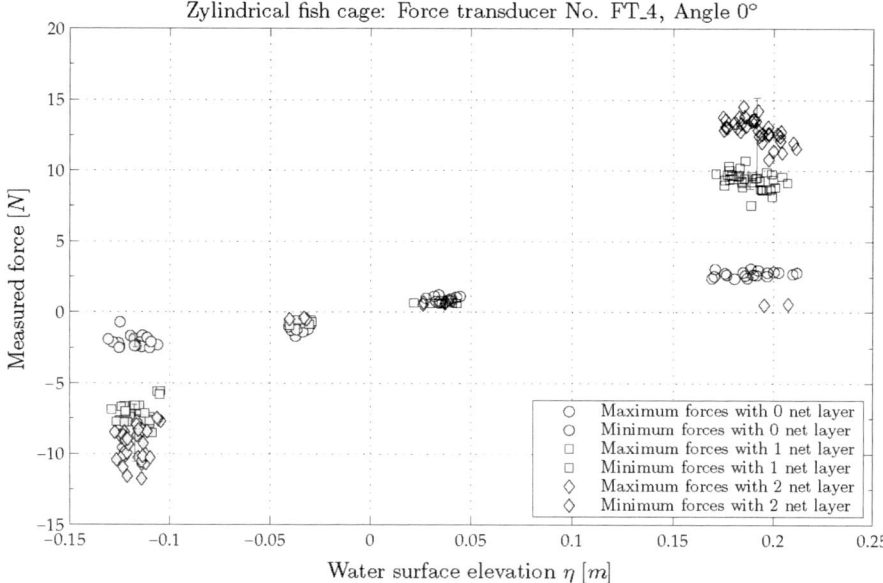

Fig. 3.13 Correlation of maximum wave heights and measured forces for cylindrical fish cages for force transducer FT_4 with respect to the number of net layers

stability reasons. Similarly, the spherical fish cage had three force transducers connecting from the equator of the sphere towards the triples plus a vertical force transducer to bear the vertical forces. Highest tension forces appear at the wave-facing tripile leg during the passage of the wave crest. In parallel, these forces initiated by the wave at the front of the cylinder are also measured at the backward force transducers but with opposite sign and vectorially separated.

In order to analyze the correlation between the wave heights and the caused reaction forces a zero-crossing method has been applied to the water surface elevation measurement which was located besides the wave-facing front tripile of the OWEC. Thereafter the maximum values of the surface elevation and the forces have been collected. Experimentally, three different flow resistances are considered. First, the fish cage is tested only with its support structure. One and two layers of net material are fixed to the fish cage in subsequent tests. Figure 3.13 shows an example of the correlation of maximum wave heights and the respective maximum force measured with respect to the number of net layers. In general it is apparent that the additional forces exerted to the tripile structure increase with increasing wave height. Though, the correlation is not linear but growth exponentially. Due to the limited number of wave heights analyzed so far it is yet not feasible to present valid regression functions. The influence of the flow resistance variation is though clearly obvious. While the increase of net layers from zero to one results in an increase of forces of more than 100%, the further increase towards two layers of net material only results in further loading of approx. 40%. The variation of let layer numbers is yet important to investigate how loadings develop under the additional resistance due to marine fouling. Up-scaled overall forces that have to be absorbed by the tripile structure are at least for stormy conditions in the mega newton range and deemed clearly design relevant for the OWEC structure. For the chosen cases, the connecting joints between the investigated fish cage variations are arranged so that bending moments are also induced. In upcoming studies it is thus recommended to lower the connecting joints further down to the sea bottom in order to greatly reduce the effects that aquaculture applications connected to the OWEC would have.

In literature a number of theoretical approaches exist to deduce the maximum scour depths at piles (Melville and Coleman 2000; Sumer and Fredsøe 2002) which nowadays depict an important piece of information for the safety assessment of OWEC. In case of complex marine constructions involving multi-use of an OWEC with multiple, complex-shaped constructional members, the analytical approaches are not easily applicable. Hence an optical method is applied which consists of sediment gauges and image capturing. A manual post-processing routine is chosen to estimate the local sea bed evolution. The local water depth of the modeled OWEC structure and the chosen monochromatic waves limit the scour tendency. Scour evolution is yet possible under the assumption that swell generated in a far field arrives at the location of interest. Therefore additional experiments with an elongated wave period of $T = 3.0$ s are conducted. In this case scour around the tripile legs can be observed clearly. Under the assumed boundary condition relative scour depth of $S/D = 0.54$ is determined at the front leg of the tripile whereas relative scour depth of $S/D = 0.36$ is found below the tower of the OWEC for a setup without a fish cage. Slightly different results are obtained under the presence of fish cages mounted to the OWEC. Relative scour depth of $S/D = 0.48$ (front leg) and $S/D = 0.25$ (central below tower of OWEC) are found. This reduction of scour observed underneath the main tower of the OWEC below the tower can be explained by a reduction of wave-induced flow velocities and shear stresses in this area.

3.5 Discussion and Conclusions

In the following, not only are conclusions drawn from the single case studies but it is highlighted how open ocean aquaculture might benefit from its transition towards a multi-use concept. Direct benefits such as cost reductions through multiple usage of anchor constructions as well as indirect ones like shared usage of supply vessels or personnel indicate that time has come to move aquaculture industry further offshore. Ongoing initiatives such as the European framework funded projects MERMAID, H2OCEAN, and TROPOS are a vital indication of the present demand for research and development in this area.

Open ocean shellfish farming is at a pivotal point at this time. Technology has reached a point where economically viable commercial production is a real prospect. Once production becomes more routine and operating under standard procedures, it is believed that companies will direct greater focus towards improving efficiency in a similar way the inshore longline system developed in New Zealand in the 80s and 90s. In further support of the open ocean mussel farm mentioned in the case study, mussel spat is being caught in commercial numbers as far as 12 km off the coast and may be accessible even further out to sea. This means the whole production chain can be handled on the same farm. It is estimated that this particular farm will deploy 350–400 km of mussel production rope per annum within the next 3–4 years. Once this farm has shown itself to be technically and economically viable, other farm owners have indicated that they will start production. The potential will then be enormous.

There are some considerations regarding production and seasonality that must be taken into account. It has been found that mussels inshore have a longer and more varied condition window than mussels grown offshore (Heasman, unpublished), i.e. shellfish conditioning at the open ocean site is more uniform across the growing area and is shorter in duration. This is an advantage in terms of processing however the shorter window leads to production peaks and troughs. For example, an offshore farm may be capable of producing 30,000 ton of mussels but if they all have to be harvested in a 6–12 week period it will require a large number of vessels and processing capability for that period. These units will sit idle or at low usage for the remainder of the year. For this reason it is important to have multiple species in production simultaneously, unless a species that can be harvested year round is utilized.

Despite the findings in New Hampshire, other open ocean locations and technologies may be more suitable for offshore farming. An example of this is Ocean Blue located 11 km offshore Panama. Here they are raising Cobia (*Rachycentron canadum*), a fast growing warm water species, in 6000 m^3 Sea Station™ cages. Last year they produced 400 MT with an expected 1200 MT by 2014. Currently they feed via day boats through pneumatic blowers into each cage. Commercial feed barges (http://www.akvagroup.com/products/cage-farming-aquaculture/feed-barges/ac-450-panorama) are available however they cannot withstand seas over 4. 5 m. These barges are common in protected water ways in Norway, Chile and Canada. Certainly if aquaculture is to move further from shore, remote feeding systems will need to be improved to withstand seas >10 m. Also important will be real time monitoring of fish behavior and environmental parameters inside and outside the cages. In this, farmers can best manage animal welfare to optimize growth and condition.

Ocean farming will slowly creep offshore as new, economical and robust technologies develop. Farms of the future must be turnkey allowing for remote operations such as feeding, environmental monitoring, fish health and harvesting of product. These platforms could ultimately produce their own energy through wave, wind and tidal turbine technologies. Ultimately, farming centers should be located close to large coastal populations to reduce the carbon footprint from exporting overseas.

For the combination of aquaculture fish cages and OWEC structures, the up-scaled experimental particle velocities at the highest position of the investigated fish cages are in the range of 1.3 m/s under prototype condition for the maximum wave height (storm conditions). This range of velocities is potentially harmful to species candidates to be grown in the North Sea and might negatively affect survival rates and thus reduce economic feasibility of multi-use approaches. Alternatively, the investigated cage systems could be modified in height (decreased) and lowered further towards the sea bottom in order to circumvent the critical particle velocities induced by storm waves. However, volume reduction of the actual fish cages also minimizes the economic potential of the fish cages as the amount of fish is decreased.

This alternative design has similar potential to reduce overall forces at the contact points with the OWEC structure. In addition, as forces grow with the degree of marine growth, net materials with growth retarding characteristics in combination with optimized maintenance cycles might decrease overall forces as well. However, the greatest challenge for further testing in prototype conditions are still the restraints from owners or designers of offshore wind energy parks which reject any additional (and yet untested) loading stemming from secondary purposes for safety reasons. Unless legitimate doubts and technological uncertainties are alleviated through further scientific effort, it seems to date still challenging to expect marine multi-use applications in the near future.

In conclusion, based on differently far developed examples it has been demonstrated that open ocean aquaculture has reasonable potential for future growth and prosperity. In future, multi-use is not only a feasible add-on to farming in open ocean conditions whenever technological or logistical challenges are resolved thoroughly but it might depict the key innovation towards economically feasible offshore farming in high energy environments. For example, it might be beneficial for the aquaculture ventures to rely on pre-existing fixed structures such as piers, foundations amongst others to attach feeding storage, supply equipment or instruments. Rather than constructing and installation of dead weight anchors in rough environments, pre-existing infrastructure is seen advantageous to the installation of various forms of aquaculture technology including fish cages, mussel ropes or net pens. In turn, infrastructure owners would be able to generate additional income through leasing out their property to aquaculture.

However, some technological or logistical challenges which have been described can only be addressed efficiently with the help of stakeholder dialog and proper incorporation of end-users or operators. A chain of development steps from the very beginning of a business idea to the final operation of this business incorporates preliminary design, feasibility studies which involves various modelling steps, a design phase to plan for a prototype development, and eventually an operational farm or multi-use deployment which has undergone further optimization steps to yield its full effectiveness under a wide range of external influences. A number of approaches to tackle those steps of the development cycle have thus been highlighted in this paper aiming to help additional projects to be launched in the future.

References

Buck, B. H., & Buchholz, C. M. (2004). The offshore-ring: A new system design for the open ocean aquaculture of macroalgae. *Journal of Applied Phycology, 16,* 355–368.

Buck, B. H., & Buchholz, C. M. (2005). Response of offshore cultivated *Laminaria saccharina* to hydrodynamic forcing in the North Sea. *Aquaculture, 250,* 674–691.

Chambers, M. D., DeCew, J., Celikkol, B., Yigit, M., & Cremer, M. C. (2011). Small-scale, submersible fish cages suitable for developing economies. *Global Aquaculture Advocate, 14,* 30–32.

Chambers, M. D., & Howell, W. H. (2006). Preliminary information on cod and haddock production in submerged cages off the coast of New Hampshire, USA. *ICES Journal of Marine Science, 63*(2)

Chambers, M. D., Langan, R., Howell, W., Celikkol, B., Watson, B., Barnaby, R., et al. (2007). Recent developments at the University of New Hampshire Open Ocean Aquaculture Site. *Bulletin-Aquaculture Association of Canada, 105*(3).

DeCew, J., Baldwin, K., Celikkol, B., Chambers, M., Frederiksson, D. W., Irish, J., et al. (2010a). Assessment of a submerged grid mooring in the Gulf of Maine. In *OCEANS 2010.* doi:10.1109/OCEANS.2010.5664025

DeCew, J., Celikkol, B., Baldwin, K., Chambers, M., Irish, J., Robinson, M. R., et al. (2012). Assessment of a mooring system for finfish aquaculture. *Journal of the World Aquaculture Society, 43,* 32–36.

DeCew, J., Frederiksson, D. W., Bugrov, L., Swift, M. R., Eroshkin, O., & Celikkol, B. (2005). A case study of a modified gravity type cage and mooring system using numerical and physical models. *IEEE Journal of Oceanic Engineering, 30,* 47–58.

DeCew, J., Tsukrov, I., Risso, A., Swift, M. R., & Celikkol, B. (2010b). Modeling of dynamic behavior of a single-point moored submersible fish cage under currents. *Aquacultural Engineering, 43,* 38–45.

Drumm, A. (2010). *Evaluation of the promotion of offshore aquaculture through a technology platform (OATP)* (p. 46). Galway, Ireland, Marine Institute. Retrieved October 1, 2016, from www.offshoreaqua.com/docs/OATP_Final_Publishable_report.pdf

FAO. (2012). *FAO yearbook. Fishery and Aquaculture Statistics 2010.* Rome, Italy: Food and Agriculture Organization of the United Nations.

Fredriksson, D. W., Swift, M. R., Eroshkin, O., Tsukurov, I., Irish, J. D., & Celikkol, B. (2004). The design and analysis of a four-cage, grid mooring for open ocean aquaculture. *Aqua Engineering, 32,* 77–94.

Fullerton, B., Swift, M. R., Boduch, S., Eroshkin, O., & Rice, G. (2004). Design and analysis of an automated feed buoy for submerged cages. *Aquacultural Engineering, 95,* 95–111.

Goseberg, N., Franz, B., & Schlurmann, T. (2012). The potential co-use of aquaculture and offshore wind energy structures. In *Proceedings of the 6th Chinese-German joint symposium on hydraulic and ocean engineering (JOINT 2012).* Keelung, Taiwan: National Taiwan Ocean University.

Hau, E., & von Renouard, H. (2006). *Wind turbines: Fundamentals, technologies, application, economics.* Heidelberg/Berlin, Germany: Springer.

Howell, W. H., & Chambers, M. D. (2005). Growth performance and survival of Atlantic halibut (*Hippoglossus hippoglossus*) grown in submerged net pens. *Bulletin-Aquaculture Association of Canada, 9,* 35–37.

Huang, C.-C., Tang, H.-J., & Liu, J.-Y. (2006). Dynamical analysis of net cage structures for marine aquaculture: Numerical simulation and model testing. *Aquacultural Engineering, 35,* 258–270.

Huang, C.-C., Tang, H.-J., & Liu, J.-Y. (2008). Effects of waves and currents on gravity-type cages in the open sea. *Aquacultural Engineering, 38,* 105–116.

Irish, J. D., Carrol, M., Singer, R. C., Newhall, A. E., Paul, W., Johnson, C., et al. (2011). *Instrumentation for open ocean aquaculture monitoring.* WHOI. doi:10.1575/1912/25

Irish, J. D., Fredriksson, D. W., & Boduch, S. (2004). Environmental monitoring buoy and mooring with telemetry. *Sea Technology, 45,* 14–19.

Kapetsky, J. M., Aguilar-Manjarrez, J., & Jenness, J. (2013). A global assessment of potential for offshore mariculture development from a spatial perspective. FAO Fisheries and Aquaculture Technical Paper 549. Rome, Italy: Food and Agriculture Organization of the United Nations.

Lacroix, D., & Pioch, S. (2011). The multi-use in wind farm projects: More conflicts or a win-win opportunity? *Aquatic Living Resources, 24,* 129–135.

Lader, P. F., & Enerhaug, B. (2005). Experimental investigation of forces and geometry of a net cage in uniform flow. *IEEE Journal of Oceanic Engineering, 30,* 79–84.

Langan, R., & Horton, C. F. (2003). Design, operation and economics of submerged longline mussel culture in the open ocean. *Bulletin of the Aquaculture Association of Canada, 103,* 11–20.

Langhamer, O., & Wilhelmsson, D. (2009). Colonisation of fish and crabs of wave energy foundations and the effects of manufactured holes—A field experiment. *Marine Environmental Research, 68,* 151–157.

Liu, T.-L., & Su, D.-T. (2013). Numerical analysis of the influence of reef arrangements on artificial reef flow fields. *Ocean Engineering, 74,* 81–89.

Marra, J. (2005). When will we tame the oceans? *Nature, 436.* doi:10.1038/436175a

Melville, B. W., & Coleman, S. E. (2000). *Bridge scour.* Littleton, Colorado: Water Resources Publications.

Raffel, M., Willert, C., & Kompenhans, J. (1998). *Particle image velocimetry—A practical guide.* Berlin/Heidelberg, Germany: Springer.

Rice, G., Stommel, M., Chambers, M. D., & Eroshkin, O. (2003). The design, construction, and testing of the university of New Hampshire feed buoy. Open Ocean Aquaculture, From Research to Commercial Reality, The World Aquaculture Society

Rillahan, C., Chambers, M., Howell, W. H., & Watson, W. H. (2009). A self-contained system for observing and quantifying the behavior of Atlantic cod, *Gadus morhua,* in an offshore aquaculture cage. *Aquaculture, 293,* 49–56.

Rillahan, C., Chambers, M. D., Howell, W. H., & Watson, W. H. (2011). The behavior of cod (*Gadus morhua*) in an offshore aquaculture net pen. *Aquaculture, 310,* 361–368.

Sumer, B. M., & Fredsøe, J. (2002). *The mechanics of scour.* Singapore: The Marine Environment, World Scientific.

Sveen, J. K., & Cowen, E. A. (2004). *PIV and Water Waves, Kapitel: Quantitative imaging techniques and their application to wavy flow.* Singapore: World Scientific.

Swift, M. R., Palczynski, M., Ketler, K., Michelin, D., & Celikkol, B. (1998). Fish cage physical modeling for software development and design applications. In Proceedings of the 26th U.S.-Japan aquaculture symposium nutrition and technical development of aquaculture (pp. 199–206), University of New Hampshire, Durham, NH.

Tsukrov, I., Eroshkin, O., Frederiksson, D., Swift, M. R., & Celikkol, B. (2003). Finite element modeling of net panels using a consistent net element. *Ocean Engineering, 30,* 251–270.

Tsukrov, I., Ozbay, M., Swift, M. R., Celikkol, B., Fredriksson, D. W., & Baldwin, K. (2000). Open ocean aquaculture engineering: Numerical modeling. *Marine Technology Society Journal, 34*(1), 29–40.

Turmelle, C., Swift, M., Celikkol, B., Chambers, M., DeCew, J., Fredriksson, D., et al. (2006). Design of a 20-ton Finfish Aquaculture feeding buoy. *Proc. Oceans06*. Boston, MA.

Chapter 4
Operation and Maintenance Costs of Offshore Wind Farms and Potential Multi-use Platforms in the Dutch North Sea

Christine Röckmann, Sander Lagerveld and John Stavenuiter

Abstract Aquaculture within offshore wind farms has been identified as one of the many possibilities of smart use of marine space, leading to opportunities for innovative entrepreneurship. Offshore areas potentially pose less conflict with co-users than onshore. At the same time, offshore areas and offshore constructions are prone to high technical risks through mechanical force, corrosion, and biofouling. The expected lifetime of an offshore structure is to a great extent determined by the risk of failures. This chapter elaborates on logistical challenges that the offshore industry faces. Operation and maintenance (O&M) activities typically represent a big part of the total costs (e.g. 25–30% of the total lifecycle costs for offshore wind farms). The offshore wind energy sector is considered an industry with promising features for the public and private sector. Large wind farms farther off the coast pose high expectations because of higher average wind speeds and hence greater wind energy yield (in terms of megawatts per capital). These conditions entail additional challenges in logistics, though. One of the main hurdles that hinders use of offshore wind energy is the high cost for O&M. The offshore wind industry will have to solve these problems in order to achieve substantial cost reduction - alone or jointly with other (potential) users. It is precisely the logistical problems around O&M where most likely synergy benefits of multi-use platforms (MUPs) can be achieved. The offshore wind energy industry is eagerly looking for technical innovations. Until now they mostly sought the solutions in their own circles. If the combination of offshore wind energy and offshore aquaculture proves to be feasible and profitable in practice, there may be an additional possibility to reduce the O&M costs by synergy effects of the combined operations. Logistic waiting times, for example, can result in substantial revenue losses, whereas timely spare-parts supply or sufficient repair capacity (technicians) to shorten the logistic delay times are beneficial. A recent study suggests that a cost reduction of 10% is

C. Röckmann (✉) · S. Lagerveld
Wageningen University and Research—Wageningen Marine Research,
P.O. Box 57, 1780 AB Den Helder, The Netherlands
e-mail: Christine.rockmann@wur.nl

J. Stavenuiter
Asset Management Control Centre, Willemsoord 29, 1781 AS Den Helder, The Netherlands

© The Author(s) 2017
B.H. Buck and R. Langan (eds.), *Aquaculture Perspective of Multi-Use Sites in the Open Ocean*, DOI 10.1007/978-3-319-51159-7_4

feasible, if the offshore wind and offshore aquaculture sectors are combined in order to coordinate and share O&M together. The presented asset management control model proves useful in testing the innovative, interdisciplinary multi-use concepts, simulating return rates under different assumptions, thus making the approach more concrete and robust.

4.1 Introduction

Aquaculture within offshore wind farms has been identified as one of the many possibilities of smart use of marine space, leading to opportunities for innovative entrepreneurship (Buck 2002; Buck and Krause 2012; Michler-Cieluch et al. 2009a, b; Michler-Cieluch and Krause 2008; Lagerveld et al. 2014; Verhaeghe et al. 2011). Offshore areas potentially pose less conflict with co-users than onshore. Large wind farms farther offshore are considered promising because of higher average wind speeds and hence greater and more constant wind energy yield in terms of megawatts per capital. Similarly, positive assets to aquaculture offshore are the often favorable conditions for growth due to water depth and hydrodynamics (e.g. quick nutrient input and waste dispersal), and less potential for disease spread, pollution and agricultural interactions. At the same time, offshore areas and offshore constructions are prone to high technical risks and entail additional logistical challenges.

This chapter elaborates on logistical challenges concerning operation and maintenance (O&M) of offshore activities. O&M activities typically represent a big part of the total costs, which are some of the main hurdles that hinder use of offshore wind energy. It is precisely the logistical problems around O&M where synergy benefits of multi-use platforms (MUPs) can most likely be achieved. The offshore wind energy industry is eagerly looking for technical innovations and until now, they mostly sought the solutions in their own circles. If the combination of offshore wind energy and offshore aquaculture proves to be feasible and profitable in practice, there may be additional possibilities to reduce the O&M costs by synergistic effects of the combined operations. Reducing logistic waiting times, for example, can result in substantial cost savings, just as timely spare-parts supply or sufficient repair capacity (technicians) are beneficial to shorten the logistic delay times. If the offshore wind and aquaculture sectors join forces, O&M activities can be coordinated and shared together and thus costs saved. The aquaculture sector—in order to be recognized as a potential partner—still has to demonstrate its feasibility offshore, though, under the wide range of rough oceanographic conditions. Aquaculture at offshore sites faces major challenges compared to coastal and land based production systems. The longer transport distances result in higher O&M costs than for coastal sites. Although technical solutions have been developed and are available (e.g. Gimpel et al. 2015; Buck et al. 2010), offshore aquaculture in the North Sea is still absent due to economic/financial reasons and a lack of cross-sector planning. A combination of an offshore wind farm with aquaculture, e.g. as proposed already in 2002 by Buck (2002), Buck et al. (2004) and recently for the Dutch North Sea by Lagerveld et al. (2014), involves both risks and benefits. The potential risks include technical failures due to corrosion and biofouling; ecological risks, such as underwater-noise disturbance of marine mammals, disturbance of the seabed sediments and seabed communities; collision risks to birds and bats above water, and attraction of invasive species. Benefits of combining an offshore wind farm with offshore aquaculture can be found in eco-facilitation and in economic/financial savings. Eco-facilitation refers to the enhancement of biological diversity and production (e.g. by offering increased food availability and shelter, thereby attracting flora and fauna). The economic/financial benefit refers to expected synergy effects through sharing and thus savings on operation and maintenance costs (Krause et al. 2011).

Section 4.2 provides an overview of relevant offshore O&M activities and an analysis of O&M costs. Section 4.3 describes the potential for cost savings in a combined offshore Multi-Use Platform (MUP), based on the example of a virtual offshore wind-mussel-farm (OWMF), and it also depicts an Asset Management Control (AMC) model, that could help to manage such a combined business. The AMC model can, for example, simulate different O&M scenarios of a virtual OWMF over 20 years. The chapter concludes with lessons learned and recommendations for pilot studies.

4.2   Offshore Operation and Maintenance Activities

4.2.1   Accessibility of Offshore Wind Farms

The offshore marine environment is characterized by harsh conditions. Project developers of offshore wind farms have to cope with many logistical and safety issues that developers of wind energy projects on land do not, or at least not to the same extent. Operation and maintenance costs make up 25–30% of the total costs of an offshore wind farm (Miedema 2012, cf. Sect. 4.2.5). This is almost as much as the cost of the wind turbines and about as much as the costs of construction and installation. Individual offshore wind turbines currently require about five site visits per year: one regular annual maintenance visit, and three to four visits in case of malfunction (cf. Noordzeewind website). With technological progress, this can potentially be reduced to three visits per year. Nonetheless, a future offshore wind farm consisting of 200 turbines of 5 MW each will therefore need some 3000 offshore visits per year. Operation and maintenance (O&M) visits are carried out by boat or helicopter, which means that the personnel performing the repair, has to climb onto the turbines. Especially in rough conditions—helicopters for example are used at wind speeds of up to 20 m/s—this is a risky undertaking. Systems need to be developed to ensure the safety of staff and to expand workability.

In the future, certain maintenance tasks may also be carried out remotely (see DOWES, Sect. 4.2.4).

Until now, O&M visits are carried out when the significant wave heights (SWH) are less than or equal to 1.5 m. According to Stavenuiter (2009), each support vessel has a certain maximum allowable significant wave height for several operations. Therefore, the availability of a vessel is correlated with the occurrence of certain significant wave heights. The cumulative occurrence of significant wave heights less than or equal to 1.5 m is 68%. Waves up to 2.0 m increases the cumulative occurrence to 83% (Stavenuiter 2009). Significant wave heights measured close to two Dutch offshore wind farm locations (OWEZ and PAWP) are almost identical, despite a distance of 40 NM between each other (Rijkswaterstaat 2009).

The number of days per month that offshore locations are accessible or not due to weather downtime varies over the year, with more no access days during the windy and stormy winter months (Stavenuiter 2009). The maximum number (22) of no access days of two offshore locations, very close to the Dutch offshore wind farms OWEZ and PAWP, in 2009 was in November. In spring and summer months, weather downtime fluctuated between 6 and 11 days between April and August 2009, and 16 in September 2009 (Stavenuiter 2009).

The possibility of larger wave heights will require new systems for safe O&M personnel transfer. If transfers are to be restricted to wave heights of 1.5 m, this will limit offshore work to about 200 days a year (Noordzeewind 2010; Miedema 2012). Noordzeewind (2010) estimated a total of approximately 218 possible access days in 2009, and the remaining time of the year was considered non-productive time ('weather downtime') in 2009. According to a study in the German North Sea, available working days were estimated to be even lower, in a range of 30–100 days/year (Michler-Cieluch et al. 2009a). Increasing the workable significant wave height from 1.5 to 2 m, could increase the accessibility of wind farms by 15% (Stavenuiter 2009). An increase of the safe working wave height to 3 m and above could increase the number of days available for transfers up to 310 days per year. Hence, increasing overall accessibility can lead to cost reduction of wind energy production. To achieve this, new ships with motion stabilizers are required to guaranty safe transfers of personnel and material. Current solutions are offshore access systems such as 'Ampelmann', a hydraulic ship-based transfer platform that compensates the vessel's motions, thereby enabling safe offshore access and operations. But even if these new systems for operating in far-offshore conditions are developed, a constant shuttling of workboats to and from the coast is impractical and costly. Therefore, developers and offshore service providers are looking for new methods, one of which is the 'mother ship' approach. A single large vessel would then service one or more offshore wind farms staying in the neighborhood of these farms for long periods of time and deploying multiple smaller craft for daily servicing.

4.2.2 Infrastructure for Cabling and Cable Repair

Up to now, there are neither standardized practices nor procedures to procure cables or share cabling equipment, ships, and all other elements necessary for a safe and speedy repair. If developers were more willing to collaborate with each other, to share facilities, vessels, and their particular knowledge, this could lead to far more efficient procedures through economies of knowledge. So far, the desire to keep cable choices and technologies confidential prevailed over the opportunity to develop a more efficient infrastructure for joint installation and maintenance or repair of cables. But these facilities will be necessary as bases for long-range offshore vessels and to service the offshore wind farms closer to the shore. Especially with future far and large offshore wind farms (FLOW), it could be a unique asset to have manufacture and dedicated repair and storage facilities for spare parts closer to the FLOW sites. Despite the benefits to be expected, it is far from certain whether developers and offshore operators are willing to pay for collective facilities that they may not need to use.

4.2.3 Trained Staff

To keep up with developments, companies will need to permanently invest in capacity building and training to ensure that sufficiently skilled O&M personnel are available. This holds even more for FLOW farms. A rough calculation suggests that one O&M job will be created for every two turbines installed. With 200 turbines of 5 MW each, this equates to a need of about 100 FTE of trained staff. Even if this calculation is conservative, and the number of staff can be reduced through greater efficiency, there will still be a huge need for skilled personnel. To meet that demand, operators and developers will have to set up offshore training centers and training programs. It would not be wise, if they do this just for their own purposes. As with the cabling sector, it is obvious that collaboration and joint financing have great advantages.

4.2.4 Dutch Offshore Wind Energy Services (DOWES)

There are three lines of intervention in a wind farm: first, scanner control with remote management; second, helicopter intervention; and third, heavy lift operations. Reactive maintenance, e.g. arranging a site visit if a turbine stops working, is always expensive and can sometimes be impossible; for instance, in bad weather conditions or if boats and crew are unavailable. This dependence on weather, crew, and boat availability increases the risk of an expensive wind generation asset being unable to produce electricity for weeks or even months. Predictive maintenance, i.e. remote surveillance, can help in constant monitoring and real time information about what is happening at a site. Key to such planned predictive maintenance is the increased deployment of sensors in offshore wind turbines. Modern offshore wind turbines, particularly those that are custom built for offshore, will contain a huge number (>1000) of sensors in key components. The ongoing Dutch Offshore Wind Energy Services (DOWES.nl) project focuses on developing an innovative information and communications technology (ICT) system to manage offshore wind parks in the Den Helder region (2008–2014). The DOWES management plan aims to lead to high wind farm availability at minimum cost. The ICT system will be capable of reading the sensors on the wind turbines using remote control, making use of the most up-to-date science.

It is possible to manage and maintain offshore wind parks in various ways. DOWES aims to safeguard offshore wind parks from a distance/on land. Constant monitoring of the state of the wind turbines can facilitate timely information of the right people. This can aid in making cost-effective choices and carrying out maintenance optimally. In the long run such systems are expected to increase the manageability of offshore wind parks and reduce maintenance costs.

4.2.5   Analysis of Operation and Maintenance Costs

To get more insight in the O&M cost structure of OWFs, the total O&M costs are split over specific O&M disciplines. It starts with the breakdown of the operational expenditures (OPEX) (Fig. 4.1).

This breakdown shows that the O&M costs represent 53% of the OPEX (15% "Operation" + 38% "Maintenance"). In the Asset Management Control (AMC) approach (Stavenuiter 2002) the discipline "Maintenance" is considered to be the combination of all technical, logistic, administrative and managerial actions during the life cycle of an asset/object, intended to retain the asset or restore it to a state, in which it can perform the required function. Therefore, the activity "Port Activities" is considered a part of "Maintenance". For the UK's seabed, the Crown Estate applies license fees. However, this aspect is not applicable for the offshore wind industry in the Netherlands. For this reason the cost for license fees is also included under "Maintenance". "Other cost" which are not specified by Board (2010; Fig. 4.3) are distributed among the O&M disciplines: 5% are placed under "Operation" and 7% under "Maintenance" since this discipline holds more variable and unspecified costs.

The next objective is to validate a realistic average annual O&M cost for offshore wind farms. For this purpose, a specific annual O&M cost analysis has been carried out. Figure 4.4 illustrates the spread of O&M cost, as applied in several reports (Board 2010; Feargal 2009; Pieterman et al. 2011; Kjeldsen 2009; Musial and Ram 2010; Rademakers and Braam 2002). The total annual O&M cost varies between 15 and 45 €/MWh. The cited reports do not mention the size of the wind farms, nor the distance to shore. It seems likely though, that these aspects have great

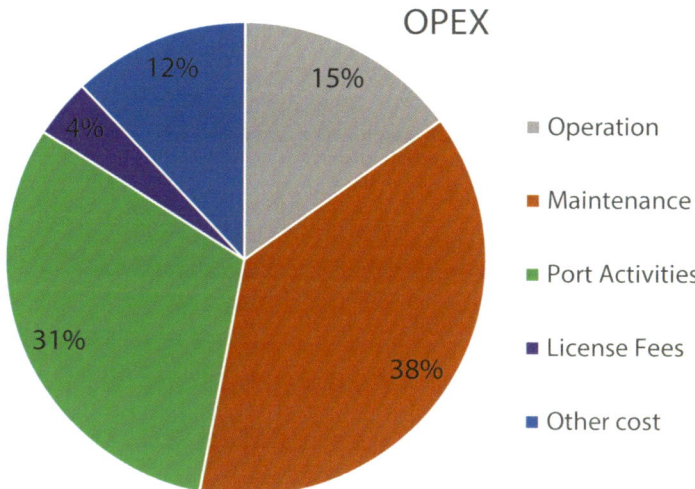

Fig. 4.1 Breakdown of operational expenditures (OPEX) of an offshore wind farm, according to Board (2010). *Source* Lagerveld et al. (2014)

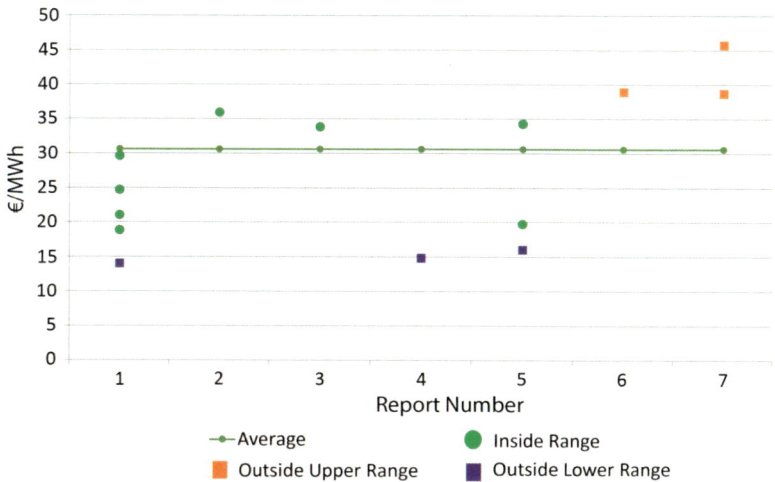

Fig. 4.2 Spread of the O&M cost of offshore wind farms of seven different studies (Miedema 2012)

influence on the O&M cost. An average (line in Fig. 4.2) for O&M cost is determined at 30 €/MWh (€ 0.03 per kWh), by calculating a boxplot based on the middle 50%, omitting the maximum and minimum outliers, which are considered as unreliable or exceptional (Miedema 2012).

To identify possibilities for synergy, a more refined O&M OPEX distribution is necessary to identify activities which can be executed more efficiently by

Table 4.1 Cost share (in % of total O&M costs) and explanation of the different O&M disciplines in the total life cycle management of offshore wind farms (Miedema 2012)

Operations (11%)	In this distribution 'Operations' purely deals with the primary process; by moving 3% to 'Life Cycle Management' (LCM) and 6% to 'Inspective Maintenance', 'Operations' (usually 20%; Fig. 4.2) is reduced to 11%
Life cycle management (7%)	'Life Cycle Management' (LCM) is used for the benefit of both operations and maintenance. LCM takes care of maintenance schedules and planning (3%) and covers activities that are normally housed under 'Maintenance' (Fig. 4.2), thereby leading to a transfer of 4% from 'Maintenance' to 'LCM'
Inspective (10%), preventive (12%), corrective (35%) maintenance	The overall activity 'Maintenance' (usually 80%; Fig. 4.2) is split up into three specific maintenance types and 'Improvement', which covers refit, overhauls and modification programs. 'Inspective Maintenance' is often seen as an operational activity or part of preventive maintenance. In this study it is recognized as a specific maintenance type with a total share of 10%, composed of 6% 'Operations' and 4% 'Maintenance' (Fig. 4.2). Although most studies apply a preventive/corrective maintenance ratio of app. 1:2, in this study it is this set at app. 1:3, because inspective maintenance is usually considered to be part of preventive maintenance[a].
Improvement (25%)	Total O&M cost includes refits, major overhauls and modifications, to maintain optimal performance of the wind farm. With a total O&M cost distribution of 21–34 €/MWh, the share of 'Improvement' O&M is set at 25%

[a]According to Rademakers and Braam (2002), the preventive maintenance cost dispersion is 3–6 €/MWh, where the corrective cost dispersion is 5–10 €/MWh

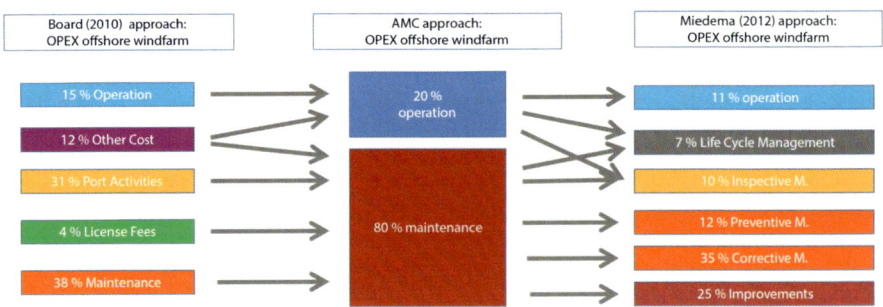

Fig. 4.3 OPEX breakdown. *Left* distribution according to Board (2010); *right* distribution used in this study; adopted from DGAME (AMC Centre 2011). *Source* Lagerveld et al. (2014)

combining wind energy production and mussel farming. For this purpose we used a cost distribution (Table 4.1), as elaborated by Miedema (2012), according to the AMC approach (Stavenuiter 2002).

Figure 4.3 presents a summary of the three consecutive approaches of allocating costs to the different operation and maintenance disciplines.

4.3   Potential for Synergy

To estimate the potential synergy through combining wind and mussel farming, the following assumptions apply.[1]

4.3.1   Operations and Life Cycle Management

For OWFs larger than 200 MW, it is common to have a control room ashore, 24 h and 7 days a week staffed by two to four people. Here, the assumption is made that with little extra effort this team can also manage the mussel farm, if it is integrated in the wind farm environment.

4.3.2   Inspective, Preventive, Corrective Maintenance and Improvement Maintenance

Previous studies and practical experiences (Thomsen 2012) have shown that in general 50% of the charged maintenance labour are non-productive time because of waiting for e.g. specific certified personnel, transport opportunities, acceptable weather windows, adequate spares, tools and equipment. It is assumed that by combining wind energy and mussel production these 'lost hours' can be reduced to at least 25% of the charged maintenance labour. This means that, when the labour cost is 60% of the total O&M cost of a wind farm, a cost reduction of 15% is attainable.

To reduce the waiting time related to O&M of wind farms, and thus reduce O&M costs, there are several logistical opportunities for synergy. For example, when a multi-purpose ship sails out for a week to transport a maintenance crew to and from the wind turbines, it can inspect the longline-installations and/or harvest

---

[1]These assumptions were formulated and agreed on in an expert workshop, held in June 2010 at the AMC Centre in Den Helder, Netherlands. Participants: Ramses Alma (AMC T&T), Nico Bolleman (Blue-H), Henk Braam (ECN), Wim de Goede (HVA), Ko Hartog (HVA), Bertrand van Leersum (ATO NH), Tom Obdam (ECN), Luc Rademakers (ECN), Hein Sabelis (Peterson), John Stavenuiter (AMC Centre) and Frans Veenstra (IMARES).

Table 4.2 Estimation of cost shares for wind farming when carried out singly and in combination with mussel farming, based on the expert workshop

O&M Disciplines	Wind farming (%)	Combination wind and mussel farming (%)
Operations	11	9
Life cycle management	7	6
Inspective maintenance	10	9
Preventive maintenance	12	11
Corrective maintenance	35	32
Improvement	25	23
Cost reduction	–	10
Total	100	100

Baseline (bl) is the O&M OPEX distribution according to Miedema (2010)

the mussels, while the crew is busy carrying out the maintenance work. When tasks are finished, the ship takes the crew on board again and brings the harvest ashore.

To achieve the pursued cost reductions, the following aspects of synergy are seen as prerequisites:

- Clusters of aquaculture integrated with, or between, clusters of wind turbines
- Combined Operations and Life Cycle Management
- Use of multi-purpose support vessels, capable to operate under significant wave-height conditions of up to 3 m
- Well-trained staff, capable to operate and maintain all installations
- No additional staff needed for the control room

The previously mentioned assumptions are expected to lead to an overall reduction of O&M costs by at least 10%. The following cost breakdown (in % of the total O&M cost of wind energy) is considered to be an adequate estimation for offshore mussel farming and the combination of offshore wind and mussel farming. The Figures derived serve as set targets and baselines or references for a first analysis in the LCA model (Table 4.2).

Although the cost breakdown for offshore wind farming is fairly well-founded, it must be taken into account that the estimations for combined wind and mussel farming are indicative and used as a first estimated baseline for running the AMC model.

4.3.3 Asset Management Control (AMC) Model

More participants in the O&M process will lead to a more complex organization and more uncertainty and financial risk for the asset owner. A model that oversees all participants and processes, involved in the O&M of OWFs, will prove to be essential to determine the cost-effectiveness of the wind and mussel farm (W&MF)

System Identification Diagram

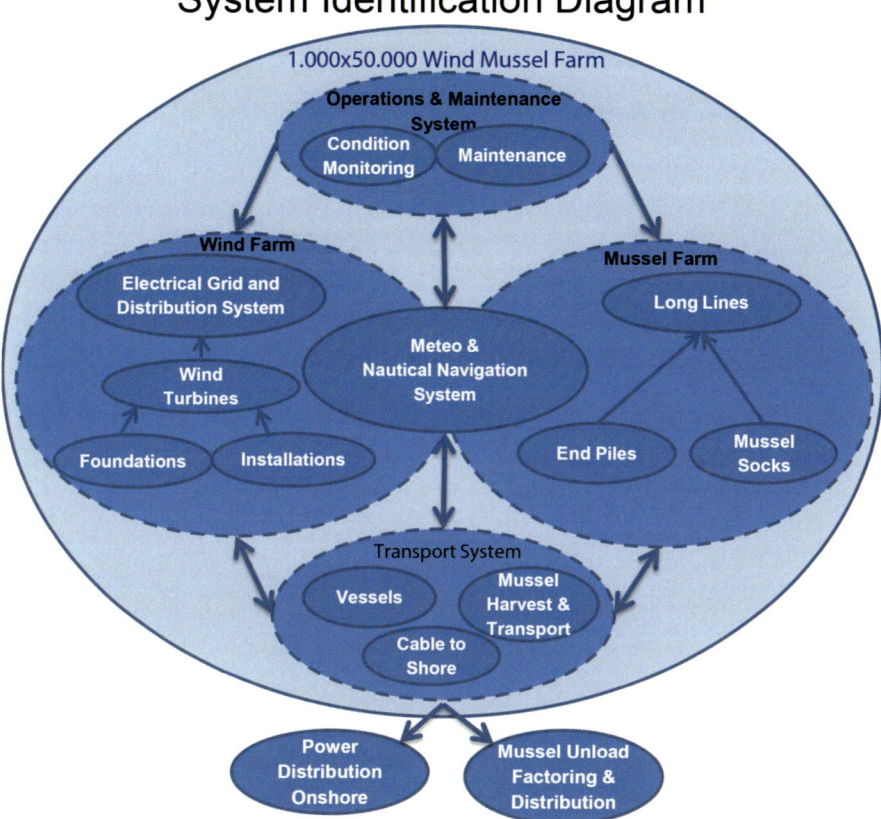

Fig. 4.4 Wind & Mussel Farm (W&MF) system identification diagram

system over the design lifecycle. A system approach is chosen that gives sufficient insight and at the same time is kept manageable. The prime operational functions, namely wind energy production and mussel farming, are the main components of the system identification diagram (Fig. 4.4). The diagram illustrates the two main systems, their support systems, and the system boundaries. The two main systems are supported by three support systems:

1. Operations & Maintenance System
2. Meteorological & Nautical Navigation System
3. Transport System (for details see Annex 1)

The physical building blocks of the systems (dashed lines) are defined as 'functional packages'. The functions: power distribution onshore, mussel unload, factoring, and distribution, are not included because it is assumed that these (sub-) systems are available in adequate capacity.

Based on the identification diagram and our assumptions on how a W&MF system like this could be realized in the North Sea, Lagerveld et al. (2014) elaborated on the following physical concepts:

- Wind & Mussel Farm outline
- Operation and Wind farm clusters, including auxiliary systems
- Auxiliary systems
- Mussel farm clusters
- Transport system
- Operation & Maintenance management system
- Economic input parameters

Here, we describe the Operations and Maintenance management system. It is assumed that a large-scale offshore wind and mussel farm should be managed and controlled with an integrated information and communication system. For that reason, the newest AMC concept in the field is adopted for this case (Van Leersum et al. 2010). This system is an integrated monitoring and control system over the value chain, called DOWES (Dutch Offshore Wind Energy System). DOWES is designed for managing optimal system performance.

Based on the physical representation of the virtual Wind & Mussel Farm case, system configuration figures are determined (see Appendix D in Lagerveld et al. 2014).

In this study an existing AMC model (Stavenuiter 2002) is expanded/elaborated, including the entities physical components, process activities and time periods (years). In order to perform the simulation runs the model has to meet the following requirements:

- the whole W&MF can be adapted by parameter settings (for long-term maintenance)
- benefits can be viewed by different parameters/key performance indicators
- different settings in O&M plans (maintenance strategies) are possible
- fluctuations in business, technical and economic parameters can be taken into account
- financial and technical balance sheets should be available per year
- price elasticity, or price changes in general, can be analyzed
- main results, such as revenues, system cost-effectiveness, market value, return on investment, over the exploitation phase, can be presented in charts

A synergy factor is introduced to model the synergy effects of combining wind and mussel farming. It is expressed as a percentage, which indicates the extent to which the O&M cost of the wind farm can be reduced by more efficient use of labor, transport and equipment. For example: if the O&M cost of a wind farm is set at 100%, and it will be 95% if it is combined with a mussel farm, then the synergy factor, as defined, is 5%.

For offshore-aquaculture the location is the most important criterion for a successful combination with offshore wind (Lagerveld et al. 2014). Previous studies have shown that the North Sea is a good habitat for mussels (Lagerveld et al. 2014, Sect. 3.2.2 and references therein). Promising locations for a 1000 MW wind farm in the North Sea seem to be Borssele, IImuiden-Ver and Schiermonnikoog. The model is based on average parameter settings, but can be tuned for specific locations.

The backbone of the AMC model used for our case, is a system model, called AMICO (Stavenuiter 2002), that is capable of modelling the physical system over the lifecycle, including a logistic process model per year period.

The AMC Blauwdruk model (Lagerveld et al. 2014) is based on these concepts, simulating and calculating the cost/performance parameters per year period. Within this model the following adjustments are possible per (sub-) system:

1. total installed capacity
2. installed capacity per cluster
3. number of clusters
4. number of (critical) installations per system
5. initial realization investment
6. interest rate
7. inflation rate
8. farm yield coefficient
9. baseline ROI discount factor (average interest (%) on investment over the years)
10. O&M cost as % CAPEX
11. starting sales prices (in MWh for wind and kg for mussels)
12. sales price increasing rate (if calculations are 'fixed')
13. mussel farming cycles (1–4 year).

In this study, two O&M plans were developed: the Base O&M plan and the related retrofit and overhaul plan, which can be budgeted by a multiply factor, called the retrofit and overhaul factor. Settings that can be entered in the LCA model by the control panel, are:

- mussel farming and wind synergy factor active? (y/n)
- fixed calculations (y/n). If yes, no uncertainty simulations will be executed
- multiyear maintenance (MYM) period (1–10 year)
- aquaculture and wind synergy factor, expressed as % of the wind farm O&M cost.

In addition, several simulation parameters are included which may vary over the years and can be adjusted by the trend diagram parameters:

- wind power sales price developments
- mussel sales price developments
- operational excellence factor (simulates windfalls and setbacks in business)

- maintenance management control factor (simulates windfalls and setbacks in system failures, based on Mean Time Between Failure (MTBF) and Mean Time To Repair (MTTR).

To simulate price changes for the wind power sales price and mussel sales price, the prices can be adjusted by parameter setting of the trend diagram (Lagerveld et al. 2014).

4.4 Conclusions

In line with Michler-Cieluch et al. (2009a), this article highlights the need for collaboration between the different disciplines relevant for combining activities offshore. We show that an asset management control model is useful for bringing the relevant offshore disciplines together. Multi-use activities offshore do have an effect on (the assessment of) risks arising from (multiple) combined O&M processes. The exact details of these processes are still unknown and hence estimations are uncertain.

Simulations with the AMC model under different assumptions allow us to estimate the benefits and costs of the interdisciplinary MUP approach, thus making the approach more concrete and robust.

In summary, we conclude that there are opportunities for all actors (the government, the wind sector, and the aquaculture sector) to achieve their different objectives by combining offshore wind energy production with offshore aquaculture. Lagerveld et al. 2014 suggest that an overall cost reduction on O&M activities of approximately 10% is feasible, if offshore wind farms and offshore aquaculture are combined.

Furthermore, running the Asset Management Control Model the return of investment (ROI) for four different scenarios was simulated. Based on the chosen economic parameter values and sales prices estimates, the model simulations show that a ROI of 4.9% should be possible in unfavourable economic conditions when synergy is absent. When 10% synergy can be achieved, a ROI of 5.5% seems possible. The ROI is significantly higher when economic conditions are favourable. Even when there is no synergy, a ROI of 8.3% should be feasible, and in case of 10% synergy the ROI is likely to reach 9.6%.

Recommendations
The mitigation of physical and chemical processes that pose a risk to the constructions should be investigated. In collaboration with all sectors involved, it should be investigated in more detail what operational processes in a multi-use setting can look like, thus enabling us to accurately quantify potential synergy benefits. Only then will we be able to assess the reliability of our input values and the robustness of the model results.

Annex 1—Transport System Details

Subsea Power Cable Subsystem

The Subsea Power Cable Subsystem provides the electrical power transport from the transmission station to shore. For this case the total length of this cable is estimated at 30 NM (55 km).

The technical specifications and cost figures used in this case were and can be extracted from the internet (e.g. http://www.nationalgrid.com/NR/rdonlyres/62196427-C4E4-483E-A43E-85ED4E9C0F65/39230/ODISAppendicesFinal_0110.pdf).

Estimated cost for a 300 kV cable 1031 MW (price level of 2013):

CAPEX (20 years depreciation): 111.120.000 Euros

OPEX (annual): 560.000 Euros

Offshore Wind and Fish Farming Support Ships

The Offshore Wind and Fish Farming Support Ships are important to attain the 10% savings on O&M costs. The design requirements of these ships include:

- capable to transport and accommodate 40 persons, working in 3 shifts for one week
- wind farm spares transport and repair capability
- mussel harvest and transport capability.

In addition this ship must be also be capable to navigate and work in harsh weather conditions. The following equipment is considered to make this possible:

- a dynamic position system (DP-2)
- a motion compensated crane
- a wide working deck.

The size, shape, weight contribution and propulsion should be tuned on such a way that this ship will be a comfortable platform to live on for one week and to work on 24/7 h with a significant wave height up to 3 m (North Sea conditions). For an illustration of a new preliminary design of a Wind & Mussel Farming Support ship which could meet these specifications, see Stavenuiter (2009).

The optimization of size, shape and displacement makes these ships stable platforms to operate in a working window up to 95% over the year. Besides this, the dynamic position system (DP2) and motion compensated crane make it possible to access wind turbines, with personnel, by man riding with a crew basket, but also with spare parts and tooling, up to 3000 kg.

As support ship for Mussel Farming it is assumed that these ships will be equipped with a cargo hold for 600 tons of mussels. For inspecting and harvesting the mussels, the dynamic position system and motion compensated crane are also considered essential. The idea is that already proven mussel harvest systems could be mounted on the motion compensated crane. With this system it should be possible to harvest approximately 5 tons/h.

Tooling and Spars Container Support System

It is supposed that the Support Ships will be designed and built in a multi-functional concept. For that reason a configuration with a wide deck and containers is chosen. In this case it is assumed that for tooling, equipment and spare parts (mainly for Wind Farm Maintenance), 18 specially prepared containers will be sufficient for serving 2 ships.

Mussel Harvest Subsystems

The assumption is made that the existing mussel harvest systems for near shore can be modified in such a way that they can be used for offshore, up to 3 m significant wave height, when combined with the motion compensated crane as one system.

References

AMC Centre. (2011). *DGAME*. Retrieved June 3, 2016, from www.amccentre.nl/dgame

Board, R. A. (2010). *Value breakdown for the offshore wind sector*. Report commissioned by UK Renewables Advisory Board.

Buck, B. H. (2002). Offshore-windparks und green aquaculture. In: K. Övermöhle & K. P. Lehmann (Eds.), *Fascination offshore—Use of renewable energies at sea*. Report 2002.

Buck, B. H., Krause, G., & Rosenthal, H. (2004). Extensive open ocean aquaculture development within wind farms in Germany: The prospect of offshore co-management and legal constraints. *Ocean and Coastal Management, 4,* 95–122.

Buck, B. H., Ebeling, M. W., & Michler-Cieluch, T. (2010). Mussel cultivation as a co-use in offshore wind farms: Potential and economic feasibility. *Aquaculture Economics & Management, 14,* 255–281.

Buck, B. H., & Krause, G. (2012). Integration of aquaculture and renewable energy systems. In: R. Meyers (Ed.), *Encyclopaedia of sustainability science and technology*. Springer Science +Business Media LLC.

DOWES (2014). *Dutch offshore windenergy service*. Accessed March 1, 2014, at www.dowes.nl

Feargal, B. (2009). *The offshore wind challenge*. Cranfield University.

Gimpel, A., Stelzenmüller, V., Grote, B., Buck, B. H., Floeter, J., Núñez-Riboni, I., et al. (2015). A GIS modelling framework to evaluate marine spatial planning scenarios: Co-location of offshore wind farms and aquaculture in the German EEZ. *Marine Policy, 55,* 102–115.

Kjeldsen, K. (2009). *Cost reduction and stochartic modelling of uncertainties for wind turbine design*. MSc Thesis. Aarlborg University.

Krause G., Griffin, R. M., & Buck, B. H. (2011). Perceived concerns and advocated organisational structures of ownership supporting 'offshore wind farm–mariculture integration'. *Intech*. doi:10.5772/15825

Lagerveld S., Röckmann C., & Scholl M. (2014). *A study on the combination of offshore wind energy with offshore aquaculture.* IMARES Report C056/14. Retrieved August 2, 2016, from http://edepot.wur.nl/318329

Michler-Cieluch, T., & Krause, G. (2008). Perceived concerns and possible management strategies for governing wind farm-mariculture integration. *Marine Policy.* doi:10.1016/j.marpol.2008.02.008

Michler-Cieluch, T., Krause, G., & Buck, B. H. (2009a). Reflections on integrating operation and maintenance activities of offshore wind farms and mariculture. *Ocean and Coastal Management, 52,* 57–68.

Michler-Cieluch, T., Krause, G., & Buck, B. H. (2009b). Marine aquaculture within offshore wind farms: Social aspects of multiple-use planning. *GAIA-Ecological Perspectives for Science and Society, 18,* 158–162.

Miedema, R. (2012). *Research: Offshore Wind Energy Operations & Maintenance Analysis.* MSc Thesis, Hogeschool van Amsterdam.

Musial, W., & Ram, B. (2010). *Large-scale offshore wind power in the United States. Assessment of Opportunities and Barriers.* NREL/TP-500-40745. Denver, CO, USA: National Renewable Energy Laboratory.

Noordzeewind (2010). *Operations report 2009.* OWEZ-R_000_20101112. Retrieved November 27, 2015, from www.noordzeewind.nl

Pieterman, R. P., Braam, H., Obdam, T. S., Rademakers, L. W. M. M., & van Zee, T. J. J. (2011). *Optimisation of maintenance strategies for offshore wind farms. A case study performed with the OMCE-Calculator.* ECN Report nr. ECN-M-11-103. Energie Centrum Nederland Petten.

Rademakers, L. W. M. M., & Braam, H. (2002). *O&M aspects of the 500 MW offshore wind farm at NL7.* DOWEC-F1W2-LR-02-080/0. Petten.

Rijkswaterstaat. (2009). *Europlatform 1985–2008.*

Stavenuiter, J. (2002). *Cost effective management control of capital assets: An integrated life cycle management approach.* the Netherlands: Asset Management Control Research Foundation, Medemblik.

Stavenuiter, W. (2009). *The missing link in the offshore wind industry: Offshore wind support ship.* Report SDPO.09.020.m. Master Thesis, Delft University of technology.

Thomsen, K. (2012). *Offshore wind—A comprehensive guide to successful offshore wind farm installation.* Elsevier.

Van Leersum, B., Stavenuiter, J., Sabelis, H., Braam, H., & van der Mijle Meijer, H. (2010). *Integrated offshore monitoring system.* DOWES. A European Fund project, ATO Den Helder, The Netherlands. Accessed December 2, 2015, at www.dowes.nl

Verhaeghe, D., Delbare, D., & Polet, H. (2011). Haalbaarheidsstudie: Passieve visserij en maricultuur binnen de Vlaamse windmolenparke? ILVO-mededeling, no. 99, Instituut voor Landbouw-en Visserijonderzoek—ILVO.

Chapter 5
Technical Risks of Offshore Structures

Job Klijnstra, Xiaolong Zhang, Sjoerd van der Putten
and Christine Röckmann

Abstract Offshore areas are rough and high energy areas. Therefore, offshore constructions are prone to high technical risks. This chapter elaborates on the technical risks of corrosion and biofouling and technical risks through mechanical force. The expected lifetime of an offshore structure is to a great extent determined by the risk of failures through such risks. Corrosion and biofouling threaten the robustness of offshore structures. Detailed and standardized rules for protection against corrosion of offshore structures are currently lacking. There is a need for an accepted uniform specification. A major technical risk of a combined wind-mussel farm is that of a drifting aquaculture construction that strikes a wind turbine foundation. We investigate two scenarios related to this risk: (1) Can a striking aquaculture construction cause a significant damage to the foundation? (2) If a drifting aquaculture construction gets stuck around a turbine foundation and thus increases its surface area, can the foundation handle the extra (drag) forces involved? A preliminary qualitative assessment of these scenarios leads to the conclusion that a drifting mussel or seaweed farm does not pose a serious technical threat to the foundation of a wind farm. Damage to the (anticorrosive) paint of the turbine foundation is possible, but this will not lead to short term structural damage. Long term corrosion and damage risks can be prevented by taking appropriate maintenance and repair actions. Contrarily to mussel or seaweed farms, the impact/threat of a drifting fish farm on structural damage to a wind foundation depends on type, size and the way of construction of the fish cages. The risk of extra drag force due to a stuck aquaculture construction relates particularly to jacket constructions because any stuck construction may lead to (strong) increase of the

---

J. Klijnstra (✉) · X. Zhang
Endures BV, Bevesierweg 1 DC002, 1781 AT Den Helder, The Netherlands
e-mail: job.klijnstra@endures.nl

S. van der Putten
TNO Structural Dynamics, Van Mourik Broekmanweg 6,
2628 XE Delft, The Netherlands

C. Röckmann
Wageningen University and Research—Wageningen Marine Research,
P. O. Box 57, 1780 AB Den Helder, The Netherlands

© The Author(s) 2017
B.H. Buck and R. Langan (eds.), *Aquaculture Perspective of Multi-Use Sites in the Open Ocean*, DOI 10.1007/978-3-319-51159-7_5

frontal surface area of the immersed jacket structure and thereby give increased drag forces from currents or waves. To ensure an optimal lifetime and lower operational costs maintenance aspects of materials for both offshore wind and aquaculture constructions should be taken into account already in the design phase of combined infrastructure.

5.1   Introduction

For the successful operation of a wind farm and the successful combination of a wind farm with aquaculture, it is essential that the expected lifetime of the constructions used is acceptable. The expected lifetime of an offshore structure is to a great extent determined by the risk of failures. These failures can be the result of many different problems.

This chapter focuses on two technical risks, typically associated with a combination of wind farming and aquaculture: damage mechanisms of corrosion and bio-fouling, and damage risks of mechanical loads. More precisely, a major technical risk of a combined wind-mussel farm is that of a drifting aquaculture construction that strikes a wind turbine foundation. We carry out a preliminary qualitative assessment of two scenarios related to this risk: (1) Can a striking aquaculture construction cause a significant damage to the foundation? (2) If a drifting aquaculture construction gets stuck around a turbine foundation and thus increases its surface area, can the foundation handle the extra (drag) forces involved?

There are additional risks, which are not dealt with in this report. The risk of collision with ships is also there, and it may even be slightly elevated, but in terms of possible damage it does not substantially differ from the single-use situation (wind farm). Impacts of foreign (drifting) objects are also not taken into account.

The findings presented in the following sections are based on literature data. We focus on risks arising from offshore wind energy production combined with offshore mussel farming. Additionally, risks arising from seaweed culture and using fish cages are also presented here because information on technical aspects of offshore structures, available in current literature, is scarce and often does not discriminate between the different types of aquaculture. Mechanical risks are described in some more detail in Janssen and van der Putten (2013).

Chapter 3 deals specifically with corrosion aspects and biofouling of offshore structures, Chap. 4 deals with mechanical risks of wind farms in the presence of offshore aquaculture, in this chapter elaborates the two technical risk scenarios, and Chap. 6 finishes with conclusions and recommendations.

5.2 Corrosion Aspects and Biofouling

5.2.1 Corrosion Mechanisms and Corrosivity Zones for Offshore Structures

In general, the same mechanisms that can damage offshore structures like wind turbines and platforms can also damage aquaculture structures that are made of the same or similar material. Offshore structures are exposed to different and varying corrosive environmental conditions.

Based on theory and practical experience with offshore structures, in total eleven different corrosion zones of offshore wind structures can be identified. The most critical zones are the splash/tidal zone and closed compartments filled with seawater (e.g. the internal of a monopile or jacket foundation structure). Design specifications for steel structures define a corrosion allowance. In case of uniform corrosion this is an applicable design tool. However, when local corrosion mechanisms like microbial corrosion (MIC), galvanic corrosion or corrosion fatigue occur, the structural integrity of the steel structure must be evaluated. The offshore wind structure design is determined by fatigue load. Local defects like pitting attack may act as initiation sites for fatigue cracking. For this reason special attention should be given to local defects in the foundation and the tower structure.

5.2.2 Corrosion Risks in Currently Used Offshore Wind Turbines

The offshore wind energy market is young, compared to the offshore oil and gas and shipment markets; the first offshore wind farm was installed in 1991. The most important lesson learned from the first generation offshore wind turbines is: wind turbines based on onshore technology are not suitable for offshore application. The first offshore wind farm, Horns Rev (D), suffered from a major coating failure of eighty wind turbine foundations. The coating on the transition pieces broke down and resulted in unexpected repair and maintenance costs. The reason was a combination of wrong coating selection and improper application of the coating. This points out to the key issue: a lack of conformity between the manufacturer, coating applicator and coating supplier.

Other corrosion related problems reported are failing cathodic protection systems, corroding boat landings by combination of wear, impact and seawater and corroding secondary structure components like ladders and railings. The impact of corrosion damage varied from increased safety risks for maintenance personnel to re-evaluating the structural integrity of the foundation structure because of local pitting attack.

Local corrosion attack by MIC has been noticed on the internal surface of different monopile foundations on different locations in the North Sea. With grouting failure repair of several monopile foundations, local corrosion attack was detected on the internal surface area of the unprotected monopile. Until then the internal area had been a black box: the hedge was sealed to reduce and stop the internal corrosion process. Nowadays, MIC processes inside monopile foundations are still not known in very much detail and require further investigation to find optimal control measures.

Specification of corrosion protection for specific offshore wind structures is still an issue. The applied standards for European offshore wind farms vary from onshore related specifications to those deriving from offshore oil and gas specifications. Based on the experiences with coating and cathodic protection failures, there is a need for an accepted uniform specification. Up to date, such a specification is lacking.

5.2.3   Biofouling on Offshore Structures

Offshore constructions are attractive to biofouling species. Biofouling may result in increased costs due to antifouling measures that have to be taken: extensive inspection and maintenance, creation of micro-environments discouraging microbial corrosion, and heightened design criteria as a consequence of the extra hydrodynamic and weight loading (Fig. 5.1).

Generally four different process stages of bio-fouling in seawater are described (Callow & Callow 2011). These may take place in different time frames. The first

(a)   **(b)**

Fig. 5.1 Biofouling on an offshore jacket foundation (**a**) and access to a wind turbine foundation for maintenance (**b**): biofouling is visible in the tidal zone and on the stairs to the platform (*Source* Windcat Workboats B.V.)

stage starts almost instantly upon immersion with the formation of a conditioning layer of dissolved organic matter such as glycoproteins and polysaccharides. Subsequently a so-called biofilm can be formed with colonizing bacteria and micro-algae. Hours to days later a more complex community may form including multicellular primary producers and grazers, for instance algal spores, marine fungi and larvae of hydroids, bryozoans, and barnacles. If time and environmental conditions allow for, such communities may evolve to diverse and sometimes very thick layers with both hard fouling organisms (barnacles, mussels, tube worms, corals, etc.) and large populations of soft fouling such as ascidians, hydroids and macro algae. However, it should be explicitly mentioned that in a natural environment the biofouling process is very variable and never follows exactly this schematic representation. The process is influenced by many abiotic factors as well, such as salinity, nutrient content, sunlight intensity and duration, currents, and temperature.

In existing wind farms, no antifouling techniques are currently applied on the foundations. In this situation, the uncoated steel subsea zone and the coating system on the transition piece are both susceptible to biofouling. Especially the boat landing area (see Fig. 5.1b) is a substructure that for safety reasons may need extra attention with regard to fouling prevention.

Biofouling on floating foundations as well as the tether ropes should be taken into account when assessing the lifetime of the construction. Calculations of design loads of offshore wind turbine foundations commonly apply a maximum biofouling layer thickness of about 200 mm for extreme load conditions. A load calculation model would also take into account weight and hydrodynamic loading (current and wave load) by biofouling. At first glance, a value of 200 mm of maximum biofouling layer thickness seems sufficient. However, in order to deduce a more reliable biofouling layer thickness depending on the location, regular checks over a twenty year period must take place. Biofouling on tether ropes can additionally influence the hydrodynamic behavior by the increased diameter of these tether ropes.

Biofouling can pose a risk to offshore wind foundations in the following cases:

- **Increased drag load**. The hydrodynamic profile of a biofouling layer strongly deviates from that of the flat surface of a foundation. Extensive growth, in the form of long trail-like colonies of mussels, algae and other soft elongated macro-organisms that move along with the current, may sometimes result in unexpectedly high drag loading. Biofouling may, however not necessarily pose a risk to the mechanical load on the foundations in moderate tidal current conditions.
- **Influence on cathodic protection**. Another effect of biofouling is coverage of anodes, which affects the function of the cathodic corrosion protection system. For visual inspection on site (weld inspection, wall thickness measurements) a biofouling layer must be removed.
- **Influence on MIC**. Biofouling creates micro-environments encouraging microbial corrosion (MIC). Knowledge on MIC processes inside monopile foundations is still scarce and needs further elaboration for proper assessment of risks on failure due to pitting corrosion.

- **Safety and accessibility**. For safety reasons biofouling must be prevented on stairs and boat landing area, to ensure safe access of maintenance personnel to the foundation and wind turbine (Fig. 5.1b).

There are several techniques that can be applied to prevent or clean biofouling on surfaces: antifouling coatings, electrochemical and physical methods for fouling control, cleaning of surfaces by robots or handheld tools. It is recommended to inspect the foundation and anodes after a period of 5–10 years. Visual inspection and quantification of fouling composition and thickness can be combined with regular cleaning of the external surface.

Considering the three types of wind turbine foundations (Fig. 5.2; Table 5.1) no clear differences in biofouling settlement and/or development are expected. The basic materials used in the foundation are equally susceptible to fouling under immersion. Fouling control coatings can be applied to all types of materials. Also cleaning techniques for removal of fouling do not substantially differ between the three types of foundation structures.

5.2.4   Potential Influence of Offshore Aquaculture on the Corrosion of Unprotected Steel Structures

Processes in seaweed farms may influence seawater chemistry. The salinity of ambient sea water at open sea is 3.0–3.6% in most cases. The pH of seawater is relatively stable whereas temperature, dissolved oxygen and nutrients may vary strongly (Bartoli et al. 2005; Mantzavrakos et al. 2007). Seawater is generally at a pH of 7.5–8.5 due to its buffering capacity with many ions and interaction with carbon dioxide and water. Oxygen levels can range from zero to over 10 ppm in temperate waters (Valdemarsen et al. 2012).

Seaweed photosynthesis increases dissolved oxygen in the water: The oxygen concentration in seaweed tanks can vary from 7.0 to 13.0 ppm, while in ambient seawater it varies from 8.0 to 10.3 (Msuya and Neori 2008). The increased level of dissolved oxygen in the water might result in an increased corrosion rate of unprotected steel structures at sea. The corrosion rate of steel under a calcite film (deposited by seawater on cathodic areas of metal) is 250% higher in the presence

Table 5.1 Typical design properties of three different wind turbine foundations

	Monopile	Jacket	Gravity based
Weight	500 tonnes	800 tonnes	5000 tonnes
Main material	Steel	Steel	Concrete
Max. water depth	30 m	30 m	40 m
Max. wave height (H_{max})	13.7 m	16.2 m	17.5
Max overturning moment at seabed	200 MNm		450 MNm

of seaweeds than without (Buzovkina et al. 1992). Seaweeds may raise the pH of the water by 0.1–0.4 pH units (Robertson-Andersson et al. 2008). This variation may have an influence on scale formation on steel structures and thereby induce or change localized corrosion processes (Beech and Campbell 2008). Careful monitoring of scale formation and appropriate maintenance measures will help to keep corrosion risks below critical levels.

Fish farms cause metal enrichment in the bottom of the sea, e.g. extreme high concentrations of Zn, Cu and Cd in sediments and pore water (Dean et al. 2007; Kalantzi et al. 2013; Loucks et al. 2012; Nordvarg and Johansson 2002). Such high concentrations may also increase the corrosion risk of steel due to higher conductivity of the electrolyte and creation of galvanic effects. Additionally, oxygen consumption because of biodegradation may create an anoxic or anaerobic environment that stimulates MIC by microorganisms such as sulfate reducing bacteria (SRB; Kawahara et al. 2008). Increase of carbon oxides and nitric oxides can also increase the corrosion of steels (Beech and Campbell 2008). On sites with substantial water currents, however, it is not very likely that these processes will have a strong effect on corrosion of materials and constructions.

No literature data have been found on effects of mussel farms on environmental parameters that can be associated with corrosion risks. A priori such risks cannot be fully excluded, depending on type of materials used in mussel farms. If similar phenomena occur as described above for fish farms, e.g. metal enrichment and/or anoxic conditions in the near environment, then potential risks may exist but again on locations with sufficient water currents these risks are probably low.

5.3   Mechanical Risks of Wind Farms Due to the Presence of Offshore Aquaculture Constructions

Offshore wind farms are constructed and developed to withstand the forces of the oceans. Wind and waves cause the highest loads on a wind turbine (tower and foundation). The presence of an offshore aquaculture may pose an additional threat to the wind farm. The research question is: What are the effects of aquaculture constructions and activities on the (mechanical) safety of offshore wind turbines?

To grow seaweed or mussels, usually nets or ropes are used (e.g. submerged longlines, cf. Buck et al. 2010; Lagerveld et al. 2014); fish farms usually apply special cages or more or less rigid characteristics. Common materials for fish cage construction are wood, steel and plastic (Burak Cakaloz 2011).

The next section discusses likely scenarios that may occur and could lead to mechanical risks to the turbine foundation when offshore aquaculture is carried out within or in close vicinity of an offshore wind farm. Because the risks can be different depending on the type of foundation, three commonly used structures and their properties are considered: monopile, jacket and gravity based (Fig. 5.2; Table 5.1).

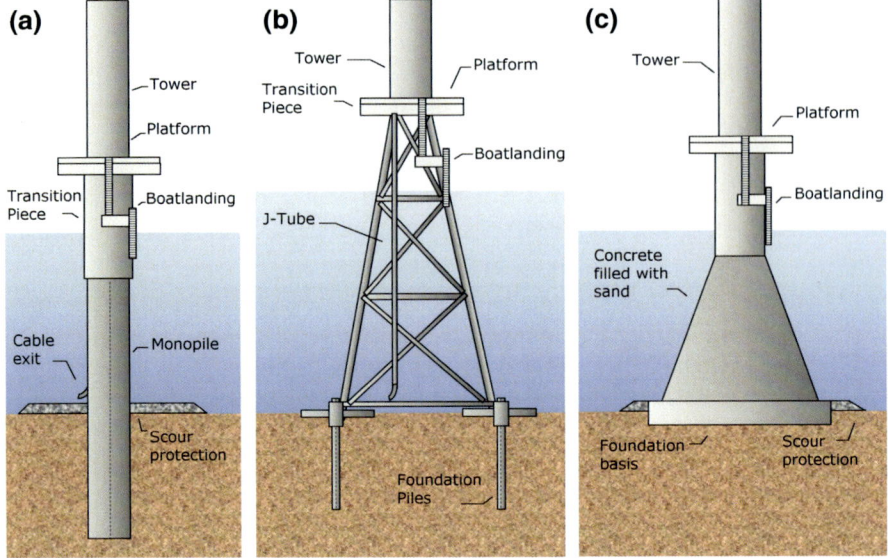

Fig. 5.2 Three types of turbine foundations: Monopile (**a**), jacket (**b**) and gravity based (**c**)

5.4   Scenario Analyses

Our analyses focus on narrative scenarios that may lead to mechanical (and corrosion) damage to the wind turbine foundation. Scenarios that could lead to damage of the aquaculture construction or the supply/maintenance vessels are not (yet) included. These risks can only be investigated at a later stage when the operational processes of maintenance and harvesting are known in detail.

Two scenarios that may occur and questions that arise are:

1. Impact. Drifting aquaculture construction **strikes** the turbine foundation.
   Is there a risk of significant damage to the foundation?
2. Extra drag force. Drifting aquaculture construction **gets stuck around** the turbine foundation, increasing its surface area.
   Can the foundation handle the extra (drag) forces involved?

The answers to these questions depend on the type of aquaculture (mussel, seaweed, fish) and corresponding constructions, and on the specific turbine foundation (i.e. monopile, jacket or gravity based). Therefore, the scenarios are presented in matrix tables. The two different scenarios and their possible risks are described below.

Scenario 1: Impact. Drifting aquaculture strikes the turbine foundation

It is possible that a drifting aquaculture (e.g. the longline construction, whether or not overgrown) strikes a turbine foundation. In such a case there are three main parameters that determine the risk of damage to the foundation:

1. the mass
2. the impact velocity
3. the deformability/robustness of the aquaculture construction

As mussel and seaweed farms mainly consist of nets and ropes, the deformability of such structures is large. In case of an accident, it is the aquaculture construction that deforms, and not the foundation. Probably this also holds for most fish cages. Elastic fish cages will not damage the foundation structure; only larger, more rigid cages have the potential to do so.

Damage to the protective coating of the foundation structures when they are hit, is possible in all cases. On a longer term, this could induce additional corrosion risks and negatively influence the safety of the construction. Inspections are required and possible repair of the coating may be necessary. Table 5.2 summarizes the effects, which do not differ for the three different foundation types.

Scenario 2: Extra drag force. Drifting aquaculture construction gets stuck around the turbine foundation, increasing its surface area

It is possible that a drifting aquaculture does not only strike, but gets stuck around a turbine foundation. In the case of a monopile or gravity based foundation, the stuck aquaculture construction will not significantly increase the frontal surface area of the structure. The frontal surface area is an important parameter in the determination of drag forces. With increasing frontal surface, drag forces due to current and surface waves increase. In the case of a jacket consisting of a lattice structure with many beams, it is possible that an aquaculture construction gets stuck around the beams and significantly increases the frontal surface area. In this case, the local force on such a beam, and the overall drag forces on the whole structure certainly increase. The effects are summarized in Table 5.3.

Possible effects of the 'worst case' scenario (grey cells in Tables 5.2 and 5.3) are preliminarily analyzed in Janssen and van der Putten (2013).

Table 5.2 Scenario 1: Drifting aquaculture strikes the turbine foundation

	Mussels	Seaweed	Fish
Monopile	No significant	No significant struc-	Damage depends on
Jacket	structural impact	tural impact damage	mass, velocity and de-
Gravity based	damage expected	expected	formability of fish cage

Grey cells indicate the worst case scenario

Table 5.3 Scenario 2: Drifting of the aquaculture

	Mussels	Seaweed	Fish
Monopile	No significant increase in loads expected		
Jacket	Increase in drag force		
Gravity based	No significant increase in loads expected		

The aquaculture is stuck around the turbine foundation. Grey cells indicate the worst case scenario

5.5 Conclusions and Recommendations

The combination of different activities offshore influences the assessment of risks arising from multi-use offshore platforms. The exact details of the processes involved in such multi-use offshore activities are still unknown and hence estimations are uncertain.

The main risk of a combined wind-mussel farm investigated here is that of a drifting aquaculture construction. Two major scenarios and related questions were investigated:

1. Is there a risk that a drifting aquaculture constructions strikes the turbine foundation and causes a significant damage to the foundation?
2. What is the risk if a drifting aquaculture construction gets stuck around the turbine foundation and thus increases its surface area? Can the foundation handle the extra (drag) forces involved?

A preliminary qualitative assessment of these scenarios yields that scenario 1 (impact between offshore aquaculture and wind turbine foundation) is not a real threat in case of mussel and seaweed farms. It is highly unlikely that aquafarm structures will be used that are heavy and rigid enough to cause significant structural damage. The (anticorrosive) paint of the turbine foundation might get damaged in case of an impact, but this will not lead to short term structural damage. In order to prevent corrosion and damage risks in the long term, appropriate actions (i.e. repair) can and should be taken. For fish farms the situation in scenario 1 may vary with the type and size of cages that are used and the way they are constructed. Potential risks of consequences of the impact should be assessed already in the design phase of such combined infrastructure.

Scenario 2 (extra drag force from currents and waves due to stuck aquaculture constructions) poses a risk especially to jacket constructions because it may lead to (strong) increase of frontal surface area of the immersed structure and thereby give increased drag forces. With monopiles and gravity based constructions the stuck aquaculture material may attach to the turbine foundation at a single point only with insignificant increase of frontal surface area and minimal increase in such drag force.

For a jacket construction, in the extreme case of a 100% coverage of its underwater surface by stuck aquaculture material during a storm, the overturning moment at the seabed could increase by 200–300 MNm (Janssen and van der Putten 2013), and eventually lead to the collapse of the wind turbine. However, this risk is merely theoretical, considering the type and construction of aquaculture materials being far less massive than the foundation itself and the unrealistic assumption of a 100% coverage. Nevertheless, appropriate methods to avoid this small risk can be investigated in the design phase of such infrastructure, for instance modular aquaculture structures that fall apart in case of drifting under severe conditions.

In severe storms with extremely high waves, an intact aquaculture structure that is physically directly connected to the turbine foundation could theoretically lead to the collapse of the turbine if the overturning moment at the seabed becomes too large. For this reason, the investigated scenarios only consider aquaculture installations that are not attached to any wind turbine foundations (Lagerveld et al. 2014). Nonetheless, if a connected wind farm-aquaculture infrastructure is considered and designed, methods to reduce and prevent high tensile forces on the turbine foundation should be taken into account. For example, use of suitable anchors to hold the aquaculture structure in place or application of so-called safety wires that break at predefined tensile forces. Although the aquaculture farm will be lost in the latter case, the turbine foundation will stay intact.

Finally, a few recommendations for the future implementation of a multi-use platform offshore, based on Noël (2015); van der Putten (2015) and Lagerveld et al. (2014):

- Appropriate measures should be taken to protect aquaculture and offshore wind constructions from corrosion attack either by selection of corrosion resistant materials or by application of suitable protective techniques or coatings.
- Type and size of aquaculture activities determine the extent of effects on water and sediment quality. In turn, water and sediment quality may affect corrosion resistance of the materials used. This aspect should be dealt with in a dedicated risk assessment for the specific location.
- Maintenance aspects of materials for both offshore wind and aquaculture constructions should be taken into account already in the design phase to ensure optimal lifetime of infrastructure.

References

Bartoli, M., Nizzoli, D., Naldi, M., Vezzulli, L., Porrello, S., Lenzi, M., et al. (2005). Inorganic nitrogen control in wastewater treatment ponds from a fish farm (Orbetello, Italy): Denitrification versus ulva uptake. *Marine Pollution Bulletin, 50,* 1386–1397.

Beech, I. B., & Campbell, S. A. (2008). Accelerated low water corrosion of carbon steel in the presence of a biofilm harbouring sulphate-reducing and sulphur-oxidising bacteria recovered from a marine sediment. *Electrochimica Acta, 54,* 14–21.

Buck, B. H., Ebeling, M. W., & Michler-Cieluch, T. (2010). Mussel cultivation as a co-use in offshore wind farms: Potential and economic feasibility. *Aquaculture Economics & Management, 14*(4): 255–281.

Burak Cakaloz, A. (2011). Fish cage construction. Presentation at the FAO Regional Training on the Principles of Cage Culture in Reservoirs. Issyk-Kul, Kyrgyzstan, 22–24 June 2011. Retrieved July 27, 2016 from http://www.fao.org/fileadmin/templates/SEC/docs/Fishery/Fisheries_Events_2012/Principles_of_cage_culture_in_reservoirs/Fish_Cage_Construction.pdf.

Buzovkina, T. B., Aleksandrov, V. A., & Shlyaga, L. I. (1992). Influence of the initial fouling on the marine corrosion of steel. *3*, 501–503.

Callow, J. A., & Callow, M. E. (2011). Trends in the development of environmentally friendly fouling resistant marine coatings. *Nature Communications,* doi:10.1038/ncomms1251. www.nature.com/naturecommunications.

Dean, R. J., Shimmield, T. M., & Black, K. D. (2007). Copper, zinc and cadmium in marine cage fish farm sediments: An extensive survey. *Environmental Pollution, 145,* 84–95.

Janssen, M. M. H. H., & van der Putten, S. (2013). *Mechanical risks involved with aqua farming on offshore wind farm sites.* TNO-MEM-2013–0100000996.

Kalantzi, I., Shimmield, T. M., Pergantis, S. A., Papageorgiou, N., Black, K. D., & Karakassis, I. (2013). Heavy metals, trace elements and sediment geochemistry at four mediterranean fish farms. *Science of the Total Environment, 444,* 128–137.

Kawahara, N., Shigematsu, K., Miura, S., Miyadai, T., & Kondo, R. (2008). Distribution of sulfate-reducing bacteria in fish farm sediments on the coast of southern fukui prefecture, Japan. *Plankton and Benthos Research, 3,* 42–45.

Lagerveld, S., Röckmann, C., & Scholl, M. (2014). *A study on the combination of offshore wind energy with offshore aquaculture.* IMARES Report C056/14. Retrieved October 12, 2015, from http://edepot.wur.nl/318329

Loucks, R. H., Smith, R. E., Fisher, C. V., & Fisher, E. B. (2012). Copper in the sediment and sea surface microlayer near a fallowed, open-net fish farm. *Marine Pollution Bulletin, 64,* 1970–1973.

Mantzavrakos, E., Kornaros, M., Lyberatos, G., & Kaspiris, P. (2007). Impacts of a marine fish farm in Argolikos Gulf (Greece) on the water column and the sediment. *Desalination, 210,* 110–124.

Msuya, F. E., & Neori, A. (2008). Effect of water aeration and nutrient load level on biomass yield, N-uptake and protein content of the seaweed *Ulva lactuca* cultured in seawater tanks. *Journal of Applied Phycology, 20,* 1021–1031.

Noël, N. (2015). *Microbial influenced corrosion (MIC): Assessing and reducing the risk.* Invited presentation at 2nd Annual Integrity and Corrosion of Offshore Wind Structures Forum, June 1–3, London, UK.

Nordvarg, L., & Johansson, T. (2002). The effects of fish farm effluents on the water quality in the Aaland Archipelago, Baltic Sea. *Aquacultural Engineering, 25,* 253–279.

Robertson-Andersson, D. V., Potgieter, M., Hansen, J., Bolton, J. J., Troell, M., Anderson, R. J., et al. (2008). Integrated seaweed cultivation on an abalone farm in South Africa. *Journal of Applied Phycology, 20,* 579–595.

Valdemarsen, T., Bannister, R. J., Hansen, P. K., Holmer, M., & Ervik, A. (2012). Biogeochemical malfunctioning in sediments beneath a deep-water fish farm. *Environmental Pollution, 170,* 15–25.

Van der Putten, S. (2015). *Joint Industry project Felosefi. Improved fatigue crack growth models taking into account load sequence effects*. Presentation at 2nd Annual Integrity and Corrosion of Offshore Wind Structures Forum, June 1–3, London, UK.

Part II
Aquaculture Governance

Chapter 6
Aquaculture Site-Selection and Marine Spatial Planning: The Roles of GIS-Based Tools and Models

Vanessa Stelzenmüller, A. Gimpel, M. Gopnik and K. Gee

Abstract Around the globe, increasing human activities in coastal and offshore waters have created complex conflicts between different sectors competing for space and between the use and conservation of ocean resources. Like other users, aquaculture proponents evaluate potential offshore sites based primarily on their biological suitability, technical feasibility, and cost considerations. Recently, Marine Spatial Planning (MSP) has been promoted as an approach for achieving more ecosystem-based marine management, with a focus on balancing multiple management objectives in a holistic way. Both industry-specific and multiple-use planners all rely heavily on spatially-referenced data, Geographic Information System (GIS)-based analytical tools, and Decision Support Systems (DSS) to explore a range of options and assess their costs and benefits. Although ecological factors can currently be assessed fairly comprehensively, better tools are needed to evaluate and incorporate the economic and social considerations that will also be critical to identifying potential sites and achieving successful marine plans. This section highlights the advances in GIS-based DSS in relation to their capability for aquaculture site selection and their integration into multiple-use MSP. A special case of multiple-use planning—the potential co-location of offshore wind energy and aquaculture—is also discussed, including an example in the German EEZ of the North Sea.

V. Stelzenmüller (✉) · A. Gimpel
Thünen-Institute of Sea Fisheries, Palmaille 9, 22767 Hamburg, Germany
e-mail: vanessa.stelzenmueller@thuenen.de

M. Gopnik
Independent Consultant, Washington DC, USA

K. Gee
Helmholtz Zentrum Geesthacht, Max-Planck-Str. 1, 21502 Geesthacht, Germany

6.1 Reconciling Ocean Uses Through Marine Spatial Planning

The last decades have witnessed an unprecedented race between different sectors for access to the sea. With interests such as offshore renewable energy, sand and gravel extraction, national security, fishing, and nature conservation all pushing for more space, exclusive uses of marine areas are being replaced by a search for more integrated solutions (Halpern et al. 2008b; Katsanevakis et al. 2011). Marine Spatial Planning (MSP) has been widely advocated as one such place-based, integrated tool for managing human activities in the marine environment (Douvere 2008; Douvere and Ehler 2010; Collie et al. 2013).

A key challenge for MSP is to make spatial choices that strike a balance between multiple ecological, economic and social objectives, typically identified through a political process (Katsanevakis et al. 2011; Jay et al. 2012; Carneiro 2013; Foley et al. 2013). Regardless of the governance framework present, or specific process selected, sustainable spatial planning should account for the cumulative effects of all human activities on the marine environment at meaningful ecological scales (Halpern et al. 2008a; Stelzenmüller et al. 2010). Tradeoff analyses using, for instance, explicit weighting criteria can improve transparency in decision-making and should form a crucial part of any MSP process (White et al. 2012; Stelzenmüller et al. 2014). These analyses should focus not only on economic and ecological values, but also on social and cultural values associated with coastal communities and the sea, many of which are extremely difficult to measure (Gee and Burkhard 2010).

A recent EU Directive (EPC 2014a); (Article 3) describes MSP as a cross-cutting policy tool, enabling public authorities and stakeholders to apply a coordinated, integrated, and trans-boundary approach "to promote sustainable development and to identify the utilization of maritime space for different sea uses as well as to manage spatial uses and conflicts in marine areas." The Directive specifically encourages nations to explore multi-purpose uses in accordance with relevant national policies and legislation, and encourages Member States to cooperate in the sustainable development of offshore energy, maritime transport, fisheries, and aquaculture. Nevertheless, existing MSP initiatives show that spatial planning remains open to very diverse interpretations.

Although they do consider multiple maritime uses, early marine spatial plans within the EU, such as the German plan for the EEZ, were often driven by specific sectoral needs (Halpern et al. 2012; Collie et al. 2013; Olsen et al. 2014), often reflecting changing political priorities, shifting prioritization among sectors, or technological advances. This has been particularly apparent with regard to the desired expansion of offshore renewable energy (Gimpel et al. 2013; Christie et al. 2014; Davies et al. 2014). Different ideas about how to *implement* MSP have also emerged. Some marine spatial plans (e.g. in the UK) favor a broad, strategic approach that sets out general guidelines for the use of sea areas, while others (e.g. in Germany) are based on more detailed zoning, creating areas that favor a particular use and other areas where certain uses are prohibited (Schultz-Zehden and Gee 2013, Jay and Gee 2014). As MSP develops from isolated initiatives and projects into statutory plans, broad strategic planning and the concept of co-location or multiple-uses of marine offshore areas are set to become more and more significant (Buck et al. 2004).

6.2 Potential Benefits of MSP to Aquaculture

Effective spatial management is being recognized as one avenue for advancing sustainable aquaculture development worldwide. Europe, keen to encourage growth in the aquaculture sector, has published Strategic Guidelines that identify improved access to space as one of four priority areas to be tackled (EPC 2014b). In the Baltic region, even countries without existing aquaculture facilities are expected to consider future operations as part of their emerging marine spatial plans (project 2013). The Finnish regional fisheries administration prepared regional aquaculture site selection plans that identify offshore areas where existing production can be concentrated and new production begun, using a participative process and Geographic Information System (GIS) mapping as a supporting tool (Olofsson and Andersson 2014).

Although aquaculture is usually mentioned in reports on Integrated Coastal Zone Management (ICZM) and MSP, they rarely focus specifically on aquaculture siting because of their multiple use orientation (Olofsson and Andersson 2014). The English East Inshore and East Offshore Marine Plans (Government 2014) do identify a range of "optimum sites of aquaculture potential" where other uses would be restricted to preserve the potential for aquaculture development. In Germany, the 2014 draft spatial plan for Mecklenburg-Western Pomerania also calls for "spatially compatible" siting of aquaculture operations to minimize environmental impacts (MEIL; Ministerium für Energie 2014).

After initial hesitation in many countries, fisheries and aquaculture stakeholders are now becoming actively engaged in MSP to secure the most suitable sites for their activities (Jentoft and Knol 2014). MSP is also seen as a possible means of resolving animal welfare issues (e.g. assessing maximum carrying capacities) which can help improve public acceptance of the sector (Bryde 2011). These considerations apply not only to existing types of aquaculture operations, but also to future trends such as large offshore installations, potentially combined with offshore wind farming, and specialized production, such as sturgeon, feed production, nutrient removal, and energy production from micro and macro algae (Wenblad 2014). While it could be argued that other place-based approaches to aquaculture siting and management might deliver similar benefits, statutory MSP brings certain strategic advantages.

To start, MSP brings a more coordinated approach to overall sea use, promising greater accountability and transparency of decision-making by including a wide range of stakeholders from all sectors. It may also increase the effectiveness of investments, reduce duplication of effort, and speed up decision-making

(FAO 2013). For example, designating appropriate aquaculture areas and then linking these areas to streamlined licensing procedures could render development less uncertain and increase investor interest (EC 2013). As a strategic tool, MSP can allocate space for aquaculture at sites with both favorable operational characteristics (economic and ecological) as well as lower potential for conflict with other sectors (FAO 2013). MSP would also allow for more structured consideration of co-location of different uses, such as aquaculture taking place around offshore wind structures, providing both a venue for the respective stakeholders to come together and a greater incentive for investment. Hence, the most important reason for aquaculture proponents to engage fully in MSP may be its emphasis on cross-sectoral dialogue and conflict resolution. A well-run MSP process can turn aquaculture from a relatively minor player in a very large debate to an equal participant at the table, able to explain and advocate for its requirements for space at sea (project 2013). The value of open, fair dialogue is particularly relevant in interactions with the environmental sector, but also in considering other uses that might restrict or conflict with aquaculture operations. A 2006 report that examined the suitability of co-locating aquaculture and offshore wind farms in the UK found that the offshore wind energy sector would resist such efforts and concluded that MSP, accompanied by semi-commercial trials, was the only viable way forward for this type of co-use in the UK (Mee 2006).

6.3   Decision Support Systems for MSP and Aquaculture Siting

The EU MSP Directive stipulates that maritime spatial plans should be based on reliable data and encourages Member States to share information and make use of existing instruments and tools for data collection (EPC 2014a); (Article 19). Given the spatial context of MSP, applications to scale economic, environmental, and social dimensions geographically are in high demand (Kapetsky et al. 2013). Spatial data are commonly handled in GIS that make it possible to translate many work-flows into a connected series of process steps (Stelzenmüller et al. 2012). Thus, from a practical perspective, sustainable MSP requires not only spatially explicit information about suitable areas but also a sound spatial assessment of the overlap of human activities (Stelzenmüller et al. 2012) and their combined impact on the marine environment (Kelly et al. 2014). Even more challenging, the identification of a suitable site for a given use does not just depend on physical, chemical and biological factors, but also on political, economic, and social criteria (Wever et al. 2015).

As a result of these challenges, flexible Decision Support Systems (DSS) that are able to consider complex interactions in a unique analytical framework are critical. DSS can be distinguished based on their relative focus on data, models, knowledge, or communication (Power 2003). Current DSS can range from simple spreadsheet models to complex software packages (Bagstad et al. 2013). One example of a DSS for use in MSP is MIMES (Multi-scale Integrated Models of Ecosystem Services), which uses GIS data to simulate ecosystem components under different scenarios defined by stakeholder input. It features a suite of models to support MSP decision making (www.afordablefutures.com/services/mimes). Further, the MMC (Multipurpose Marine Cadastre) is an integrated, online marine information system for viewing and accessing authoritative legal, physical, ecological, and cultural information in a common GIS framework (www.marinecadastre.gov). Another example is MaRS (Marine Resource System), which is a GIS-based DSS designed to enhance marine resource analysis and ultimately identify areas with potential for development in UK waters by The Crown Estate (www.thecrownestate.co.uk/mars-portal-notice). It assists in identifying areas of opportunity and constraint, by identifying how different activities would interact in a particular area and providing statistics showing the value of the area to a competing industry.

6.3.1 The Importance of Spatial Data in the Planning Process

In general, GIS-based data and robust spatial analyses help collate and harmonize data for use at different stages of the planning process (Jay and Gee 2014; Shucksmith et al. 2014), including scoping, development, and evaluation of planning options (Stelzenmüller et al. 2010, 2013a). The use of GIS to support aquaculture development and planning has a long tradition (Kapetsky et al. 1990) and the identification of suitable sites for aquaculture has been among the most frequent applications of GIS (Fisher and Rahel 2004). In recent years, the use and relevance of spatial data in supporting informed decision making in MSP has been increasingly demonstrated (Caldow et al. 2015). For instance, the development of GIS data layers to inform MSP includes: the mapping of sensitivity of seabirds to offshore wind farms (Bradbury et al. 2014); the assessment of potential whale interactions with shipping (Petruny et al. 2014); the mapping of offshore (Campbell et al. 2014) and inshore (Breen et al. 2015) fishing activity; or the mapping of ethnographic information (Sullivan et al. 2015). Decision makers and planners are most likely to require spatial data layers at the development stage of a MSP process, enabling them to explore the data, and develop and evaluate planning scenarios (Stelzenmüller et al. 2012).

6.3.2 DSS for Aquaculture Siting

The combination of DSS into one GIS-based system that can support sustainable aquaculture development has been identified as an important future need (Ferreira et al. 2012; Filgueira et al. 2014). Suitability modeling refers to the spatial overlay of geo-data layers to identify suitable aquaculture sites by identifying, for instance, favorable environmental factors or constraints. Such studies can determine the suitability for aquaculture development at various intensities (Longdill et al. 2008) or can distinguish suitability by type of aquaculture cage (Falconer et al. 2013).

GIS-based suitability modeling is one of the most frequent DSS applications used to evaluate potential aquaculture sites. The first applications of these techniques date back to 1985 when the siting of aquaculture and inland fisheries using GIS and remote sensing was conducted by the FAO. Until the mid-1990s, most studies continued to target data-rich, small-scale environments (Gifford et al. 2001). Early studies looked primarily at coastal or land-based aquaculture related to oysters (shellfish) and shrimps and by using simple siting models (Fisher and Rahel 2004). The main drawback of applying suitability models in offshore environments was a lack of fine-scale data with the necessary temporal and spatial resolution (Fisher and Rahel 2004). Since GIS applications and models have become significantly more complex, and the resolution and quality of data has greatly improved (Fisher and Rahel 2004), one might expect the focus to have shifted to offshore areas. However, as yet, the majority of GIS-based site selection efforts are still focused on shrimp aquaculture in coastal areas around Asia, while studies in offshore environments remain rare.

Once spatially resolved data are available, a GIS-based Multi-Criteria Evaluation (GIS-MCE)—also referred to as area weighted rating (Malczewski 2006)—can be used as a flexible and transparent DSS for potential aquaculture siting. Applications of GIS-MCE in offshore areas were undertaken in a study by Perez et al. (2005) in which suitable sites were modelled for offshore floating marine fish cage aquaculture in Tenerife, Canary Islands. The untapped potential of offshore mariculture is addressed in a global assessment wherein all spatial analyses of suitability and constraints were conducted with the help of GIS (Kapetsky et al. 2013).

Recently, the combination of GIS and dynamic models to identify suitable sites, as done by Silva et al. (2011) for shellfish aquaculture, is becoming more popular. Superimposed models such as FARM (Farm Aquaculture Resource Management; www.farmscale.org) aim to support the siting process with detailed analyses of production, socio-economic outputs, and environmental effects (Silva et al. 2011). In Nunes et al. (2011), the implementation of an ecosystem approach to aquaculture has been advanced by testing various complementary analytical tools. The tools were used to assess multiple aspects of blue mussel cultivation in Killary Harbour, Ireland at different spatial scales (farm- to system-level), times (seasonal to annual to long-term analyses) and levels of complexity (from simple to complex process-based modelling). The selected tools included a system-scale, process-based ecological model (EcoWin 2000; www.ecowin2000.com), a local-scale carrying capacity and environmental effects model (FARM), and a management level eutrophication screening model (ASSETS). Further examples of combining different ecosystem tools for decision support for aquaculture is presented in Filgueira et al. (2014).

In terms of advanced, web-based interactive DSSs, AkvaVis includes site selection, carrying capacity, and management monitoring modules (Ervik et al. 2008, 2011). With suitable adaptations, it appears to be a promising tool for estimating offshore aquaculture potential at national levels and for managing its subsequent development. Filgueira et al. (2014) proposed dynamic, fully-spatial modeling, scenario-building, and optimization tools such as PEST (model-independent Parameter ESTimation, www.pesthomepage.org) as an ideal combination of tools for effective MSP. In the context of MSP decision support, other tools such as MaxEnt (Maximum Entropy modelling) or MARXAN have been utilized in combination with GIS to identify trends, opportunities and concerns related to sustainable management and farm locations (http://dspace.stir.ac.uk/handle/1893/19465) or to identify fisheries areas (Schmiedel and Lamp 2012). ARIES (ARtificial Intelligence for Ecosystem Services) assists in mapping service flows of the ecosystem such as aquaculture benefits (ariesonline.org/docs/ARIESModelingGuide1.0.pdf) and InVEST (Integrated Valuation of Ecosystem Services and Tradeoffs) enables the user to evaluate how aquaculture can affect production and value of marine ecosystem services (www.naturalcapitalproject.org/models/models.html).

As described in Ferreira et al. (2012), the data requirements for DSS expand with the scale of the aquaculture operation. Thus, it will be challenging to use aquaculture-specific DSS in a broader spatial planning system, such as MSP or ICZM, where a large ecosystem scale is required. Indeed, as might be expected, and as articulated by Gifford et al. (2001), there are no "ideal" sites for aquaculture and compromise will always be required. Fortunately, a range of GIS-based DSS exist already to help to find this compromise and to support MSP using transparent data management and advanced visualization.

Several tools have been developed to help assess conflicts and synergies between fisheries, aquaculture, and other marine sectors and to advance practical applications based on that knowledge (Stelzenmüller et al. 2013b). Given the multiple-use context of many sea areas, the identification of suitable sites for aquaculture will depend on location-specific understanding of conflicts and synergies between various proposed types of sea use. While conflicts should be minimized, the discovery of synergies can help identify areas suitable for co-location (Stelzenmüller et al. 2013b; Griffin et al. 2015).

6.4 The Co-location Scenario: Combining Offshore Wind Energy and Aquaculture

Both English and German marine plans encourage the combination of aquaculture with other uses. In the UK, a strong national policy statement calls for consideration of the "significant opportunities for co-existence of aquaculture and other marine activities" (Government 2011, 3.9.6). The UK's East Inshore and Offshore Marine

Plans also stipulate that co-location opportunities should be maximized wherever possible, and that "proposals for using marine areas should demonstrate the extent to which they will co-exist with other existing or authorized activities and how this will be achieved" (Government 2014, p. 106). Identifying opportunities for, and the technical feasibility of, co-location becomes all the more important for supporting decision-making (Christie et al. 2014; Hooper and Austen 2014).

6.4.1 Co-location as an Opportunity for Spatial Planning?

The co-location of offshore infrastructure and aquaculture has been a particular focus of research (Buck et al. 2004; Lacroix and Pioch 2011; Wever et al. 2015), with "infrastructure" typically referring to offshore wind energy facilities. During the past ten years there has been growing interest among policy makers, scientists, the aquaculture industry, and other stakeholders in implementing pilot studies to demonstrate the feasibility of such co-location. In the southern North Sea and German Bight, the potential co-location of offshore wind and aquaculture has gained momentum due to the allocation of large areas for offshore wind, including approximately 35% of the German EEZ of the North Sea, and the resulting loss of space for other sectors, such as fisheries (Stelzenmüller et al. 2014).

Based on an extensive stakeholder consultation process, Wever et al. (2015) identified future research needs to support implementation of the co-location concept. One of these needs was the development of site-selection criteria that include environmental, economic, socioeconomic, and technological parameters. A recent study by Benassai et al. (2014) used a GIS-MCE DSS to evaluate suitable areas for the co-location of offshore wind and aquaculture at a large spatial scale, using only environmental criteria. At a much finer resolution, Gimpel et al. (2015) assessed the potential for coupling offshore aquaculture and wind farms in the German EEZ of the North Sea based on environmental and infrastructure criteria. In the following section we provide a brief summary of the methods, key criteria, and results of this case study.

6.4.2 Case Study in the German Bight

In order to evaluate different spatial co-location scenarios for the coupling of offshore Integrated Multi-Trophic Aquaculture (IMTA) systems and wind farms, possible aquaculture candidates (seaweed, bivalves, fish and crustaceans) were identified. Those have been selected accounting for their native occurrence in the German North Sea, their resistance to hydrodynamic conditions in offshore environments as well as their economic potential for the EU market. The study area comprised the German EEZ of the North Sea with a surface area of 28.539 km^2 (Fig. 6.1). A vector grid was superimposed to the study area with a grid size

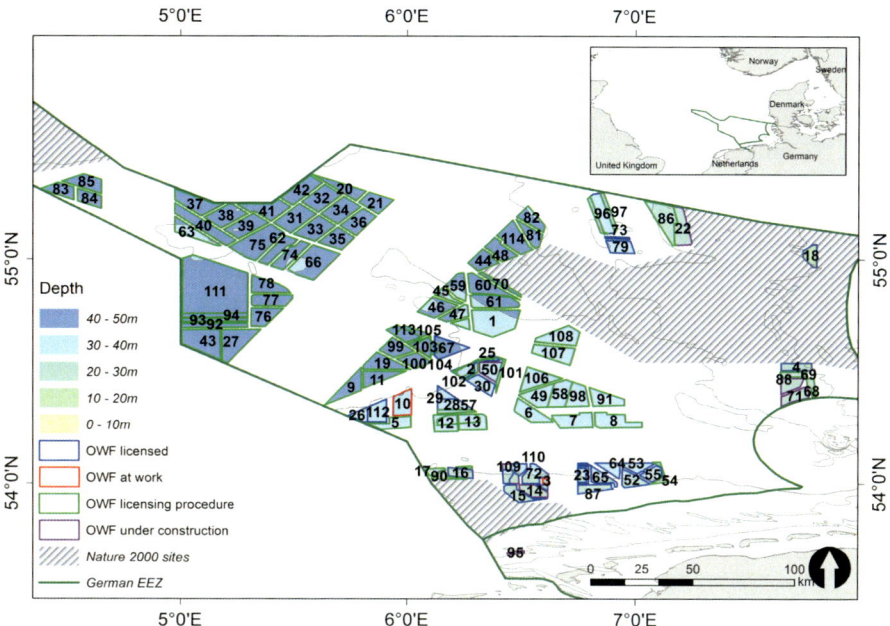

Fig. 6.1 Map of Offshore Wind Farm (OWF) areas in the German EEZ of the North Sea, numbered, coloured per depth level and framed per status. *Shaded* districts show the Nature 2000 areas. Note that depth, the OWF areas (effective from December 2013; BSH) and the Nature 2000 sites constituted a physical constraint applied, limiting suitable sites for co-use with aquaculture. OWF 18, 80 and 95 have not been considered during this study, as they appear within the 12 nm zone or in Nature 2000 sites (redrawn from Gimpel et al. 2015)

resolution of 9.26 km². A GIS-MCE technique was applied to index suitable co-sites. In order to provide all criteria needed (Fig. 6.2), hydrographic data were extracted, analysed and interpolated to derive depth stratified mean values per quarter of the year. Further all data were standardised using fuzzy membership functions with control points to guarantee comparability among factors, whereby the choice of function and control points was based on expert knowledge and literature research. With the pairwise comparison method of the Analytical Hierarchy Process (AHP) all factors were weighted by priority for all grid cells. Also a range of weighting designs was modelled using an Ordered Weighted Average (OWA) approach to address the uncertainty in prediction results. If one grid cell appeared to be unsuitable during OWA weighting, it had been excluded from further assessments. The final weighting of the factors was based on expert judgement and focused on the optimal growth under farmed conditions. Using this weighting scheme the GIS-MCE resulted in a series of geo-referenced aquaculture suitability layers comprising the whole German EEZ of the North Sea. In a next step, an offshore co-location suitability index was developed by accounting for overlaps between the aquaculture sites and referenced offshore wind farms provided

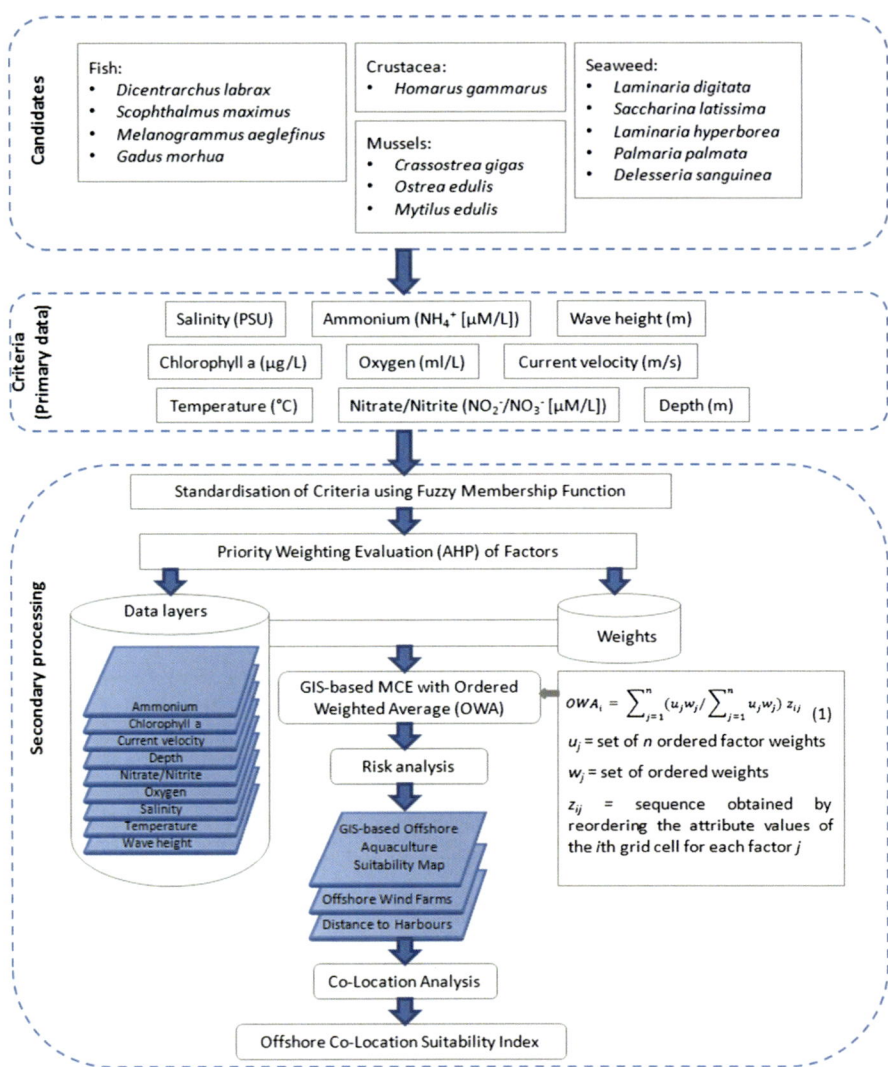

Fig. 6.2 Overall methodological approach used to index potential co-use locations of offshore wind farms in combination with offshore aquaculture, redrawn from (Ouma and Tateishi 2014) (taken from Gimpel et al. 2015)

by the Federal Maritime and Hydrographic Agency (BSH), excluding the wind farms located in existing nature conservation sites or within the German territorial waters (18, 56, 82 and 95). Further, the co-location sites were examined concerning their suitability for IMTA techniques. The overall methodological approach is shown in Fig. 6.2.

While the conditions for fish proved to be highly suitable during summer, the mussels and algae revealed peak suitability in spring. Still, when examining suitable sites at 10–20 m depth for spring, haddock (*Melanogrammus aeglefinus*) showed highest suitability of all aquaculture candidates closest to the coast (Fig. 6.3). Though fish can be cultured offshore the whole year around, but they require a high degree of care (feeding, clearing of cages etc.). Therefore, due to logistic constraints a cultivation approach closer to the coast is preferred. In contrast, oarweed (*Laminaria digitata*) presented the highest suitability scores at wind farm areas located further offshore. When seaweed is seeded elaborately on the rope and transferred at sea at a juvenile stage, holdfasts will not be dislodged and cauloids will not break leading to a resistance to harsh conditions. As they require a very low level of maintenance, a cultivation approach offshore is forthcoming. In general, if seaweed is part of an IMTA approach and also a candidate to be sold on the EU market, it has to be harvested latest by the end of spring. If the seaweed is cultivated within a bioremediation concept and is only used to extract nutrients from the water column, it can be on-site year around.

Results showed several wind farms were de facto suitable sites for IMTA systems combining fish species, bivalves and seaweeds. As Laminaria species (*L. digitata, Laminaria hyperborea*) cultured near fish farms bring along better growth rates, a candidate set of Oarweed (*L. digitata*), Pacific oyster (*Crassostrea gigas*)

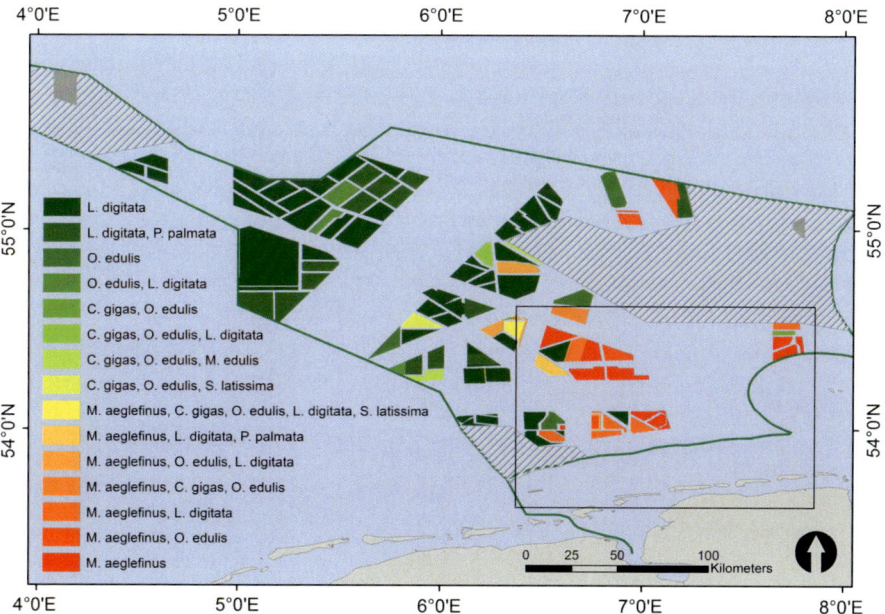

Fig. 6.3 Map of suitable co-location sites in the German EEZ of the North Sea, colored per aquaculture candidate featuring the highest suitability per wind farm area. Results are shown for spring at 10–20 m depth with Nature 2000 areas indicated as shaded areas

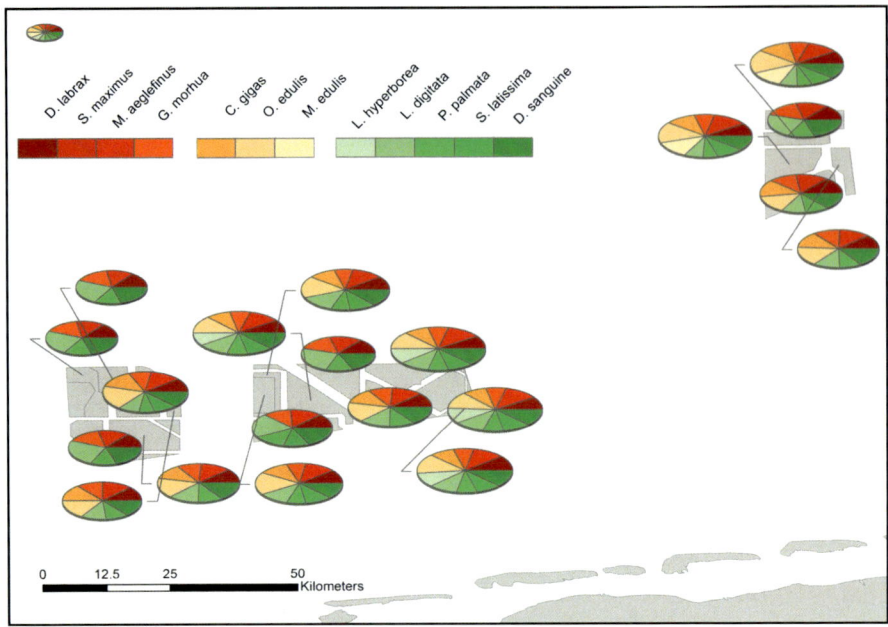

Fig. 6.4 Co-location sites for aquaculture candidates, suitable for possible IMTAs close to the German coast (North Sea) are shown as an example. Results presented depict the conditions given at 10–20 m depth during spring time. The size of the pies reflects the height of the relative suitability scores

and *M. aeglefinus* could be of interest regarding IMTA, especially if cultured in spring near the coast (Fig. 6.4). If it is about an IMTA candidate set which could be on site year around, Atlantic cod (*Gadus morhua*), blue mussel (*Mytilus edulis*) and sea beech (*Delessaria sanguine*) can be mentioned. Though, the here presented suitability for co-locations does not account for economic viability analyses for the respective candidates. Nevertheless, the case study example illustrated how competing needs might be balanced by strategic planning for the needs of sectors, offshore wind energy and offshore IMTA. This might offer guidance to stakeholders and assist decision-makers in determining the most suitable sites for pilot projects using IMTA techniques.

6.5   Conclusions and Future Needs

As highlighted by the here presented examples, some form of strategic spatial planning will be critical in advancing sustainable offshore aquaculture. Olofsson and Andersson (2014) describe a successful planning process conducted in the Baltic Sea Region to site sustainable aquaculture farms for finfish and mussels.

GIS tools were used to evaluate geographical data and identify suitable areas, fish carrying capacity was calculated, mussel settling in the selected areas was estimated, and a public consultation process was carried out in both regions. Hence, such a spatial planning process will be crucial for aquaculture development as it lowers the threshold for new entrepreneurs, minimizes the risk of appeals, makes the business more environmentally safe, and lowers the risk of social conflicts.

But, increasingly, aquaculture siting will be conducted in a broader, multiple-use context where tradeoffs will be more complex. As stated by Lovatelli et al. (2014), "meeting the future demand for food from aquaculture will largely depend on the availability of space [and] 'MSP' is needed to ensure [that] allocation of space." Although a variety of GIS-based tools and DSS are currently available to assist in the planning process, the allocation of space in the ocean remains a complex, contentious process unlikely to be fully resolved by even the most sophisticated mathematical calculations. Successful MSP, including the co-location of compatible activities, relies as much on the willingness of relevant stakeholders to become involved as it does on tools and techniques for identifying optimum areas (Gopnik 2015). The integration of relevant actors is a "complex and controversial issue" (Buck et al. 2008) which depends on a multitude of factors, including inclusiveness, transparency of the process and of decision-making, timing, credibility of the data and science, and impartial mediation.

Despite these caveats, GIS-based DSS will continue to play an important role in planning and spatial decision-making because of their ability to evaluate the results of many different spatial scenarios. Ideally, this should include assessments of the economic and socio-cultural impacts of different siting decisions, which can be the main sources of conflict and are too often overlooked (ICES 2013). Socio-economic data integrate publicly-held values into the decision-making processes. Primary data on socio-cultural values—such as the importance people give to cultural identity and the degree to which that is related to the ecosystem (de Groot et al. 2010)—are usually not available. Surveys on secondary data as well as their spatial analysis still remain complex tasks. In general, the spatial aggregation of socio-economic data in a GIS framework is difficult, involving close collaboration with the respective sectors (Ban et al. 2013). Most progress can be found with regard to the mapping of fleet-specific fisheries activities due to technical advances in combining Vessel Monitoring System (VMS) and logbook information (Bastardie et al. 2010; Lee et al. 2010; Hintzen et al. 2012). Finally, GIS-based DSS should be flexible enough to respond to shifting circumstances, such as changes in environmental conditions, environmental targets, growth expectations in the aquaculture sector, or policy environments. Like all analytical tools, GIS-based DSS are only as good as the quality and thoroughness of the data they are based on, and their strengths and limitations should be clearly explained to stakeholders during the planning process.

Acknowledgements The German Federal Office for Agriculture and Food (Bundesanstalt für Landwirtschaft und Ernährung, BLE) supported the contribution of AG as part of the project Offshore Site Selection (OSS) (313-06.01-28-1-73.010-10). Case study data were freely provided by National Oceanographic and Atmospheric Administration (NOAA), Helmholtz-Zentrum

Geesthacht—Centre for Materials and Coastal Research (HZG) and the German Federal Maritime and Hydrographic Agency (Bundesamt für Seeschifffahrt und Hydrographie, BSH), Hamburg (Germany) in raw, uninterpreted form.

References

Bagstad, K. J., Semmens, D. J., Waage, S., & Winthrop, R. (2013). A comparative assessment of decision-support tools for ecosystem services quantification and valuation. *Ecosystem Services, 5*, 27–39.

Ban, N. C., Bodtker, K. M., Nicolson, D., Robb, C. K., Royle, K., & Short, C. (2013). Setting the stage for marine spatial planning: Ecological and social data collation and analyses in Canada's Pacific waters. *Marine Policy, 39*, 11–20.

Bastardie, F., Nielsen, J. R., Ulrich, C., Egekvist, J., & Degel, H. (2010). Detailed mapping of fishing effort and landings by coupling fishing logbooks with satellite-recorded vessel geo-location. *Fisheries Research, 106*, 41–53.

Benassai, G., Mariani, P., Stenberg, C., & Christoffersen, M. (2014). A sustainability index of potential co-location of offshore wind farms and open water aquaculture. *Ocean and Coastal Management, 95*, 213–218.

Bradbury, G., Trinder, M., Furness, B., Banks, A. N., Caldow, R. W. G., & Hume, D. (2014). Mapping seabird sensitivity to offshore wind farms. *PLoS ONE, 9*(9), e106366. DOI 10.1371/journal.pone.0106366

Breen, P., Vanstaen, K., & Clark, R. W. E. (2015). Mapping inshore fishing activity using aerial, land, and vessel-based sighting information. *ICES Journal of Marine Science, 72*, 467–479.

Bryde, M. (2011). Marine spatial planning and aquaculture—A norwegian perspective. *Presentation at the conference on competitive and sustainable aquaculture*, October 4–5, 2011, Helsinki. Retrieved September 4, 2015 from http://www.rktl.fi/www/uploads/pdf/Aquaculture%20Forum/theme_1_3_sustainability_bryde.pdf

Buck, B. H., Krause, G., Michler-Cieluch, T., Brenner, M., Buchholz, C. M., Busch, J. A., et al. (2008). Meeting the quest for spatial efficiency: progress and prospects of extensive aquaculture within offshore wind farms. *Helgoland Marine Research, 62*, 269–281.

Buck, B. H., Krause, G., & Rosenthal, H. (2004). Extensive open ocean aquaculture development within wind farms in Germany: The prospect of offshore co-management and legal constraints. *Ocean and Coastal Management, 47*, 95–122.

Caldow, C., Monaco, M. E., Pittman, S. J., Kendall, M. S., Goedeke, T. L., Menza, C., et al. (2015). Biogeographic assessments: A framework for information synthesis in marine spatial planning. *Marine Policy, 51*, 423–432.

Campbell, M. S., Stehfest, K. M., Votier, S. C., & Hall-Spencer, J. M. (2014). Mapping fisheries for marine spatial planning: Gear-specific vessel monitoring system (VMS), marine conservation and offshore renewable energy. *Marine Policy, 45*, 293–300.

Carneiro, G. (2013). Evaluation of marine spatial planning. *Marine Policy, 37*, 214–229.

Christie, N., Smyth, K., Barnes, R., & Elliott, M. (2014). Co-location of activities and designations: A means of solving or creating problems in marine spatial planning? *Marine Policy, 43*, 254–261.

Collie, J. S., Vic-Adamowicz, W. L., Beck, M. W., Craig, B., Essington, T. E., Fluharty, D., et al. (2013). Marine spatial planning in practice. *Estuarine, Coastal and Shelf Science, 117*, 1–11.

Davies, I. M., Watret, R., & Gubbins, M. (2014). Spatial planning for sustainable marine renewable energy developments in Scotland. *Ocean and Coastal Management, 99*, 72–81.

de Groot, R. S., Alkemade, R., Braat, L., Hein, L., & Willemen, L. (2010). Challenges in integrating the concept of ecosystem services and values in landscape planning, management and decision making. *Ecological Complexity, 7*, 260–272.

Douvere, F. (2008). The importance of marine spatial planning in advancing ecosystem-based sea use management. *Marine Policy, 32,* 762–771.

Douvere, F., & Ehler, C. N. (2010). The importance of monitoring and evaluation in adaptive maritime spatial planning. *Journal of Coast Conservation* (Online First)

EC. (2013). Strategic guidelines for the sustainable development of EU aquaculture—COM/2013/229 communication to the european parliament, the council, the european economic and social committee and the committee of the regions (April 29, 2013).

EPC. (2014a). Directive 2014/89/EU of the European parliament and of the council of July 23, 2014 establishing a framework for maritime spatial planning. *Official Journal of the European Union, L, 257,* 135–145.

EPC (2014b) Aquaculture. October 3, 2015, retrieved from http://ec.europa.eu/fisheries/cfp/aquaculture/index_en.htm

Ervik, A., Agnalt, A. L., Asplin, L., Aure, J., & Bekkvik, T. C. (2008). *AkvaVis—Dynamic GIS-tool for siting of fish farms for new aquaculture species: Environmental quality requirements for new aquaculture species and Atlantic salmon.* Bergen, NO: Fisken og Havet.

Ervik, A., Agnalt, A. L., Asplin, L., Aure, J., Bekkvik, T. C., Døskeland, I., et al. (2011). *AkvaVis decision support system.* Bergen, NO: Fisken og Havet.

Falconer, L., Hunter, D. C., Scott, P. C., Telfer, T. C., & Ross, L. G. (2013). Using physical environmental parameters and cage engineering design within GIS-based site suitability models for marine aquaculture. *Aquaculture Environment Interactions, 4,* 223–237.

FAO. (2013). Applying Spatial Planning for Promoting Future Aquaculture Growth. COFI: AQ/VII/2013/6. Retrieved October 12, 2014, from www.fao.org/cofi/31155–084a1eedfc71131950f896ba1bfb4c302.pdf

Ferreira, J. G., Aguilar-Manjarrez, J., Bacher, C., Black, K., Dong, S.L., Grant, J., et al. (2012). Progressing aquaculture through virtual technology and decision-support tools for novel management. In R. P. Subasinghe, J. R. Arthur, D. M. Bartley, S. S. De Silva, M. Halwart, N. Hishamunda, C.V. Mohan, et al. (Eds.), *Farming the Waters for People and Food. Proceedings of the Global Conference on Aquaculture 2010.* Phuket, Thailan, September 22–25, 2010. FAO, Rome and NACA, Bangkok.

Filgueira, R., Grant, J., & Strand, Ø. (2014). Implementation of marine spatial planning in shellfish aquaculture management: Modeling studies in a Norwegian fjord. *Ecological Applications, 24,* 832–843.

Fisher, W. L., & Rahel, F. J. (2004). Geographic information systems in fisheries. *Journal of Fish Biology, 66,* 290–291.

Foley, M. M., Armsby, M. H., Prahler, E. E., Caldwell, M. R., Erickson, A. L., Kittinger, J. N., et al. (2013). Improving ocean management through the use of ecological principles and integrated ecosystem assessments. *BioScience, 63,* 619–631.

Gee, K., & Burkhard, B. (2010). Cultural ecosystem services in the context of offshore wind farming: A case study from the west coast of Schleswig-Holstein. *Ecological Complexity, 7,* 349–358.

Gifford, J. A., Benetti, D. D., & Rivera, J. A. (2001). *National marine aquaculture initiative: Using GIS for offshore aquaculture siting in the U.S. Caribbean and Florida.* Final Report NOAA-National Sea Grant, Rosenstiel School of Marine and Atmospheric Science, University of Miami.

Gimpel, A., Stelzenmüller, V., Cormier, R., Floeter, J., & Temming, A. (2013). A spatially explicit risk approach to support marine spatial planning in the German EEZ. *Marine Environmental Research, 86,* 56–69.

Gimpel, A., Stelzenmüller, V., Grote, B., Buck, B. H., Floeter, J., Núñez-Riboni, I., et al. (2015). A GIS modelling framework to evaluate marine spatial planning scenarios: Co-location of offshore wind farms and aquaculture in the German EEZ. *Marine Policy, 55,* 102–115.

Gopnik, M. (2015). *Public lands management and marine spatial planning.* Oxon, UK: Routledge.

Government HM. (2014). *East inshore and east offshore marine plans.* London: Department for Environment, Food and Rural Affairs.

Government HM, Northern Ireland Excecutive, Scottish Government & Welsh Assembly Government. (2011). *UK marine policy statement*. London: HM Government, Northern Ireland Executive, Scottish Government, Welsh Assembly Government.

Griffin, R., Buck, B., & Krause, G. (2015). Private incentives for the emergence of co-production of offshore wind energy and mussel aquaculture. *Aquaculture, 436,* 80–89.

Halpern, B. S., Diamond, J., Gaines, S., Gelcich, S., Gleason, M., Jennings, S., et al. (2012). Near-term priorities for the science, policy and practice of coastal and marine spatial planning (CMSP). *Marine Policy, 36,* 198–205.

Halpern, B. S., McLeod, K. L., Rosenberg, A. A., & Crowder, L. B. (2008a). Managing for cumulative impacts in ecosystem-based management through ocean zoning. *Ocean and Coastal Management, 51,* 203–211.

Halpern, B. S., Walbridge, S., Selkoe, K. A., Kappel, C. V., Micheli, F., D'Agrosa, C., et al. (2008b). A global map of human impact on marine ecosystems. *Science, 319,* 948–952.

Heinimaa, S., & Rahkonen, R. (2012). Conference on competitive and sustainable aquaculture. In *Dimensions and tools of competitive and sustainable aquaculture in northern Europe.* Copenhagen. TemaNord 2012, p. 518. Conference Proceedings.

Hintzen, N. T., Bastardie, F., Beare, D., Piet, G. J., Ulrich, C., Deporte, N., et al. (2012). VMS tools: Open-source software for the processing, analysis and visualisation of fisheries logbook and VMS data. *Fisheries Research, 115–116,* 31–43.

Hooper, T., & Austen, M. (2014). The co-location of offshore windfarms and decapod fisheries in the UK: Constraints and opportunities. *Marine Policy, 43,* 295–300.

ICES. (2013). *Report of the joint HZG/LOICZ/ICES workshop: Mapping cultural dimensions of marine ecosystems (WKCES), 17–21 June 2013, ICES CM 2013/SSGHIE: 07*. Germany: Helmholtz Zentrum Geesthacht.

Jay, S., Ellis, G., & Kidd, S. (2012). Marine spatial planning: A new frontier? *Journal of Environmental Policy & Planning, 14,* 1–5.

Jay, S., & Gee, K. (Eds.) (2014). *TPEA good practice guide: Lessons for cross-border MSP from transboundary planning in the European Atlantic*. Liverpool: University of Liverpool

Jentoft, S., & Knol, M. (2014). Marine spatial planning: risk or opportunity for fisheries in the North Sea? *Maritime Studies, 13,* 1–16.

Kapetsky, J. M., Aguilar-Manjarrez, J., & Jenness, J. (2013). *A global assessment of potential for offshore mariculture development from a spatial perspective*. FAO Fisheries and Aquaculture Technical Paper 549. Rome, Italy: FAO.

Kapetsky, J. M., Hill, J. M., Worthy, L. D., & Evans, D. L. (1990). Assessing potential for aquaculture development with a geographic information system. *Journal of the World Aquaculture Society, 21,* 241–249.

Katsanevakis, S., Stelzenmüller, V., South, A., Sørensen, T. K., Jones, P. J. S., Kerr, S., et al. (2011). Ecosystem-based marine spatial management: Review of concepts, policies, tools, and critical issues. *Ocean and Coastal Management, 54,* 807–820.

Kelly, C., Gray, L., Shucksmith, R. J., & Tweddle, J. F. (2014). Investigating options on how to address cumulative impacts in marine spatial planning. *Ocean and Coastal Management, 102,* 139–148.

Lacroix, D., & Pioch, S. (2011). The multi-use in wind farm projects: More conflicts or a win-win opportunity? *Aquatic Living Resources, 24,* 129–135.

Lee, J., South, A. B., & Jennings, S. (2010). Developing reliable, repeatable, and accessible methods to provide high-resolution estimates of fishing-effort distributions from vessel monitoring system (VMS) data. *ICES Journal of Marine Science, 67,* 1260–1271.

Longdill, P. C., Healy, T. R., & Black, K. P. (2008). An integrated GIS approach for sustainable aquaculture management area site selection. *Ocean and Coastal Management, 51,* 612–624.

Lovatelli, A., Aguilar-Manjarrez, J., & Cardia, F. (2014). FAO supported the 5th offshore mariculture conference 2014. *FAO Aquaculture Newsletter, 52,* 8–9.

Malczewski, J. (2006). GIS-based multicriteria decision analysis: A survey of the literature. *International Journal of Geographical Information Science, 20,* 703–726.

Mee, L. (2006). Complementary benefits of alternative energy: Suitability of offshore wind farms as aquaculture sites. In *Inshore fisheries and aquaculture technology innovation and development, SEAFISH—Project Ref: 10517*. Plymouth, UK: Marine Institute – University of Plymouth.

MEIL (Ministerium für Energie, Infrastruktur und Landesentwicklung M-V). (2014). Entwurf Landesraumentwicklungsprogramm. Retrieved December 2, 2015, from http://www.regierung-mv.de/Landesregierung/em/Raumordnung/Landesraumentwicklungsprogramm/aktuelles-Programm/.

Nunes, J. P., Ferreira, J. G., Bricker, S. B., O'Loan, B., Dabrowski, T., Dallaghan, B., et al. (2011). Towards an ecosystem approach to aquaculture: Assessment of sustainable shellfish cultivation at different scales of space, time and complexity. *Aquaculture, 315,* 369–383.

Olofsson, E., & Andersson, J. (2014). *Spatial planning guidelines for baltic sea region aquaculture*. (AQUABEST report 3/2014). Helsinki: FI: Finnish Game and Fisheries Research Institute

Olsen, E., Fluharty, D., Hoel, A. H., Hostens, K., Maes, F., & Pecceu, E. (2014). Integration at the round table: Marine spatial planning in multi-stakeholder settings. *PLoS ONE. 9*(10), e109964. doi 10.1371/journal.pone.0109964

Ouma, Y., & Tateishi, R. (2014). Urban flood vulnerability and risk mapping using integrated multi-parametric AHP and GIS: Methodological overview and case study assessment. *Water, 6,* 1515–1545.

Perez, O. M., Telfer, T. C., & Ross, L. G. (2005). Geographical information systems-based models for offshore floating marine fish cage aquaculture site selection in Tenerife, Canary Islands. *Aquaculture Research, 36,* 946–961.

Petruny, L. M., Wright, A. J., & Smith, C. E. (2014). Getting it right for the North Atlantic right whale (*Eubalaena glacialis*): A last opportunity for effective marine spatial planning? *Marine Pollution Bulletin, 85,* 24–32.

Power, D. (2003). *Categorizing decision support systems: A multidimensional approach*. Decision Making Support Systems: Achievements, Trends and Challenges for the New Decade: 20–27 project P (2013) Summary report of the workshop on aquaculture/new uses of marine resources. Retrieved January 9, 2015, from http://www.partiseapate.eu/dialogue/aquaculture-new-uses/

Schmiedel, J., & Lamp, J. (2012). *Case study: Site selection of fisheries areas for maritime spatial planning with the help of tool "Marxan with Zone" in the pilot area pomeranian bight*. BaltSeaPlan Report 30, Rostock, January 2012. Retrieved January 12, 2015, from www.baltseaplan.eu

Schultz-Zehden, A., & Gee, K. (2013). *BaltSeaPlan findings—Experiences and lessons from BaltSeaPlan*. Retrieved June 18, 2016, from http://www.baltseaplan.eu/index.php/Reports-and-Publications;809/1

Schultz-Zehden, A., & Matczak, M. (Eds.). (2012). *SUBMARINER compendium. An assessment of innovative and sustainable uses of Baltic marine resources*. Gdansk, PL: Maritime Institute in Gdansk.

Shucksmith, R., Gray, L., Kelly, C., & Tweddle, J. F. (2014). Regional marine spatial planning—The data collection and mapping process. *Marine Policy, 50,* 1–9.

Silva, C., Ferreira, J. G., Bricker, S. B., DelValls, T. A., Martín-Díaz, M. L., & Yáñez, E. (2011). Site selection for shellfish aquaculture by means of GIS and farm-scale models, with an emphasis on data-poor environments. *Aquaculture, 318,* 444–457.

Stelzenmüller, V., Breen, P., Stamford, T., Thomsen, F., Badalamenti, F., Borja, T., et al. (2013a). Monitoring and evaluation of spatially managed areas: A generic framework for implementation of ecosystem based marine management and its application. *Marine Policy, 37,* 149–164.

Stelzenmüller, V., Fock, H. O., Gimpel, A., Rambo, H., DIekmann, R., Probst, W. N., et al. (2014). Quantitative environmental risk assessments in the context of marine spatial management: Current approaches and some perspectives. *ICES Journal of Marine Science*. doi:10.1093/icesjms/fsu206

Stelzenmüller, V., Lee, J., South, A., Foden, J., & Rogers, S. I. (2012). Practical tools to support marine spatial planning: A review and some prototype tools. *Marine Policy, 38,* 214–227.

Stelzenmüller, V., Lee, J., South, A., & Rogers, S. I. (2010). Quantifying cumulative impacts of human pressures on the marine environment: A geospatial modelling framework. *Marine Ecology Progress Series, 398,* 19–32.

Stelzenmüller, V., Schulze, T., Gimpel, A., Bartelings, H., Bello, E., Bergh, Ø … Verner-jeffreys, D. (2013b). *Guidance on a better integration of aquaculture, fisheries and other activities in the coastal zone.* From tools to practical examples. COEXIST project 2013. Printed.

Sullivan, C. M., Conway, F. D. L., Pomeroy, C., Hall-Arber, M., & Wright, D. J. (2015). Combining geographic information systems and ethnography to better understand and plan ocean space use. *Applied Geography.* doi:10.1016/j.apgeog.2014.11.027

Wenblad, A. (2014). *MSP: Principles and future developments in relation to aquaculture.* Presentation given at the PartiSEApate workshop on Aquaculture and new uses of marine resources, April 15–16, 2013, Gdansk. Retrieved January 9, 2015, from http://www.partiseapate.eu/dialogue/aquaculture-new-uses/

Wever, L., Krause, G., & Buck, B. H. (2015). Lessons from stakeholder dialogues on marine aquaculture in offshore wind farms: Perceived potentials, constraints and research gaps. *Marine Policy, 51,* 251–259.

White, C., Halpern, B. S., & Kappel, C. V. (2012). Ecosystem service tradeoff analysis reveals the value of marine spatial planning for multiple ocean uses. *Proceedings of the National Academy of Sciences of the United States of America, 109,* 4696–4701.

Chapter 7
Governance and Offshore Aquaculture in Multi-resource Use Settings

Gesche Krause and Selina M. Stead

Abstract The notion of the sea as a seemingly endless source of resources has long dominated marine governance. This is despite that different perceptions and valuation systems underlie the institutional structures that govern and manage marine systems. Socio-political considerations cover the whole range of stakeholders and their type of involvement in the establishment and operation of multi-use offshore systems. However, within the vast variety of regulations inside the EU, the EU Member States as well as in North America, their implementation for offshore multi-use settings is as yet incipient and examples of best practice in multi-use scenarios are needed. These need to combine different knowledge systems (e.g. authorities, decision-makers, local communities, science, etc.) to generate effective insights into the management of multiple uses of ocean space and to complement risk-justified decision-making. Pre-existing social networks can provide significant political leverage for governance transformations as required for the move offshore. That said, a range of organizational and social challenges related to the collective use of a defined ocean territory have to be taken into account. For instance, the creation and compliance with defined responsibilities and duties or the introduction of cross-sectoral management lines, such as an offshore co-management, that integrates the different demands and practices of the involved parties within an operational scheme that is practical on a day to day manner are in case in point. Indeed, how people perceive and value marine environments and the resources they provide determines individual and collective preferences, actions and strategies in the marine realm. Thus, for the effective

G. Krause (✉)
Alfred Wegener Institute Helmholtz Centre for Polar
and Marine Research (AWI), Earth System Knowledge Platform (ESKP),
Bussestrasse 24, 27570 Bremerhaven, Germany
e-mail: gesche.krause@awi.de

G. Krause
SeaKult—Sustainable Futures in the Marine Realm,
Sandfahrel 12, 27572 Bremerhaven, Germany

S.M. Stead
School of Marine Science and Technology, Newcastle University,
Newcastle NE1 7RU, UK

© The Author(s) 2017
B.H. Buck and R. Langan (eds.), *Aquaculture Perspective of Multi-Use Sites in the Open Ocean*, DOI 10.1007/978-3-319-51159-7_7

implementation of sustainable marine resource management, the public has to be included in the knowledge production in order to understand processes that take place in our economies, environment and societies which in turn will affect the outcomes of management actions. In the following chapter, Marine Spatial Planning (MSP) approaches, linkages between site-selection criteria's, GIS and modelling towards the multi-use of offshore areas to marine governance are discussed in more detail.

7.1 Introduction

Increasing demand by marine resource users for access to offshore marine environments, especially as a potential solution to improve food security, highlights an urgent need for action to ensure sustainable development of offshore areas is supported by effective management and policy. This should be prioritised by maritime nations worldwide since demand for offshore user rights is likely to grow especially when considering the global aquaculture industry is one of the fastest-growing food producing sectors (FAO 2016). As a commercial sector, marine aquaculture is a relatively young sector which has seen rapid changes particularly in technological advances over the last 4–5 decades (Duarte et al. 2007). Public sector policies play a critical role in the development of this sector (Hishamunda et al. 2014). To illustrate, Hishamunda et al. (2014) credit China's expansion of aquaculture and current prominence to specific policies introduced by central government authorities that promote stewardship among those involved in this sector. Aquaculture policies provide a vision and broad qualitative goals that can, if effectively implemented, support growth of the sector and further frame the roles of the key actors that are to be involved in aquaculture development (Brugere et al. 2010). In this respect, corresponding governance systems need to be considered because success or failure of achieving policy goals cannot be judged on the basis of their theoretical or technical attributes without considering the institutional, political and cultural context in which they are applied (Araujo et al. 2004).

Thus, the role of effective governance in supporting sustainable development of marine aquaculture especially offshore is receiving increasing interest. This is especially valid in countries and regions failing to meet sector growth targets through governments prioritising research and development on production and technology, as is the case in Europe (Stead 2005, 2015). With advances in research and development attracting growing interest in offshore aquaculture as a potential viable investment opportunity it is timely to examine the governance and policy needs required to support its growth in a sustainable way. So far, the Atlantic salmon, *Salmo salar* L., is the most intensely farmed commercial finfish in seawater and considered a suitable candidate for offshore aquaculture. Interest in culturing other marine species offshore, such as aquatic plants like seaweed, molluscs and various finfish, is growing, unlike the availability of appropriate inshore coastal space for marine aquaculture. Thus arguably, the offshore aquaculture sector could contribute more significantly to the overall aquaculture sector's growth, offering alternative or supplementary solutions for meeting food security challenges and provide a source of income generation, to mention only a few of the possible benefits not always fully realised.

Globally, it has been well recognized that there is an increased need to conserve ocean ecosystems and to use ocean space as efficiently as possible, thus requiring planning for multiple uses of compatible activities, and the development of strategies to promote, enhance, and optimize the multiple uses in order to protect ocean ecosystems and conserve ocean space (Buck et al. 2004; Mee 2006). That said, all of these activities planned in the offshore realm require strong policy backing with effective governance arrangements and clear multi-use management goals in place. Traditionally, the notion of the sea as an endless source of resources has long dominated marine resource management, especially in the fisheries sector where weak governance is recognised to have contributed to unsustainable exploitation of fishery resources. Historically, an absence of institutions governing the sea was highlighted as early as 1968 by Hardin through his seminal article 'the tragedy of the commons'. Furthermore, in the recent efforts in tailoring contemporary institutional structures that govern and manage marine ecosystems, a lack of understanding about the different perceptions and value systems of marine resource users and how these influence human behaviour has surfaced. However, such socio-political considerations over the whole range of stakeholders and their type of involvement in the establishment and operation of multi-use offshore systems has received little attention, thus this chapter is timely in highlighting the role of governance in multi-use offshore aquaculture development.

In the late 1970s, the introduction of the UNCLOS (United Nations Convention on the Law of the Sea) led to successive efforts on governing the offshore commons in a more streamlined approach. Despite not all countries officially ratifying the UNCLOS to date, most have endorsed the concept of the rights and duties pertaining to Exclusive Economic Zone definitions (Buck et al. 2004). Nonetheless, managing the complexities surrounding marine governance issues including equitable access to offshore marine resources requires policies that take account of the structures and principles relevant to decision-making. This chapter provides a broad overview of the role good governance can play in supporting development of the offshore aquaculture sector in an increasingly contested marine environment where multiple use demands robust governance systems.

7.2 Defining Governance, Management and Policy

Managing aquaculture, like much of marine resources management, is often dealt with independently to other marine sectors competing for use of similar areas, for example, fisheries and offshore renewable energy sectors (Stead 2005, 2015). Good governance can help integrate consideration of impacts from decisions made about different sectors. Good governance is defined herein simply as the process in which decisions are made formally, e.g. government policies and regulations and/or informally codes of conduct from an industry/community and health of marine ecosystems perspective (Stead 2015).

In this chapter we employ the definition by Olsen (2003), who differentiates between <u>management</u> as the process by which human and material resources are harnessed to achieve a known goal within a known institutional structure and <u>governance</u>, which sets the stage in which management occurs by defining—or redefining—the fundamental objectives, policies, laws and institutions by which societal issues are addressed. By fundamental objectives, we refer to the fundamental human rights, e.g. access to a healthy environment. This governs the interest to engage in the various aspects thereof, e.g. maintenance of biodiversity and their related ecosystem services. The concept of governance and management thus focuses on norms, institutions (laws and regulations), organizational structures and processes in the (marine) human-nature context. A lack of understanding about linkages between decision-making by different actors and stakeholders and between different levels at which decisions are made (local industry, national jurisdiction authority and international conventions/treaties) can lead to fragmented and weak governance systems. Indeed, governance is often confused with government and has many meanings depending on the disciplinary view or context in which the term is being applied. The political scientist, Roderick Rhodes (1996) developed a concept of governance which was adopted by the European Commission in their White Paper on European Governance where "European governance" refers to rules, processes and behaviour affecting the way powers are exercised at a European level (EU 2001). Herein, the qualities of good governance are referred to as the 5 "principles of good governance":

(1) coherence, (2) openness, (3) participation, (4) accountability, and (5) effectiveness.

For the purpose of this chapter, it is useful to tailor the definitions of governance, policy and management to the specifics of aquaculture, as these terms are often used interchangeably leading to constraints in implementing sustainable aquaculture development strategies (Table 7.1).

In summary, management of offshore aquaculture is often dealt with independently to marine governance, and even more distant from multi-use approaches. To date, insufficient effort has been directed towards considering and integrating both management and governance in the development of offshore aquaculture sector policy. The hypothesis for debate in this chapter is 'before the offshore aquaculture sector can optimise production regardless of the target species and technology employed, a better understanding of the links between governance structure and offshore multi-use management implementation is required to facilitate growth of this sector'. The next section shares lessons from development of marine governance models in a large marine ecosystem context based on the work by Fanning et al. (2007, 2009) and Mahon et al. (2008, 2011) to illustrate the opportunities or supporting development of the offshore aquaculture sector associated with adopting a multi-level governance framework

Table 7.1 Overview of working definitions for governance, management, policy and ecosystem approach to policy making (*Source* by authors)

Governance	Describes a social function centred on efforts or incentives to steer the actions of humans toward achieving desirable outcomes and avoiding undesirable ones. It covers the fundamental goals, institutional processes and structures which are the basis for planning and decision-making, and sets the stage within which management occurs
Policy	Refers to a set of basic principles and associated guidelines derived from governance that defines the process by which human and material resources are viewed to achieve a defined goal within a known institutional structure. Policy is driven by broader societal issues, for example, food security and/or economic development through offshore aquaculture
Management	Refers to the process of how human and material resources are used to achieve a known goal within a known institutional structure
Ecosystem approach to policy making	Viewed as a potential solution to managing and valuing complex marine ecosystems and measuring policy impact of different sectors such as offshore aquaculture on humans and economic growth. However research methods must consider different response variables when examining relationships between different factors and by using integrated modelling techniques: e.g., Bayesian Belief Networks—which can deal with directly and indirectly related aquaculture data. Ecosystem services sometimes negatively relate to each other and so marine management decisions about offshore aquaculture need to consider interactions and changing values for now and in the future

7.3 Developing a Multi-level Governance Framework for Offshore Aquaculture

In contrast to terrestrial systems, knowledge on marine ecosystems is more limited, e.g. pertaining to the existence of spatial and temporal data, species distribution and data format issues (Martin et al. 2014). Furthermore, ocean governance is still in its infancy as it confronts political, legal and economic development options for ocean use as well as efforts to restore marine ecosystems and the services they provide (WBGU 2013). The recognition of the latter gap is reflected by the significant efforts to develop and reshape various concepts over the last 30 years. These address the objective of influencing marine environments positively and purposively through human and societal engagement by management and governance means. Integrated Coastal Zone Management (e.g. Cicin-Sain and Knecht 1998) and more recently also Coastal and Ocean Governance (Olsen 2003; WBGU 2013) are the most prominent and influential examples (Stead et al. 2002; Stead 2005). With respect to management measures such as marine protected areas, for example, Heylings and Bravo (2007) heralded the benefits of developing a co-management regime based on strategic vision, participation, and consensus building, while Grafton and Kompas (2005) called for a governance system that uses socio-economic criteria in the development of management goals as well as the physical design of the conservation areas themselves. With regard to the complexities implicit in larger-scale, social-ecological systems (SESs), Norgaard et al. (2009) reported on the lack of clarity from policy-makers and lawmakers concerning setting objectives and accountability. Further, these authors cautioned that because professionals often participate more as individuals than as representatives, the knowledge lines are very fuzzy among different groups thus broader representative views may not always be captured. Also, because adaptive management often does not go beyond theory and demonstration of successes in practice in marine sustainable multi-use management are limited for aquaculture, especially where there is a multiplicity of perspectives confounding interpretation, then initiatives to strengthen interactions among scientists, policy-makers, stakeholders, and the public could help improve benefits associated with adoption of adaptive management for the offshore aquaculture sector.

Thus arguably marine governance is in a nascent stage, and even more so in regard of endorsing offshore aquaculture in a multi-use setting. Thus, for any offshore aquaculture initiative, ocean governance will set the stage in influencing the rate of development for this industry. Governance effectiveness, as defined herein, is influenced by a set of fundamental objectives, followed by a suite of policies, laws and institutions that exist on multiple levels (e.g. local, national and international). Therefore, to debate our hypothesis stated earlier, this section illustrates the importance of having a governance framework to facilitate decision-making. It illustrates this through marine governance research done to support sustainable marine resources management in the Caribbean, using a policy cycle based approach (Fanning et al. 2007; Turner et al. 2014). An adaptive governance framework was useful in helping different countries to identify their strengths and weaknesses in governance structures which can be achieved using a policy cycle exercise to analyse existing arrangements for offshore aquaculture (Fanning et al. 2007). The Regional Ocean Governance Framework (ROGF) developed for the Wider Caribbean for the Caribbean Large Marine Ecosystem (CLME) by Mahon et al. (2012) illustrates the conceptual 'architecture' to guide identification at multiple scales and multiple levels and when populated with relevant data shows who are and who are not involved in the multi-use policy processes. Figure 7.1 helps visualise hypothetically linkages vertically (local, national, sub-regional, regional and global) and horizontally (e.g., between sectors such as offshore aquaculture and offshore renewable energy) of governance.

Policy cycles occur at several levels from local, national, sub-regional, regional and global (Fig. 7.1). Within each of these levels there may be many policy cycles. The concept underpinning the policy cycle (Fig. 7.2) is that to achieve effective governance, there must be a complete policy process for decision-making at any level (Fig. 7.1). For an efficient system of governance, the policy cycles need to be complete and there needs to be communication, not only between the different

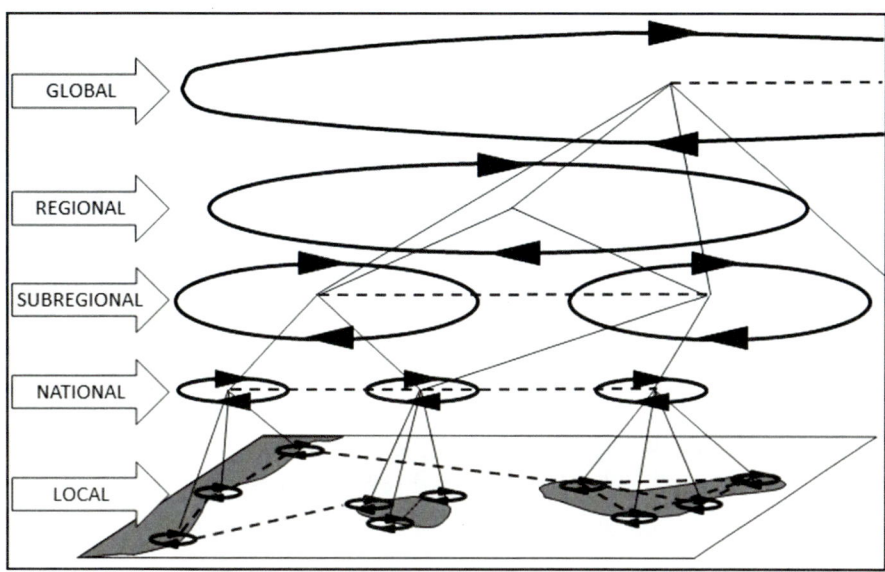

Fig. 7.1 Schematic of linkages between (*vertical*) and within (*horizontal*) levels of governance needed for effective governance. The conceptual Regional Ocean Governance Framework (ROGF) is based on two fundamental parts of the governance system: (i) complete policy processes that are linked *vertically* (nationally and regionally) and *horizontally* (aquifers, lakes, oceans and fresh water systems; sectors such as fisheries and tourism) as shown in Fig. 7.1, and (ii) the policy cycle (see Fig. 7.2) (Fanning et al. 2007)

Fig. 7.2 The 5 stages of the generic policy cycle (redrawn from Fanning et al. 2007). The concept underpinning the policy cycle is to achieve effective governance, there must be a complete policy process. This includes the ability to (i) take up data and information, (ii) generate advice, (iii) make decisions, (iv) implement decisions, and (v) review all aspects of the process

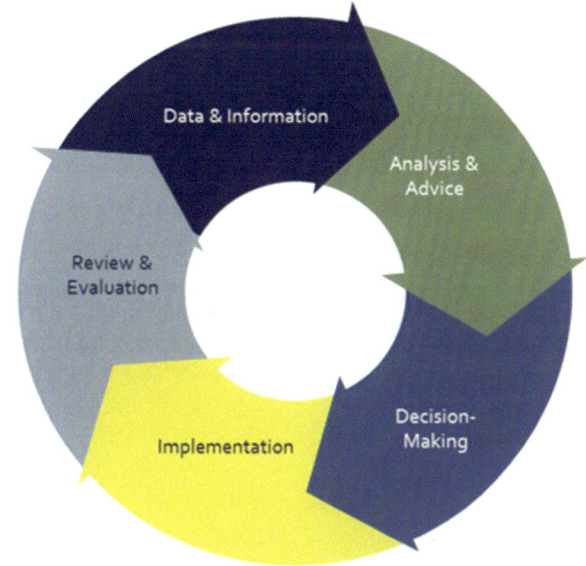

levels of governance (vertical connections in Fig. 7.1) but also across the policy cycles at each level (horizontal connections in Fig. 7.1). There is also an assumption that communication pathways should be two-way allowing for sharing of data and information and improvements in governance using ideas from both higher and lower levels.

National level policy cycle review exercises for the offshore aquaculture sector in a multi-use setting should be undertaken to identify involvement of various actors and players in management and policy. The processes underpinning the different stages in the policy cycle are to ensure (1) appropriate data and information leads to (2) analysis and provision of advice that informs (3) decision-making, which then gets (4) implemented, and subsequently is (5) reviewed and evaluated to determine effectiveness of decisions (Fig. 7.2). The policy cycle review process is useful for: (i) Identifying the government, non-government and private sector stakeholder groups involved in formal and informal governance structures that exist and govern marine resource use both directly and indirectly to the offshore aquaculture sector; (ii) Identifying groups involved in the governance policy cycle; and (iii) Identifying strengths and weaknesses in the policy cycle.

7.4   Knowledge and Information Gaps in Offshore Aquaculture Multi-use Governance

Promoting new forms of governance is by no means the sole responsibility of one institution alone. It is the responsibility of all levels of public authority, private undertakings and organised civil society because good governance—openness, participation, accountability, effectiveness and coherence—are what the public expects at the beginning of the 21st century (EU 2001).

To date, economic viability, technical and environmental barriers remain principal research topics where knowledge is still required to aid creation and exploitation of new multi-use ventures, such as offshore wind farms or open ocean aquaculture (Michler-Cieluch and Krause 2008). In recent years, studies have also started to consider public or specific stakeholder groups' perceptions in relation to the fledgling offshore wind industry and/or to aquaculture development in the open ocean (examples given in Nichols et al. 2003; Robertson and Carlsen 2003; James and Slaski 2006). It has been recognized that powerful stakeholder groups, in particular those directly involved in or affected by innovations, exert a great influence on new developments: they can impact negatively on progress of projects (Tango-Lowy and Robertson 2002) but also contribute positively to the improvement of management processes (Dalton 2006; Apt and Fischhoff 2006).

Two stakeholder analysis by Krause (2003) and Michler-Cieluch (2009) for the North Sea area of Germany revealed that there are different types of actors involved in the offshore realm as compared to nearshore areas. Different types of conflicts, limitations and potential alliances surface. These are rooted in the essential differences in the origin, context and dynamics of nearshore- versus offshore resource uses (Krause 2003). Whereas in nearshore waters historically well-rooted social networks with traditional use patterns exist, the offshore waters are dominated by large, often international operating companies with limited social networks and engagement with each other. The latter can be viewed as holding "pioneer" mentalities, since the offshore development was initiated with the onset of technical developments in ship building and platform construction. These fundamental differences between the diverse stakeholders in nearshore and offshore waters make a streamlined approach to support multi- use management very difficult (Krause 2003; Krause et al. 2011). This finding was reflected by the recent assessment by the WBGU (2013) that revealed major deficits in the status of ocean governance, which have yet to catch up with the pace of technological advancements which allow an intensification of offshore multi-use resource utilisation. This is in part due to the lack of redundant multi-level institutions in place that are able to communicate local and regional findings in a bottom-up fashion as well as a lack of instruments in place that support good communication vertically and horizontally (Fig. 7.1).

To date, the offshore wind farm operators, perceiving themselves as pioneers, hold "client" ties with the decision-makers, in which other users (such as offshore aquaculture, fisheries, etc.) and their interests are not included in development considerations. Employing good governance would help to find stakeholder-led solutions which could be perceived as "win-win" for multiple stakeholders in the offshore setting, the wind energy operator may improve their public perception (Gee 2010) and engender greater participation in future decision-making processes (Turner et al. 2014).

In practice, undertaking the policy cycle exercise is stakeholder-led and provides a way for actors and players in the offshore aquaculture sector to assess how complete the policy cycle is in a particular country. The exercise provides a basis to inform what, if any, gaps exist at different stages of the policy cycle and what the implications are for effective governance of multi-use approaches to offshore aquaculture operations.

Thus, identification of actors involved in a policy cycle, as well as its strengths and weaknesses, can lead to building awareness of the many organisations that can potentially be involved in offshore aquaculture multi-use governance. Fostering improved communication and cooperation between these organisations at different levels should improve the management of the offshore aquaculture sector at a national level and regionally. Hereby pre-existing social networks can provide significant political leverage for governance transformations as research for example from Chile (Gelich et al. 2010) and the Caribbean (Turner et al. 2014) has shown. Applying good governance principles that includes decisive legislative bodies to determine the specific constitutional rules to be used, possibly through marine spatial planning (MSP), can be useful in understanding the institutional framework required for effective decision-making in multiple-marine resource user settings.

However, the "social embeddedness" (Granovetter 1985) and the role of discreet informal social networks must also be considered alongside formal governance mechanisms to address the probability of collectively concerted action in the off-shore realm. Indeed, if viewing offshore aquaculture as a common-property activity within a common-pool resource system, then two issues commonly need to be addressed:

1. cost/benefits and challenges of managing exclusion or the control of access to potential users, and;
2. if each user is capable of subtracting from the welfare of all other users, then to what extent are marine resource users willing to trade-off negative impacts with potential benefits?

In the context of offshore aquaculture, if we consider Hardin's (1968) quote that "this dilemma of the tragedy of the commons has no technical solution"—then in debating our hypothesis we should look more into the social dimensions of offshore aquaculture in a multi-use setting. Therefore, to fully address our hypothesis —'before the offshore aquaculture sector can optimise production regardless of the target species and technology employed, a better understanding of the links between governance structure and offshore management implementation is required'—in practice, workshops should be conducted in each country with an interest in offshore aquaculture. These events should be stakeholder-led and seek to identify the efficacy of current governance processes and governance reform needed to improve sustainable development of this sector.

Focussed workshops of this kind can help governments identify likely structural issues associated with current governance arrangements and determine reforms needed for policy and associated management measures to meet identified con-straints for future development of the offshore aquaculture sector (Wever et al. 2015). To address the respective technical, economic, social, and political chal-lenges of offshore as well as established inshore aquaculture production, gover-nance processes must also include the relevant stakeholders in a triple-loop manner as proposed by Pahl-Wostl (2009) over time (Fig. 7.3).

Increasing use of ocean territory demands from the respective actors that they move beyond the narrow focus of business entrepreneurship and consider a more complex picture of the multiple-use challenges. In this sense, comprehensiveness implies to take a wider scope of competing demands on multiple marine resources and stakeholders potentially develop a holistic view of all issues, including those of the other parties involved. Marine spatial planning can be helpful in ensuring information from different sectors and actors is assimilated, which goes beyond a single individual or organizations' priorities. Additionally, an insight into the existing underlying offshore aquaculture sector ideas, competing interests and normative considerations has to be generated to understand complex problems and to overcome misunderstandings. This demands governance processes that are either more results-oriented (e.g. for integrating technical knowledge of the participating sectors) or more process-oriented (e.g. for establishing new linkages between

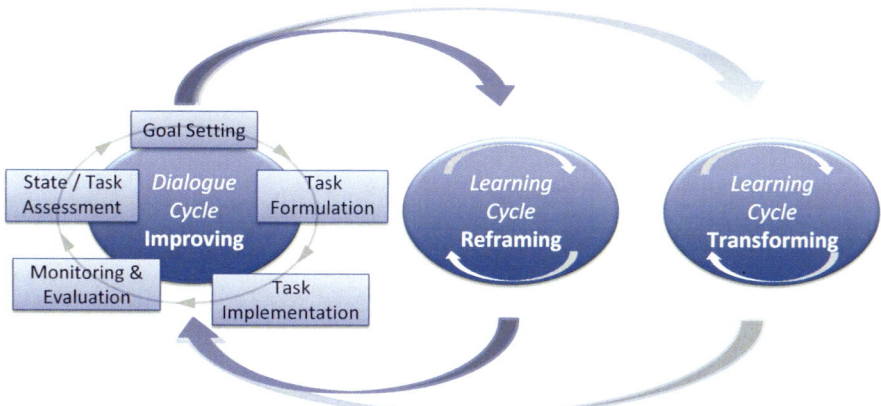

Fig. 7.3 The concept of "triple-loop learning" applied to a multi-use oriented governance regime for the implementation of offshore aquaculture. It is assumed that the different stages of learning among participating stakeholder groups leads to a change in the composition of actor groups and thus effects a change in the institutional framework (Pahl-Wostl 2009)

different groups) than many of today's management systems support (Michler-Cieluch and Krause 2008). How to operationalise highly context-specific processes is a current gap in knowledge for building the institutional arrangements needed to support growth of the offshore aquaculture sector. In this context, it must be stressed that stakeholders are important sources of information which is useful for identifying and contributing to the socio-economic drivers underpinning decision-making in multiple level governance systems which exist offshore.

7.5   Outlook

The strong expansion of offshore wind farms in marine environments, such as in the North Sea, increases the potential competition and likely conflicts among marine resource users especially in sea areas that have formerly been used for other purposes, such as for fishing or shipping activities, or that have been free of human activity and possibly protected. Hence, reconciling the demands of multi-use activities with livelihood and conservation needs will be difficult to balance in practice. Risk and uncertainty associated with starting new businesses are particularly high in such emerging and innovative industrial sectors like the offshore aquaculture industry. Thus a common understanding of uncertainty factors impacting a prospected offshore multiple-use site for aquaculture particularly when combined with renewable energy systems is a first step towards turning some uncertainty factors into more measurable and "controllable" elements. Putting in place multi-use governance systems that can address these complex issues vertically and horizontally can help stakeholders to consider ways of dealing with anticipated risks.

However, a wide range of *organizational and social* challenges related to the collective use of a defined ocean territory have to be taken into account and preferably during the infancy of a new sector. Governance frameworks can help define responsibilities and duties thus supporting cross-sectoral management associated with offshore co-management that integrates the different demands and practices of the involved parties. It can be postulated that policy drivers will change over time and inform management actions for the future, thus complexity in decision-making is likely to increase where there are demands from multiple users of ocean space. Concomitant changes to the natural, marine, and social ecosystems should ideally be considered in advance of offshore aquaculture developments so that governance systems can be developed to tailor the context-specific needs of different locations. The trajectory of anticipated changes identified through policy workshops (Fig. 7.2) could result in offshore aquaculture businesses providing a more vibrant, innovative, marine economy with compatible management measures supported by appropriate governance systems. To fully address our hypothesis —'before the offshore aquaculture sector can optimise production regardless of the target species and technology employed, a better understanding of the links between governance structure and offshore management implementation is required', then one approach could be to implement national policy cycle workshops to identify the relevant stakeholders and get their buy-into co-develop effective multi-use governance arrangements that will support social, economic and environmental development of the offshore aquaculture sector.

References

Apt, J., & Fischhoff, B. (2006). Power and people. *The Electricity Journal, 19*, 17–25.

Araujo, M., Acosta, A., Linan, A., & Saiegh, S. (2004). *Political institutions, policy making and policy outcomes in Ecuador: Latin American research network*. Inter-American Development Bank.

Buck, B., Krause, G., & Rosenthal, H. (2004). Extensive open ocean aquaculture development within wind farms in Germany: The prospect of offshore co-management and legal constraints. *Ocean and Coastal Management, 47*, 95–122.

Brugère, C., Ridler, N., Haylor, G., Macfadyen, G., & Hishamunda, N. (2010). Aquaculture planning: policy formulation and implementation for sustainable development. FAO Fisheries and Aquaculture Tech. Paper No. 542. Rome: FAO.

Cicin-Sain, B., & Knecht, R. W. (1998). *Integrated coastal and ocean management, concepts and practices*. Washington: Island Press.

Dalton, T. M. (2006). Exploring participants' views of participatory coastal and marine resource management processes. *Coastal Management, 34*, 351–367.

Duarte, C. M., Marbá, N., & Holmer, M. (2007). Rapid domestication of marine species. *Science, 316*, 382–383.

EU. (2001). European Governance – A white paper. COM (2001) 428 final. Brussels.

Fanning, L., Mahon, R., & McConney, P. (2009). Focusing on living marine resource governance: the caribbean large marine ecosystem and adjacent areas project. *Coastal Management, 37*, 219–234.

Fanning, L., Mahon, R., McConney, P., Angulo, J., Burrows, F., Chakalall, B., et al. (2007). A large marine ecosystem governance framework. *Marine Policy, 31*, 434–443.

FAO. (2016). *State of world fisheries and aquaculture 2014*. Rome: FAO.

Gee, K. (2010). Offshore wind power development as affected by seascape values on the German North Sea coast. *Land Use Policy, 27*, 185–194.

Gelich, S., Hughes, T. P., Olsson, P., Folke, C., Omar, D., Fernandez, M., et al. (2010). Navigating transformations in governance of Chilean marine coastal resources. In *Proceedings of the National Academy of Sciences of the United States of America (PNAS) Sustainability Science Early Edition*.

Grafton, R. Q., & Kompas, T. (2005). Uncertainty and the active adaptive management of marine reserves. *Marine Policy, 29*, 471–479.

Granovetter, M. (1985). Economic action and social structure: The problem of embeddedness. *American Journal of Sociology, 91*(3), 481–510.

Hardin, G. (1968). The tragedy of the commons. *Science, 162*, 1243–1248.

Heylings, P., & Bravo, M. (2007). Evaluating governance: a process for understanding how co-management is functioning and why, in the Galapagos marine reserve. *Ocean and Coastal. Management, 50*, 174–208.

Hishamunda, N., Ridler, N., & Martone, E. (2014). Policy and governance in aquaculture: Lessons learned and way forward. FAO Fisheries and aquaculture technical paper no. 577. Rome: FAO.

James, M., & Slaski, R. (2006). *Appraisal of the opportunity for offshore aquaculture in UK waters*. Report of project FC0934, commissioned by Defra and Seafish from FRM Ltd.

Krause, G. (2003). Spannungsfelder im Aquakultursektor in Deutschland: Inshore versus Offshore. *Deutsche Gesellschaft für Meeresforschung, 3–4*, 7–10.

Krause, G., Griffin, R. M., & Buck, B. H. (2011). Perceived concerns and advocated organisational structures of ownership supporting 'offshore wind farm – mariculture integration'. In G. Krause (Ed.), *From turbine to wind farms: Technical requirements and spin-off products* (pp. 203–218). InTECH.

Mahon, R., Cooke, A., Fanning, L., & McConney, P (2012). *Governance arrangements for marine ecosystems of the Wider Caribbean Region. Centre for Resource Management and Environmental Studies (CERMES)*. CERMES technical report no. 61. Barbados: University of the West Indies, Cave Hill Campus.

Mahon, R., Fanning, L., & McConney, P. (2011). Principled ocean governance for the Wider Caribbean Region. In L. Fanning, R. Mahon, & P. McConney (Eds.), *Towards marine ecosystem-based management in the Wider Caribbean* (pp. 27–38). Amsterdam: Amsterdam University Press.

Mahon, R., Fanning, L., McConney, P., & Toro, C. (2008). Governance for Caribbean marine resources: Seeking a path. *Proceedings of the Gulf & Caribbean Fisheries. Institute, 60*, 3–7.

Martin, C. S., Fletcher, R., Jones, M. C., Kaschner, K., Sullivan, E., Tittensor, D. P., et al. (2014). *Manual of marine and coastal datasets of biodiversity importance*. May 2014 release. Cambridge (UK): UNEP World Conservation Monitoring Centre.

Mee, L. (2006). *Complementary benefits of alternative energy: Suitability of offshore wind farms as aquaculture sites*. Report to Seafish project ref no: 10517.

Michler-Cieluch, T. (2009). Co-management processes in integrated coastal management—The case of integrating marine aquaculture in offshore wind farms (PhD University Hamburg).

Michler-Cieluch, T., & Krause, G. (2008). Perceived concerns and possible management strategies for governing wind farm-mariculture integration. *Marine Policy, 32*, 1013–1022.

Nichols, R. B., Robertson, R. A., & Lindsay, B. E. (2003). Northern New England commercial fishermen and open ocean aquaculture: An analysis of how commercial fishermen perceive the government, fishing, and their way of life. In C. J. Bridger & B. A. Costa-Pierce (Eds.), *Open ocean aquaculture: From research to commercial reality*. Baton Rouge, USA: World Aquaculture Society.

Norgaard, R. B., Kallis, G., & Kiparsky, M. (2009). Collectively engaging complex socio-ecological systems: Re-envisioning science, governance, and the California Delta. *Environmental Science & Policy, 12,* 644–652.

Olsen, S. B. (Ed.). (2003). *Crafting coastal governance in a changing world.* Kingston, USA: University of Rhode Island Coastal Resource Center.

Pahl-Wostl, C. (2009). A conceptual framework for analysing adaptive capacity and multi-level learning processes in resource governance regimes. *Global Environmental Change, 19,* 354–365.

Rhodes, R. A. W. (1996). The New Governance: Governing without Government. *Political Studies, 44*(4), 652–667.

Robertson, R. A., & Carlsen, E. (2003). Knowledge, relevance, and attitudes towards open ocean aquaculture in Northern New England: Summary of five sample surveys. In C. J. Bridger & B. A. Costa-Pierce (Eds.), *Open ocean aquaculture: From research to commercial reality.* Baton Rouge, LA, USA: World Aquaculture Society.

Stead, S. M. (2005). A comparative analysis of two forms of stakeholder participation in European aquaculture governance: Self-regulation and integrated coastal zone management. In T. S. Gray (Ed.), *Participation in fisheries governance.* Berlin: Springer.

Stead, S. M. (2015). Mariculture: Aquaculture in the marine environment. In H. D. Smith, J. L. Suárez De Vivero, & T. S. Agardy (Eds.), *Routledge handbook of ocean resources and management.* London: Taylor and Francis.

Stead, S. M., Burnell, G., & Goulletquer, P. (2002). Aquaculture and its role in integrated coastal zone management. *Aquaculture International, 10,* 447–468.

Tango-Lowy, T., & Robertson, R. A. (2002). Predisposition toward adoption of open ocean aquaculture by northern New England's inshore commercial fishermen. *Human Organization, 61,* 240–251.

Turner, R. A., Fitzsimmons, C., Forster, J., Peterson, A., Mahon, R., & Stead, S. M. (2014). Measuring good governance for complex ecosystems: Perceptions of coral reef-dependent communities in the Caribbean. *Global Environmental Change, 29,* 105–117.

WBGU. (2013). *World in transition: Governing the Marine Heritage.* Flagship report 2013. German Advisory Council on Global Change.

Wever, L., Krause, G., & Buck, B. H. (2015). Lessons from stakeholder dialogues on marine aquaculture in offshore wind farms: Perceived potentials, constraints and research gaps. *Marine Policy, 51,* 251–259.

Chapter 8
The Socio-economic Dimensions of Offshore Aquaculture in a Multi-use Setting

Gesche Krause and Eirik Mikkelsen

Abstract Decision-making within the marine realm is a complex process, which endorses ecological, societal and economic needs and they must therefore be managed jointly. Much of the formerly "free oceans" is nowadays subject to intensive uses, thus making the need to optimise the management of the resources within a multifunctional and multi-use(r) context apparent. The high competition for functions and uses of inshore and nearshore waters has given strong incentives to investigate the opportunities of moving industrial activities offshore. The current raise of offshore aquaculture is one prominent example of this. However, our understanding of the social dimensions and effects of offshore aquaculture is yet incomplete. We need to consider also how different multi-use settings for offshore aquaculture affect the socio-economic outcomes on various levels. During the development of offshore aquaculture, this multifunctional perspective has emerged especially for the combination with offshore wind farms. This synergy of two different stakeholders, the so-called multifunctional utilisation of marine areas, can be viewed as a new concept by the implementation of integrated, consensus-based resource planning conditions. We suggest a typology of social dimensions of marine aquaculture, based on the literature of "traditional" nearshore aquaculture. Based on this typology we discuss the current level of knowledge on the socio-economic dimensions of multi-use offshore aquaculture and point to further research needs.

G. Krause (✉)
Alfred Wegener Institute Helmholtz Centre for Polar and Marine Research (AWI), Earth System Knowledge Platform (ESKP), Bussestrasse 24, 27570 Bremerhaven, Germany
e-mail: gesche.krause@awi.de

G. Krause
SeaKult—Sustainable Futures in the Marine Realm, Sandfahrel 12, 27572 Bremerhaven, Germany

E. Mikkelsen
Departement of Social Science, Norut Northern Research Institute AS, Postboks 6434, 9294 Tromsø, Norway

B.H. Buck and R. Langan (eds.), *Aquaculture Perspective of Multi-Use Sites in the Open Ocean*, DOI 10.1007/978-3-319-51159-7_8

8.1 Background

Aquaculture has been widely employed for a long time, i.e. traditional fishpond aquaculture in Asia has been a significant landscape element for centuries. The last decades have, however, seen a marine "Neolithic revolution". About 430 (97%) of the species presently in aquaculture have been domesticated since the start of the twentieth century, and 106 species have been domesticated over the past decade alone (Duarte et al. 2007). Aquaculture is posed to get a prominent role to address one of the major global challenges at the start of the twenty first century, in providing alternative sources for marine food proteins, supply and food-security. Yet, many challenges remain, and not only the numerous technical and biological issues, but also regarding the social, cultural and economic character of future development. Indeed, vis á vis the impressive growth in production volumes over the last decades, with aquaculture expanding from practically being negligible compared to capture fisheries, to constituting over 40% of total global marine production (FAO 2014), this development has had manifold socio-economic repercussions on various levels. This recent rise of aquaculture and accompanying socio-economic relevance has been coined as the so-called "Blue Revolution" (Krause et al. 2015). At the same time the growth in capture fisheries seen over the last 50 years seem to have stagnated (FAO 2014).

Most of this rather recent global growth in aquaculture production has taken place in inland and coastal areas (FAO 2014). However, there are major obstacles to accommodate further growth into existing marine resource use patterns, which would increase conflicts along coastal areas. This is partly due to stakeholder groups growing in numbers or prominence (Buanes et al. 2005), but also due to the risk of spread of diseases and parasites between aquaculture farms, which limits farm densities in coastal areas. Although the typical size of fish farms inshore has grown strongly the last 10–20 years, the potential for creating very large aquaculture production facilities inshore appears limited. In contrast, in the offshore realm, the size of aquaculture plants can be much larger, thus targeting at more cost-efficient scales of production. Hence, moving aquaculture facilities offshore seems a promising way to try to tackle these challenges and limitations, and the technology for offshore aquaculture is emerging now (Buck et al. 2008).

The high and rapidly increasing demand for offshore space for different purposes, such as installations for the production of energy from renewable sources, oil and gas exploration and exploitation, shipping and fishing, nature conservation, the extraction of raw materials such as sand and gravel, aquaculture installations and underwater cultural heritage, as well as the multiple pressures on coastal resources, require an integrated planning and management approach (EU 2014). Indeed, since the offshore move is rather risky and expensive, a multi-use approach is favored, that is the integrated production of marine species with other resource uses, such as offshore wind farms (Buck and Krause 2012). In the case of such multi-use offshore concepts, the typical practical procedure of looking for the most suitable site will be confined to those sites where offshore wind farms are planned or already in place.

This is due to the fact that aquaculture acts as "secondary newcomer", since the current momentum of moving activities offshore stems from the political will to enforce renewable energy systems in the first place (Buck et al. 2003). Therefore, the typical site-selection criteria catalogue applicable for aquaculture can usually not be implemented. Hence criteria for the selection process must be tailored to capture the relevant local parameters of the conditions around and within an offshore wind farm. Offshore equipment will need to be adapted to co-exist with the other uses to which the platforms may be put. For instance in the case of aquaculture, equipment has been developed for more benign environments and as such is still in the redesign-phase for harsher conditions. It must be noted that several projects are working to realize offshore aquaculture farm designs independent of renewable energy production facilities. In Norway the problems of salmon lice in the fish farms and their transfer to wild salmon populations are also a strong incentive to enable offshore aquaculture. In other parts of the world independent offshore aquaculture farms have been in operation for some time, i.e. in the Caribbean also in real open ocean environments (Ryan et al. 2004).

Additionally, and maybe most importantly, the socio-economic framing conditions must be assessed. They can either promote or hamper such offshore multi-use concepts. While the density and variety of stakeholders and interests affected by inshore aquaculture in general seems much higher than offshore aquaculture could be in the relatively near future, it would be naïve to assume that socio-economic issues can be ignored when going offshore. Inshore as well as offshore aquaculture production activities are subject to dispute and conflict when management regimes have not been established properly, as the participating stakeholder groups have different and sometimes opposing interests (Krause et al. 2011; Wever et al. 2015). Additionally, the flow of costs and benefits and end-consumer preferences vary a lot from place to place (Griffin et al. 2015). This can change an initial local acceptance to strong opposition against the instalment of aquaculture in coastal rural landscapes. Resolving this requires additional input from social, economic and political sciences (Michler-Cieluch and Krause 2008). This has over the years lead to an increasing awareness to the social dimensions of aquaculture production (Krause et al. 2015).

Marine Spatial Planning (MSP), which now has expanded out to offshore areas, as for example in Norway's Integrated management plans for the Barents Sea, the North Sea, and the Norwegian Sea (Anon 2008–2009; Anon 2010–2011; Anon 2012–2013), attempts to combine governance of stakeholders and their interests with the needs and limitations inherent in ecological sustainability. So far, stakeholder participation in marine spatial planning has been less than in typical integrated coastal zone management (ICZM) processes, but the ecosystem component of MSP management appear to be stronger—the current EU Efforts to define and reach "Good Environmental Status" of Marine Waters by 2020 (Marine Strategy Framework Directive) is a case in point. However, with increasing interest for the use of offshore areas, marine spatial planning must include socio-economic aspects and stakeholder participation to a stronger degree to be successful.

In this chapter we consider the social dimensions of multi-use offshore aquaculture, and the complexities involved in managing the ecological, societal and economic aspects of offshore aquaculture in an integrated and systematic manner. We suggest a first typology of socio-economic dimensions for aquaculture and outline their implications for offshore aquaculture development in multi-use contexts. The chapter closes with a discussion on the current level of knowledge on the socio-economic dimensions of multi-use offshore aquaculture, and point to further research needs.

8.2   Socio-economic Dimensions of Aquaculture—A First Typology

Over the past decades, scientists and policymakers have become increasingly aware of the complex and manifold linkages between ecological and human systems. Social-ecological systems are understood to be complex adaptive systems where social and biophysical agents are interacting at multiple temporal and spatial scales (Janssen and Ostrom 2006). This has stimulated researchers across multiple disciplines to look for new ways of understanding and responding to changes and drivers in both systems and their interactions (Zurek and Henrichs 2007). In this contextual setting, Krause et al. (2015) showed that most socio-economic analysis to date deal mainly with the effects of salmon or shrimp farming, and to a lesser extent with e.g. Pangasius and Tilapia, as well as filter feeders (such as Crassostrea gigas) and seaweeds (such as Kappaphycus alvarezii and K. striatum as well as Eucheuma denticulatum) (see Buanes et al. 2004; Barton and Fløysand 2010; Fröcklin et al. 2012; Stonich and Bailey 2000; Joyce and Satterfield 2010; Sievanen et al. 2005; Buchholz et al. 2012).

In order to promote the sustainable co-existence of uses and, where applicable, the appropriate apportionment of relevant uses in the offshore realm, a framework should be put in place that consists at least of the establishment and implementation by Member States of maritime spatial planning, resulting in plans. Such a planning process should take into account land-sea interactions and promote cooperation among Member States. Its main purpose is to promote sustainable development and to identify and encourage multi-purpose uses, in accordance with the relevant national policies and legislation (EU 2014).

Thus, the management of marine offshore areas is complex and involves different levels of authorities, economic operators and other stakeholders. However, questions pertaining to the inter-relationships between community impacts, right of access, ownership, taxation, liabilities for the negative repercussions from the environmental effects on society, and ethical issues, to name but a few, have remained largely untackled in a comprehensive, integrated manner (Krause et al. 2015). As a result, the socio-economic consequences of aquaculture operations are often poorly understood and repercussions such as poaching not fully anticipated (see examples given in Barrett et al. 2002; Bunting 2004; Fröcklin et al. 2012; Isaksen and Mikkelsen 2012; Jentoft and Chuenpagdee 2013; Sandberg 2003; Sievanen et al. 2005; Varela 2001). In many cases the omission of relevant stakeholders and social concerns in aquaculture development projects has contributed to inequity, social conflicts and violence (Martinez-Alier 2001; Nagarajan and Thiyageasan 2006; Varela 2001). The unavoidability of feedbacks between largely structural and technical interventions and the socio-economic systems within which they are embedded, highlights the need for employing more systematic (or ecosystem) approaches to analyse cause and effect relationships and to explore future sustainable, efficient and equitable development scenarios (Hopkins et al. 2011; Belton and Bush 2014).

This raises the question, what processes are needed to include issues and concerns that are not currently promoted by active and resourceful stakeholders (Buanes et al. 2004). This is especially important in offshore areas under multi-use conditions, since this is a novel line of resource use which lacks yet experiences on which stakeholder group's work well together under which governance and management conditions and what type of socio-economic outcomes can be expected. Indeed, more detailed and context-specific socio-economic dimensions of aquaculture operations include many important factors which need to be understood: gender, employment and income, nutrition, food security, health, insurance, credit availability, human rights, legal security, privatization, culture/identity, global trade and inequalities, as well as policies, laws and regulations, macro-economic context, political context, customary rules and systems, stakeholders, knowledge and attitudes, ethics, power, markets, capital and ownership (Hishamunda et al. 2009). Certainly, the lack of consensus on the social dimension is striking when compared with the universally accepted general definitions that exist for the biological and economic dimensions for sustainable ecosystem management (Krause et al. 2015). This is even more so the case of offshore marine management.

Based on the existing literature from coastal aquaculture (see references below) we identify these major types of socio-economic dimensions of aquaculture which are universally applicable:

(a) Attitudes to and perceptions of aquaculture and its effects
(b) Organization of and participation in planning for aquaculture
(c) Direct benefits of aquaculture and their distribution
(d) Negative effects of aquaculture production activities and conflicts with other interests
(e) Effects on the wider economic and innovation system
(f) Effects on cultural fabric and other social aspects

As the list indicates, the social implications of aquaculture are multi-dimensional and affect multiple levels. As a starting point of this analysis, we capture what we account as social dimension by stating what aspects of aquaculture we have excluded: The internal organization and efficiency of the aquaculture production units, and the ecological effects of aquaculture production (including biological, physical, and chemical). These aspects will, however, affect some of the social dimensions, like the number of jobs provided by aquaculture, the economic benefits to society, and possible conflicts with other stakeholders. Further, the different social dimensions (a)–(f) will typically interact. In the following sub-chapters, we elaborate in more detail on the suggested typology and their implications especially for offshore aquaculture.

8.2.1   Attitudes to and Perceptions of Aquaculture

How attitudes to aquaculture vary and correspond between social groups must be taken into account by planning authorities. However, capturing these varying stakeholder attitudes are also of relevance for businesses, NGOs and other stakeholders (Mazur and Curtis 2008; Freeman et al. 2012; Ladenburg and Krause 2011). Studies from different countries have explored groups' views on how aquaculture impacts, i.e. on the environment (e.g. by chemical pollution, effecting local fish stocks and wildlife, visual pollution of coastal landscape), on possible job-creation and economic benefits of mostly rural marginal areas, how it interferes with tourism, fishing or recreation, how it contributes to food security, and how the regulation of aquaculture should be.

This has been done for a number of countries, including Australia (Mazur and Curtis 2006, 2008), Spain (Bacher et al. 2014), Thailand (Schmitt and Brugere 2013), US and Norway (Chu et al. 2010), Greece (Katranidis et al. 2003), New Zealand (Shafer et al. 2010), Germany (Krause et al. 2011; Wever et al. 2015), Israel (Freeman et al. 2012), and Scotland (Whitmarsh and Palmieri 2009). Many of the studies concentrate on local stakeholder groups related to aquaculture (e.g. Bacher et al. 2014), several include tourists (e.g. Katranidis et al. 2003), some consider the general public (e.g. Mazur and Curtis 2008), and some also try to consider how the attitudes and risk perceptions expressed correlate with personal characteristics (e.g. lifestyle behaviour, Freeman et al. 2012), or even with community situation (e.g. employment or income deprivation, Whitmarsh and Palmieri 2009).

Bergfjord (2009) asked fish farmers in Norway what they see as the greatest risks to their business. The majority of respondents raised biggest concerns pertaining to market conditions for their product and towards the outbreak of diseases. Socially related dimensions do, however, also appear on their list, including sea area access, changes to the license system and of environmental regulations, which are all relatively high up on the list, while public repugnance to fish farms (aesthetics, environment) is ranked lower.

Slater et al. (2013), investigate how personal characteristics correlate with willingness towards choosing aquaculture as a livelihood in a developing country setting. Their model allows policy makers to consider the influence of socio-economic factors on the success of introducing aquaculture in different local contexts.

There is a need for further research on attitudes towards aquaculture and their formation, both for inshore and offshore aquaculture. The central theme to date on attitudes is how aquaculture under different settings and circumstances leads to conflicts. Questions that remain to be covered are whether the large increase in aquaculture production will reduce social acceptance, and if attitudes depend on the degree of exposure to prior experience of inshore aquaculture, as a study by Ladenburg and Krause (2011) has indicated for the expansion of wind turbines. They could show that prior experience plays an important role in affecting positive or negative perceptions of renewable energy systems. It can be assumed that this effect may also be relevant for offshore aquaculture development.

8.2.2   Organization of and Participation in Planning for Aquaculture

The omission of relevant stakeholders and social concerns in aquaculture development projects has, more often than not, contributed to inequity, social conflicts and violence (Krause et al. 2015). Therefore, the analysis and documentation of who the relevant stakeholder groups are is important for achieving a good planning process. This may include assessment of their legitimacy and power (Buanes et al. 2004). The design and execution of the planning process for aquaculture, or for the broader MSP process, is important both for the efficiency of the process and for the social sustainability of the outcomes. Qualities like transparency, representativeness and fairness matter for i.e. the legitimacy and support for the outcome (Buanes et al. 2004), for the wider trust in authorities, for development/maintenance of democracy, and for sustainable development (Krause et al. 2015).

Who to include in the planning process, and how, can be difficult to decide upon, balancing the ideals above with the ambition for an effective planning process, which typically also is embedded in a larger governance system. Whilst addressing the interactions and feedbacks between issues (e.g. economic, social and environmental consequences) in a MSP context, it becomes evident that many of these play out over time (i.e. in past, present and future contexts) and space (i.e. at local, regional and ecosystem/global scale)—these are referred to as 'cross-scale' or 'multi-scale' processes (Krause et al. 2015). What time-scale to consider in the planning process, and thus also what geographical scale, is also something that must be decided.

Discussions of the availability, usefulness and ease of use of knowledge is also important, e.g. on aquaculture's value creation and its distribution (Isaksen and Mikkelsen 2012). For example, with the development of feed-aquaculture turning to more land-based production of feeds, agriculture can also be seen as a new emerging stakeholder in aquaculture production (Costa-Pierce 2010). Indeed, when new groups of stakeholders are included in planning processes they may often bring about new knowledge to the process. However, the authorities managing the planning process are likely to influence specific knowledge realms that will be used with MSP.

The relative position of different stakeholder groups in MSP, either as holding perceived historical rights by decision-makers or de facto allocated property rights, also shapes future planning. Being a relatively new activity, aquaculture has often had a weaker position than traditional activities, according to some authors (Burbridge et al. 2001; Wever et al. 2015). On the other hand, the lack of appropriate governance systems in some developing countries, and specifically the presence of corruption, can lead to unwanted privatization by aquaculture entrepreneurs (Cabral and Aliño 2011).

8.2.3   Direct Benefits of Aquaculture, and Their Distribution

The provision of food and nutrition security, jobs and income are the main reasons for promoting and employing aquaculture. In addition to these rather straightforward benefits, aquaculture makes a further contribution as the consumption of animal source food facilitates uptake of nutrients from dietary components of vegetable origin (Leroy and Frongillo 2007). This role of addressing the hidden hunger problem, that is the lack of certain nutrients in everyday accessible foodstuff, is particularly important in countries such as i.e. Bangladesh, Cambodia, Ghana, Nigeria, and the Pacific Islands, where many people are impoverished and fish is by far the most frequently consumed animal-source food (Belton et al. 2011; Hortle 2007; Biederlack and Rivers 2009).

The documentation of the economic and employment benefits of aquaculture is often done in either official statistics (like for Norway: http://ssb.no/en/fiskeoppdrett) (Goulding et al. 2000), or in reports from aquaculture industry associations to strengthen support for the industry (e.g. see Sandberg et al. 2014), or as commissions from regional authorities who want to understand how different industries contribute to regional development (e.g. see for Troms county in Norway Robertsen et al. 2012).

In contrast to these Western country examples, cases from Thailand and Bangladesh suggest that, whilst the financial and employment gains generated by this sort of activity may appear substantial on paper (incomes from fish more than doubled, etc.), when placed in the context of the overall livelihood portfolio of practicing households, they are generally fairly modest. Thus, whilst economic gains in aquaculture production achieved in Thailand were superficially impressive, the already relatively affluent households involved subsequently abandoned techniques required to sustain them, because they could not compete with alternatives such as selling labor in a buoyant non-farm economy (Belton and Little 2011). The retention by project participants in rural Bangladesh of similar techniques capable of generating similar production increases may be indicative of the generally more severe nature of rural poverty there, which is itself linked to more limited opportunities for well-remunerated diversification of economic activities beyond the farm (Belton et al. 2011).

Assessing the benefits of aquaculture can be difficult, even focusing solely on financial and economic aspects, as exemplified by Burbridge et al. (2001) and Isaksen and Mikkelsen (2012). While aquaculture is seen as important for alleviating protein deficiency in diets in developing countries (FAO 2014), some have questioned the total effect of fish protein supply if aquaculture does not manage to reduce wild fish inputs in feed and adopt more ecologically sound management practices (Naylor et al. 2000). Belton et al. (2011) point out that global fish supply is undergoing a fundamental transition as capture fisheries is succeeded by aquaculture. This transition is however far from uniform. Even among major aquaculture producing nations, capture fisheries is yet crucial to food and nutrition security. For numerous countries outside this 'elite' group, wild capture fisheries remain the dominant supplier (Hall et al. 2011). Thus, who will benefit from aquaculture in what ways and on what level remains yet unresolved. These issues are even more difficult to assess in offshore aquaculture conditions, by which the initial monetary input is much higher and thus the benefits play out on much different levels and dimension than small-scale subsistence farming.

8.2.4 Negative Effects of Aquaculture Production Activities and Conflicts with Other Interests

The literature on attitudes and perceptions about aquaculture referred earlier can well be used as documentation of actual external effects and conflicts. Indeed, the conflicts between aquaculture and other interests are typically due to negative environmental effects from aquaculture, or competition over marine areas (Primavera 2006). Changes in environmental conditions due to aquaculture can be measured technically (Grigorakis and Rigos 2011) and thus rather objectively.

Assessing the impact or severity of a change in the natural environment due to aquaculture, and even more so in offshore environments, is however more difficult. It will vary with the local environment, and the local use of the area by humans, as shown by for example Primavera (2006) and Paul and Vogl (2011). Asking stakeholders or the general public to give their input for this has its challenges. However, instead of asking end-consumers to what degree aquaculture is or creates problems, one should rather investigate their use of and valuation of ecosystem services (Millenium Ecosystem Assessment 2005) affected by aquaculture, as well as the impact of aquaculture on these services. Media analyses as a source of information on negative effects of aquaculture and conflicts associated with it also has its problems (Tiller et al. 2012). This difficulty in assessing negative effects was reflected in a recent end-consumer study in Germany by Feucht and Zander (2014). Their focus group outcomes revealed an obvious lack of knowledge among the participants concerning aquaculture in general. It was found that consumers are mostly unfamiliar with aquaculture (Aarset et al. 2004). The image of aquaculture seems to be created by comparing it to agricultural systems and by contrasting it with fishing, whilst, at the same time the image of aquaculture being an unsustainable, antibiotic-driven production activity prevailed. Further negative conflicts of aquaculture pertain to the rights to utilize certain marine areas, more often than not having roots in legal, ethical, economic, historical and social aspects of marine use (Joyce and Satterfield 2010).

8.2.5   Effects on the Wider Economic and Innovation System

The establishment of aquaculture businesses in a region can influence the availability of input factors like skilled labor, specialized suppliers, education programs, and other infrastructure. Competition for input factors in limited supply may hamper the development of other industries, particularly if they are not as profitable as aquaculture. Over time, the development and growth of the aquaculture sector can however stimulate other more or less related industries to grow in the area (Ørstavik 2004), i.e. through the contribution to the development of regional innovation systems (Asheim and Coenen 2005), including developing knowledge intensive service activities (Aslesen and Isaksen 2007).

However, some countries governments have reacted to this situation and placed demands on large producers (Huemer 2010). In the case of Scotland, this has led to significant investment in infrastructures like improved roads, schools and other facilities (Georgakopoulos and Thomson 2008).

In Norway, there has been frustration in many coastal communities in the latter years, as increased ownership concentration and centralization of production have excluded many of them from the benefits generated by the aquaculture activities and production chain (Huemer 2010; Sandberg 2003; Isaksen and Mikkelsen 2012).

Thus, the rules of aquaculture that evolved over the past decades, based on notions of 'managing' marine resources for aquaculture practices, were almost all oriented toward determining who could gain access to a certain marine area and how much they would be taxed (see e.g. salmon aquaculture tax practices in Norway, Isaksen and Mikkelsen 2012).

8.2.6   Effects on Cultural Fabric and Other Social Aspects

Socially-sound aquaculture development relies on the understanding of two fundamental aspects: (A) the conditions that aquaculture operates under and (B) the mechanisms and channels by which aquaculture affects the social fabric. The latter term encapsulates the social context-specific setting in a particular ecosystem with its respective people and their attributes, e.g. knowledge holders, right holders, access to power holders, gender and institutions, among others (Krause et al. 2015). Indeed, all across human history and geography, people have perceived, lived, used and explored marine resources in coastal lagoons and bays, estuaries and shores, which conducted to changes in those habitats and their natural populations. Consequently, coastal communities changed their values and perceptions of the sea, as well as their way of living and of using natural resources. Thus, advancing change within any civilization does not occur in a vacuum, but rather must evolve out of the given circumstances and discourses that prevail. For instance in western civilization, we cannot think of culture without considering the context in which products and goods are produced, mediated, and consumed over time. Thus, aquaculture as culture exists in relationship to broader societal discourses that evolve across different scales, from interpersonal and group relationships to mainstream media discourses (Bell-Jordan 2008; Fiske 1987; Rosteck and Frentz 2009).

New aquaculture industries, especially in rural areas, should strive to integrate into the cultural fabric of the local community (Burbridge et al. 2001). This is paramount consideration, i.e. if an aquaculture industry develops successfully in a region or community that has previously been dominated by other types of industries, it can alter the very image of what the region, community and its inhabitants "is". This can be because the base economic activity has been a fundamental factor for the identity of the community and its inhabitants, be it fisheries, tourism, agriculture or something else. It can be because aquaculture introduces new and very visible landscape elements, and a third possibility is that in-migration substantially alters the cultural mix in the community population. Furthermore, gender issues, like the opportunities for increased women participation and responsibility in the labor market through aquaculture development may gain more prominence. However, this seems to primarily have been an issue for developing countries (Veliu et al. 2009; Ndanga et al. 2013).

How aquaculture development and planning processes are organized may also affect learning among and between stakeholder groups. Leach et al. (2013) examine qualities in aquaculture partnerships in the US that enhance knowledge acquisition and belief change, and these include procedural fairness, trustworthiness among participants, diverse participation and the level of scientific certainty. Their work also indicate that knowledge acquired through collaborative partnerships make the participants primed to change their opinions on science or policy issues. Stepanova (2015) find that knowledge integration and joint learning are crucial for conflict resolution over coastal resource use.

8.3 Current Knowledge on Socio-economic Effects of Offshore Aquaculture

Having briefly presented the general types of socio-economic effects of aquaculture, we will review the current knowledge of the socio-economic effects of offshore aquaculture. Hereby we specifically review existing, rather limited, knowledge on the socio-economic dimensions of offshore multi-use. However, we believe that it is likely not be a very large difference of the socio-economic categories mentioned above pertaining to the relevant aspects for offshore aquaculture in a multi-use setting.

For instance, summarizing several stakeholder analyses for offshore aquaculture, Krause et al. (2003, 2011) and Wever et al. (2015) showed that there are different types of actors involved in the offshore realm than in near-shore areas. Due to this, different types of conflicts, limitations and potential alliances surface. These root in the essential differences in the origin, context and dynamics of near-shore—versus offshore resource uses. For instance, the near-shore areas in Germany have been subject to a long history of traditional uses through heterogeneous stakeholder groups from the local to national level (e.g. local fisheries communities, tourism industry, port developers, military, etc.), in which traditional user patterns emerged over a long time frame. In contrast, the offshore areas have only recently experienced conflict. This can be attributed to the relatively recent technological advancements in shipping and platform technology, both of which have been driven by capital-strong stakeholders that operate internationally. Whereas there is a well-established organizational structure present among the stakeholders in the near-shore areas in terms of social capital and trust, as well as tested modes of conduct and social networks, these are lacking in the offshore area. Indeed, for the latter, a high political representation by stakeholders is observed, that possess some degree of "client" mentality towards decision-makers in the offshore realm. It implies in this context that financial powerful and political influential «newcomers», such as offshore wind farm operators, effect the political and economic environment in providing favorable operation conditions (Krause et al. 2011; Griffin et al. 2015).

These fundamental differences between the different stakeholders in near-shore and offshore waters make a streamlined approach to multiple use or conflict management difficult. The results of the survey of stakeholders in Krause et al. (2011) indicate the importance of the social context for how various mariculture-wind farm integration processes go forward, specifically regarding the various forms of ownership and management such a venture might take.

Thus, the effects covered in the offshore aquaculture studies mentioned above pertain mainly to the socio-economic typology realms of *Attitudes to and perceptions of aquaculture and its effects*, the *Effects on the wider economic and innovation system* as well as on the *Organization of and participation in planning for aquaculture* and, closely related to these, the *Effects on cultural fabric and other social aspects*.

A number of studies have considered the economic performance of various types of offshore aquaculture primarily from the farm or business side. Some also provide more general frameworks or analyses to the profitability of offshore aquaculture. All these contributions fits into the social typology category of *Direct benefits of aquaculture and their distribution*. Jin et al. (2005) provides a risk-assessment model of open-ocean aquaculture, and present a case study of farming of Atlantic cod farming in New England, USA. Knapp (2008a) provides an analytical approach to the economic potential of offshore aquaculture in the United States of America, while Knapp (2008b) consider the potential employment and income which might be created, directly and indirectly, from U.S. offshore aquaculture. Kim et al. (2008) investigate the investment decision of offshore aquaculture under risk, for Rock Bream aquaculture in Korea.[1] Kam et al. (2003) gave an early case study of offshore Pacific Threadfin aquaculture in Hawaii. Kim et al. (2008) considered the economic viability of offshore Rock Bream aquaculture in Korea, and later (Kim and Lipton 2011) provided a comparative economic analysis of inshore and offshore Rock Bream aquaculture in Korea. Kim (2012) provides an economic feasibility study of mackerel offshore aquaculture production in Korea (see footnote 1).

A recent study of the economic performance of Italian offshore mariculture by Di Trapani et al. (2014) evaluated the economic performance of an offshore production system for sea bass compared to an inshore one, based on interviews with actual, "representative" farmers. They compared the net present value, discounted payback time, and the internal rate of return. They found better economic profitability of offshore farming than inshore, even if sensitivity analysis revealed that financial indicators of both aquaculture production systems have been very sensitive to market condition changes. They also ran Monte Carlo simulations to test the robustness of their analysis. They concluded that an "offshore production system could represent an opportunity for fish farmers to increase their profitability, obtaining a more sustainable production and avoiding possible conflicts with other human activities in coastal areas." The authors have, however, not specifically investigated possible conflicts with other activities, only the financial performance.

A few scientific contributions have considered the legislative side required for offshore aquaculture in various countries, fitting into the social dimensions type *Organization of and participation in planning for aquaculture*. Stickney et al. (2006) considered the interest in open ocean aquaculture in the USA, the regulatory environment, and the potential for sustainable development. They concluded that in the time of their study there was little interest in commercial offshore aquaculture, largely because of the lack of a formal regulatory structure. Cicin-Sain et al. (2005) in a technical report considered requirements and proposed a legislative framework for offshore aquaculture in the USA, including planning and site assessment, leasing and permitting for sites, and monitoring and compliance and enforcement. They briefly also considered the economic potential and possible environmental effects. Forster (2008) tried to answer the question "What new law, if any, is needed

---

[1]Paper in Korean, abstract in English.

to enable private farming in marine public lands?" with the USA as the empirical reference. Cha et al. (2009) studied legislation and planning for offshore aquaculture in Korea (see footnote 1).

It seems common to assume that moving aquaculture further offshore will reduce the conflicts with other users. Some papers have tried to investigate this, in particular to fisheries. These papers consider many more interactions than just competition over ocean space. Knapp (2008b) considers, for the USA only, potential impacts of market-driven changes from offshore aquaculture prices and production volumes of wild and farmed fish, and the subsequent changes in net economic benefits to fishermen, and also fish farmers and consumers.

Valderrama and Anderson (2008) include several other mechanisms for interaction, and point to a number of ways that modern coastal aquaculture has affected fisheries, and through which also offshore aquaculture could make an impact. They believe that the largest influence of aquaculture on wild fisheries has probably occurred through international trade and the market, having: "(a) influenced prices negatively through increased supply and positively through the development of new markets; (b) changed consumer behavior; (c) accelerated globalization of the industry; (d) increased concentration and vertical integration in the seafood sector; (e) resulted in the introduction of new product forms; and (f) significantly changed the way seafood providers conduct business." They find that the growth of aquaculture has stimulated the traditional wild fisheries sector to improve product quality, and in some cases also wild fisheries management to improve. But the success of the aquaculture sector has also lead to attacks from the fisheries sector—and environmental organizations—which Valderrama and Anderson (2008) link to the establishment of international trade restrictions, for example for salmon, shrimp and catfish. While some of the interactions and effects fit into the social typology category of *Negative effects of aquaculture production activities and conflicts with other interests*, some also relate to *Effects on the wider economic and innovation system*.

Valderrama and Anderson (2008) also consider environmental interactions. They sum up that "aquaculture has: (a) directly influenced fish stocks through its use of wild fish stocks for inputs, such as feed and juveniles; (b) influenced fish stocks through intentional releases (salmon stock enhancement) or through unintentional escapes; (c) displaced wild fish through its use of habitat and, in some cases, enhanced fisheries habitat (e.g., some oyster operations); and (d) influenced and been influenced by wild fish stocks through transmission of diseases and parasites." This mainly fits into the social typology categories of *Negative effects of aquaculture production activities and conflicts with other interests*.

In a recent study, Tiller et al. (2013) used systems thinking and Bayesian-belief networks approach to investigate how offshore aquaculture developments in California can impact commercial fishermen. The scientists arranged a workshop with 10 commercial fishermen where they presented 7 pre-selected drivers for how offshore aquaculture could affect the fishermen. These drivers, e.g. such as "the quantity of farmed seafood released to the market", were defined as "variables that influence other variables, but are typically not affected themselves, within the stakeholder's sector" (Tiller et al. 2013). During the workshop the participants nominated Marine Spatial Planning (MSP) as an additional driver. They even coined it a "super" driver, and felt it necessary as a pre-condition for the discussion as a whole. The exercise Tiller et al. (2013) conducted was thus to explicitly identify what attitudes and risk perceptions stakeholders have for offshore aquaculture. The concrete risks or effects they identified were, among others, related to loss of income, to extra costs incurring due to loss of gear, and to being excluded from the fishing grounds. All the effects they considered fit in the social typology category of *Negative effects of aquaculture production activities and conflicts with other interests*.

Forster (2008) sets out to tackle some broader issues related to offshore aquaculture, including how the potential of offshore aquaculture fit into the bigger picture of global food supply, how the assessed long term potential of offshore aquaculture may be important in evaluating current efforts to get it developed, and how offshore aquaculture should be judged in comparison to other methods of food production.[2] He looked into the anticipated future need and demand for food and hence seafood, but also into other ways of making more productive use of the sea, including for energy production and animal feed. He further discussed some of the criteria to assess the sustainability of offshore aquaculture, as well as some of the major substantial issues. Thus, his deliberations point out to the typology item *Effects on the wider economic and innovation system*.

Lastly, a recent study by Ferreira et al. (2014) in southern Portugal investigated interactions between inshore and offshore clam aquaculture through a modelling framework. This enabled them to consider production volumes in the two contrasting aquaculture settings, as well as the environmental effects and disease interactions between them. They could show that whilst the inshore aquaculture activity targets clams of high value, a substantial part of the primary production which is food for the clams originates from the offshore. The offshore area has one of the world's first offshore aquaculture parks, 3.6 nm from the coast. The park has 60 leases for aquaculture production, 70% for finfish cage culture and 30% for bivalve longline culture, covers 15 km^2 and is at 30–60 m depth. Ferreira et al. (2014) found that the bivalve offshore production has caused a decrease of clam yields inshore. While this is replaced by the yields offshore, it is a source of stakeholder conflict. The authors' modelling of potential disease spread between the offshore and inshore systems made it possible to develop a risk exposure map. The authors argue that such quantitative models of interactions, including reduced yields for inshore stakeholder, demonstrate a need for "strong governance to offset disease risks", and they stress "the need to go beyond the conventional spatial planning toolset in order to ensure an ecosystem approach to aquaculture." These findings fit into the social typology categories of *Direct benefits of aquaculture and their distribution*; *Negative effects of aquaculture production activities and conflicts with other interests*.

---

[2]He also considered legislative issues, as noted earlier.

8.4 Implications for Assessing the Socio-economic Effects of Offshore Aquaculture in a Multi-use Setting

Combining offshore aquaculture with other activities, such as offshore wind farming, is an opportunity to share stakeholder resources and can lead to greater spatial efficiency in the offshore environment. However, next to the apparent questions on the biological and technical nature of how to link these activities, these prospected challenges endorse that the participating social actors will have to negotiate agreements and management regulations for elaborating and coordinating their individual tasks (Michler-Cieluch and Krause 2008).

So far, the main motivation to undertake this multi-use efforts in the offshore aquaculture sector can be attributed to the assumed positive socio-economic effects that such an approach, in which resources and activities are shared and managed jointly, may hold. Indeed, the decision to partner with mariculture firms may be primarily motivated by cost considerations for wind energy firms (Reith et al. 2005; Griffin and Krause 2010). Thus, in an offshore setting where many users are competing for space, allowing the concurrent use of a wind farm for mariculture may provide a dual benefit to wind energy producers and aquaculture (Krause et al. 2011). The benefits can be large when firms coordinate core skills to form an alliance with unique capabilities that neither partner could efficiently provide alone. For instance, Michler-Cieluch and Krause (2008) showed that there is scope for such wind farm-mariculture cooperation in terms of operation and maintenance activities (Table 8.1).

This assessment points out to the strong role of attitudes towards offshore aquaculture that act as building blocks to engage and act in a multi-use context. These relate directly to the question of acceptance, the prevailing mental-mind models of stakeholders and the probability of joint action. Indeed, pre-existing

Table 8.1 SWOT-Matrix on potentials and constraints factors of multi-use approaches in offshore aquaculture and wind farms (modified from Michler-Cieluch and Krause 2008)

	Potential	Constraint
	Strengths	Weaknesses
Internal	Development of a flexible, collective transportation scheme Sharing of high-priced facilities Shortening of adaptive learning process by making use of available experiences and knowledge	Little to no interest in joint planning process Little willingness to engage into new fields of activity Ambiguous assignment of rights and duties
	Opportunities	Threats
External	Available working days coincide Availability of a wide range of expertise Lack of legislation in EEZ favors implementation of innovative concepts	Lack of regulatory framework supporting co-management arrangements No access rights within wind farm area for second party Unsolved problems of liability

social networks can provide significant political leverage for governance transformations as required for the move offshore. Moving beyond sole offshore operation and maintenance aspects of multi-use offshore aquaculture, our proposed first typology of socio-economic effects of aquaculture points to this aspect.

This typology aims to capture the socio-economic consequences and effects of aquaculture, and more specifically offshore aquaculture. We have analyzed the current knowledge of such effects for offshore aquaculture, including in a multi-use setting. The socio-economic dimensions of offshore aquaculture in the existing literature relate primarily to *Attitudes and perceptions of stakeholders*, *Organization of and participation in planning for aquaculture*, *Direct benefits of aquaculture*, and *Negative effects of aquaculture production activities and conflicts with other interests*.

In regard to our proposed typology, we can detect considerable knowledge gaps in all the different socio-economic dimensions. Most noteworthy are the gaps to the specific multi-use issues of offshore aquaculture. Questions on who will benefit on which level to which degree, who takes the burden of the associated risk, and who can be made liable if a multi-use concept affects others in the offshore realm remain to be addressed in more detail. Next to the yet emerging body of literature on the multi-use of offshore waters for aquaculture, the current situation demonstrates clearly that within the vast variety of regulations inside the EU, the EU Member States as well as in North America, their implementation is as yet incipient and examples of best practice in multi-use settings are needed (Krause et al. 2011). These need to combine different knowledge systems (e.g. authorities, decision-makers, local communities, science, etc.) to generate novel insights into the management of multiple uses of ocean space and to complement risk-justified decision-making.

Hence, social and regulatory issues will play a significant role in fostering or hindering collaboration (Christie et al. 2014) for offshore aquaculture in both single and multi-use settings. Given the significant volume of subsidies already used to promote wind energy and smarter use of offshore resources, relatively modest technical or financial support for co-production could provide the catalyst to more fully scope this idea and hopefully move the focus of marine spatial planning a little closer to collaborative solutions (Griffin et al. 2015). In this regard, building partnerships amongst actors and increasing 'social capital' can be a way forward in multi-use offshore aquaculture. Localizing activities in marine spatial planning involves organizing a knowledge base of particular social, cultural, ecological and economic values related to the context of each marine activity. As most offshore aquaculture in a multi-use setting will take place beyond national jurisdictions (although still in the EEZ), a debate on who decides on the future of the sea and what criteria are used to take such decisions remains to be worked out. Indeed, unresolved issues of ownership of the process, i.e. which stakeholders are involved in the consent procedure and their relative influence appear to be crucial (Krause et al. 2015). Furthermore, socio-economic dimensions in aquaculture operation, e.g. emotional ownership of the sea/coastal area by the local residents/stakeholders and the social values that drive this ownership are difficult to capture in such remote offshore settings.

In addition to the issues of how to undertake a streamlined socio-economic assessment based on the suggested typology, scale issues of the effects of multiple-use activities need to be addressed. Indeed, the appropriate scales to analyses the effects and interactions of offshore aquaculture naturally depends on how far different effects extend and how they interact in a multi-use setting. Engaging in offshore aquaculture production, larger scales are required to understand the *context* in which the activity works and the smaller scales support our understanding of the underlying *mechanisms* of the respective aquaculture operation. The necessary interconnectedness of the different scales and time frames needs therefore to be captured by a multi-layered approach (Krause et al. 2015).

8.5   Outlook

Especially for offshore aquaculture in a multi-use setting, we need to be more specific to identify and to articulate societal choices and their related values that build the foundation of the decision to move offshore in a sustainable manner in the first place. This directly frames the socio-economic effects of such activity. The typology of socio-economic dimensions of aquaculture presented in this chapter can be regarded as a first step to capture these societal values and thus remains to be worked out in more detail for their implications for offshore aquaculture. It must be verified with the emerging body of insights and growing experience on how this new form of marine resource may develop in the future.

The present practice involves the political allocation of ocean space for specific purposes only, which leads to a complex mix of ownership, associated commons and private property. Depending on the activity, these contain very different customary and statutory rules and regulations, in which we can detect a "failure of understanding" the socio-economic effects on offshore marine resource use by and large. Indeed, questions as "how *should* the socio-economic dimensions of offshore aquaculture be captured and interpreted?", and How *can* it be managed, and what are the major challenges for efficient sustainable management?" point to existing knowledge gaps and opens up the arena for discussion on what is required to address the socio-economic dimensions of offshore aquaculture.

This current gap between oceans as common and oceans as private property as well as diverging views and pictures leads to a contested sea space. In multi-use settings, these are especially important to address at the interface of policy, research and practice. There is a high risk of failing in integrating offshore aquaculture within the emerging marine management regime. What is at odds is the balanced management of the politically powerful vs. newcomers. Critical for dealing with the whole breadth of socio-economic issues in offshore aquaculture are the further development of suitable and robust methods. These are necessary for analyzing and assessing the cultural fabric and other social aspects that may be impacted through offshore production systems, as well as generating insight on what effects we can expect on the wider economic and innovation systems involved therein.

Acknowledgements We thank Maximilian Felix Schupp for his critical review of an earlier draft of this manuscript. The members of the ICES Working Group on Social and Economic Dimensions of Aquaculture (WGSEDA) are thanked for inspiring workshops over the course of the years that helped to sharpen the social typology presented here.

References

Aarset, B., Beckmann, S., Bigne, E., Beveridge, M., Bjorndal, T., Bunting, J., et al. (2004). The European consumers' understanding and perceptions of the "organic" food regime. *British Food Journal, 106*(2), 93–105.

Anon. (2009). White paper No. 37 to the Storting (2008–2009) Integrated Management of the Marine Environment of the Norwegian Sea. N. M. o. t. Environment. Oslo.

Anon. (2011). White paper 10 (2010–2011) to the Storting. First update of the Integrated Management Plan for the Marine Environment of the Barents Sea–Lofoten Area. N. M. o. t. Environment. Oslo.

Anon. (2013). White paper 37 to the Storting (2012–2013) Integrated Management of the Marine Environment of the North Sea and Skagerrak (Management Plan) N. M. o. t. Environment.

Asheim, B. T., & Coenen, L. (2005). Knowledge bases and regional innovation systems: Comparing Nordic clusters. *Research Policy, 34*(8), 1173–1190.

Aslesen, H. W., & Isaksen, A. (2007). New perspectives on knowledge-intensive services and innovation. *Geografiska Annaler: Series B. Human Geography, 89*, 45–58.

Bacher, K., Gordoa, A., & Mikkelsen, E. (2014). Stakeholders' perceptions of marine fish farming in Catalonia (Spain): A Q-methodology approach. *Aquaculture, 424–425*, 78–85.

Barrett, G., Caniggia, M., & Read, L. (2002). There are more vets than doctors in Chile: Social and community impact of the globalization of aquaculture in Chile. *World Development, 30*(11), 1951–1965.

Barton, J. R., & Fløysand, A. (2010). The political ecology of Chilean salmon aquaculture, 1982–2010: A trajectory from economic development to global sustainability. *Global Environmental Change, 20*, 739–752.

Bell-Jordan, K. E. (2008). Black. White. And a survivor of the real world: Constructions of race on reality TV. *Critical Studies in Media Communication, 25*(4), 353–372.

Belton, B., Haque, M. M., Little, D. C., & Sinh, L. X. (2011). Certifying catfish in Vietnam and Bangladesh: Who will make the grade and will it matter? *Food Policy, 36*, 289–299.

Belton, B., & Little, D. C. (2011). The social relations of catfish production in Vietnam. *Geoforum, 42*(5), 567–577.

Belton, B., & Bush, S. R. (2014). Beyond net deficits: New priorities for an aquacultural geography. *The Geographical Journal, 180*(1), 3–14.

Bergfjord, O. J. (2009). Risk perception and risk management in Norwegian aquaculture. *Journal of Risk Research, 12*(1), 91–104.

Biederlack, L., & Rivers, J. (2009). *Comprehensive Food Security & Vulnerability Analysis (CFSVA): Ghana.* United Nations World Food Programme.

Buanes, A, Jentoft, S., Karlsen, G. R., Maurstad, A., & Søreng, S. (2004). In whose interest? An exploratory analysis of stakeholders in Norwegian coastal zone planning. *Ocean and Coastal Management, 47*(5–6), 207–223.

Buanes, A., Jentoft, S., Maurstad, A., Søreng, S., & Karlsen, G. R. (2005). Stakeholder participation in Norwegian coastal zone planning. *Ocean and Coastal Management, 48*(9–10), 658–669.

Buchholz, C., Krause, G., & Buck, B. H. (2012). Seaweed and man. In C. Wiencke & K. Bischof (Eds.), *Seaweed biology: Novel insights into ecophysiology, ecology and utilization* (pp. 471–493). Heidelberg: Springer.

Buck, B. H., Krause, G., Rosenthal, H., & Smetacek, V. (2003). Aquaculture and environmental regulations: The German situation with the North Sea. In A. Kircher (Ed.), *International marine environmental law: Institutions, implementation and innovation* (pp. 211–229). Kluwer Law International 64.

Buck, B. H., Krause, G., Michler-Cieluch, T., Brenner, M., Buchholz, C. M., Busch, J. A., et al. (2008). Meeting the quest for spatial efficiency: Progress and prospects of extensive aquaculture within offshore wind farms. *Helgoland Marine Research 62*, 269–281.

Buck, B. H., & Krause, G. (2012). Integration of aquaculture and renewable energy systems. In R. A. Meyers (Ed.), *Encyclopaedia of sustainability science and technology* (p. 23). Berlin: Springer.

Bunting, S. W. (2004). Wastewater aquaculture: Perpetuating vulnerability or opportunity to enhance poor livelihoods? *Aquatic Resources, Culture and Development, 1*, 51–75.

Burbridge, H., Roth, E., & Rosenthal, H. (2001). Social and economic policy issues relevant to marine aquaculture. *Journal of Applied Ichthyology, 17*(4), 194–206.

Cabral, R. B., & Aliño, P. M. (2011). Transition from common to private coasts: Consequences of privatization of the coastal commons. *Ocean and Coastal Management, 54*(1), 66–74.

Cha, C.-P., Lee, K.-N., & Kim, M.-J. (2009). A study on a legislation plan for introduction of offshore aquaculture fisheries regime. *Journal of Fisheries and Marine Sciences Education, 21* (3), 335–346.

Christie, N., Smyth, K., Barnes, R., & Elliot, M. (2014). Co-location of activities and designations: A means of solving or creating problems in marine spatial planning? *Marine Policy, 43*, 254–261.

Chu, J., Anderson, J. L., Asche, F., & Tudur, L. (2010). Stakeholders' perceptions of aquaculture and implications for its future: A comparison of the USA and Norway. *Marine Resource Economics, 25*(1), 61–76.

Cicin-Sain, B., Bunsick, S. M., Corbin, J., DeVoe, M. R., Eichenberg, T., Ewart, J., et al. (2005). An operational framework for offshore marine aquaculture in U.S. federal waters. Technical Report. Center for Marine Policy, University of Delaware.

Costa-Pierce, B. A. (2010). Sustainable ecological aquaculture systems: The need for a new social contract for aquaculture development. *Marine Technology Society Journal, 44*(3), 88–112.

Di Trapani, A. M., Sgroi, F., Testa, R., & Tudisca, S. (2014). Economic comparison between offshore and inshore aquaculture production systems of European sea bass in Italy. *Aquaculture, 434*, 334–339.

Duarte, C. M., Marbá, N., & Holmer, M. (2007). Rapid domestication of marine species. *Science, 316*, 382–383.

EU. (2014). Directive 2014/89/EU of the European Parliament and of the Council of 23 July 2014 establishing a framework for maritime spatial planning. Brussels.

FAO. (2014). *The state of World Fisheries and aquaculture—opportunities and challenges* (p. 240). Rome: FAO Food and Agricultural Organisation of the UN.

Ferreira, J., Saurel, C., Lencart e Silva, J. D., Nunes, J. P., & Vazquez, F. (2014). Modelling of interactions between inshore and offshore aquaculture. *Aquaculture, 426*, 154–164.

Feucht, Y., & Zander, K. (2014). Consumers' knowledge and information needs on organic aquaculture. *Building Organic Bridges, 2*, 375–378.

Fiske, J. (1987). *Television culture*. London: Routledge.

Forster, J. (2008). Broader issues in the offshore fish farming debate. In M. Rubino (Ed.), *Offshore aquaculture in the United States: Economic considerations, implications & opportunities* (pp. 245–260). U.S. Department of Commerce; Silver Spring, MD; USA. NOAA Technical Memorandum NMFS F/SPO-103.

Freeman, S., Vigoda-Gadot, E., Sterr, H., Schultz, M., Korchenkov, I., Krost, P., et al. (2012). Public attitudes towards marine aquaculture: A comparative analysis of Germany and Israel. *Environmental Science and Policy, 22*, 60–72.

Fröcklin, S., de la Torre-Castro, M., Lindström, L., Jiddawi, N. S., & Msuya, F. E. (2012). Seaweed mariculture as a development project in Zanzibar, East Africa: A price too high to pay? *Aquaculture, 356–357*, 30–39.

Georgakopoulos, G., & Thomson, I. (2008). *Risk, risk conflicts, sub-politics and social and environmental accounting and accountability in scottish salmon farming*. Netherlands: Amsterdam Business School Research Institute.

Goulding, I., & Hallam, D., Harrison-Mayfield, L., Mackenzie-Hill, V., & da Silva, H. (2000). Regional socio-economic studies on employment and the level of dependency on fishing. Regional Profiles. Lot. Brussels, Belgium, Commission of the European Communities, Directorate-General for Fisheries, 15.

Griffin, R., Buck, B. H., & Krause, G. (2015). Private incentives for the emergence of co-production of offshore wind energy and mussel aquaculture. *Aquaculture, 436,* 80–89.

Griffin, R., & Krause, G. (2010). Economics of Wind Farm-Mariculture Integration. Working Paper, Department of Environmental and Natural Resource Economics, University of Rhode Island.

Grigorakis, K., & Rigos, G. (2011). Aquaculture effects on environmental and public welfare— The case of Mediterranean mariculture. *Chemosphere, 85*(6), 899–919.

Hall, S. J., Delaporte, A., Phillips, M. J., Beveridge, M., & O'Keefe, M. (2011). *Blue frontiers: Managing the environmental costs of aquaculture*. Penang, Malaysia: The World Fish Center.

Hishamunda, N., Cai, J., & Leung, P. S. (2009). Commercial aquaculture and economic growth, poverty alleviation and food security. FAO Fisheries and Aquaculture Tech. Paper, no. 512, Rome.

Hopkins, T. S., Bailly, D., & Støttrup, J. G. (2011). A systems approach framework for coastal zones. *Ecology and Society, 16*(4), 25.

Hortle, K. G. (2007). Consumption and the yield of fish and other aquatic animals from the Lower Mekong Basin. *MRC Technical Paper, 16,* 1–88.

Huemer, L. (2010). Corporate social responsibility and multinational corporation identity: Norwegian strategies in the Chilean aquaculture industry. *Journal of Business Ethics, 91,* 265–277.

Isaksen, J. R., & Mikkelsen, E. (2012). Økonomer i kystsonen: Kan kunnskap om verdiskaping gi bedre arealforvaltning? In B. Hersoug & J. P. Johnsen (Eds.), *Kampen om plass på kysten: interesser og utviklingstrekk i kystsoneplanleggingen* (pp. 159–178). Oslo: Universitetsforlaget.

Janssen, M. A., & Ostrom, E. (2006). Governing social-ecological systems. In L. Tesfatsion & K. L. Judd (Eds.), *Handbook of computational economics*, Vol. 2 (pp. 1465–1509). North Holland.

Jentoft, S., & Chuenpagdee, R. (2013). Concerns and problems in fisheries and aquaculture— exploring governability. In M. Bavinck, R. Chuenpagdee, S. Jentoft, & J. Kooiman (Eds.), *Governability of fisheries and aquaculture* (pp. 33–44). Netherlands: Springer.

Jin, D., Kite-Powell, H., & Hoagland, P. (2005). Risk assessment in open-ocean aquaculture: A firm-level investment-production model. *Aquaculture Economics and Management, 9,* 369–387.

Joyce, A. L., & Satterfield, T. A. (2010). Shellfish aquaculture and First Nations' sovereignty: The quest for sustainable development in contested sea space. *Natural Resources Forum, 34*(2), 106–123.

Kam, L. E., Leung, P., & Ostrowski, A. C. (2003). Economics of offshore aquaculture of Pacific Threadfin (*Polydactylus sexfilis*) in Hawaii. *Aquaculture, 223*(2), 63–87.

Katranidis, S., Nitsi, E., & Vakrou, A. (2003). Social acceptability of aquaculture development in coastal areas: The case of two Greek Islands. *Coastal Management, 31*(1), 37–53.

Kim, D.-H., Choi, J.-Y., & Lee, J.-U. (2008). A study on the investment decision of offshore aquaculture under risk. *The Journal of Fisheries Business Administration, 39*(2), 109–123.

Kim, D., & Lipton, D. (2011). A comparison of the economic performance of offshore and inshore aquaculture production systems in Korea. *Aquaculture Economics & Management, 15*(2), 103–117.

Kim, D.-H. (2012). An economic feasibility study of mackerel offshore aquaculture production system. *The Journal of Fisheries Business Administration, 43*(3), 23–30.

Knapp, G. (2008a). Economic potential for U.S. offshore aquaculture: An analytical approach. In: M. Rubino (Ed.), *Offshore aquaculture in the United States: Economic considerations, implications & opportunities (pp. 15–50). U.S. Department of Commerce; Silver Spring, MD; USA. NOAA Technical Memorandum NMFS F/SPO-103.*

Knapp, G. (2008b). Potential economic impacts of U.S. offshore aquaculture. In M. Rubino (Ed.), *Offshore aquaculture in the United States: Economic considerations, implications & opportunities (pp. 161–188). U.S. Department of Commerce; Silver Spring, MD; USA. NOAA Technical Memorandum NMFS F/SPO-103.*

Krause, G., Brugere, C., Diedrich, A., Ebeling, M. W., Ferse, S. C., Mikkelsen, E., et al. (2015). A revolution without people? Closing the people–policy gap in aquaculture development. *Aquaculture, 447,* 44–55.

Krause, G., Griffin, R. M., & Buck, B. H. (2011). Perceived concerns and advocated organisational structures of ownership supporting 'offshore wind farm—mariculture integration'. In G. Krause (Ed.), *From turbine to wind farms: Technical requirements and spin-off products* (pp. 203–218). InTECH Book.

Krause, G., Buck, B. H., & Rosenthal, H. (2003). Multifunctional use and environmental regulations: Potentials in the offshore aquaculture development in Germany. *Proceedings of the multidisciplinary scientific conference on sustainable coastal zone management "Rights and Duties in the Coastal Zone",* 12–14 June 2003. Stockholm, Sweden.

Ladenburg, J., & Krause, G. (2011). Local attitudes towards wind power: The effect of prior experience. In G. Krause (Ed.), *From turbine to wind farms: Technical requirements and spin-off products* (pp. 3–14). InTECH.

Leach, W. D., Weible, C. M., Vince, S. R., Siddiki, S. N., & Calanni, J. C. (2013). Fostering learning through collaboration: Knowledge acquisition and belief change in marine aquaculture partnerships. *Journal of Public Administration Research and Theory, 24*(3), 591–622.

Leroy, J. L., & Frongillo, E. A. (2007). Can interventions to promote animal production ameliorate undernutrition? *The Journal of Nutrition, 137*(10), 2311–2316.

Lipton, D. W., & Kim, D. H. (2007). Assessing the economic viability of offshore aquaculture in Korea: An evaluation based on rock bream (*Oplegnathus fasciatus*) production. *Journal of the World Aquaculture Society, 38*(4), 506–515.

Martinez-Alier, J. (2001). Ecological conflicts and valuation: Mangroves versus shrimps in the late 1990s. *Environment and Planning C: Government and Policy, 19*(5), 713–728.

Mazur, N. A., & Curtis, A. L. (2006). Risk perceptions, aquaculture, and issues of trust: lessons from Australia. *Society and Natural Resources, 19*(9), 791–808.

Mazur, N. A., & Curtis, A. L. (2008). Understanding community perceptions of aquaculture: Lessons from Australia. *Aquaculture International, 16*(6), 601–621.

Michler-Cieluch, T., & Krause, G. (2008). Perceived concerns and possible management strategies for governing "wind farm–mariculture integration". *Marine Policy, 32*(6), 1013–1022.

Millenium Ecosystem Assessment. (2005). Ecosystems and Human Well-Being. Synthesis. Washington, DC: Island Press.

Nagarajan, R., & Thiyageasan, K. (2006). The effect of coastal shrimp farming on birds in Indian Mangrove forests and tidal flats. *Acta Zoologica Sinica, 52,* 541–548.

Naylor, R. L., Goldburg, R. J., Primavera, J. H., Kautsky, N., Beveridge, M. C., Clay, J. et al. (2000). Effect of aquaculture on world fish supplies. *Nature, 405*(6790), 1017–1024.

Ndanga, L. Z., Quagrainie, K. K., & Dennis, J. L. (2013). Economically feasible options for increased women participation in Kenyan aquaculture value chain. *Aquaculture, 414,* 183–190.

Ørstavik, F. (2004). Knowledge spillovers, innovation and cluster formation: the case of Norwegian aquaculture. In C. Karlsson, P. Flensburg, & S. Å. Hörte (Eds.), *Knowledge spillovers and knowledge management.* Edvard Elgar Publishing.

Paul, B. G., & Vogl, C. R. (2011). Impacts of shrimp farming in Bangladesh: Challenges and alternatives. *Ocean and Coastal Management, 54*(3), 201–211.

Primavera, J. (2006). Overcoming the impacts of aquaculture on the coastal zone. *Ocean and Coastal Management, 49*(9), 531–545.

Reith, J. H., Deurwaarder, E. P., Hemmes, K., Curvers, A. P. W. M., & Kamermans, P. (2005). BIO-OFFSHORE Grootschalige teelt van zeewieren in combinatie met offshore windparken in de Noordzee. ECN Windenergie. Report ECN-C-05-008. Retrieved September 13, 2015, from http://www.ecn.nl/docs/library/report/2005/c05008.pdf

Robertsen, R., Iversen, A., & Andreassen, O. (2012). Havbruksnæringens ringvirkninger i Troms (Ripple effects of aquaculture industry in Troms). Nofima Rapport. Tromsø, Norway, Nofima: 67.

Rosteck, T., & Frentz, T. S. (2009). Myth and multiple readings in environmental rhetoric: The case of an inconvenient truth. *Quarterly Journal of Speech, 95*(1), 1–19.

Ryan, J., Mills, G., & Maguire, D. (2004). *Farming the deep blue*. Dublin, Ireland: Marine Inst.

Sandberg, A. (2003). Commons for whom? New coastal commons on North-Norwegian Coasts. In A. Sandberg, E. Berge, & L. Carlsson (Eds.), *Landscape, law & justice: Proceedings from a workshop on old and new commons* (pp. 135–140). Trondheim, Norway: Dept. of Sociology and Political Science, Norwegian University of Science and Technology.

Sandberg, M. G., Henriksen, K., Aspaas, S., Bull-Berg, H., & Stokka, A. (2014). Verdiskaping og sysselsetting i norsk sjømatnæring - en ringvirkningsanalyse med fokus på 2012 (Value creation and employment in Norwegian Seafood industry—a ripple effect analysis with focus on 2012), SINTEF Fiskeri og havbruk.

Schmitt, L. H. M., & Brugere, C. (2013). Capturing ecosystem services, Stakeholders' preferences and trade-offs in Coastal aquaculture decisions: A Bayesian belief network application. PLoS ONE. doi:10.1371/journal.pone.0075956

Shafer, C. S., Inglis, G. J., & Martin, V. (2010). Examining residents' proximity, recreational use, and perceptions regarding proposed aquaculture development. *Coastal Management, 38*(5), 559–574.

Sievanen, L., Crawford, B., Pollnac, R. B., & Lowe, C. (2005). Weeding through assumptions of livelihood approaches in ICM: Seaweed farming in the Philippines and Indonesia. *Ocean & Coastal Management, 48*, 297–313.

Slater, M. J., Mgaya, Y. D., Mill, A., Rushton, S. P., & Stead, S. M. (2013). Effect of social and economic drivers on choosing aquaculture as a coastal livelihood. *Ocean and Coastal Management, 73*, 22–30.

Stepanova, O. (2015). Conflict resolution in coastal resource management: Comparative analysis of case studies from four European countries. *Ocean and Coastal Management, 103*, 109–122.

Stickney, R. R., Costa-Pierce, B., Baltz, D. M., Drawbridge, M., Grimes, C., Phillips, S., et al. (2006). Toward sustainable open ocean aquaculture in the United States. *Fisheries, 31*, 583–610.

Stonich, S. C., & Bailey, C. (2000). Resisting the blue revolution: Contending coalitions surrounding industrial shrimp farming. *Human Organization, 59*(1), 23–26.

Tiller, R., Gentry, R., & Richards, R. (2013). Stakeholder driven future scenarios as an element of interdisciplinary management tools; the case of future offshore aquaculture development and the potential effects on fishermen in Santa Barbara, California. *Ocean and Coastal Management, 73*, 127–135.

Tiller, R., Brekken, T., & Bailey, J. (2012). Norwegian aquaculture expansion and Integrated Coastal Zone Management (ICZM): Simmering conflicts and competing claims. *Marine Policy, 36*(5), 1086–1095.

Valderrama, D., & Anderson, J. (2008). Interactions between capture fisheries and aquaculture. In M. Rubino (Ed.), *Offshore aquaculture in the United States: Economic considerations, implications & opportunities* (pp. 189–206). U.S. Department of Commerce; Silver Spring, MD; USA. NOAA Technical Memorandum NMFS F/SPO-103.

Varela, J. (2001). The human rights consequences of inequitable trade and development expansion: The abuse of law and community rights in the Gulf of Fonseca, Honduras. In D. Barnhizer (Ed.), *Effective strategies for protecting human rights: Prevention and intervention, trade and education*. Dartmouth: Ashgate.

Veliu, A., Gessese, N., Ragasa, C., & Okali, C. (2009). *Gender analysis of aquaculture value chain in Northeast Vietnam and Nigeria*. Washington, DC: The World Bank.

Wever, L., Krause, G., & Buck, B. H. (2015). Lessons from stakeholder dialogues on marine aquaculture in offshore wind farms: Perceived potentials, constraints and research gaps. *Marine Policy, 51*, 251–259.

Whitmarsh, D., & Palmieri, M. G. (2009). Social acceptability of marine aquaculture: The use of survey-based methods for eliciting public and stakeholder preferences. *Marine Policy, 33*(3), 452–457.

Zurek, M. B., & Henrichs, T. (2007). Linking scenarios across geographical scales in international environmental assessments. *Technological Forecasting and Social Change, 74*(8), 1282–1295.

Chapter 9
Regulation and Permitting of Standalone and Co-located Open Ocean Aquaculture Facilities

John S. Corbin, John Holmyard and Scott Lindell

Abstract Aquaculture will be the dominant producer of global seafood in the 21st Century and the emerging open ocean farming sector must play an important role. Current concepts for utilizing the ocean for aquaculture in national Exclusive Economic Zones, include; standalone structures for growing shellfish and finfish and aquaculture production facilities co-located with other, compatible ocean uses, e.g., offshore oil platforms and wind energy towers. There may also be potential for multiple use of marine space incorporating aquaculture at different trophic levels together with artificial reefs, recreational angling and diving and commercial fishing using static gear. This chapter examines the global status of open ocean aquaculture (OOA), stand alone and co-located projects, in national ocean jurisdictions. Regulatory regimes and experiences permitting OOA shellfish and finfish farming in the United States and shellfish farming in the United Kingdom are shared. Examples of the permitting process from several U. S. states are also cited. Finally, conclusions and recommendations are offered to assist nations in formulating and implementing regulatory systems to effectively facilitate commercial aquaculture development in the open ocean environment.

9.1 Introduction

The current world population of 7.2 billion is projected to increase by 1 billion over the next 12 years and reach 9.6 billion by 2050 (UNFAO 2014). With this steadily rising population pressure on the earth's resources, as well as its political systems, social structures, and food production alternatives, there are many challenges to

J.S. Corbin (✉)
Aquaculture Planning and Advocacy LLC, Kaneohe, HI, USA
e-mail: jscorbin@aol.com

J. Holmyard
Offshore Shellfish Ltd., Brixham, Devon, UK

S. Lindell
Biology Department, Woods Hole Oceanographic Institution,
MS 34, 104 Redfield Bldg., Woods Hole, MA, USA

© The Author(s) 2017
B.H. Buck and R. Langan (eds.), *Aquaculture Perspective of Multi-Use Sites in the Open Ocean*, DOI 10.1007/978-3-319-51159-7_9

reaching the goal of a sustainable planet (Brown 2009; Friedman 2008). Two of the most important challenges are sufficient food and sufficient energy; both from sustainable sources.

9.1.1 Cultured Seafood Trends

Fish provide 17% of world protein and for nearly 3 billion people and for 4.3 billion people 15% of their consumption (UNFAO 2014). It is well established that yields from capture fisheries have leveled off and future increases in seafood production must come from aquaculture. In 2012, humans consumed 136.2 Mmt of seafood and aquaculture (excluding aquatic plants) provided 66.6 Mmt or 48%. Current cultured supplies are dominated by freshwater species (carps, tilapias, cyprinids, catfishes) and the marine species component (finfish, shellfish and crustaceans) makes up 36% (UNFAO 2014).

Projections of future seafood demand pose a daunting target for producers in both developing and developed countries. Notably, besides relentless population growth driving increased demand, rising income levels are contributing to rising world per capita yearly seafood consumption, which has increased from 9.9 kg in the 1960s to 19.2 kg in 2012 and this trend will continue (UNFAO 2014). Recent World Bank projections indicate "the most plausible scenario" for aquaculture expansion is by 2030 production will reach 93 Mmt, which is an increase over the 2012 value of 66 Mmt of 27 Mmt in only 15 years (World Bank 2013). Notably, aquaculture could supply more than 60% of fish for direct human consumption by this date, however the huge challenge can be portrayed as expanding production by more than the current size (1.3 Mmt) of Norway's highly successful 30 year old salmon industry each year going forward (Corbin 2010).

While there are many challenges for mariculture development to overcome, these seafood trends are focusing greater attention on expansion of coastal, offshore and open ocean farming and its environmental and resource advantages over land-based and inland water production (Lovatelli et al. 2013; Kapetsky et al. 2013). The vision for marine aquaculture, is becoming clear, but the legal and regulatory environment for offshore industry development is consistently cited as a major hurdle to expansion (Knapp 2013; Percy et al. 2013). Experts note the most important factors affecting successful growth will be the extent to which enabling regulatory frameworks establish clear, stable, and timely processes for permitting and regulating large-scale, ocean farming (Knapp 2013).

9.1.2 Ocean Energy Trends

The offshore environment of the continental shelf, including the deep ocean beyond the neritic/sub-littoral zones, is the location and potential location of important energy generation technologies; namely oil and natural gas drilling platforms and wind farms. As with the trend in marine aquaculture, ocean energy generation facilities are moving further offshore and offer opportunities for co-location of fish and shell fish farming.

Drilling for oil and gas in nearshore waters has been going on since the late 1800s and as technologies improved and market forces increased demand, companies moved further and further offshore. Present day technologies are capable of operating in water depths up to 3000 m (AOGHS 2014). Worldwide as of February 2014 there were 1501 offshore oil and gas rigs deployed, with the highest concentrations in the Gulf of Mexico (229) and in the North Sea (180) (Statista 2014). While a projection of numbers of future drilling rigs is not available, it is forecast that the annual investment in rigs will grow from approximately $80 billion dollars to over $120 billion dollars in 2020.

Wind energy farms have moved offshore as technology has become available to support renewable energy production in the harsh, exposed ocean environment. Today 4420 MW of offshore wind power have been installed globally as of June 2012, representing 2% of total installed world wind power capacity. More than 90% of this capacity is installed off northern Europe, in the North Sea, Irish Seas and the English Channel. Expectations are that major new developments will be occurring in China, Japan, Korea, the United States, Canada, and Taiwan in the near future (GWEC 2014).

Current plans for locating more ocean wind farms around the world indicate the future of this form of renewable energy is very bright. According to the more ambitious projections a total of 80 GW of offshore wind power could be installed world-wide by 2020, with three quarters located in Europe (GWEC 2014).

9.2 Status of Commercial Offshore Aquaculture

Aquaculture is the fastest growing food production technology in the world, though rates have been decreasing in recent years (UNFAO 2014). As a greater diversity of marine species became available from advances in husbandry techniques and sturdier growout technologies (e.g., anchored cages, net pens, and long lines), the industry has begun to utilize the offshore environment (Forster 2013; Jeffs 2013). But, what is a widely accepted definition of offshore aquaculture?

Recently the UNFAO Aquaculture Office held a workshop to comprehensively review the status and potential of offshore aquaculture and addressed this question among others (Lovatelli et al. 2013). Marine aquaculture in the ocean was operationally classified into three categories based on site characteristics; Coastal, Off-the-Coast, and Offshore. They can be defined as follows:

(1) Coastal aquaculture—Operations less than 500 m from the coast and in protected or semi-protected waters—less than 90° exposure to the open ocean. Water depth is less than 10 m. These facilities can be accessed 100% of the

time. This category is not the main focus of this discussion, though contrasts between this category and the others will be made.

(2) Off-the-Coast aquaculture—Operations that are 500 m to 3 km from the coast. Generally, they are within the Continental Shelf zone or open ocean and with water depths of 10–50 m and wave heights of less than 3–4 m. These locations are exposed greater than 90° to the open ocean and include some automation to operate.

(3) Offshore aquaculture—Operations located greater than 2 km or out of sight from the coast, in water depths greater than 50 m, with wave heights of 5 m or more, ocean swells, variable winds and strong currents. These locations are exposed equal to or greater than 180° to the open sea and there is a great need for automation of these remote operations (Lovatelli et al. 2013).

These discussions focus on off-the-coast and offshore locations for standalone and co-located aquaculture farms. The term open ocean will be used when both locations are being discussed.

9.2.1 Standalone Aquaculture Projects

Standalone commercial aquaculture in off-the-coast and offshore environments is an emerging activity in many countries that have a history of land-based and coastal marine aquaculture development. As a result, off-the-coast farming technology is evident in many developed countries and is being touted in developing countries for expansion of fish production. Moreover, while offshore technology has been proven technically feasible, it's wide spread application has been slow due to the significant logistical, technical and economic challenges of operating a farm at a site exposed to open ocean winds, waves and currents (Lovatelli et al. 2013).

A recent study of the status and potential of global offshore mariculture from a spatial perspective (suitable area) utilizing nations with existing industry is instructive. Between 2004 and 2008, 93 nations produced an average yearly production of 30 Mmt, with a mean of 15 tonnes/km of coastline and a median value of 1 tonne/km, indicating a few large producers and many small ones. The maximum observed value was 520 tonnes/km. About half the 93 producing nations have production of less than 1 tonne/km of coastline (Kapetsky et al. 2013).

Indicative of potential, 80 of these producing nations have a total coastline length of 1,472,111 km, while at present 44% of maritime nations with 0.3 million km of coast line are not practicing mariculture. Nations with the greatest long-term development potential based on coastline length are Canada, Indonesia, and Chile. Nations with the greatest long-term potential based on area of their EEZ are Russia, the United States and France (Kapetsky et al. 2013).

Most of temperate water marine fish production in ocean locations is located in Northern Europe (40%), followed by South America (27%); with Norway and Chile dominating. Atlantic salmon is the dominant fish, with Norway the largest producer followed by Chile. The next most important species are Sea Bream and Sea Bass, with China being the largest producer followed by Mediterranean Sea countries, with Greece as the largest producer. For shellfish (largely oysters and mussels), the dominant producers are the Republic of Korea, Japan, and Spain, followed by much smaller production from France, Chile, and the United States (Holmer 2013).

The majority of tropical offshore aquaculture activity is concentrated in two regions, Asia and tropical locales of the Caribbean and Hawaii-USA. Off-the-coast and offshore aquaculture in Asia are dominated by China, with 900 cages growing cobia, amberjack, sea bream, and other economically important species. Deep water offshore cages were initiated in China in the late 1990s and expansion has continued as a government priority in the coastal provinces with suitable climate (Angel and Edelist 2013).

Hawaii-U.S.A. and the tropical locales of Puerto Rico, the Bahamas, Panama, Belize, Mexico, and Costa Rica all have had large-scale research, pilot projects and several commercial scale projects (Panama currently boosts the largest open ocean farm in the world), with cobia as the most popular species in the Gulf of Mexico and the Caribbean (Alston et al. 2005; Angel and Edelist 2013; Open Blue Sea Farms 2014). In October 2014, Hubbs Sea World Research Institute (HSWRI) began the permit process to locate a 5000 tonne fish farm for Pacific yellowtail 4.5 miles off Southern California, USA; which if successful will be the first commercial fish farm in federal waters (Leschin-Hoar 2014).

9.2.2 Aquaculture Co-located with Platforms

The concept of co-locating commercial farms with established and emerging open ocean private sector activities, that is ocean wind farms and oil and gas drilling, has come to the fore front in recent years (Bridger 2004; Buck et al. 2004). A perceived major advantage of co-location is that the much needed energy generating activities have an existing regulatory framework that may be utilized for siting aquaculture and that permitting/leasing mechanisms are lacking or poorly developed for open ocean aquaculture in most countries (Holmer 2013).

More practical considerations in the case of utilizing wind farms include: the massive platforms to attach aquaculture equipment and the unused space between the turbines for aquaculture that could have restricted public access and enhanced security. In the case of oil and gas drilling structures, active or decommissioned, there is a large structure to attach farm equipment and more importantly with decommissioned rigs, there is unused deck space on the platform for fish tanks and support activities. Also important, with oil and gas platforms, the energy company would not have to go to the expense and could avoid the environmental disruption of dismantling the platform by fostering an ongoing beneficial use.

While strong analytical arguments have been made for incorporating aquaculture into the planning of wind farms and to some extent oil and gas platforms, actual demonstrations are few and small scale (Buck et al. 2004; Bridger 2004; Mee 2006). Cultivation of seaweed and blue mussels have been demonstrated to be biologically and technically feasible in the high energy ocean environment of the North Sea (Buck et al. 2004, 2008). In addition, conceptualization of aquaculture systems that can utilize wind farm infrastructure and adjacent areas is well along (AWI 2014; Lagerwald et al. 2014).

In the US, a pilot demonstration shellfish farm associated with a large, offshore wind farm off Cape Cod, State of Massachusetts, was recently approved and is discussed in the following section. In Europe, in recent years, numerous successful co-location trials have been carried out using blue mussels and assorted other species (Lagerwald et al. 2014).

Co-location of aquaculture research and pilot demonstration projects with oil and gas ocean infrastructure began in the US in the mid-1990s. The first aquaculture facility associated with the industry was the SeaFish project in 1998, located 34 miles off of the State of Texas. SeaFish operated for one year, growing red drum, but ended in 1999 when Shell Oil decided to reactivate the platform.

The second major co-location project was planned for Platform Grace in federal waters 10.5 miles off Ventura, California, USA which stopped producing oil and gas in 1997. Recognizing the opportunity, HSWRI, a prominent US aquaculture research institute, leased the facility in August, 2003 for a project to culture shellfish and a variety of native finfish (HSWRI 2003; Krop and Polefka 2007). In 2004, the agreement with HSWRI was not renewed by Venoco and the company resumed oil and gas production shortly thereafter.

With the rising interest in offshore aquaculture, in 2007 the Minerals Management Service (MMS) that manages all offshore U.S. oil and gas leasing (now the Bureau of Ocean Energy Management, Regulation and Enforcement), approved a program for the outer continental shelf to issue leases, easements, and rights of way for alternative uses of offshore oil and gas production platforms. MMS could authorize individual projects and offshore aquaculture is identified as one of the activities.

9.3   Case Studies on Permitting and Regulation

9.3.1   Introduction

Commercial offshore aquaculture will not fully develop unless governments create a supportive political climate and resulting regulatory conditions. More specifically, a recent study of the offshore industry from an economic perspective concluded:

> Offshore aquaculture will develop only where there is an enabling regulatory framework which allows investors to undertake projects with a reasonable expectation that their investments in the farm and their fish will be secure and a reasonable degree of certainty about how the operation will be regulated (Knapp 2013).

A wide variety of government policies may affect the development of offshore aquaculture. However, required investment will not occur without policies which give fish farmers the opportunity and the incentive to invest (Knapp 2013). Some of the most important policies can be grouped broadly under two categories: Regulatory and Leasing (Table 9.1). Under regulatory policies the most important concerns for industry are: number and complexity of the regulations; predictability of the process of obtaining permits; and costs of compliance. Under leasing policies the most relevant concerns are: complexity and predictability of the leasing process; how legally secure is the site in terms of use, exclusive use and property rights; and the available length of lease and predictability/stability of the cost (rent/royalties) (Percy et al. 2013).

Percy et al. (2013) provides a succinct discussion of the purposes of a permit/license/lease system for marine aquaculture (Table 9.2). The overarching principle is no person can carry on ocean farming without first obtaining a permit, license or lease from the government, be it national, state or local. The requirement confirms by law the government has the right to regulate aquaculture and prosecute those that violate the imposed requirements.

In particular, the requirement enables the government to directly regulate the operator of a facility to: enforce the accepted management concepts of aquaculture, protect the environment, restrict the location and number of facilities, and obtain public input on development (Table 9.2). These purposes should be kept in mind when reviewing the Case Studies.

International law distinguishes generalized sea areas and jurisdictions that serve as the framework for government's declaring control, ownership and authority to regulate off-the-coast and offshore ocean waters (Fig. 9.1). Moving out from the shore, there exists the following categories (Lovetalli et al. 2013).

Table 9.1 Selected government policy issues affecting offshore aquaculture development

Category	Policy issues
Regulatory policies	• What regulations does government impose? • How costly are the regulations? • What is the process for developing regulations? • How stable and predictable are the regulations? • What are the objectives of the regulations? • How cost efficient are the regulations?
Leasing policies	• Is there a process to lease offshore sites? • How predictable is the process? • How long does the process take? • How legally secure are sites? • How flexible are permitted uses? • Can sites be transferred? • What do sites cost?

Source adapted from Knapp 2013

Table 9.2 Purposes of requirements for permits, licenses and leases to establish a regulatory framework for marine aquaculture	• Enable the government to assess the capacity of the applicant to carry out the project, e.g., aquaculture experience
	• Show in advance how the applicant will meet all regulatory requirements
	• Enable the government and others to identify all the other facilities by maintaining a register
	• Provide a means of enforcing the basic rules applicable to an operation through the attachment of conditions
	• Enable the government to control the numbers of operations issued to avoid excessive facilities concentration and supervise geographical distribution
	• Use the application/approval process to obtain public input on the proposed operations

Source adapted from Percy et al. (2013)

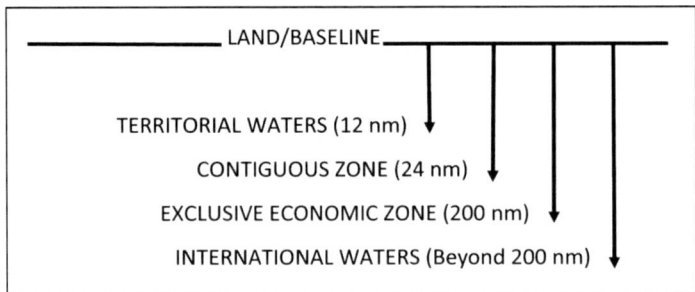

Fig. 9.1 Generalized sea areas and jurisdiction in international rights

- Baseline—the low-water line along the coast of a State (Country).
- State's Territorial Sea—extends up to 22 km (12 nm) from its baseline. The State has sovereignty over its territorial sea.
- Contiguous Zone—is a band of water extending from the outer edge of the territorial sea to up to 44 km (24 nm) from the baseline, within which the State can exert limited control as in the territorial sea.
- Exclusive Economic Zone (EEZ)—extends from the outer limit of the territorial sea to a maximum of 370.4 km (200 nm) from the territorial sea baseline and inclusive of the contiguous zone. A coastal State has control of all economic resources within the EEZ.
- International Waters (High Seas or Area Beyond National Jurisdiction (ABNJ))—are oceans, seas, and waters outside of national jurisdiction, i.e., beyond the EEZ.

These Case Study discussions of permitting and regulation will address off-the-coast and offshore aquaculture activities in the Territorial Sea, Contiguous Zone, and the EEZ; with the focus being the situations in the United States and United Kingdom.

In the U.S., the ocean jurisdictions are defined somewhat differently with coastal states generally having jurisdiction from shore out to 3 nm (and a few states out to 12 nm), so called State waters, and the national government having principal jurisdiction from 3 nm (or 12 nm) out to 200 nm, so called Federal waters. In the United Kingdom ocean jurisdictions reflect centuries of government ownership and administration of so called Crown Estate land and sea bed resources out to 12 nm, however management in the nation's EEZ is not clear. By law Crown Estate resources belong to the reigning monarch, however they are not the monarch's private property, but are managed (e.g., leased for offshore aquaculture) by an independent entity, the Crown Estate to maintain and enhance the value and generate revenue for the UK treasury (Crown Estate 2014).

9.3.2   Regulating Finfish Aquaculture in the U.S. EEZ, a Regional Approach

The U.S. has had a National Aquaculture Plan since 1980, however growth has been primarily in land based, fresh water species, until the last decade (Anderson and Shamshak 2008). The 2012 US Aquaculture Census put the industry at 269,550 tonnes, valued at $1.233 B. Mariculture species represent 37% of that value, and consist of primarily near shore, ocean culture of salmon and various mollusks, grown entirely in state waters (NMFS 2013).

US seafood imports has steadily risen to 91% of annual consumption, causing numerous government and public calls for expanding domestic supplies from aquaculture and particularly from marine aquaculture and farming the EEZ (e.g., PEW Ocean Comm. 2003; U.S. Commission on Ocean Policy 2004; U.S. Ocean Action Plan 2004; NOAA 10-year Plan for Marine Aquaculture 2007; PEW Marine Aquaculture Task Force 2007; NOAA Aquaculture Policy 2011; Department of Commerce Aquaculture Policy 2011). Clearly, there is a well-documented national policy to encourage commercial aquaculture in the EEZ to increase domestic supplies and help address the growing seafood security issue. Central to this policy is establishing the regulatory and legal framework to permit and manage aquaculture sustainably in federal waters (e.g., a permit and site administration process that provides for management, environmental monitoring, and enforcement and site tenure, exclusive use, and property rights and protections for the commercial projects) (Fletcher 2004).

From publication of the first National Aquaculture Plan to the present, regulatory constraints have been frequently cited as a major obstacle to U.S. industry expansion (Firestone et al. 2005; Stickney et al. 2006; GAO 2008; NOAA 2008). The regulatory framework for aquaculture in the EEZ is the least developed ocean jurisdiction and can best be described as nascent or a work in progress. There are two emerging approaches extant:

(1) A region by region permitting process for certain federally regulated finfish species in the region that nominally exists, but is currently being fleshed out from an operational standpoint and remains untested, and

(2) An individual lead federal agency permitting process for shellfish (and possibly finfish) that is emerging and may be used to locate a commercial farm anywhere in federal waters (the next case study), though questions remain regarding its long-term viability.

This section addresses the regional approach to permitting commercial aquaculture in the EEZ. As background, the recent Congressional history of attempts to legislate an overarching national framework to site aquaculture in federal waters is described. Initial actions by the Regional Fishery Management Councils to plan and develop fish farming in the EEZ are highlighted. The permitting process formulated by the Gulf of Mexico Regional Fishery Management Council is described in detail. Finally, permitting and leasing processes successfully used in state waters by Maine and Hawaii for more than 15 years are briefly described to contrast with the federal effort.

9.3.2.1   Background

Legislation to establish a federal regulatory system that provides a comprehensive nationwide approach (not regional) for permitting and management of EEZ aquaculture was introduced by previous Administrations and members of Congress on numerous occasions over the past 10 years. Bills were introduced in the 2005, 2007, 2009, 2011 sessions and a bill was in preparation for the 2013 session, but was halted with the death of the author. In all instances, the National Oceanic and Atmospheric Administration (NOAA), U.S. Department of Commerce (USDOC) was the designated lead agency for national marine aquaculture research and development and specifically for implementing a permit process for commercial farming in the entire U.S. EEZ.

The general purpose of all the bills was three fold: (1) support the development of a sustainable marine offshore industry, (2) safeguard the marine environment and coastal communities, and (3) support research and development to further industry expansion. All the bills were comprehensive in structure and virtually all the measures included sections on: Definitions, Administration, Office of Sustainable Marine Aquaculture, Marine Aquaculture Research and Development, Offshore Aquaculture Permits, Prohibitions by Coastal States, Recordkeeping and Access to Information, Protections for Marine Environment, Unlawful Activities, Enforcement Provisions, Civil Enforcement and Permit Sanctions, Criminal Offenses, Forfeitures, Severability, and Judicial Review, and Authorization of Appropriations. For a bill example see the National Offshore Aquaculture Act of 2007 (GCFMC 2014b). None of these measures passed due largely to opposition by environmental groups (Non-Governmental Organizations) and sympathetic members of Congress (Jeffs 2013; Angel and Edelist 2013).

9.3.2.2 A Path Forward

In the context of this highly political struggle to promulgate comprehensive, over-arching legislation, the federal government improvised a path forward. The approach was to use America's cornerstone national legislation for regionally managing fisheries in federal waters, the Magnuson-Stevens Fishery Conservation and Management Act (MSA), though the Act does not specifically mention aquaculture. In 1993, NOAA legally defined marine aquaculture as fishing, under the authority of MSA, which allowed three possible approaches to developing fish farming in each of the eight regions defined in the Act: (1) Issuing an Exempt Fishing Permit of limited scope and duration, (2) Amending an existing Fisheries Management Plan (FMP) for a regulated species or ecosystem to include aquaculture, and (3) Developing a specific FMP for regulating offshore mariculture in the region.

Several of the Regions embraced NOAA's definition and increasing advocacy for marine aquaculture. The New England Regional Fishery Management Council established evaluation criteria for open ocean aquaculture proposals that dictated the use of best management practices aimed at reducing environmental and fishery impacts-one commercial scallop project has been permitted to date (Upton and Buck 2010). The Western Pacific Council developed and passed a detailed Aquaculture Policy Statement to encourage projects that comply with a set guidelines-NOAA has issued permits for two short-term research projects testing floating and deep water, anchored cages have been approved off Hawaii and they prompted a court case instigated by an out-of-state environmental group that up held the 1993 legal definition opinion. In addition, the Western Pacific Council has begun a process to amend all its Fishery Management Plans to permit issuance of permits for commercial aquaculture.

The Gulf of Mexico Council has been the most active in putting in place a detailed regulatory process for selected managed fish species in federal waters, but progress has been very slow. In 2003 the Council adopted an aquaculture policy for the Gulf EEZ. Recommendations were made in six key areas: (1) allowable species, (2) habitat protection, (3) research, (4) location and design, (5) water quality, and (6) health management and disease control (GMFMC 2014a).

This led the Council to undertake preparation of a Management Plan for aquaculture in federal waters of the Gulf, inclusive of a comprehensive Federal Programmatic Environmental Impact Statement (PEIS), Regulatory Flexibility Analysis and Regulatory Impact Review (GMFMC and NOAA 2009). The massive document, after public review, was adopted by the Council in January 2009 and was sent, as required, to the Secretary of Commerce for final review and approval. In an unprecedented move, the Secretary and NOAA allowed the Plan and PEIS to become law without agency approval because, "Offshore aquaculture activities should be governed by a comprehensive national policy rather than by regional regulatory frameworks." At the time NOAA indicated they would work on an overarching national policy with the Councils and Congress to ensure a smooth transition to a national approach for regulating offshore aquaculture (Balsinger 2009).

National marine aquaculture policy statements by the Department of Commerce and NOAA emerged and were adopted in 2011. It is notable that the NOAA policy changed direction and supported a regional approach (rather than a national approach) and provided extensive regulatory and operational guidance to the Fishery Councils (NOAA 2011). Unfortunately, it took until 2014 for NOAA to draft and put out Rules to implement the Gulf Plan that the Agency has touted as the model for the other Councils. Many in the industry attribute this long delay to the key leadership positions in NOAA having environmental concerns with marine aquaculture in federal waters(not supported by science), as well as lobbying by anti-aquaculture NGO groups (Angel and Edelist 2013).

The comment period on the Proposed Rules closed on October 27, 2014 and at this writing no firm time table currently exists for adoption. Therefore, the Proposed Rules will be used as the basis for describing the regional regulatory framework and process being considered by the U.S. to regulate aquaculture in the EEZ. (Note to the Reader: During the preparation of this book NOAA adopted the Proposed Rule discussed in this Chapter on January 11, 2016. On Febuary 16, 2016 a broad coalition of environmental and fishery interest groups filed a law suit challenging the Rule, which stopped further implementation and will be adjudicated in 2017. See the website of the Gulf of Mexico Fishery Management Council for an update on the situation.)

9.3.2.3   The Gulf Council Permit Process

Adoption of the Proposed Rules for the Gulf Council Plan will put in place a permit process to allow commercial aquaculture of species regulated by the Council in the federal waters of the Gulf of Mexico. To highlight the major aspects of the proposed process, the following are discussed: core terms, application description, process description, consultation requirements, and operational and monitoring requirements. Those readers interested in more detail should go to the original documents (GMFMC and NOAA 2009; GMFMC 2014a).

Core Terms

There was much deliberation of the Plan and permit content by the Gulf Council and input from the public; namely the core terms of placing limits on annual production, species covered, the initial permit term and the renewal process, the permit costs, and the requirements to obtain an assurance bond sufficient for removal of the project.

These extended discussions were largely prompted by concerns raised by the environmental community and the Gulf fishing industry. Notably, it was recognized by the Council that changes would be needed as experience was gained and they included in the Plan an adaptive management mechanism that could allow changes to be made, e.g., to core Plan terms and the Rules (GMFMC and NOAA 2009). Core terms are highlighted:

- Limits on Production—The limits placed on the amount of fish grown annually, termed Maximum Sustainable Yield and Optimum Yield to be consistent with fisheries terminology, was 29 M kg round weight. The amount represented the average landings of all marine species in the Gulf. The Council set production limits for individual farms at 5.8 M kg so there was a diversity of producers.
- Species Grown—Utilizing MSA to regulate offshore aquaculture requires the species cultured must be in the portfolio of native species managed by the Council. Native species not under Council management may be cultured without a NOAA permit. Both the Rules and the Plan expressly prohibit culture of genetically modified and transgenic animals.
- Site Size—There is no limit on the size of a farm site. But, sites have to be twice as large as the combined area of the allowable aquaculture systems (e.g., cages and net pens) to allow for best management practices such as the rotation of systems for fallowing.
- Method of Site Disposition and Terms—The disposition process uses a permit to gain site access. The term for an initial aquaculture permit, which is issued by NOAA, is 10 years. Renewal periods for projects in good standing are in 5 year increments with no limits on number.
- Cost of Permit—The permit cost was determined through a government formula that is aimed at administrative cost recovery. The initial permit fee is $10,000 and a $1000 fee would be assessed annually. Each 5 year renewal would cost $5000. A permit is transferrable to a new owner.
- Bond for Facility Removal—The applicant is required to obtain an assurance bond sufficient to cover the costs associated with removing all components of the facility, including the stock. This is a precautionary step to reduce the potential for navigational hazards and environmental impacts if structures remain after the operation has terminated.

Description of the Application

- Required information includes: business, applicant and hatchery contact information; documentation of U.S. citizenship or resident alien status; a baseline environmental assessment of the proposed site (content requirements in preparation); a description of the geographic location and dimensions of the aquaculture facility and site; a description of equipment, allowable aquaculture systems, and methods to be used for growout; a list of species to be cultured and estimated production levels; a copy of an emergency disaster plan; and copies of currently valid Federal permits applicable to the proposed operation.
- The applicant is required to obtain an assurance bond sufficient to cover costs associated with removing all components of the facility, including cultured animals.
- The applicant is required to provide a document certifying that all broodstock or progeny of such broodstock were originally harvested from U.S. waters the Gulf of

Mexico and were from the same population or sub-population where the facility is located, and no genetically modified or transgenic animals would be stocked.

- The applicant is required to provide a copy of the contractual agreement with a certified aquatic animal health expert. An expert is defined as a licensed doctor of veterinary medicine or a person who is certified by the American Fisheries Society, Fish Health Section, as a Fish Pathologist or Fish Health Inspector.

Description of the Permit Process

- Once the NOAA Regional Administrator (RA) has determined an application is complete, notification of receipt of the application is published in the Federal Register. Interested persons are given 45 days to comment and comments would be requested during public testimony at a Council meeting. The RA would notify the applicant in advance of any Council meeting to offer an opportunity to appear to support the application.
- After public comment ends, the RA would notify the applicant and the Gulf Council in writing of the decision to issue or deny the permit. Reasons the RA may deny a permit might include: failing to disclose material information; falsifying statements of material facts; the project poses a significant risk to marine resources, public health or safety; and issuing the permit would result in conflicts with established or potential oil and gas infrastructure, safe transit to and from infrastructure and future geological and geophysical surveys.
- The RA may consider revisions to the application made by the applicant in response to public comment before approving or denying the permit.

Consultations

- The RA will consult with the Bureau of Ocean Energy Management, Regulation and Enforcement, the Bureau of Safety and Environmental Enforcement, and other Federal agencies as appropriate, to address and resolve any conflicts in the use of the Outer Continental Shelf (OCS), with special emphasis on energy programs.
- Other federal agencies that will require consultation are: U.S. Fish and Wildlife Service, National Coastal Zone Management Program, National Historic Preservation Council, and the military (Navy and Coast Guard). In addition, adjacent states and selected state agencies may need to be contacted.

Operational and Monitoring Requirements

- Start Up—To ensure permits are used, permittees are required to place 25% of allowable aquaculture systems approved for use at a specific facility in the water at the permitted site within 2 years of permit issuance and cultured fish would be stocked at the site within 3 years of permit issuance.

- Source Hatchery—Fingerlings and other juvenile animals obtained for growout at an aquaculture facility in the EEZ should only be obtained from a hatchery located in the U.S. All broodstock used for spawning at a hatchery supplying fingerlings or other juvenile animals should be certified by the hatchery owner as having been marked or tagged (e.g., dart or internal wire tag).
- Health Certificate—Prior to stocking fish the applicant should provide NMFS a copy of an animal health certificate signed by an aquatic animal health expert certifying the fish have been inspected and are visibly healthy and the source population tests negative World Organization of Animal Health (OIE) pathogens specific to the cultured species or additional pathogens identified in the National Aquatic Animal Health Plan. The process should be repeated with each new stocking event.
- Broodstock Collection—At least 30 days before each time a permittee or the permittee's designee intends to harvest broodstock from the Gulf, including state waters, they are required to submit a request for harvest to the RA. The request should include information on the number, size and species to be harvested, the methods, gear, and vessels used in capturing, holding, and transporting broodstock, the date and specific location of intended harvest, and the location of where the broodstock would be delivered. If the harvest is approved, the permittee would be notified by the RA and required to submit a report within 15 days of the date of harvest summarizing the number, size, and species harvested and the location where the broodstock were captured.
- Stock Genetics—NMFS may sample cultured animals to determine genetic lineage. If cultured animals. If cultured animals are determined to be genetically modified or transgenic, then NMFS would order the removal of all these cultured animals.
- System Evaluation—Aquaculture systems (e.g., cages or net pens) used for growing fish would be evaluated by the RA on a case-by-case basis. The structural integrity and ability of proposed systems to withstand physical stresses associated with major storm events (e.g., hurricanes) would be reviewed by the RA, using engineering analyses, computer and physical oceanographic models, or other required documentation.
- Risk Assessment—The RA will evaluate the potential risks of proposed aquaculture systems to Essential Fish Habitat, endangered or threatened species, marine mammals, wild fish stocks, public health, or safety. Any system approved for use should be marked with a minimum of one properly functioning locating device (e.g., GPS device) in the event that the allowable system is damaged or lost. In addition, the U.S. Coast Guard requires structures be marked with lights and signals to ensure compliance with private aids to navigation.
- Facility Separation—No aquaculture facility should be sited within (3 km) of another facility to minimize transmission of pathogens between facilities.
- Restricted Zone—A Restricted Zone will be established for each aquaculture facility to afford additional protection to an operation's equipment and systems, and increase safety by reducing potential encounters with fishing vessels. Restricted access zone boundaries should be clearly marked with a floating

device, such as a buoy. No recreational or commercial fishing, other than aquaculture may occur in the restricted zone.

- Facility Access—Permittees are required to provide NMFS personnel and authorized officers access to their aquaculture facility and records in order to conduct inspections and determine compliance with applicable regulations.
- Facility Inspection—Permittees should inspect the aquaculture systems for entanglements or interactions with marine mammals, protected species, and migratory birds. The frequency of inspections will be specified by NMFS as a condition of the permit.
- Biologics, Pesticides, Drugs—Farms have to comply with all applicable U.S. Department of Agriculture (USDA), EPA, and Food and Drug Administration (FDA) requirements. Use of aquaculture feeds should be conducted in compliance with EPA feed monitoring and management guidelines.
- Notification of Harvest—NMFS should be notified 72 h prior to the intended time of landing. The landing notification would include the time, date, and port of landing.
- Crop Landing—Permittees participating in the aquaculture program are allowed to offload fish at dealers only between 6 a.m. and 6 p.m., local time. All fish landed would have to be maintained whole with heads and fins intact. Any cultured fish harvested and being transported should be accompanied by the applicable bill of lading through landing ashore and first point of sale.
- Reporting Requirements—Permittees are required to report to NMFS the following events: major escape events, e.g., escape of 10% of fish from one cage; findings of reportable pathogens; and entanglements or interactions with marine mammals, protected species or migratory birds. All these events have to be reported within 24 h of discovery. Permittees must also report any change in the hatchery source of fingerlings or use of a new animal drug.
- Compliance with Permits—Permittees have to monitor and report environmental assessment data to NMFS in accordance to procedures specified by NMFS. Applicants also have to comply with all monitoring and reporting requirements specified in their Environmental Protection Agency (EPA) and the U.S. Army Corps of Engineers (ACOE).

Sanctions and Denials

- Permit Revocation, Suspension and Modification—An aquaculture permit issued pursuant to the Gulf Plan and Proposed Rule may be revoked, suspended, or modified in accordance with general procedures governing enforcement-related permit sanctions found in sub part D of 15 Code of Federal Regulations part 904. Further denial of submissions of initial permits and renewals are covered by this part of the law (CFR 2014).
- Review Process—Briefly, the process includes the following steps: (1) Notice of permit sanction or denial, (2) Opportunity for a hearing before the NOAA

Regional Administrator, and (3) Final Administrative decision. Suspended permits may be reinstated when a sanction has been reinstated. In emergency situations to protect the marine resource, a Judge can order suspension of farming activities or other actions (CFR 2014).

9.3.2.4   Other Required Permits

In addition to the NOAA permit described above, there are two other major permits required to locate and operate a fish farm in federal waters: the Department of the Army (DA) Section 10 Permit issued by the ACOE District Office and the National Pollution Discharge Elimination System (NPDES) Permit, issued by the Regional Office of the EPA. According to the Gulf Plan and Proposed Rules, the sequence for the applicant securing all permits is to obtain the DA Section 10 Permit and the EPA NPDES Permit, then these documents must be presented to NOAA as the final step for the agency to issue the NOAA permit.

The ACOE and the EPA permits are briefly highlighted below and for more details see the appropriate references (APA 2011; IWGA 2014).

DA Section 10 Permit, ACOE-

- Purpose—The DA permit program was established to ensure that the navigational characteristics of U.S. coastal waters were not adversely affected by development. The jurisdiction originally extended only to structures in navigable waters, however with the passage of amendments to the Clean Water Act in 1972, the authority of ACOE was extended to include issuing permits for discharge of dredge or fill material in U.S. waters, including all federal waters.
- Applicability—Navigable waters are defined as those waters that are subject to the ebb and flow of the tides and/or are presently used, or were used in the past, or may be susceptible to use in interstate or foreign commerce. The ACOE can issue an individual permit an aquaculture facility in federal waters. Individual permits include standard permits and for less complex projects a Letter of Permission.
- The Section 10 DA permit is required for structures or work in navigable waters and such work includes: piers, boat docks, breakwaters, permanent moorings, and anchored aquaculture facilities. Maintenance and repair of existing structures is generally exempt from a permit.
- Issuance of a permit is based on a "public interest review" which evaluates the probable impacts of a project, including its cumulative impacts. The decision process requires a balancing of the reasonable benefits which can be expected to accrue from an activity against its reasonable detriments. The permit will have conditions governing the execution of the project.

- Information Requirements—The DA permit requires the proposed work be described in sufficient detail so its potential impact on the affected environment can be fully evaluated. The information required is diverse and besides addressing navigation, it includes: consideration of such areas as: conservation, economics, historic values, water quality, aesthetics, coastal zone management, environmental concerns, safety, food production, fish and wildlife values, and the general needs and welfare of the people.
- Both a complete narrative description and detailed plans and drawings are needed. Strictly, an Environmental Assessment is not required, though a large-scale project may need one. Consultations are required with NOAA, the U.S. Fish and Wildlife Service, the Historic Preservation Council and the Coastal Zone Management Office of the nearest state. There is a public comment period of 30 days on the application.

National Pollution Discharge Elimination System (NPDES), EPA

- Purpose—The purpose of the NPDES permit is to protect the water quality of the U.S., and by regulating the discharge of wastewater from fixed point sources into surface waters, including wetlands, coastal waters and the EEZ. Fixed point sources include pipes, ditches, channels, etc. from which pollutants may be discharged. Ponds, tanks, offshore fish cages, or other similar aquaculture infrastructure are generally considered fixed point sources.
- Applicability—The NPDES permits applies to aquaculture systems, generally categorized as Concentrated Aquatic Animal Production Facility (CAAP). A CAAP means a hatchery, land-based fish farm, offshore fish farm or other facility which contains, grows or holds aquatic animals. There are statutory exemptions available for fish farms growing less than 9091 kg of cold water animals or 45,454 kg of warm water animals. Also, the EPA can designate any size facility a CAAP on a case by case basis.
- In general, EPA may create farm effluent limitations for a point source in federal waters by reference to three types of standards set in law: Effluent Limitation Guidelines (ELG), Water Quality Standards (WQS), and Ocean Discharge Criteria (ODC). Regardless, managing aquaculture in federal waters has not been an agency priority and EPA has not issued WQS for aquaculture that apply to federal waters to date and the current ODC's provide little guidance for aquaculture discharges (HLS et al. 2012)
- In 2004, EPA did establish ELG's for aquaculture, including net-pen facilities. The CAAP ELG's include only "narrative effluent limitations" requiring implementation of effective operational measures to achieve reduced discharges of solids and other materials in the CAAP waste waters, i.e., not numerical effluent limitations. To illustrate with two narrative limitations: (a) the employment of efficient feed management and feed input to the minimum amount reasonably necessary to achieve production goals and sustain targeted

rates of aquatic animal growth and (b) that the production system be inspected on a routine basis in order to identify and promptly repair any damage and regular maintenance be employed in order to ensure that it is properly functioning.

- Facilities subject to the ELG's are required to develop and maintain a Best Management Practices (BMP) plan describing how they will achieve these and other requirements. In the absence of distinct ELG for the project, the law allows the permit writer to use "best professional judgment" to issue the permit (HLS et al. 2012). A permit is issued for a set time period of no more than five years and is renewable and transferrable.

- Information Required—In general, a project application must describe the proposed activity, including location, species, and number of animals, daily flow of effluent and infrastructure making up the facility. The physical-chemical nature of the effluent, including pH, temperature, dissolved oxygen, nitrogen, and phosphorus, must be characterized. Further, it describes the receiving water and water source and total kilograms of food being fed during the calendar month of maximum feeding. Tests on the existing quality of the receiving water, a survey of the receiving water's ecosystem, and an analysis of the prevailing water currents may also be required.

9.3.2.5   Successful State Permit/Leasing Processes

Several U.S. states have supported offshore aquaculture in state waters for many years as a tool for economic development and diversification, particularly in rural, coastal areas. Two states among the leaders are Maine and Hawaii. It is beyond the scope of this chapter to detail the permitting processes to secure an offshore site in Maine and Hawaii, which include most of the same considerations as the regional permitting process under MSA, i.e., environmental impacts, multiple use conflicts, public participation, consultations with other agencies, and public safety, etc. These considerations allow the lead state agency to grant a use permit/lease, the ACOE District to grant a Section 10 permit and the state's water quality agency to grant an NPDES permit. Readers are referred to the appropriate web sites for details (APA 2011; IWGA 2014).

However, an important difference between the regional permit process as defined previously and these state processes is the property management vehicle to grant access to an aquaculture site is a lease and not a permit. Legal convention in the U.S. is a lease for property is needed to grant exclusive use, property rights and protections and tenure and is also important for the leasee to qualify for financing and insurance (Fletcher 2004; Firestone et al. 2005; Callies 2010; NSGLC 2012). The core aspects of the offshore aquaculture leasing programs in Maine and Hawaii are highlighted to contrast with the regional permit approach for federal waters being implemented by the Gulf. The aspects discussed are: limits on production, species that can be grown, size of the site, terms of the lease, cost of the lease, and bond requirements.

Maine

Maine began leasing sheltered State offshore waters for salmon aquaculture in 1970. Shellfish was the primary species initially and then cage culture of salmon became a target. There are three different access vehicles to State waters for different purposes, described below (State of Maine 2014). Today Maine boosts one of the most valuable aquaculture industries in the U.S. valued at $57 M in 2013. The Core Terms are highlighted as follows:

- Limits on Production—No limits are placed on volume of production, but there are limits on the size of the site that can be leased.
- Species Grown—For the Experimental Lease and the Standard Lease, any species already present in Maine. Bringing in out-of -state stock requires a State Department of Marine Resources import permit. With a Limited Purpose Aquaculture License (LPA) the species cultured can be blue mussels, clams, American and European oysters, sea and bay scallops, green sea urchins and marine algae.
- Site Size—Maximum size for a Standard Lease is 40 ha and for an Experimental Lease is four acres. For the LPA license it is up to 400 sq. ft.
- Method of Site Disposition and Terms—A lease is the site disposition vehicle for an experimental project that can be up to 3 years and the large-scale, long-term commercial project that can be up to 10 years. A license mechanism is used for the short-term, one calendar year LPA project. Renewals are possible for the Standard Lease and LPA license for similar terms.
- Cost of the Lease—The Standard Lease costs $1500 for shellfish and $2000 for finfish, plus $100 per acre per year rent. Renewal fees after 10 years are $1000 for shellfish and $1500 for finfish.
- Bond for Facility Removal—Both the Standard Lease and Experimental Lease require a bond for removal of the facility and crop if the lease is terminated. The LPA license does not require a bond.

Hawaii

Hawaii began leasing State marine waters for commercial aquaculture in 2001, with the first farm in the U.S. to be sited in open ocean conditions and operated totally submerged growing the Pacific Threadfin fish. This was followed shortly after by a second farm growing Amberjack (Sims 2013). Today Hawaii has one offshore fish farm and several in the permit process. Its total aquaculture industry was valued at $58 M in 2013.

Obtaining an offshore site requires a two-step process. First the company must obtain a Conservation District Use Permit (CDUP) from the Department of Land and Natural Resources (DLNR), which is a conditioned permit that describes the site and the aquaculture use (State of Hawaii 2014). The CDUP application requires an Environmental Assessment be carried out. Upon approval of the use by DLNR, the applicant can then negotiate the terms of a long-term lease from DLNR for State marine waters (APA 2011; Sims 2013). The Core Terms are highlighted as follows:

- Limits of Production—No limits are placed on the volume of production, but the oceanographic characteristics of the site must support the proposed project.
- Species Grown—Any species native to Hawaii may be grown. If the species is subject to State regulations then a license for procession of the species must be obtained.
- Site Size—There is no maximum size for a site, but the applicant must justify the size of the site requested from both the business and environmental perspectives.
- Method of Site Disposition and Terms—A long-term lease is the disposition vehicle for commercial projects. Terms could go up to 65 years, however State policy has been for the initial period to be 15 years, with possibility of renewal in up to 15 year increments.
- Cost of the Lease—The CDUP processing fee for State marine waters is 2.5% of the project cost, with a limit of $2500. Lease rents are a flat per acre fee of $100 per acre per year or a percentage of gross sales ($1\frac{1}{4}\%$), whichever is higher.
- Bond for Facility Removal—The commercial lease requires a bond to remove the farm infrastructure and crop should the lease be terminated. The amount shall be sufficient to protect the public interest with the removal of all structures and stock.

9.3.2.6 Discussion

Evolution of the Regional Approach

Clearly the establishment of a regulatory mechanism for siting aquaculture in the U. S. EEZ has been controversial and a struggle, with members of the Congress, the current Administration and the public being on both sides of the issue, for and against. The U.S. has had a National Aquaculture Plan since 1980, but emphasis on open ocean aquaculture has only come to the forefront in the past 10 years, being fueled by: availability of mass culture hatchery technology for marine species, development of cage and mooring technology for open ocean conditions, growing seafood security concerns with growing reliance on imports (currently at 91%), advances in understanding and monitoring environmental impacts of ocean farms, and well publicized successes with offshore culture in Europe and the Mediterranean.

Numerous efforts to establish a national regulatory system for the U.S. EEZ via national legislation have been frustrated by successful lobbying by environmental groups and actions by sympathetic members of Congress, as well as the U.S. industry being fragmented and not having a unified voice. A steady stream of public and private sector planning and policy studies over the past 10 years state it is in the best interests of the American people (the public interest) to develop open ocean aquaculture, yet only limited action has occurred.

Advantages of an overarching national regulatory system include: direct agency implementation of nationwide plans and policies, designation of a lead agency to issue and manage permits for all federal waters, opportunity for close coordination with all stakeholder agencies involved in the EEZ, and placement of the management entity in the nation's ocean management agency for efficient application of the best science to decisions. There appear to be no major disadvantages with this comprehensive, overarching approach.

Due to the political road blocks in Congress for national legislation, agency proponents utilized an obscure legal opinion that aquaculture is fishing and therefore the nation's principle fisheries management legislation, the MSA, as amended, could be used to move forward. MSA utilizes a regional approach through eight agency-stakeholder managed councils that prepare plans and policies and manage the fisheries in their jurisdiction, while working closely with NOAA. This community-based approach allows each region to develop a plan (or not) for aquaculture in federal waters and work with NOAA to promulgate rules to manage it.

The advantages of this regional approach include some of the same achieved by a national approach: the lead agency for fisheries would also be the lead agency for aquaculture, NOAA would still issue and manage site permits for the EEZ, and coordination would have a more regional focus.

Apparent disadvantages to this approach include: regionalizing planning and policy development could lead to differing approaches to implementation for each region (NOAA does have final approval on all plans and policies), plans and projects have to go through the "filter" of a diverse public-private management body (e.g., includes national and state agencies, commercial and recreational fishers, ocean recreation interests, cultural and environmental interests, etc.), the Councils are currently not staffed to manage commercial aquaculture in the EEZ and implementation and placement under the Fishery Councils adds another layer of bureaucracy to navigate.

The Gulf Plan Rules, Concerns

The adopted Gulf of Mexico Offshore Aquaculture Plan and pending Rules have been subject to intense stakeholder and public comment using Internet posting. Comments from a wide variety of stakeholders note a number of concerns and potential flaws in the Rules as currently written. The major concerns expressed by the aquaculture industry and potential investors are briefly highlighted below. Fortunately, if these concerns are realized as serious impediments to development, the Gulf Plan includes change processes to amend both the Plan and the Rules, hence the approach could be considered an Adaptive Management Approach to establishing a regulatory system for the U.S. EEZ.

Core Terms, Concerns

- Production limitations—Placing limitations on total farm production for the region and on production for an individual farm could act as a disincentive to investment. The environmental characteristics should be used to establish the carrying capacity for an area as with U.S. states.
- Size limitations—Though there is not limit on farm size, dictating that a farm site must be twice as large as the combined area of allowable aquaculture systems to facilitate fallowing is inappropriate. Farms that want to use separate fallowing sites at their discretion would be penalized with excessive costs.
- Site disposition—There are concerns with the proposed method of site disposition, a permit rather than a lease. Legal questions have been raised as to whether a permit under the MSA enabling legislation is equivalent to a lease and can provide property rights and protections, tenure, exclusive use and capacity to secure financing and insurance.
- Terms of the permit—The concern is that the initial permit term of 10 years and a 5 year renewal are too short. Industry believes that these terms could be problematic for a large-scale commercial operation to fully build out and achieve profitability, as well as provide sufficient stability in rental costs and key terms to attract the large investment required. More reasonable and attractive terms would be and initial minimum period of 20 years and a minimum renewal period of 15 years, similar to several U.S. states.

The Application, Concerns

- Certified Broodstock—The applicant is required to certify that all broodstock or progeny of such broodstock were originally harvested form the Gulf of Mexico from the same population or sub-population where the facility is located. There are concerns that there is not sufficient reliable fish species distribution information for the Gulf to implement this provision and that the genetic sequencing is costly.

Operating and Monitoring Requirements, Concerns

- Time limits—There is a requirement to deploy 25% of allowable aquaculture systems approved for a site within 2 years and have fish stocked within 3 years of receiving the permit. Industry believes this time period should be at least 4 years for deployment and stocking, due to the many complexities of starting up both a large-scale hatchery and open ocean farm.
- Systems evaluation—Aquaculture systems utilized for growing fish would have to be evaluated by NOAA on a case-by-case basis. While the general

requirement is reasonable in concept considering the frequency of storms in the Gulf of Mexico, the concern is it suggests the applicant may need to purchase expensive engineering analyses if not available from the vendor and construct computer and physical oceanographic models, and other complex documentation requiring extensive data collection over time, if not readily available from government sources.

- Offloading fish—Permittees are allowed to off load fish at Gulf dealers only between 6 a.m. and 6 p.m., local time. All fish landed must be maintained whole with heads and fins intact. The concern is this restriction will affect the farmer's ability to serve market needs and take full advantage of aquaculture's being able to offer predictable supplies.
- Other permits—In addition to the NOAA permit, two other permits are required to operate a fish farm in the EEZ; and ACOE Section 10 permit and an EPA NPDES permit. The concern is the rule requires applicants to secure these permits first before securing the NOAA permit, yet it is perceived the NOAA permit will have the most impact on the final structure, operation and maintenance of the farm. Several states issue the permit for the site first and subject to receiving the ACOE and NPDES permits, so that the major characteristics of the proposed farm are fairly firm before these other permit processes are concluded.

Comparison with State Processes

As with the regional aquaculture permit, siting of open ocean farms in Maine and Hawaii requires site disposition from the lead agency, as well as a Section 10 and NPDES permit. In contrast to the proposed regional approach, these successful states handle some of the core terms differently. To illustrate:

(a) The regional approach places limits on regional and individual farm production, while Maine has no limits on production but limits site size. Hawaii has no limits on production or site size, but the scale of a project must be justifiable and suitable for the oceanographic conditions.

(b) The regional approach uses a permit as the site disposition vehicle to provide the farmer some specified degree of site access and control. While Maine and Hawaii utilize a lease to clearly convey terms, tenure, property rights and protections and exclusive use for a public benefit of what was a public resource.

(c) The regional approach limits species to those managed by the Gulf Council. While Maine's experimental and standard leases have no native species restrictions and likewise there are no native species restrictions for Hawaii open ocean farms.

(d) The regional approach does not limit total site size, but requires a site be twice the size of that required by the aquaculture facility (cages, net pens, moorings) to allow for fallowing portions. While Maine has limits of 40 ha for its

Standard lease and 1.62 ha for its experimental lease and Hawaii has no limits on lease size but scale must be justified.

(e) The regional approach utilizes a permit of 10 years duration and 5 years for each renewal period. While in Maine a lease uses terms of 3 years for the Experimental and 10 years for the Standard, and renewals are possible. In Hawaii a lease is also used that could go up to 65 years, though recent policy has been 15 years and 15 years for renewals.

(f) The regional approach costs a flat fee of $10,000 and $1000 annual fee, with each 5 year renewal costing $5000. While in Maine, the Standard lease costs $1500 or shellfish and $2000 for finfish, plus $100 an acre annual rent; with renewals of $1000 for shellfish and $1500 for finfish. In Hawaii the CDUP processing fee is 2.5% of the project cost, with a limit of $2500. Lease rents are a flat per acre fee of $100 per acre per year or a percentage of gross sales ($1\frac{1}{4}\%$), whichever is greater.

(g) The regional approach requires a bond for facilities removal. Both Maine and Hawaii have a similar requirement.

It remains to be seen if this as yet unapproved and untried regional approach to allow commercial aquaculture into the U.S. EEZ will be embraced by private industry, as the permitting and leasing processes have been in Maine and Hawaii.

9.3.3   Case Study—Shellfish Farming in the Northeastern and West Coasts of the U.S., Recent Examples

The regulatory regimes for offshore shellfish farming in the US vary depending on locale and the agencies involved. The regulatory process is quite new for most applicants and authorities so there is little precedent from which to build. Some states (e.g. Rhode Island and Maine) have established a "Limited Production Application" process which takes as little as 6 months, and allows a proponent to establish a small-scale demonstration of their intended farming design and practices. Other states do not have such provisions and require a substantial investment of time and money for permitting regardless of the size of the operation being proposed. Small offshore shellfish farms, mostly growing mussels have been permitted in the state waters of New Hampshire, Massachusetts, Rhode Island and Connecticut in the last 5 years.

Despite these promising developments in state waters (defined as within 3 nm of shore), recent experiences to permit shellfish farms in federal waters (3–200 nm from shore) off Massachusetts and California show that site selection, regulatory uncertainty, and monitoring requirements are a substantial challenge to offshore aquaculture development in the US. The following case histories illustrate that user conflicts and associated permitting requirements have resulted in substantial downsizing, relocation, and redesign of proposed activities despite the best efforts of project proponents to identify and avoid potential conflicts in advance.

9.3.3.1   Catalina Sea Ranch—First Farm Permitted in Federal Waters

In California, Catalina Sea Ranch (CSR) engaged with the US Coast Guard and devoted substantial efforts towards understanding marine mammal avoidance when identifying its initial 405 hectare site for its mussel farm. Subsequently it was required to downsize to 40 ha and engage more fully with the State of California Coastal Commission (CCC), fishing interests, and the offshore oil and gas industry interests before identifying a site that effectively minimized conflicts with other offshore users.

CSR was issued a provisional Army Corps of Engineers (ACOE) permit under Section 10 of the Rivers and Harbors Act on July 17, 2012 (see also Sect. 3.2.4). The Rivers and Harbors Act was designed to permit marine construction projects while avoiding hazards to navigation. This provisional permit included a variety of expected permit conditions. A surprising and unprecedented permit requirement for ACOE approval and State Coastal Zone Management (CZM) agency concurrence was a monitoring plan for the seawater filtration effects, biodeposition, and changes in the abundance/distribution of non-native fouling organisms.

The permit was termed provisional because it was not valid and did not authorize the project to be carried out because, by Federal law, no permit can be issued until the State agency in charge of CZM consistency certification (the CCC) has concurred with the permit. To meet the concurrence requirements, it took another 18 months (January 2014) for the California Coastal Commission and CSR to agree to terms for a monitoring plan. Provisional permits may be also be issued until an applicant can satisfy a State's Section 401 Water Quality Certification. This was not necessary in this case as it is generally agreed that filter-feeding shellfish aquaculture produce a net benefit for water quality.

Catalina Sea Ranch's Offshore Mariculture Monitoring Program

Prior to commencement of construction (slated for early 2015) CSR must submit a Revised Offshore Mariculture Monitoring Program for approval by the Executive Director of the CCC. The Offshore Mariculture Monitoring Program is required to meet unprecedented environmental study, carried out by an independent entity approved by and reporting to the CCC for at least 5 years and funded by CSR. The monitoring must evaluate:

  (i) Quantity, type, and distribution of biological materials from the shellfish facility (such as feces and pseudofeces, shell material, and fouling organisms) that accumulate on the seafloor below and in the vicinity of the facility, and it must evaluate any biochemical changes in the sediment and changes in biomass, diversity of benthic infaunal and epifaunal communities.
 (ii) Response of fish, seabird, and marine mammal populations in the project area to the presence of the facility infrastructure, include estimates of the species

     diversity and abundance of the water column biota, including phytoplankton, zooplankton, and meroplankton.

(iii) Water quality in and around the facility including analysis of phytoplankton and particulate material filtration by the cultivated shellfish and release of nutrients such as ammonia nitrogen and phosphorus.

(iv) Type and amount of commercial and recreational fishing activity that occurs at and around the facility as well as compilation of all reports of lost or damaged fishing gear or catch (for which CSR may be libel) that occurs as a result of contact with the facility.

(v) Production of eggs and larvae from the cultivated non-native species, the regional dispersion of this reproductive material, and its contribution to the regional presence, persistence, and expansion of populations of these non-native species outside of cultivation.

(vi) Diversity and abundance of fouling organisms that establish on the shellfish cultivation facility, including its ropes, buoys, cables, cultivation structures, and cultivated shellfish.

These extraordinary monitoring measures contrast with any other existing shellfish farm in State waters in California. For example, a mussel farm, Santa Barbara Mariculture, has been operating 1.6 km offshore in California for over 10 years without any monitoring requirement or suspected environmental impact. No example of documented negative impact of shellfish farms in open or offshore waters are known anywhere else in the world.

9.3.3.2 Massachusetts Case Studies—A Tale of Two Projects

Cape Ann Mussel Farm

In 2012, a mussel farming project was proposed in federal waters 8 miles off Cape Ann by Salem State University, and funded by NOAA as a research demonstration project. One of objectives was to advance and further define the permitting process in Federal waters, and another was engage and train displaced fishermen in a new venture. In this case, 2 sites were chosen but one was abandoned due to conflicts identified early with the Stellwagen Bank National Marine Sanctuary nearby and its management plan that does not permit aquaculture within its boundaries.

Despite substantial collaboration and engagement with the capture fishing industry to identify the preferred site, there was potential conflict with commercial vessel traffic requiring that the longlines be deployed at 15.2 m below the ocean surface instead of 9.1 m. The potential for marine mammal entanglement was a big issue, in both the CA and MA cases. Project proponents were unaware of these potential conflicts and/or mandatory permitting or consultation requirements that might require remediation or project alteration—in part due to a lack of data or prior permitting experience that would help predict and avoid these hurdles.

Another issue that arose during the Massachusetts permitting process is that federal waters, in contrast to more coastal state waters, lack routine monitoring of Harmful Algal Blooms (HABs) that can result in offshore waters closed to shell-fishing or farming due to potential for Paralytic Shellfish Poisons (PSP). In 2013, re-opening of the Northern and Southern Temporary PSP Closure Areas was not considered because there was no research or controlled experimental harvests conducted in these areas that would provide samples to indicate that the PSP toxin levels are below the regulatory limit. However, it was recently announced that as of January 2015, the Massachusetts Division of Marine Fisheries will be conducting testing of shellfish in the closed areas to re-open the fishing for gastropods (whelks, conchs, snails) and whole and roe-on scallops, and farmed mussels. Finally, after making the same concessions as described below a permit for the project was issued in January 2015.

Nantucket Sound Mussel Farm

In July of 2013, Santoro Fishing Corporation submitted an application to ACOE for a permit to commercially farm mussels in an 11.5 ha site 9.6 km offshore just east of the permitted Cape Wind farm in the Federal waters of Nantucket Sound. After a protracted permitting process described below, a permit was issued August 21, 2014 to deploy longlines for growing blue mussels (*Mytilus edulis*) and sugar kelp (*Saccharina latissima*). The principal reason that the permitting was prolonged hinged on perceived risks to protected species that migrate through the area.

NOAA Protected Resources Division (PRD) decided that they needed further information. The ACOE called a meeting in November 2013 of the permit appli-cants and the various NOAA regulatory interests (Essential Fish Habitat, Marine Sanctuaries, Protected Resources, Aquaculture, etc.) to foster exchange of infor-mation, concerns, and promote better understanding of the Nantucket Sound and Gloucester project proposals. Similar longline mussel operations have been oper-ating in fully-exposed offshore state waters of California and New England for over 15 years without apparent impact or interference with protected species, such as whales and sea turtles. Despite this history, NOAA's PRD decided that they could not make a determination of "No effect" or "Not Likely to Adversely Affect" without reviewing a full biological assessment that addressed possible threats to protected species by the specific projects.

NOAA's PRD staff provided a guidance document for submitting a Biological Assessment. This started a confusing 9-month process involving exchanging drafts of the biological assessment with NOAA staff thru the permit coordinator at ACOE. Some of the confusion stemmed from the determination process of "effect", "not likely to adversely affect", or "no effect". It is the action agency (in this case ACOE) that makes this determination in order to initiate informal consultation. This was not clear to ACOE nor the applicant, and apparently was only made clear by PRD late in the process, after ACOE threatened to close the file because PRD's review process was taking too long.

After numerous correspondence through ACOE and concessions by the applicant, PRD indicated that the Nantucket Sound draft Biological Assessment contained elements with which they believed could lead to a determination of "Not Likely to Adversely Affect". Specifically, to prevent possible entanglement, the permit application was revised to include stiff sheathing on the vertical lines that connect the surface corner buoys to the submerged headrope. Also, the project would be developed in stages, with no expansion beyond the first stage, only 3 longlines, without further review by NOAA PRD and ACOE. Finally, NOAA PRD asked ACOE to initiate an "informal consultation" with the issuance of a permit letter and the final biological assessment to NOAA. Evidently a "formal consultation" under Section 7 of the Endangered Species Act is only initiated when there is likely to be an effect, and perhaps disagreement between parties about the suitability of a project. The confusing terminology surrounding the permit process's communication and consultation in its various forms (pre-informal, informal and formal) needs an explicit framework from which all parties can work through with transparency.

9.3.3.3 Discussion

The formal permitting process from time of filing the application to approval for the projects above varied from 13 months to more than 2 years. The timing of actual preparation of the permit application (consulting with various local, State and Federal constituents) may constitute another 3 to 6 months.

Early consultation is recommended with responsible agencies and data portals like the Northeast Ocean Data Viewer for maps and data for ocean site planning in the northeastern United States (Northeast Ocean Data Viewer 2014).

Applicants should meet with State and Federal fisheries resource managers and commercial fishing organizations and other ocean users to avoid potential conflicts with other marine activities. Applicants should also request a meeting with the Division heads (Protected Species, Aquaculture) of their regional NOAA fisheries office to discuss the need for and scope of a Biological Assessment. If necessary, a Biological Assessment template available on line will help to understand the process (IWGA 2014).

The permit application process for future mussel farms of similar designs should be much simpler to replicate for other sites based on the examples described above. As the government regulatory system becomes more familiar with the environmental impacts of mussel farms and well informed site selection decisions are made, uncertainties encountered by permit applicants should be reduced.

9.3.4   Case Study: Mussel Farming Off the English Coast; One Farmer's Experience

9.3.4.1   Introduction

The following case study describes the practical procedures and requirements experienced in order to obtain the various permits and leases needed to establish a large scale, suspended culture mussel farm off the coast of South Devon, England.

It should be noted that the experience described is specific to shellfish aquaculture and it only relates to the legal framework in place in England during the application period of 2008–2010. Legislation has been modified since that date although the general procedure is still broadly similar.

The procedures for licensing finfish aquaculture in the UK have significant differences to those for shellfish. Aquaculture licensing within the 12 mile limit is a devolved matter in the UK and there are fundamental differences in procedures between the separate administrations of England, Scotland, Wales and Northern Ireland.

The license and lease requirements described here were those relevant to suspended culture mussel farming and differ from those that would be needed for seabed cultivation of mussels or other shellfish.

9.3.4.2   Description of Farm

Permission was granted in 2010 for a suspended culture mussel farm that will cover a total area of 1540 ha of Lyme Bay off the coast of South Devon, England. The farm will be divided into 3 separate areas; two of 600 ha and one of 340 ha. All three of the areas lie between 3.5 and 9 km from the adjacent coast.

The exposure of the farm areas to open sea conditions varies, with the distance to the nearest shore being 3.5 km to the North, 300 km to the South East and 7000 km to the South West. The depth of water varies gradually between 20 and 28 m. The seabed is generally mud and shell.

The farm equipment specified in the permit and lease application consisted of 790 longlines spread evenly between the 3 areas. Each longline will be individually moored to the seabed at each end and measure 250 m from mooring to mooring with 150 m of each headline used to support culture ropes. The longlines will be moored using helical screw anchors with the headline supported 2–3 m below the surface using vertical spar shaped floats. The culture ropes will be the continuous looped New Zealand system with loops hanging up to 9 m below the headline.

The corners of each area will be marked by a yellow special mark navigation buoy topped with a St Andrews cross and a yellow flashing light visible at 2 miles.

Construction of the bulk of the farm will take 3–4 years and is being preceded by the operation of two small but commercial scale trial areas of 4 ha, each of which were established in 2013.

9.3.4.3 Legislatory Framework

The legislation covering suspended culture shellfish farming in England provides for applications for farms in waters up to 12 nautical miles (nm) beyond the coast. Historically, the few farms that have been established beyond the limits of estuarine waters have been small and close enough to the coast to come under the jurisdiction of Harbor Authorities (local authorities over close inshore waters that exist is some locations). For waters beyond 12 nm, i.e. the EEZ, it is not clear what the legal mechanism would be for granting permits or leases and so far the process has not been tested in practice by the lodging of an application.

The case study describes the first application in English waters for what could be considered a fully offshore farm located in open ocean conditions. The leases, licenses and permits that were required for the establishment of this farm and the agencies involved were as follows:

- Consent to deposit equipment on the seabed, Marine and Fisheries Agency.
- Lease of the seabed, The Crown Estate.
- Authorization as an Aquaculture Production Business, Fish Health Inspectorate.
- Shellfish Harvesting Areas Classification, Food Standards Agency.

9.3.4.4 Application Process

Informal Consultation

As this application was the first of its type and was novel in terms of both its scale and location, the first stage of the application was to discuss the possibility of establishing the farm with the owners of the seabed, which in this case was the Crown Estate. The discussions established the likely terms and conditions of any lease and a list of the various third parties that would potentially be interested and/or affected by the establishment of the farm. The parties were then approached and the proposed development of the farm was discussed on an informal basis. This informal process took place over approximately one year.

Coast Protection Act 1949; Section 34

The deposit of the materials and equipment on the seabed to construct the farm required government consent under Section 34 of the Coast Protection Act 1949, which is principally concerned with hazards to navigation. This consent was administered by the Marine and Fisheries Agency (MFA) which was an executive agency of the UK government. (Note: In 2010 the MFA was replaced by the Marine Management Organization (MMO) and the need for a CPA Section 34 consent was replaced by a Marine License issued by MMO.)

The application for consent gave full details of the location of the proposed farm areas, the equipment to be deposited and the design of the equipment, the schedule of construction, the operation methods and management procedures for the farm, and the likely gross cost of the project.

Details of the application were then circulated for individual consideration to statutory bodies with responsibility for:

- Natural environment—Natural England
- Historic and cultural environment—English Heritage
- Defense—Ministry of Defense
- Navigation—Maritime and Coast Guard Agency
- Local government—East Devon District Council, Devon County Council
- Fisheries—Department of Environment, Fisheries and Rural Affairs
- Fish and shellfish health—Fish Health Inspectorate
- Food standards—Food Standards Agency
- Water quality—Environment Agency
- Ownership of the seabed—The Crown Estate.

The application was also circulated to interested non-statutory bodies such as fishermen's organizations, local environment groups, angling clubs and sailing clubs for comment. The availability of the application was published in local newspapers and copies of the application were lodged at the East Devon District Council local government office to give the general public access to the proposal.

Responses from non-statutory consultees were collated after the 6 week time limit allowed for public comment and an extension for the consultation period was granted to the statutory bodies to enable an in-depth consideration of the implications for the natural environment.

Once all the responses were received by the MFA the applicant was given the opportunity to address any issues raised by the consultees and to make adjustments to the application where appropriate.

The issues raised included:

- Restrictions to navigation within the farm areas
- Restriction of access to commercial fishing within the farm areas
- The potential for environmental impacts and benefits within and beyond the farm areas
- The socio-economic effects on the regional and national economy
- Security of supply of seafood at UK and European Community level.

The MFA then considered all responses from the consultees and the applicant and made the judgement that the Secretary of State should issue a conditional consent. The process of application, consultation, consideration and decision took approximately 11 months.

Seabed Lease

The area of seabed required for the three farms is the property of the Crown Estate. The Crown Estate issues leases on its property and charges a rent for its use but they also have a stewardship role and an obligation to manage and enhance the value of their coastal assets. This means that the Crown Estate encourages the development of sustainable aquaculture at appropriate locations. There was therefore a need to obtain a lease from the Crown Estate who reviewed potential impacts of the project to the natural environment and to other users of the marine resource.

The application for a seabed lease followed a similar pattern to the application for a CPA Section 34 Consent. The same groups were consulted and a similar period was given for responses to be submitted. The applicant was then given the opportunity to address any of the issues raised by the consultees and to make adjustments to the application where appropriate. The issues raised included the potential effects of increased sedimentation below the farm structures, the primary productivity and carrying capacity of the water body, the structural integrity of the equipment and the provision of a decommissioning bond.

The Crown Estate then considered all the responses from the consultees and the applicant and made the judgement to issue a conditional lease for the trial phases of the project with an agreement to issue a lease for the full project on successful completion of the trials. The process of lease application, consultation, consideration and decision took approximately 10 months.

Aquaculture Production Business

Under European Council Directive 2006/88/EC all member states must maintain a register of Aquaculture Production Businesses for the purposes of preventing the spread of the disease and controlling movements of aquaculture animals and products. In England and Wales the register is maintained by the FHI (Fish Health Inspectorate) which is a unit within CEFAS (Centre for Environment, Fisheries & Aquaculture Science).

The FHI required the applicant aquaculture business to draw up a biosecurity plan, to maintain a record of movements on and off the farm premises, and to report and maintain a record of unusual mortalities. The business was then registered under Annex II of the Public register of Aquaculture Production Businesses in England and Wales.

Shellfish Harvesting Area Classification

Classification of harvesting areas is required and implemented in England and Wales under European Regulation 854/2004 before live bivalve shellfish are permitted to be placed on the market. The coordination of the shellfish harvesting area classification and monitoring programme in England and Wales is carried out by CEFAS on behalf of the Food Standards Agency (FSA).

Classification of a harvesting area is determined by the extent to which the shellfish are contaminated by *Escherichia coli*. The sampling of the shellfish must take place from a position that represents the production area and the samples should represent the shellfish that will be placed on the market. For a production area with no history of production and no existing population of market sized shellfish, classification cannot take place until market sized shellfish have been produced. In this case study the farm was offshore and used suspended culture which meant that the farm had to be established and shellfish had to be grown in situ before classification could take place.

9.3.4.5   Discussion of Current Licensing Process

The case described above shows that in theory the basic procedure for gaining permission to establish and operate a suspended culture mussel farm in the offshore zone in England was relatively straightforward and no different to that for an inshore farm. In practice there was a great deal of difference and at nearly three years, the full process took considerably longer than would normally be experienced for an inshore application.

The key delays were brought about by the lack of experience and knowledge within the statutory agencies regarding the potential environmental and socio-economic impacts of a large-scale offshore mussel farm. This is not surprising as this application was the first of its kind, but the result was a lack of capability to arrive at timely and relevant decisions which contributed to extended delays in granting permission.

Environmental Impact

The issue of potential environmental impact centered on concerns over the effects of increased sedimentation and organic enrichment directly below the culture ropes and to what extent the sedimentation footprint of the farm would extend beyond the boundaries of the farmed areas.

Whilst there was no particular concern about the seabed below most of the farm areas, which had previously been impacted by towed fishing gear, there was concern about the potential impact on a marine Special Area of Conservation (mSAC) that is adjacent to one of the three sites. There was very little experience to provide reference on these issues as there were no UK precedents for a farm of this scale built in an offshore high energy environment, and few comparable operations were to be found elsewhere in the world.

A review was made of all the information available from other studies and this was coupled with assumptions on production rates, stocking densities and management practices on the farms. Sinking rates of mussel faeces and pseudo-faeces, depths, current velocities, wave induced turbulence, sediment re-suspension and assimilation rates were estimated.

The review and estimates were presented to Natural England (NE) who acted as advisors on the natural environment to the MFA. This information enabled the applicant and NE to arrive at an adaptive management plan which entailed the applicant carrying out an environmental baseline study prior to development. The study covered a comprehensive range of parameters including biodiversity, species abundance, sediment and water chemistry, redox potential, sediment particle size distribution and hydrodynamics at a local and regional scale. The baseline study will be followed by further monitoring to detect any changes to the environment during development. This study will be carried out on the trial sites and on the first stages of development of the full commercial scale farm not adjacent to the mSAC. Results from these studies would then inform a decision on whether permission would be granted to develop the third site that is adjacent to the mSAC.

Socio-Economic Impact

The nature of the suspended culture mussel farm meant that a number of other activities would be physically excluded from the area, some would be curtailed and some would be enabled and possibly enhanced.

The principle marine activities in the area were:

- Fishing with mobile gear such as trawls and dredges
- Fishing with static gear such as pots, hand lines and set nets
- Leisure uses, i.e. angling, scuba diving and navigation for leisure vessels.

It was apparent that mobile gear would necessarily be excluded from the area, while fishing with static gear such as hand lines and angling, and scuba diving could continue to take place and could possibly be enhanced by the reef effect of the mussel ropes in the water column in an area where there was previously very little physical complexity and relief.

In reaching their positive decisions, the MFA and the Crown Estate reviewed the existing level of economic activity at the proposed areas and compared this with what could continue to take place on site together with the new economic activity that would be generated by the farm in terms of onshore, offshore and downstream employment.

The UK and the wider European Community currently relies heavily on imports and so also taken into account was the strategic need to improve security of supply of seafood and the fact that the establishment of this pioneering farm would provide information that would inform future offshore developments.

9.3.4.6   Discussion—Future Regulation and Co-location of Offshore Aquaculture

Marine aquaculture in most of its various forms is a relatively new activity that has to compete for limited space in the marine environment with more long term, established activities such as fishing, transport, leisure boating, aggregate extraction, power and communication cabling and defense, as well as the more hard to quantify services such as scenic amenity and nature conservation.

The current system of licensing marine aquaculture in England operates on a case by case basis where there is essentially a presumption against development and each proposal has to justify its application in terms of how it will either not affect existing activities or how it would be an improvement on existing activities. This has led to the alternative idea that licensing should be determined by marine plans with predetermined zones where there is a presumption for each activity.

Both of these approaches have their advantages and disadvantages and for the inshore zone where the needs of aquaculture are well defined and understood, either of these approaches can be made to work. However for the offshore zone, aquaculture is still in its infancy and the techniques, equipment, management practices and economics are not well developed and are likely to evolve over coming years. This means that the criteria by which offshore aquaculture is presently judged against other activities are likely to change and the rationale for placing aquaculture in a particular zone may prove to be inadequate in future years.

Co-location is often presented as a solution to this problem and there have been proposals and trials to co-locate aquaculture with a number of different activities worldwide. These include; the co-location and integration of different types of aquaculture at different trophic levels; co-location with reconstructed oyster reefs to enhance fisheries and biodiversity; co-location with structures such as marina pontoons, shore defenses or artificial reefs; and co-location with renewable energy installations or redundant oil and gas rigs. All of these have potential but all of them require an in-depth understanding of the individual needs of each activity before co-location could become a practicality.

Offshore mussel farming using suspended cultivation is regularly cited as a potential candidate for co-location with renewable energy structures, but currently there are no commercial examples in operation. The reasons for this are: partly because of the lack of an economic driver; partly because the technology of offshore mussel production is new and undeveloped; and partly because of the lack of a clear regulatory framework and operational protocols that would provide security for both of the co-location partners.

The economics of offshore mussel farming become more challenging the further offshore the farm is situated. The advantages are that the potential space resource for production increases, growth rates can improve, and diseases and anthropogenic pollution can decrease. The disadvantages are the increased cost of storm resistant equipment, the distance from shore base, the need for large expensive seagoing vessels and the lack of access during bad weather coupled with the difficulty of maintaining markets that require regular service. The production economics can also be rapidly altered by changes in markets for the mussels or changes in major costs such as fuel.

The technology of offshore mussel production is relatively simple in terms of equipment with most installations around the world consisting of a simple longline anchored at both ends and buoyed up to, or near to, the surface. The chief differences result from site specific conditions which determine such things as line geometry, mooring tensions, stocking density, rope spacing, seeding, harvesting and adjustment of buoyancy. The difficulty of perfecting each of these operations increases with exposure to open ocean conditions and it is likely that they would be further complicated by the need to integrate with the operational needs of a wind farm which are by their nature windy places.

Co-location of suspended culture offshore mussel farms with wind farm structures seems unlikely to be a viable option for mussel farmers in the UK under current economic conditions and would only seem to be likely if there was a regulatory imperative. This would need to be in the form of a compulsion, such as no development being allowed other than in wind farms, or in the form of an incentive such as development and operating costs being subsidized or long term price guarantees being given for product, as is the case for wind farm developers.

9.4 Recommendations for Developing a Regulatory System

Based on the review of seafood and energy trends, the status of standalone and co-located aquaculture and the Case Study experiences, recommendations are offered for consideration by countries, nation-states and local governments that have or want to establish a supportive regulatory system for developing commercial open ocean aquaculture. It is important to note in considering these recommendations that moving aquaculture into open ocean environments is in the early stages and initial approaches to regulation will evolve as technology improves and industry experience is gained.

9.4.1 Planning a Regulatory System

- Government aquaculture plans and policies should include exploration and encouragement of commercial-scale standalone and co-located with ocean energy production aquaculture farming in nearshore and open ocean locations. Successful plans, policies and regulation of nearshore projects can eventually lead to open ocean developments.
- While suitable species and technologies are evident for open ocean farming in some countries, the emerging interest in moving further offshore needs greater emphasis and encouragement through increased investment in research, development and commercial-scale demonstration projects (R, D, & D). Ultimately, government sponsored R, D, &D should foster engagement of the private sector to innovate and expand open ocean aquaculture, e.g., provide financial incentives to encourage private investment.

- Ocean wind farms in open ocean environments are the best near-term opportunity for significant future co-location of aquaculture production. In particular, planned location of wind farms off major port cities, would also convey market advantages to aquaculture. However, the mutual benefits of co-location need to be successfully demonstrated.
- Government should incorporate long-term aquaculture and ocean energy development and site identification objectives and capabilities into emerging Marine Spatial Planning Programs. For example, both the United Kingdom and the U.S. are increasing efforts in this area (UKL 2014; NOC 2014).
- Government should examine the advantages and disadvantages of regulating open ocean aquaculture on a regional basis verses a nationwide, state or local basis.
- Government should give consideration to encouraging co-location of aquaculture with other activities that benefit society and the nearby communities; for example reconstructed oyster reefs to enhance fishing and biodiversity, marina pontoons, and shore defenses and artificial reefs.

9.4.2 The Regulatory System

- Government should establish a regulatory system for open ocean aquaculture that is reasonable, science-based, predictable, and cost efficient and does not act as a disincentive to development. An adaptive management approach should be used, so experience can foster changes and in particular allow for aquaculture as a co-located use of oil and gas and wind energy platforms.
- Government should establish a lead agency and highly placed designated office for aquaculture development, which includes open ocean farming, to advocate and facilitate development (Corbin and Young 1997; Cincin-Sain et al. 2005). The office would be the first point of contact for potential projects and help farmers and investors understand the regulatory system, as well as, be responsible for education of other agencies and the public about aquaculture science, business, and societal benefits.
- Government should work to reduce overlapping and redundant permit requirements and streamline regulatory processes where possible, for example when information requirements are the same for two required permits. Also consider establishing a One Stop Location where all the permits needed for open ocean aquaculture can be secured/tracked and applicants can work with one agency.
- Government should provide dedicated applicant assistance with site identification and help rule out sites that have major environmental and multi-use conflicts early on in the process. Also it should facilitate access by applicants to oceanographic and other environmental and economic information for site and business planning.

- Government should emphasize stakeholder participation and input into the permit/license/lease process to identify and help applicants resolve conflicts early in the process. For example, government could facilitate stakeholder information meetings (e.g., informal and required public hearings) and use the Internet to promote widespread information availability for stakeholders.
- Government should consider the advantages and disadvantages of placing limits on the size and the volume of production for an open ocean site. While limits may seem appropriate initially to manage perceived risk, there should be process in the regulations to, in the future, increase or decrease or remove such limits, once site carrying capacity is demonstrated.
- Whatever the site disposition vehicle, permit, license, or lease, the governing legislation should clearly provide that the recipient has exclusive ownership of the stock and use of the area for farming, has clearly defined property rights and protections, and is not at a disadvantage in securing project financing and insurance.
- The regulatory system should provide permit, license, and lease terms (i.e., length of time) that are long enough for the aquaculture project to build out to an operational full scale and reach profitability. The suggested initial time period for an open ocean project is 15–20 years, with optional renewal time periods for projects in good standing of 15 years. Timeframes of this length will give the project a stable tenure and greatest opportunity to be successful, profitable and pay off any long-term loans.
- The regulatory system should identify allowable species for culture and incorporate economically important native species, and those introduced species that are established in the environment. Introduction of non-native species for aquaculture purposes should only be considered after extensive study of environmental risks and a public process to solicit input from stakeholders.
- Fees for permits or licenses and rents for leases should be set as a balance between cost recovery for administration of the regulatory process and a reasonable charge for the use of the public resource, so that establishing a farm is attractive to private investment. Governments could consider putting these revenues in a specially designated fund that supports further open ocean aquaculture research and development, e.g., Hawaii, U.S.A.
- Government should consider creating a separate regulatory process for open ocean aquaculture research and large-scale demonstration projects considering the emerging status of the industry. The information requirements and review process for these R, D, and D projects in the EEZ should be easier, quicker, and less complex than for a commercial project because these projects are smaller in scale, temporary, and short term.
- Government should require baseline environmental studies as part of the application process that are realistic, based on science, and can be carried out in a timely fashion at a reasonable cost. The likely inherent stability of the open ocean environment at any site should be considered in designing baseline studies and the potential positive impacts of the aquaculture project should also

be reported, such as enhancement of recreational fishing by fish naturally aggregating around the farm.

- Government should craft environmental monitoring requirements for aquaculture projects to foster sustainable development that are based on accepted science and essential parameters and reasonable sampling frequencies. Further they should not be overly burdensome to a farm located in a highly exposed, turbulent, open ocean setting. When uncertainty exists for the potential project impacts and rigorous monitoring is initially required, farmers should have the opportunity to modify monitoring plans (e.g., sampling frequency and parameters required) when sample data, are found, for example, to not be changing at a high rate or are having insignificant impact on ambient conditions.

References

Alfred-Wegener Institute. (2014). *Marine aquaculture technologies.* Accessed November 3, 2014 from http://awi.de/en/reserch/new_technologies/marime_aquaculture_maritime_technologies_and_iczm/

Alston, D., Cabarcas, A., Capella, J., Bennetti, J., & Cortes, R. (2005). *Environmental and social impact of sustainable offshore Cage culture production in Puerto Rican Waters* (p. 206). Final Report. Washington, DC: NOAA.

American Oil and Gas Historical Society. (2014). *History of offshore oil and gas drilling.* Retrieved September 28, 2014, from http://aoghs.org

Anderson, J., & Shamshak, G. (2008). Future markets for aquaculture products. In: R. Michael (Ed.), *Offshore aquaculture in the United States: Economic considerations, implications, and opportunities* (pp. 231–244). NOAA Technical Memorandum. Silver Springs, USA: USDOC

Angel, D., & Edelist, D. (2013). Sustainable development of marine aquaculture off-the-coast and offshore—A review of environmental and ecosystem issues and future needs in the tropical zones. In A. Lovatelli, J. Anguillar-Manjarrez, & D. Soto (Eds.), *Technical Workshop Proceedings: Expanding mariculture farther offshore: Technical, environmental, spatial, and governance challenges* (pp. 173–200). Rome, Italy: FAO.

Aquaculture Planning & Advocacy. (2011). Hawaii Dept. of Agriculture (124 p). Retrieved September 15, 2014, from http://hdoa.hawaii.gov/ai/files/2013/03/Permits-and-Regulatory-Requirements-For-Aquaculture-in-Hawaii-2011-Final.pdf

Balsiger, J. (2009). Letter to Dr. Robert Shiff, Chairman, Gulf of Mexico Fishery Management Council from James Balsiger, Acting Assistant Administrator for Fisheries, NMFS, NOAA. September 3, 2009. Retrieved September 3, 2014, from http://sero.nmfs.noaa/sustainable_fisheries/gulf_fisheries/aquaculture/

Bridger, C. J. (Ed.). (2004). *Efforts to develop a responsible offshore aquaculture industry in the Gulf of Mexico.* Ocean Springs, MS: Mississippi-Alabama SG Consortium.

Brown, L. (2009). *Plan B 4.0: Mobilizing to save civilization.* New York: W.W. Norton & Comp. Inc.

Buck, B. H., Krause, G., & Rosenthal, H. (2004). Extensive open ocean aquaculture development within wind farms in Germany: The prospect of offshore co-management and legal constraints. *Ocean and Coastal Management, 47,* 95–122.

Buck, B. H., Krause, G., Michler-Cieluch, T., Brenner, M., Buchlolz, C., Busch, J., et al. (2008). Meeting the quest for spatial efficiency progress and prospects of extensive aquaculture within offshore wind farms. *Helgoland Marine Research, 62,* 269–281.

Bush Administration. (2004). *U.S. Ocean action plan.* Washington, DC: NOAA.

Callies, D. L. (2010). *Regulating paradise: Land use controls in Hawaii*. Honolulu, Hawaii: Univ. of Hawaii Press.

Cincin-Sain, B., Bunsick, S., Corbin, J., DeVoe, M., Eichenberg, T., Ewart, J., et al. (2005). Recommendations for an operational framework for offshore aquaculture in U.S. Federal Waters. Technical Report. Gerard J. Mangone Center for Marine Policy, University of Delaware.

Coastal Response Research Center. (2010). *Technical readiness of ocean thermal energy conversion (OTEC)*. Durham, N.H.: University of New Hampshire.

Code of Federal Regulations. (2014). *Permit, suspension, modification and denial*. Retrieved November 2, 2014, from http://www.gpo.gov/fdsys/granule/CFR-2009-title15-vol3/CFR-2009-title15-vol3-part904

Corbin, J. S. (2010). Sustainable U.S marine aquaculture, a necessity. *Marine Technology Society Journal, 44*, 7–21.

Corbin, J. S., & Young, L. (1997). Planning, regulation, and administration of sustainable aquaculture. In J. Bardach (Ed.), *Sustainable aquaculture* (pp. 201–233). New York: John Wiley and Sons.

Crown Estate. (2014). *What we do*. Retrieved November 2, 2014, from http://www.thecrownestate.co.uk/coastal/what-we-do/

FAO. (2014). *The State of World Fisheries and Aquaculture—Opportunities and challenges*. Rome, Italy: FAO.

Firestone, J., Kempton, W., Krueger, A., & Loper, C. (2005). Regulating offshore wind power and aquaculture: Messages from land and sea. *Cornell Journal of Law and Public Policy, 14*, 72–111.

Fletcher, K. (2004). Law and offshore aquaculture: A true hurdle or a speed bump. In C. Bridger (Ed.), *Efforts to develop a responsible offshore aquaculture industry in the Gulf of Mexico: A compendium of offshore aquaculture consortium research* (pp. 23–34). Ocean Springs, MS: Mississippi-Alabama Sea Grant Consortium.

Forster, J. (2013). A review of opportunities, technical constraints, and future needs of offshore mariculture—Temperate waters. In A. Lovatelli, J. Anguillar-Manjarrez, & D. Soto (Eds.), *Technical Workshop Proceedings: Expanding mariculture farther offshore: Technical, environmental, spatial, and governance challenges* (pp. 77–100). Rome, Italy: FAO.

Friedman, T. (2008). *Hot, flat, and crowded*. New York: Straus and Giroux.

Global Wind Energy Council. (2014). *Current status global offshore wind farms*. Retrieved July 27, 2014, from http://www.gwec.net/global-offshore-current-status-future-products

GMFMC. (2014a). *Offshore mariculture rule*. Retrieved September 28, 2014, from http://www.gulfcouncil.org/index.php

GMFMC. (2014b). *National offshore aquaculture Act 2007*. Retrieved November 2, 2014, from http://www.gulfcouncil.org/fishery_management_plans/aquaculture_management.php#Legislation

GMFMC & NOAA. (2009). *Final fishery management plan for regulating offshore marine aquaculture in the Gulf of Mexico*. GMFMC and NOAA,

Harvard Law School Emmett Environmental Law and Policy Clinic, Environmental Law Institute & The Ocean Foundation. (2012). *Offshore aquaculture regulation under the clean water act*. Retrieved September 5, 2014, from http://www.eli.org/sites/default/files/docs/cwa-aquaculture.pdf

Holmer, M. (2013). Sustainable development of marine aquaculture off-the-coast and offshore—A review of environmental and ecosystem issues and future needs in temperate zones. In A. Lovatelli, J. Anguillar-Manjarrez, & D. Soto (Eds.), *Technical Workshop Proceedings: Expanding mariculture farther offshore: technical, environmental, spatial, and governance challenges* (pp. 135–172). Rome, Italy: FAO.

Hubbs Sea World Research Institute. (2003). *The grace mariculture project*. Carlsbad, CA: HSWRI.

Interagency Working Group on Aquaculture. (2014). *Guide to federal aquaculture programs and services*. Retrieved November 20, 2014, from https://www.whitehouse.gov/sites/default/files/microsites/ostp/NSTC/federal_aquaculture_resource_guide_2014.pdf

Jeffs, A. (2013). A review on the technical constraints, opportunities, and needs to ensure the development of mariculture sector worldwide—Tropical zone. In A. Lovatelli, J. Anguillar-Manjarrez, & D. Soto (Eds.), *Technical Workshop Proceedings: Expanding mariculture farther offshore: Technical, environmental, spatial, and governance challenges* (pp. 101–134). Rome, Italy: FAO.

Kapetsky, J. M., Aguilar-Mayarrez, J., & Jenness, J. (2013). *A global assessment of potential far offshore mariculture development from a spatial perspective.* FAO Fisheries and Aqua. Tech. Paper No. 549. Rome, Italy: FAO.

Knapp, G. (2013). The development of offshore aquaculture: An economic perspective. In A. Lovatelli, J. Anguillar-Manjarrez, & D. Soto (Eds.), *Technical Workshop Proceedings: Expanding mariculture farther offshore: Technical, environmental, spatial, and governance challenges* (pp. 201–244). Rome, Italy: FAO.

Krop, L., & Polefka, S. (2007). *Open ocean aquaculture in the Santa Barbara Channel: An emerging challenge for the Channel Islands National Marine Sanctuary.* Santa Barbara, CA: Environmental Defense Fund.

Lagerveld, S., Rockmann, C., Scholl, M. M., Bartelings, H., van den Burg, S. W. K., Jak, R. G., et al. (2014). *Combining offshore wind energy and large-scale mussel farming: background & technical, ecological and economic considerations.* Wageningen, NL: IMARES.

Leschin-Hoar, C. (2014). Meeting San Diego's aquaculture cowboy, Don Kent of Hubbs Sea World Research Institute forges ahead with plans for large-scale, open ocean fish farm. *Fish Farming News, 5,* 10–13.

Lovetalli, J., Anguilar- Manjarrez, J., & Soto, D. (Eds.). (2013). *Expanding mariculture farther offshore: technical, environmental, spatial, and governance challenges.* Rome, Italy: FAO.

Mee, L. (2006). *Complementary benefits of alternative energy: Suitability of offshore wind farms as aquaculture sites.* Plymouth, UK: Marine Institute—University of Plymouth.

National Marine Fisheries Service. (2013). *Fisheries of the United States 2012—Current fishery statistics No. 2012.* Silver Spring, USA: NOAA.

National Ocean Council. (2014). *Marine spatial planning.* Retrieved December 3, 2014, from http://www.whitehouse.gov/administration/eop/oceans/marine-planning

National Sea Grant Law Center. (2012). *Offshore mussel culture operations: Current legal framework and regulatory authorities.* Oxford, USA: National Sea Grant Law Center, University of Mississippi.

NOAA. (2008). *Offshore aquaculture in the United States: Economic considerations, implications and opportunities.* Washington, DC: NOAA.

NOAA. (2011). *NOAA aquaculture policy.* Washington, DC: NOAA. Retrieved August 14, 2014, from http://www.nmfs.noaa.gov/aquaculture/docs/policy/noaa_aquaculture_policy_2011.pdf

Northeast Ocean Data Viewer. (2014). *Maps and data for ocean planning.* Retrieved September 28, 2014, from http://www.northeastoceandata.org/viewer/

Open Blue Sea Farms. (2014). *World's largest open ocean fish farm.* Retrieved December 4, 2014, from http://www.openblue.com/

Percy, D., Heshamunda, N., & Kuemlangan, B. (2013). Governance in marine aquaculture: the legal dimension. In A. Lovatelli, J. Anguillar-Manjarrez, & D. Soto (Eds.), *Technical workshop proceedings: Expanding mariculture farther offshore: technical, environmental, spatial, and governance challenges* (pp. 245–262). Rome, Italy: FAO.

PEW Marine Aquaculture Task Force. (2007). *Sustainable marine aquaculture: Fulfilling the promise, managing the risks.* Takoma Park, MD: PEW Charitable Trust.

PEW Oceans Commission. (2003). *America's living oceans: Charting a course for sea change.* Takoma Park, MD: PEW Charitable Trust.

Sims, N. (2013). Kona Blue Water Farms case study: Permitting, operations, marketing, environmental impact, and impediments to expansion of global open ocean mariculture. In A. Lovatelli, J. Anguillar-Manjarrez, & D. Soto (Eds.), *Technical workshop proceedings: Expanding mariculture farther offshore: technical, environmental, spatial, and governance challenges* (pp. 263–296). Rome, Italy: FAO.

State of Hawaii. (2014). *CDUA marine waters application*. Retrieved October 10, 2014, from http://dlnr.hawaii.gov/occl/forms-2/

State of Maine. (2014). *Marine aquaculture in Maine*. Retrieved November 1, 2014, from www.maine.gov/dlnr/aquaculture/

Statista. (2014). *Offshore oil and gas rigs in 2014*. Retrieved May 28, 2014, from http://www.statista.com/rigs-worldwide-by-region/

Stickney, R., Costa-Pierce, B., Baltz, D., Drawbridge, M., Grimes, C., Phillips, S., et al. (2006). Towards sustainable open ocean aquaculture in the United States. *Fisheries, 3*, 607–610.

U.S. Commission on Ocean Policy. (2004). An ocean blueprint for the 21st Century. Washington, DC: USCOP

U.S. Dept. of Commerce. (2007). *NOAA 10 year plan for marine aquaculture*. Washington, DC: NOAA, USDOC

U.S. DOC. (2011). *U.S. Dept. of Commerce Aquaculture Policy*. Retrieved August 17, 2014, from www.nms.noaa.gov/aquaculture/docs/policy/doc_aquaculture_policy_2011.pdf

U.S. General Accounting Office. (2008). *Offshore marine aquaculture: Multiple administrative and environmental issues need to be addressed in establishing a U.S. regulatory framework*. Washington, DC: U.S. Government Accountability Office.

United Kingdom Legislation. (2014). *Marine and coastal access Act, 2009*. Retrieved December 2, 2014, from http://www.leislation.gov.uk/ukpga/2009/23/notes/contents

Upton, H., & Buck, E. (2010). *Open ocean aquaculture—CRS Report to Congress 7-5700*. Washington, DC: Congressional Research Service

World Bank. (2013). *Fish to 2030: Prospects for fisheries and aquaculture*. Agriculture environmental discussion paper No. 3. Washington, DC: World Bank.

Part III
Aquaculture Economics

Chapter 10
Economics of Multi-use and Co-location

Hauke L. Kite-Powell

Abstract Under the right circumstances, multi-use of a marine site through co-location of complementary activities can result in more efficient use of ocean space. We explore the economic dimension of multi-use and co-location, using the general example of an aquaculture operation co-located within an ocean wind farm. Co-locating aquaculture operations and wind farms can produce both public and private benefits (cost savings). The public benefits arise from the fact that an aquaculture operation co-located within the boundaries of a wind farm does not negatively affect the ecosystem services derived from the ocean area it would otherwise have occupied. The private benefits are cost savings that arise from shared permitting, infrastructure, and logistics efforts and systems. The economic value associated with these benefits depends on the scale, location, and nature of the co-located ventures and the natural resources they affect. For locations in open ocean and relatively low-value coastal waters that are candidates for wind farm or aquaculture sites in most countries, the public benefit is likely to be on the order of 500−3,000/year per hectare of area occupied by the aquaculture operation, and the private benefits are likely to be less than $50–100/ton of aquaculture operation output.

10.1 Introduction

Under the right circumstances, multi-use of a marine site through co-location of complementary activities can result in more efficient use of ocean space. In this chapter, we explore the economic dimension of multi-use and co-location, using the general example of an aquaculture operation co-located within an ocean wind farm.

We begin with a review of ocean space as a potentially scarce resource that can be an input to a range of productive economic and ecological processes. In some

H.L. Kite-Powell (✉)
Marine Policy Center, Woods Hole Oceanographic Institution,
MS 41, Woods Hole, MA 02543, USA
e-mail: hauke@whoi.edu

© The Author(s) 2017
B.H. Buck and R. Langan (eds.), *Aquaculture Perspective of Multi-Use Sites in the Open Ocean*, DOI 10.1007/978-3-319-51159-7_10

233

cases, competing uses may be incompatible in a given location (for example, dragging for wild scallops cannot easily be co-located with on-bottom or floating moored aquaculture operations). In other cases, multi-use may be feasible (for example, floating finfish cages or shellfish longlines in the open spaces between wind farm towers). It is in the interest of society, and of marine resource managers in particular, to identify and promote the most valuable use (or combination of uses) of each piece of ocean space. To do this, it is necessary to understand the economic and ecological production functions for which ocean space is an input.

We then consider the way in which co-location of activities can result in private and public economic benefits. Private benefits may arise from cost savings due to shared use of planning and permitting investments, infrastructure, logistics, supply lines, communication facilities, etc. Public benefits may arise from the accommodation of more economic activity within a smaller footprint, which leaves more ocean space for other uses and reduces losses in ecosystem service value from incompatible uses of ocean space.

We conclude with a summary of what is the likely scale of economic benefits from co-location, and summarize results from a case study of mussel culture and wind farms in the Netherlands.

10.2 Ocean Space as an Input to Economic Production

Ocean space is a heterogeneous resource that serves as an input to numerous economic and ecological production processes. Particularly near the coast, where biological and ecological production is concentrated and where economic and recreational/aesthetic uses are most intense, ocean space may be in short supply and heavily contested. Marine spatial planning activities in many regions of the world are undertaken to manage competing uses of ocean space and promote efficient use of scare marine resources, particularly in coastal areas. Examples include ongoing regional ocean planning activities in the United States under the auspices of the US National Ocean Council, such as the Northeast Regional Ocean Plan (Northeast Regional Planning Body 2016), and the German marine spatial plan (Bundesamt für Seeschifffahrt und Hydrographie 2016) (Fig. 10.1).

A number of transitory and non-transitory uses can compete for ocean space. Transitory uses include:

- Commercial fishing (trawls, seines, dredges, longlines, gillnets, handlines)
- Recreational fishing
- Military operations
- Maritime shipping
- Yachting and recreational boating
- Whale watching
- Diving

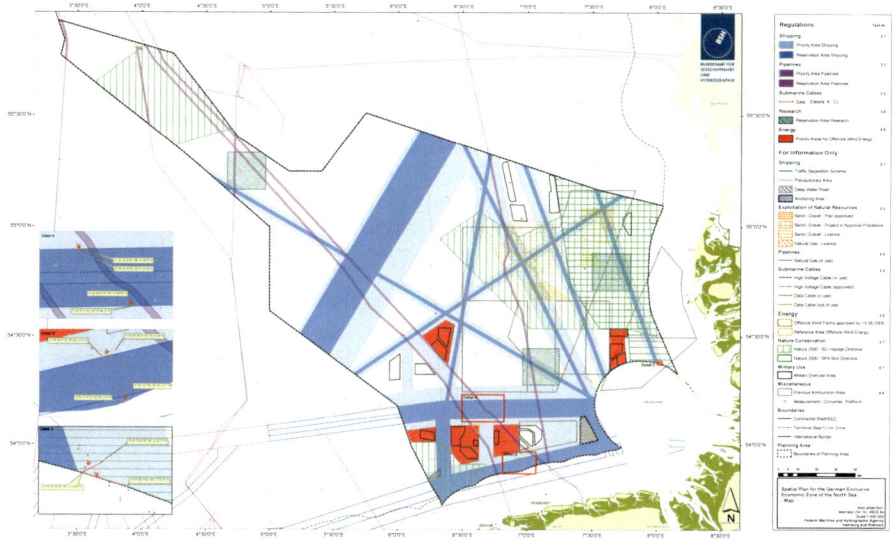

Fig. 10.1   Marine spatial plan for Germany's North Sea waters (Bundesamt für Seeschifffahrt und Hydrographie 2016)

Non-transitory uses include:

- Commercial fishing (traps/pots, weirs)
- Navigation channels
- Deepwater ports
- Wind farms
- Aquaculture sites
- Ocean dumping sites
- Shipwrecks
- Pipelines
- Marine protected areas
- Aesthetic viewsheds

The characteristics of a given area of ocean space make it more or less valuable to different uses. For example, the characteristics that determine the value of an area of ocean for marine aquaculture include distance from shore facilities (ports), nutrient fluxes, plankton densities, water temperature ranges, wind and wave conditions, current flows and flushing rates, prevalence of predators and parasites, marine contaminants, etc.

Because ocean space is of varying quality in this sense, it will produce a gradient of (potential) resource rents from alternative uses. At the same time, those uses may result in external effects if resources are degraded due to pollution, for example, or if opportunity costs arise due to forgone other uses of the space.

Historically, ocean space has not been allocated through private markets, but has instead been assigned to competing uses by pre-existing policies, implied legal interests based on tradition, and public trust considerations, among others. Historical (especially transitory) uses may have a priority in these processes. Allocation decisions are typically spread across multiple agencies. Multi-use allocation decisions will have to be made in the context of these established mechanisms.

10.2.1 Production Function

In making that happen, an economic framework is likely to be useful. The coastal and marine resources and coastal infrastructure of a geographic region can be thought of as inputs to ecological processes and economic activities that generate the value (wealth) that flows from that region's ocean resources. Understanding this "production function"—how resources and infrastructure contribute to economic value—is important if ocean resources are to be managed and used for maximum sustained benefit. A common way to measure that benefit is to quantify (estimate) the economic value generated by the economic and ecological processes that use the resources. In the following pages, we review briefly the state of knowledge about the links between marine resources and value generation. This information can suggest the scale of certain economic benefits that may flow from co-location.

Economic value exists only in the context of human populations and societies. One important determinant of economic value, therefore, is the people who participate in and receive benefits from the economic activity. The market and non-market value generated from marine resources is, in part, a function of how many people live, work, and play in the coastal and ocean areas in question, and how many visitors and tourists come to the region. There are some exceptions to this, especially in the more basic categories of ecosystem service values. For example, the value of carbon dioxide (CO_2) uptake by coastal and ocean waters is largely independent of the population living near those waters. But most categories of value will rise and fall with the number of participants; and that number can change because of population trends, changes in tourism, changes in recreational preferences, changes in wealth distribution, and other socioeconomic factors.

Value arises from marine resources because they provide ecosystem services and products and services, some of which are traded in markets. The Millennium Ecosystem Assessment (MEAB 2003) framework suggests the following classification of ecosystem services derived from coastal and marine resources:

• Provisioning Services

   – Food (fisheries, aquaculture)
   – Sea water
   – Biochemical and genetic resources
   – Minerals and other physical resources

- Regulating and Supporting Services

  - Climate regulation (CO_2 uptake, heat exchange)
  - Water purification (filtration, dilution)
  - Flood/storm protection
  - Erosion control
  - Waste assimilation
  - Nutrient cycling
  - Primary production

- Cultural Services

  - Beach recreation and coastal access
  - Recreational boating, fishing, diving
  - Aesthetic, spiritual, and cultural uses of the coast and ocean
  - Existence/bequest value of local species (value attributed by people to knowing that species exist, and will survive for future generations)

Figure 10.2 illustrates how different subsets of marine resources and infrastructure contribute to economic value generated in different economic activities and ecosystem service functions (climate regulation, water purification, and storm surge regulation) that are not captured by market data. The table is not exhaustive, but illustrates two important points. First, each natural resource and infrastructure component typically supports value generation in a variety of economic sectors and ecological functions. And second, different ocean economy sectors depend on different combinations of resources and infrastructure.

Although we know in principle which resources are used as inputs to which categories of ecosystem service and value, as suggested by Fig. 10.2, our ability to predict how changes in resources and infrastructure might affect value generation is, in most cases, incomplete at best. That is because the relationship between inputs (natural resources, infrastructure) and outputs (e.g., seafood, or recreation days) and the value of those outputs is often complicated. For some economic activities, the simple existence of access to a category of resources is sufficient: for example, the maritime transport industry needs port infrastructure and access to coastal and ocean waters to generate value; but that value does not increase, as a rule, when coastal water quality is improved. Furthermore, different areas of the ocean may have different levels of value to the maritime transportation sector, depending on their location relative to preferred shipping routes. On the other hand, the value generated by activities such as commercial fishing, aquaculture, and recreational boating and fishing depends both on the quantity and quality of coastal and ocean water resources (Hanemann et al. 2005; Keeler et al. 2012).

	Natural Resources/Habitats						Infrastructure					
	Ocean waters	Coastal waters/bays	Beaches	Wetlands/estuaries	Living resources	Cultural/archaeological resources	Shoreline structures	Commercial ports	Commercial real estate	Naval and Coast Guard facilities	Marinas	Residential real estate
Commercial fishing	X	X		X	X			X				
Aquaculture		X			X			X				
Seafood processing								X	X			
Seafood markets									X			
Recreational boating & fishing	X	X	X	X	X	X					X	
Beach recreation			X									
Tourism	X	X	X	X				X	X		X	X
Maritime transportation	X	X						X	X			
Ship- & boat building/repair	X							X	X	X		
Marine construction & manufacturing							X	X	X	X	X	X
Ocean energy	X	X						X				
Research and education	X	X	X	X	X	X		X	X			
National security	X	X						X		X		
Climate regulation	X	X		X	X							
Water purification		X		X								
Storm surge regulation		X	X	X			X					

Fig. 10.2 Mapping resources to economic sectors and value generation (adapted from Kite-Powell et al. 2016)

10.2.2 Unit Values

In general, the economic value of a resource or infrastructure component is best estimated at the margin, that is, in the context of a question such as "what is the value of an additional square kilometer of coastal wetlands to a region's seafood or coastal tourism industries," or "what is the value of an additional kilometer of beach to a region's coastal recreation benefits"? The value per unit area of an incremental piece of marine habitat, for example, depends not only on the location and characteristics of that piece, but also on how much of that kind of habitat already exists in the regional ecosystem. For these reasons, estimates of unit value (dollars per square kilometer, or dollars per year per square kilometer) for natural resources should generally be treated with caution.

The economic value of some marine ecosystem services, including the provision of seafood and other marine products, and certain recreational amenities, can be estimated from prices and quantities of goods and services traded in markets ("market values"). Other ecosystem services generate values that affect human wellbeing but are not observable from market transactions. These include the non-market or intrinsic values derived from walking on a beach, for example. There is some overlap between ecosystem service values and market values: for example, the primary production that supports biological populations of food fish is an ecosystem service, and its value is (partially) reflected in the commercial fisheries landings data.

Most ecosystem service values cannot be observed from prices in markets, and therefore must be estimated by quantifying the ecological service produced (for example, tons of CO_2 absorbed by the ocean waters of the North Atlantic each year) and then applying a unit value (in this case, the cost imposed by adding a ton of CO_2 to the atmosphere—see US EPA (2016). Published estimates of ecosystem service value from marine environments around the world span a very wide range, from near zero to more than \$100 million/year/km^2 (\$1 million/year/ha), depending on the location and the specific values included and assumptions used in the estimation. Using ecosystem service values in any particular planning context requires careful attention to the ways in which resources are used and valued, and the consequences of incremental management actions (Johnston and Russell 2011). Ecosystem service value estimates are broadly indicative of orders of magnitude for ecosystem services, but, as planning tools, they should be used with care.

In estimating the economic value derived from ocean space (habitat) in a particular location, it is necessary to take into account the particular qualities of local coastal and marine resources, and the socio-economic context of nearby populations. In general, this requires a location-specific analysis. Where no such analysis has been carried out, it is sometimes possible to develop rough estimates of economic value using a "benefit transfer" technique (see below).

Initial efforts to integrate results of ecosystem valuation studies from different locations can be found in Costanza et al. (1997). More recently, researchers have developed several databases, including the Ecosystem Services Valuation Database (Plantier-Santos et al. 2012; Texas A&M University 2016) for the US Gulf of Mexico, that summarize results of valuation studies from many locations. Table 10.1 presents summaries of values for different types of ecosystems from the meta-analysis by de Groot et al. (2012). Note that the point estimates summarized in Table 10.1 are typically derived from several studies conducted for different locations and different value categories.

Shoreline areas, and in particular beaches, provide significant value as aesthetic and recreational resources. Pendleton (2008) summarized studies suggesting that beaches along the coast of the United States see a total of approximately 800 million beach-visit-days per year. Those beach visits provide total recreational value of \$7—35 billion/year (2014 US\$), or about \$0.25 M/km^2/year for the average coastal land area (assuming approximately 9300 km of US coastline, excluding Alaska, and an average width of coastal recreation area of 100 m, or 0.1 km^2/km of coastline). The unit value of popular beaches is significantly higher

Table 10.1 Summaries of value estimates from ecosystem valuation studies

	Ecosystem service value, 2014 US$M/km²/year			
	Marine waters (open ocean)	Coral reefs	Coastal systems (incl. sea grass)	Coastal wetlands (incl. mangroves)
Provisioning services	0.01	6.41	0.28	0.34
Food	0.01	0.08	0.27	0.13
Water				0.14
Raw materials	<0.01	2.48	<0.01	0.04
Genetic resources		3.80		<0.01
Medicinal resources				0.03
Ornamental resources		0.05		
Regulating and habitat services	0.01	21.58	3.01	21.69
Climate regulation	0.01	0.14	0.06	0.01
Disturbance moderation		1.95		0.62
Waste treatment		0.01		18.64
Erosion prevention		17.62	2.92	0.45
Nutrient cycling				0.01
Nursery services			0.02	1.22
Genetic diversity	<0.01	1.86	0.02	0.75
Cultural services	0.04	12.52	0.03	0.25
Esthetic information		1.31		
Recreation	0.04	11.07	0.03	0.25
Spiritual experience			<0.01	
Cognitive development		0.13	<0.01	
All services Mean values	0.05	40.51	3.33	22.29
Standard deviation	0.09	76.89	0.58	44.18
Minimum values	0.01	4.23	3.01	0.03
Maximum values	0.19	244.85	4.84	102.10
Number of estimates (de Groot et al. 2012)	N = 14	N = 94	N = 28	N = 139

Based on data from de Groot et al. (2012); adapted from Kite-Powell et al. (2014)

than this, since they make up only a fraction of total coastline, and account for a much greater density of visits.

Table 10.2 summarizes the mean annual ecosystem service value estimates reported by de Groot et al. (2012) and other relevant studies for marine natural resources. These studies include Atkinson et al. (2012), Barbier (2012), Barbier et al. (2011), Brander et al. (2006, 2007), Costanza et al. (2008), Engle (2011), Ghermandi et al. (2008), Grabowski et al. (2012), Rao et al. (2014), Rönnbäck 1999, and Woodward and Wui (2001).

The provisioning service values are dominated by food production services, except for coral reefs, where 95% of estimated value derives from genetic resources (species with potential medical applications, etc.) and other raw materials. The dominant source of regulating/habitat value is erosion control (for coral reefs and

Table 10.2 Mean annual value (in 2014 US$M/km^2) of ecosystem services based on a review of studies from a variety of geographic settings

Habitat/biome	Ecosystem service value, 2014 US$M/km^2/year			
	Provisioning Services	Regulating and habitat services	Cultural services	Total value
Coral reef	6.410	21.100	12.500	40.000
Coastal systems (incl. sea grass)	0.276	3.020	0.035	3.330
Coastal wetlands (incl. mangroves)	0.345	21.700	0.252	22.300
Open ocean	0.012	0.008	0.037	0.057
Beach/shoreline			0.250	0.250

Adapted from Kite-Powell et al. (2014); see text above for sources

coastal systems/sea grasses) and waste treatment (filtering of runoff water) for coastal wetlands. The main source of cultural value is recreational amenities. These are useful starting points for rapid assessment at various locations using benefit transfer concepts; but care must be taken to consider carefully the similarities and differences between study locations and benefit transfer locations before adopting specific ranges of estimates.

Benefit transfer involves the application of non-market valuation models or results developed for one location to another. This method is often used when there are no data on, and insufficient resources to conduct, non-market valuation at the site in question. Benefit transfers comprise the utilization of specific values, such as mean willingness-to-pay (WTP or demand) for ecosystem services or the transfer of functions (estimated for resources in other areas) that can be used to predict WTP on the basis of environmental or socioeconomic characteristics. Boyle and Bergstrom (1992) identify three criteria for benefits transfer: (i) the resources and resource quality conditions should be similar at the two areas; (ii) the socioeconomic characteristics of the relevant populations should be similar in the two areas; and (iii) the specific nonmarket valuation methodologies, models, or estimation techniques used at the studied site should be the same as those that would be applied at the site in question. If these three criteria are not met, a benefits transfer is likely to be biased (this bias is referred to as a "transfer error"). Accepting some level of transfer error may be preferable to the only alternative: assuming that the resource or use at issue has no economic value (Johnston et al. 2014). Where good arguments can be made that criteria (i) and (ii) have been met, it is often sensible to begin with measures of central tendency (means or medians) for ecosystem service values, and to use estimates of confidence intervals, if available, to identify feasible value ranges. As an example, Rosenberger and Loomis (2001) use summary statistics from three decades of studies of recreation in the United States to develop benefit transfer estimates (means and 95% confidence intervals) of consumer surplus per person per day of $32 ± $13 for recreational swimming and $54 ± $10 for recreational fishing (2014 dollars). Additional examples of benefit transfer applications and discussion of limitations can be found in Allen and Loomis (2008),

Bergstron and Taylor (2006), Hoehn (2006), Lindhjem and Navrud (2008), and Plummer (2009).

As another example, the value of ocean areas for seafood production from commercial fishing in the New England region of the United States averages about $1200/year/km^2 ($12/year/ha), but ranges widely from near zero to more than $50,000/year/km^2 ($500/year/ha) for specific locations (Kite-Powell et al. 2016). Estimates of ecosystem service value associated with (hypothetical) open ocean aquaculture operations in that region range from $1 million to $100 million/year/km^2 ($10,000 to $1 million/year/ha).

10.3 Public Benefits from Multi-use and Co-location

When a wind farm and aquaculture operation are co-located within the footprint of the wind farm, the two activities together take up less ocean space (habitat space) than they do when they are not co-located. The public benefit from co-location in this case is the avoided opportunity cost associated with ecosystem services that would have been foregone if the aquaculture operation had been located outside the wind farm footprint.

The categories of ecosystem service value most commonly displaced by wind farms and marine aquaculture operations are seafood production and recreational boating and fishing. Both wind farms and aquaculture sites typically designate a buffer zone around the site in which commercial fishing is not permitted. Recreational boating and fishing may be permitted within a wind farm site, and may or may not be compatible with an aquaculture operation, depending on its physical "footprint" in the water column.

Since neither wind farms nor aquaculture operations are usually located in extremely high value ocean areas like coral reefs or mangrove wetlands, as a rule it makes sense to assume that the values at stake will be representative of "Open Ocean" or "coastal ocean" areas as estimated in Tables 10.1 and 10.2. If an aquaculture operation that would displace cultural and provisioning services in any location is sited within a wind farm footprint that does the same, this implies an avoided cost to the public of $0.05 to $0.3/km^2/year—of $500–$3000/ha/year—for the effective footprint of the aquaculture operation.

10.4 Private Benefits from Multi-use and Co-location

Potential private cost savings from co-locating aquaculture operations with wind farms arise from reductions in combined permitting expenses, structural (mooring) and/or site marking costs, costs associated with possible power and/or data links to shore, logistics costs, and/or lease payments. If the aquaculture operation has to bear these costs in their entirety on a site that is separate from the wind farm, and can share costs or services with the wind farm if they are co-located, the resulting avoided cost amounts to a private benefit accruing to the aquaculture operation. This benefit can in practice be shared with the operators of the wind farm.

Permitting expenses and related legal costs arise from the need to conduct studies of fish stocks, benthic resources, cultural resources, endangered species, and other aspects of the proposed site before permits to construct and operate the wind farm or aquaculture operation can be obtained. These studies tend to be site-specific and generally amount to less than 5% of total project start-up and installation costs, or less than 1% of total annualized project costs (Kite-Powell et al. 2003a, b).

If the aquaculture operation can make use of structures (moorings, pilings, etc.) that are installed as part of the wind farm infrastructure (including marker buoys demarking the boundaries of the exclusion zone around the wind farm), or of site-to-shore power cables or data links, this implies reduced installation and maintenance costs for the aquaculture operation. Like permitting costs, these typically represent less than 5% of total annual aquaculture project expenses (Jin et al. 2003; Kite-Powell et al. 2003a, b).

Logistics associated with maintaining the aquaculture infrastructure, bringing seed/fingerlings and feed to the arm site, and transporting harvested product back to shore, typically account for about 5% of the total annual operating cost of an ocean aquaculture operation. The logistics needs of an aquaculture operation tend to be much greater than those of a wind farm, but if the two can share logistics infrastructure and services, this may also represent cost savings on the order of 1% for the aquaculture operation.

The effect on combined leased payments will depend heavily on the lease payment structure imposed by the authorities responsible for permitting wind and seafood farms in the area under consideration. It may be that lease payments are assessed on the basis of output, in which case co-location may not produce any net private savings. If combined lease payments are based strictly on the area occupied by the two facilities, net savings may be achieved by co-locating.

In aggregate, the private cost savings potentially realized by the aquaculture operation from co-locating with the wind farm amount to perhaps 5% of total annualized installation and operating expenses. With cultured seafood farm production valued at $1000–$2000/ton, that is equivalent to about $50–100/ton of production per year. As noted, these private benefits would likely be shared in some fashion between the wind farm and the aquaculture operation, since both must collaborate to realize these benefits.

10.5   Case Study: Mussel Culture and Wind Farms in the Netherlands

In the Dutch debate on multi-use, the emphasis has been on combinations of either bivalve or seaweed production within offshore wind energy parks. From a market perspective, the mussel sector is the most logical choice: there is an existing market for mussels. The market recently has had troubles with the supply of mussel spat due to governmental restriction on mussel spat collection in the Wadden Sea.

Following the identification of potential for bivalve culture in the North Sea, a case-study was carried out to investigate the economic benefits for the co-production mussels and offshore wind energy in the North Sea, summarized here based on information from Sander van den Burg. This case study focused on spatial distribution and examined if vacant space in offshore wind parks can be used for mussels production. It was assumed that mussel production facilities are not physically attached to the wind turbines or their foundations. Within large offshore wind energy farms, the wind turbines are placed in clusters, leaving space between the clusters to avoid wind-free zones. If this vacant space is used for aquaculture, a relatively large and accessible area is created. Within a 1000 MW wind parks, the expected vacant area is 4000 ha (see Fig. 10.3).

This case-study included (i) a simple linear optimization model to analyze how vacant space can be used in the best manner, and (ii) sensitivity analyses to examine the effects of changes in input parameters.

For this exercise, a hypothetical offshore mussel farm was designed in consultation with experts. The proposed mussel farm uses long-lines attached to monopole foundations. The production systems is a long line for mussel spat collection, for first growing and for second growing (in a ration 1:4:16). Size of the system is set at 2.4 km of lines/ha. Based on earlier experiences from a German Case Study (see Chap. 11), expected production is 10,000 kg mussels per km (Buck et al. 2010). The costs for offshore mussel production are estimated using Buck et al. (2010). Fixed costs per ha are estimated at 4255/ha/year, variable costs 1762 €/ha/year. Price for mussels is set at 1.21 €/kg (Buck et al. 2010). Due to the increased production of mussels the price of mussels is expected to decline slightly from 0.95 to 0.94 € kg^{-1}.

Fig. 10.3 Modelled mussel farm co-located within an offshore wind farm (*Image* Sander van den Burg)

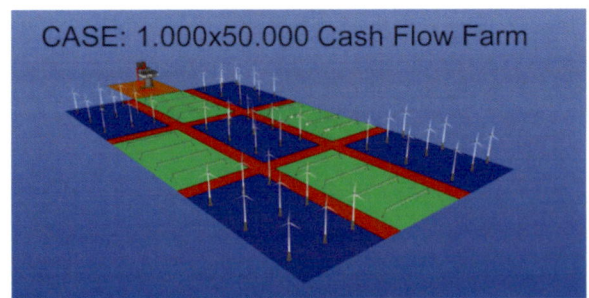

In the model, it is assumed that synergies exist between aquaculture and wind farms, especially related to transport and labor costs. This synergy is set at a relatively low percentage, only expecting benefits from combining transport, operations and maintenance. Thus not the installations, as this would increase (perceived) risks. For wind farms, we set costs for labor and transport (all concerning only O&M) at respectively 759 and 429/ha/year (note that one wind turbine takes up 78 ha (500 m security circle)). The expected synergy is set at 5% of these costs (Lagerveld et al. 2014).

Based on these input parameters, the model shows that the proposed production system for offshore mussel production in wind farms is profitable. Given the estimated costs, price and production, the overall profit to be made is €38 million if the full 4000 ha is allocated to mussel production. This is based on a production of 170,000 tons, revenue of €159 million and a total cost of €121 million.

Another option investigated was 'to do nothing' with the vacant space within the wind parks. This option was included to assess the "costs" of single-use. It is important to note that doing nothing means that some of the potential synergy is not made us of. This is calculated at 71 €/ha/year (roughly €5500 per turbine per year).

As offshore cultivation of mussels is not yet an established practice in the North Sea, the input parameters are subject to a certain degree of uncertainty. Sensitivity analysis was therefore performed to shed light on the economic consequences of changes in (i) lower base price for mussels, (ii) lower mussel yield, and (iii) higher cost for mussel production. It was shown that growing mussels is no longer profitable if the price of mussels drops below 0.70 € kg^{-1}, if the production drops to 30.5 tonnes ha^{-1}, if fixed costs increase to 31,000 €, or if transport costs increase to 14,500 € (Fig. 10.4). All sensitivity analyses showed a linear pattern and the results suggests that the model is quite robust as a reduction or increase of the input variables of −25%, −26%, 22% or 25% for price, yield, fixed costs and transport costs respectively still results in a profitable mussel cultivation system.

When it comes to multi-use in the North Sea, the combination of offshore mussel production and wind energy is considered to be the most promising combination. In this case-study, we assessed the economic feasibility of this combination.

Based on the available information, we estimated input parameters. Model results confirm that a good business case is achievable. There is a lot of uncertainty about the data but the sensitivity analysis shows that within the present business case, there is room for higher costs or lower yields.

In this sort of setting, synergies are lost in a single-use scenario. The analysis shows that the achievable synergies are—due to great differences in turnover—relatively high for the aquaculture sector but relatively low for the wind energy sector. A challenge remains to convince the wind energy sector that the synergies are worth the effort and risks.

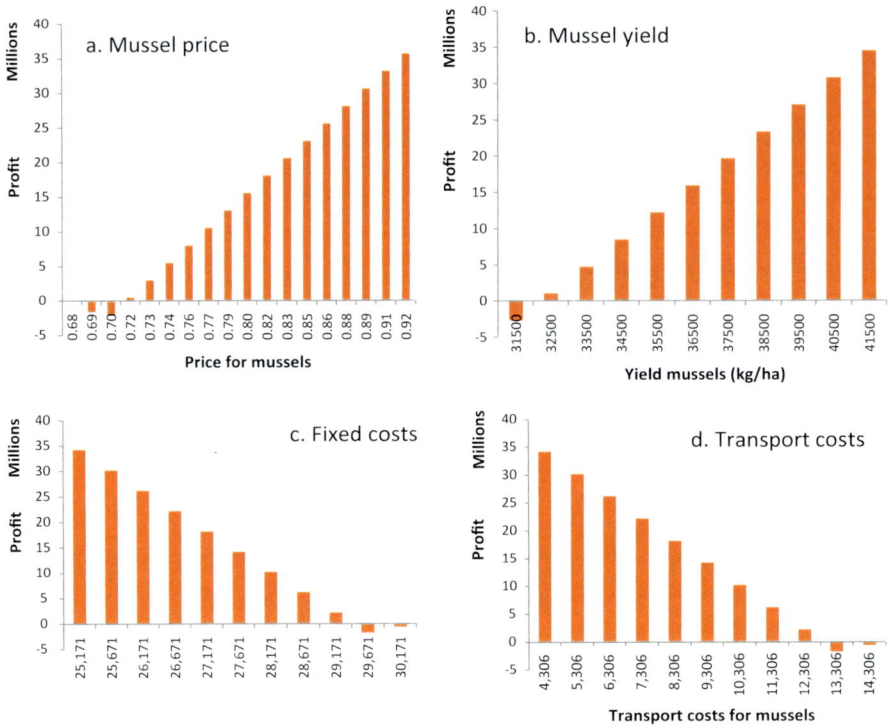

Fig. 10.4 Changes in total profit; sensitivity to changes in price (**a**), yield (**b**), fixed costs (**c**) and transport costs (**d**)

10.6 Conclusions

Co-locating aquaculture operations and wind farms can produce both public and private benefits (cost savings). The public benefits arise from the fact that an aquaculture operation co-located within the boundaries of a wind farm does not negatively affect the ecosystem services derived from the ocean area it would otherwise have occupied. The private benefits are cost savings that arise from shared permitting, infrastructure, and logistics efforts and systems.

The economic value associated with these benefits depends on the scale, location, and nature of the co-located ventures and the natural resources they affect. For locations in open ocean and relatively low-value coastal waters that are candidates for wind farm or aquaculture sites in most countries, the public benefit is likely to be on the order of $500–$3000/year/ha of area occupied by the aquaculture operation, and the private benefits are likely to be less than $50–100/ton of aquaculture operation output.

Because co-location can offer public benefits, it makes sense for local, regional, and national regulation of the siting of energy (e.g. wind farm), seafood production (e.g. aquaculture), and similar ocean uses should encourage developers to consider co-locating operations. This can take the form, for example, of more favorable lease terms for co-located facilities. Co-location is not likely to change the fundamental economics of ocean spatial use, but at the margins, it can make economic activities in the ocean more efficient.

References

Allen, B. P., & Loomis, J. B. (2008). The decision to use benefit transfer or conduct original valuation research for benefit-cost and policy analysis. *Contemporary Economic Policy, 26,* 1–12.

Atkinson, G., Bateman, I., & Mourato, S. (2012). Recent advances in the valuation of ecosystem services and biodiversity. *Oxford Review of Economic Policy, 28,* 22–47.

Barbier, E. B. (2012). A spatial model of coastal ecosystem services. *Ecological Economics, 78,* 70–79.

Barbier, E. B., Hacker, S. D., Kennedy, C., Koch, E. W., Stier, A. C., & Silliman, B. R. (2011). The value of estuarine and coastal ecosystem services. *Ecological Monographs, 81,* 169–193.

Bergstron, J. C., & Taylor, L. O. (2006). Using meta-analysis for benefits transfer: Theory and practice. *Ecological Economics, 60,* 351–360.

Boyle, K. J., & Bergstrom, J. C. (1992). Benefit transfer studies: Myths, pragmatism, and idealism. *Water Resources Research, 28,* 657–663.

Brander, L. M., Florax, R. J. G. M., & Vermaat, J. E. (2006). The empirics of wetland valuation: A comprehensive summary and a meta-analysis of the literature. *Environmental & Resource Economics, 33,* 223–250.

Brander, L. M., van Beukering, P., & Cesar, H. S. J. (2007). The recreational value of coral reefs: A meta-analysis. *Ecological Economics, 63,* 209–218.

Buck, B. H., Ebeling, M., & Michler-Cieluch, T. (2010). Mussel cultivation as a co-use in offshore wind farms: Potentials and economic feasibility. *Aquaculture Economics and Management, 14,* 1365–7305.

Bundesamt für Seeschifffahrt und Hydrographie. (2016). Spatial planning in the German EEZ: maritime spatial plans. Retrieved on August 4, 2016, from http://www.bsh.de/en/Marine_uses/Spatial_Planning_in_the_German_EEZ/index.jsp

Costanza, R., d'Arge, R., de Groot, R., Farber, S., Grasso, M., Hannon, B., et al. (1997). The value of the world's ecosystem services and natural capital. *Nature, 387,* 253–260.

Costanza, R., Perez-Maqueo, O., Martinez, M. L., Sutton, P., Anderson, S. J., & Mulder, K. (2008). The value of coastal wetlands for hurricane protection. *AMBIO: A Journal of the Human Environment, 37,* 241–248.

de Groot, R., Brander, L., van der Ploeg, S., Costanza, R., Bernard, F., Braat, L., et al. (2012). Global estimates of the value of ecosystems and their services in monetary units. *Ecosystem Services, 1,* 50–61.

Engle, V. D. (2011). Estimating the provision of ecosystem services by Gulf of Mexico wetlands. *Wetlands, 31,* 179–193.

Ghermandi, A., van den Bergh, J. C. J. M., Brander, L. M., de Groot, H. L. F., & Nunes, P. A. L. D. (2008). The economic value of wetland conservation and creation: a meta-analysis. Fondazione Eni Enrico Mattei, Sustainability Indicators and Environmental Valuation. Retrieved July 1, 2016, from http://www.feem.it/Feem/Pub/Publications/WPapers/default.htm

Grabowski, J. H., Brumbaugh, R. D., Conrad, R. F., Keeler, A. G., Opaluch, J. J., Peterson, C. H., et al. (2012). Economic valuation of ecosystem services provided by oyster reefs. *BioScience, 62,* 900–909.

Hanemann, M., Pendleton, L., & Mohn, C. (2005). *Welfare estimates for five scenarios of water quality change in Southern California: A report from the Southern California Beach Valuation Project*. Washington, DC: National Oceanic and Atmospheric Administration.

Hoehn, J. P. (2006). Methods to address selection effects in the meta regression and transfer of ecosystem values. *Ecological Economics, 60,* 389–398.

Jin, D., Hoagland, P., & Kite-Powell, H. L. (2003). A Model of the optimal scale of open-ocean aquaculture. In C. Bridger & B. Costa-Pierce (Eds.), *Open-ocean aquaculture: From research to reality*. Charleston, SC: World Aquaculture Society.

Johnston, R. J., & Russell, M. (2011). An operational structure for clarity in ecosystem service values. *Ecological Economics, 70,* 2243–2249.

Johnston, R. J., Sanchirico, J., & Holland, D. S. (2014). Measuring social value and human well-being. In R. E. Bowen, M. H. Depledge, C. P. Carlane, & L. E. Fleming (Eds.), *Oceans and human health: Implications for society and well-being* (pp. 113–137). New York: Wiley.

Keeler, B. L., Polasky, S., Brauman, K. A., Johnson, K. A., Finlay, J. C., O'Neill, A., et al. (2012). Linking water quality and well-being for improved assessment and valuation of ecosystem services. *PNAS, 109,* 18619–18624.

Kite-Powell, H. L., Hoagland, P., Jin, D., & Murray, K. (2003a). Economics of open ocean growout of shellfish in New England: sea scallops and blue mussels. In C. Bridger & B. Costa-Pierce (Eds.), *Open-ocean aquaculture: From research to reality* (pp. 293–306). Charleston, SC: World Aquaculture Society.

Kite-Powell, H. L., Hoagland, P., Jin, D., & Murray, K. (2003b). Open ocean growout of finfish in New England: a bioeconomic model. In C. Bridger & B. Costa-Pierce (Eds.), *Open-ocean aquaculture: From research to reality* (pp. 319–324). Charleston, SC: World Aquaculture Society.

Kite-Powell, H. L., Hoagland, P., & Jin, D. (2014). *Marine resource values for Saudi Aramco project planning*. Report to the Environmental Protection Department, Saudi Aramco. Dhahran, Saudi Arabia.

Kite-Powell, H. L., Colgan, C., Hoagland, P., Jin, D., Valentine, V., & Wikgren, B. (2016). Northeast ocean planning baseline assessment: marine resources, infrastructure, and economics. Northeast Regional Planning Body. Accessed November 21, 2015, at http://neoceanplanning.org/projects/baseline-assessment/

Lagerveld, S., Röckmann, C., & Scholl, M. (Eds.). (2014). *Combining offshore wind energy and large-scale mussel farming: Background & technical, ecological and economic considerations*. Ijmuiden, NL: Imares Wageningen UR.

Lindhjem, H., & Navrud, S. (2008). How reliable are meta-analyses for international benefit transfers? *Ecological Economics, 66*(2–3), 425–435.

Millennium Ecosystem Assessment Board (MEAB). (2003). *Ecosystems and human well-being: A framework for assessment*. Washington DC: Island Press.

Northeast Regional Planning Body. (2016). Draft Northeast Ocean Plan. Retrieved August 4, 2016, from http://neoceanplanning.org/plan/

Pendleton, L. H. (2008). The economic and market value of coasts and estuaries: What's at stake? Arlington, VA: Restore America's Estuaries (RAE).

Plantier-Santos, C., Carollo, C., & Yoskowitz, D. W. (2012). Gulf of Mexico Ecosystem Service Valuation Database (GecoServ): Gathering ecosystem services valuation studies to promote their inclusion in the decision-making process. *Marine Policy, 36,* 214–217.

Plummer, M. L. (2009). Assessing benefit transfer for the valuation of ecosystem services. *Frontiers in Ecology and the Environment, 7,* 38–45.

Rao, H., Lin, C., Kong, H., Jin, D., & Peng, B. (2014). Ecological damage compensation for coastal sea area uses. *Ecological Indicators, 38,* 149–158.

Rönnbäck, P. (1999). The ecological basis for economic value of seafood production supported by mangrove ecosystems. *Ecological Economics, 29,* 235–252.

Rosenberger, R. S., & Loomis, J. B. (2001). Benefit transfer of outdoor recreation use values: A technical document supporting the Forest Service Strategic Plan (2000 revision). Gen. Tech. Rep. RMRS-GTR-72. Fort Collins, CO: U.S. Department of Agriculture, Forest Service,

Rocky Mountain Research Station. Retrieved on August 4, 2016, from http://www.fs.fed.us/rm/pubs/rmrs_gtr072.pdf

Texas A&M University. (2016). Ecosystem services valuation database for the Gulf of Mexico. Retrieved on August 4, 2016, from http://www.gecoserv.org/

US Environmental Protection Agency (EPA). (2016). The social cost of carbon. Retrieved on August 4, 2016, from https://www3.epa.gov/climatechange/EPAactivities/economics/scc.html

Woodward, R. T., & Wui, Y.-S. (2001). The economic value of wetland services: A meta-analysis. *Ecological Economics, 37,* 257–270.

Part IV
Case Studies

Chapter 11
The German Case Study: Pioneer Projects of Aquaculture-Wind Farm Multi-Uses

Bela H. Buck, Gesche Krause, Bernadette Pogoda, Britta Grote, Lara Wever, Nils Goseberg, Maximilian F. Schupp, Arkadiusz Mochtak and Detlef Czybulka

Abstract Most studies on multi-use concepts of aquaculture and wind farms explored cultivation feasibility of extractive species, such as seaweed or bivalves. However, recent studies also included the cultivation of crustaceans or fish culture in the vicinity of wind turbines. Consequently, new approaches combine fed and extractive species in integrated multi-trophic aquaculture (IMTA) concepts for offshore multi-use to reduce nutrient output and the overall environmental impact of aquaculture operations. In this chapter the findings of a series of mussel and oyster cultivation experiments over several seasons are presented, which were conducted at different offshore test sites in the German Bight. Sites were selected within future offshore wind farm areas for an explicit multi-use perspective. Results have demonstrated successful growth and fitness parameters of these candidates and

B.H. Buck (✉) · B. Pogoda · B. Grote · M.F. Schupp
Alfred Wegener Institute Helmholtz Centre for Polar and Marine Research (AWI),
Marine Aquaculture, Maritime Technologies and ICZM, ZMFE Zentrum Für Maritime
Forschung und Entwicklung, Bussestrasse 27, 27570 Bremerhaven, Germany
e-mail: Bela.H.Buck@awi.de

B.H. Buck · B. Pogoda
Faculty of Applied Marine Biology, University of Applied Sciences Bremerhaven,
An der Karlstadt 8, 27568 Bremerhaven, Germany

G. Krause
Alfred Wegener Institute Helmholtz Centre for Polar and Marine Research (AWI),
Earth System Knowledge Platform (ESKP), Bussestrasse 24, 27570 Bremerhaven, Germany

G. Krause
SeaKult—Sustainable Futures in the Marine Realm, Sandfahrel 12, 27572 Bremerhaven,
Germany

L. Wever
Forschungszentrum Jülich GmbH, 52425 Jülich, Germany

N. Goseberg
Ludwig-Franzius-Institute for Hydraulic, Estuarine and Coastal Engineering,
Leibniz University Hannover, Nienburger Str. 3, 30167 Hannover, Germany

A. Mochtak · D. Czybulka
University of Rostock, 18051 Rostock, Germany

© The Author(s) 2017   253
B.H. Buck and R. Langan (eds.), *Aquaculture Perspective of Multi-Use Sites
in the Open Ocean*, DOI 10.1007/978-3-319-51159-7_11

therefore definitely proved the suitability of these bivalve extractive species for open ocean aquaculture. Another approach for multi-use in offshore wind farms is its use as marine protected area or even for reinforcement or restoration of endangered species, which need the absence of any fisheries activity for recovery. Current projects are testing this perspective for the native European oyster *Ostrea edulis* and the European lobster *Homarus gammarus*. From the technological point of view there are many options on how to connect aquaculture devices, such as longline and ring structures as well as different cage types, to the foundations as well as to install it in the centre of the free area between wind turbines. Next to the system design also experiments on drag forces originating from the aquaculture structure on the foundation and vice versa were investigated. Complementary to the biological, environmental end technical aspects, a number of studies were specifically targeted to address and include stakeholders, their attitudes, their interests and concerns over time. By this approach, the inclusion of stakeholders into the research process from its very beginning until today, co-production of knowledge could be fostered. Next to joint identification of the major impediments and concerns of offshore aquaculture under multi-use conditions, new issues and research questions were identified. Primary focus on the economic potential of aquaculture in offshore wind farms was shown for consumption mussels. The production of mussels using longline technology is sufficiently profitable even under the assumption of substantial cost increases. This is especially true, if existing capacities could be used. Last but not least, the EEZ is a special area—it is not a state territory even if a coastal state has its sovereign rights and jurisdiction. It is an area where three legal systems come together: international law, law of the European Union and national law. There are no mariculture projects in the German EEZ and no approval procedure has been completed so far. Some sites are not suitable for mariculture, especially because of nature conservation and shipping.

11.1   Introduction

The development of "offshore aquaculture" or "open ocean aquaculture" has often been described as the new challenging frontier of the "Blue Revolution", which puts current aquaculture efforts on the same level as the advances made in agriculture during the so-called "Green Revolution" in the onset of the 1950s (Krause et al. 2015). The motivation to foster this type of marine development roots against the backdrop of the rising concerns that the over-exploitation and economic inefficiencies in the capture-fisheries sector are leading to widespread severe decline of global food security. However, to date, the search for resilient solutions in the aquaculture sector to meet production, income, community development and food supply and security needs remains critical.

In contrast to nearshore aquaculture development, offshore aquaculture faces several additional challenges, one of which is that in Germany it is located in the EEZ, the Exclusive Economic Zone. The latter is defined by United Nations

Convention on the Law of the Sea (UNCLOS) as a 200 nautical mile zone off the coast of a State, over which it has special rights regarding the exploration and utilization of marine resources. Thus, the EEZ is a special area—it is not a state territory even if a coastal state has there somewhat sovereign rights and jurisdiction, which refers to the coastal state's rights below the surface of the sea. As a case in point, in Germany, the EEZ is an area where three legal systems come together: international law, law of the European Union and national German law.

Indeed, the ocean space of the EEZ has been regulated or allocated in a number of different ways, but most importantly, this has been done predominantly within individual economic sectors. Obvious examples of "sectoral zoning" include waterways for shipping, disposal areas, military security zones, concession zones for mineral extraction, aquaculture sites, and most recently marine protected areas. Although the context and outcomes are different because of the dynamic and three-dimensional nature of marine environments, the land use planning concepts and techniques can be translated to the marine environment, thus leading to a contemporary rise in the development of Marine Spatial Planning (MSP) efforts. At present, there are few frameworks that facilitate integrated strategic and comprehensive planning in relation to all activities taking place in marine areas. This current stage of development of spatial planning of marine waters can be viewed somewhat in parallel to the past land use planning efforts. The latter arose in response to specific social and economic problems and later environmental problems that were triggered by the industrial revolution at the end of the 19th century.

Next to the problematic nature of marine spatial planning, the offshore realm is a place far off the coast, which bears challenges to the technological development of aquaculture systems that are able to withstand the harsh offshore conditions, vis á vis supporting feasible operation and maintenance demands. Candidate species must be tolerable to these exposed conditions whilst holding sufficient economic potential. Last, but not least, stakeholders and their interests, concerns, rights and duties must be taken into consideration.

In the following we illustrate these different aspects and challenges of offshore aquaculture in multi-use settings employing Germany's North Sea EEZ region as a case study. We draw here on research and experience from various research efforts in this areas since early 2000 until today (see Buck et al. 2008a). Figure 11.1 shows a summary of the projects conducted in the German Bight all led by the *Alfred Wegener Institute Helmholtz Centre for Polar and Marine Research* (AWI), which were related to the four main topics: (A) biology, (B) technology, (C) social science as well as (D) economics.

All projects are displayed in a chronological order by taking the scientific coherence into account. Project No. 1, the feasibility study, gave the basis for the following projects and the entire suite of aquaculture-wind farm multi-use research. The *Coastal Futures Project* (No. 4) acted as key node project, in which the other projects either have contributed to or have been stimulated by its transdisciplinary approach. Due to their problem-focused approaches and early and continuous integration of multiple stakeholder groups, these projects initiated and accompanied several activities and outcomes outside academia. For instance, it called for wind

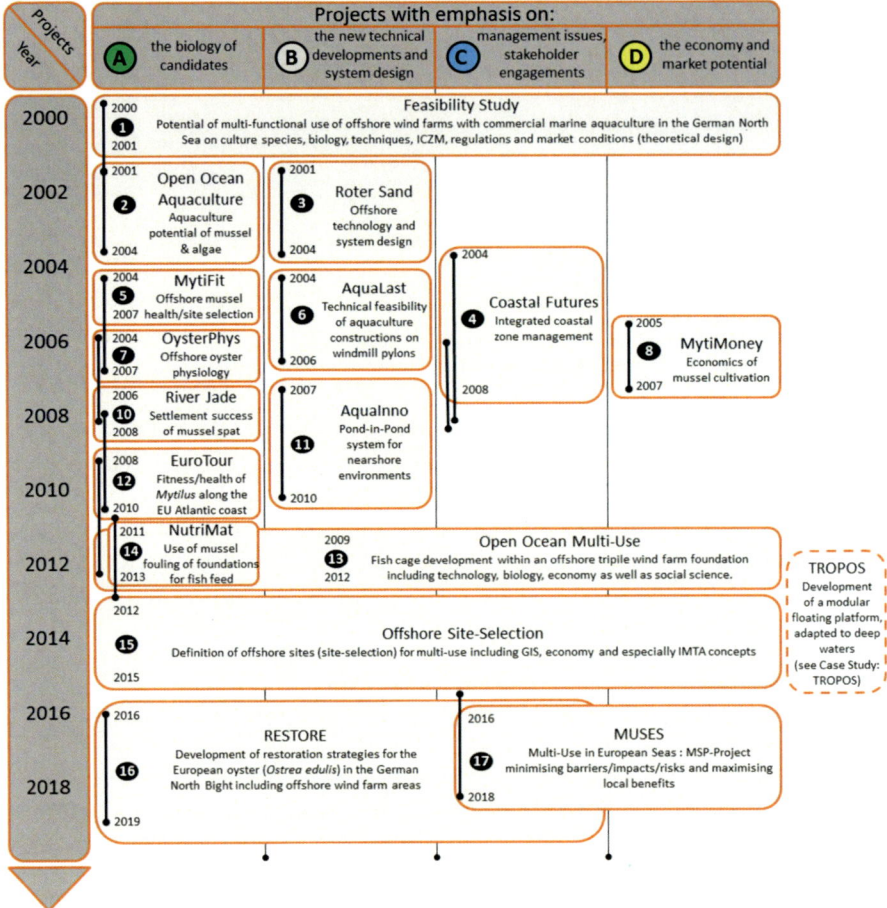

Fig. 11.1 Offshore aquaculture multi-use projects conducted in the German Bight under the guidance or participation of AWI from early 2000 ongoing. Next to the chronological order, all projects are configured in a scientific coherence manner. The project "TROPOS" is marked in a *dotted line* as it focused only partly on the German Bight. Additionally, there is a separate chapter in this book where this project is described in more detail. Project No. 11 "AquaInno" has only slight overlaps with multi-use aspects and therefore is not mentioned in detail. Project No. 12 "EuroTour" deals with site-selection and health status of offshore environments along the European Atlantic Coast. Project No. 16 focuses on restoration for nature conservation purposes. The Project No. 17 "MUSES" just started after submission of this book. Modified and updated after Buck et al. 2008a

farm planners attention to offshore aquaculture, included authorities and fishermen into the planning process for site-selection-criteria of appropriate aquaculture sites as well as stimulated the involvement of offshore engineers and wind farm operators into the technical part of an offshore aquaculture enterprise. Furthermore, it introduced fishermen to the co-management idea and appraised the potential economics of mussel cultivation, supplied authorities with maps and tools to limit regional stakeholder conflicts, and established an inshore reference station to support the data assessed offshore. All research undertakings over the entire period were conducted as follow-up projects resulting from addressing the issues of social acceptance, the demand for practical technical solutions and to fill some of the knowledge gaps in biology, social science and other related topics.

In the following sub-chapters, the central findings of these projects are discussed in more detail and the development of this contextual knowledge generation over time is synthesized. This will be done by starting with the potential species and their respective biological characteristics, which determine their feasibility for offshore aquaculture in the first place. In the successive sub-chapters, the prospective technologies to harness their potential as aquaculture products are discussed, as well as their likely economic potential for this contextual setting in the German Bight. The chapter closes with an overview over the existing stakeholder characteristics and their respective attitudes, perceptions and concerns and the respective regulatory framework.

11.2 The Beginning

The feasibility study (Project No. 1) (Buck 2000, 2001, 2002) aimed to ascertain the economic feasibility of an offshore marine aquaculture structure for breeding of marine organisms in the North Sea, predominantly in combination with an offshore wind farm. In the year 2000 several thousand wind-generators were planned for installation in the North Sea before the end of the decade. However, at that time, after completion of this feasibility study, no offshore wind farm was in use or in preparation. The major goal of this combination was to use environmentally-friendly wind-driven power generation, or so-called renewable energy production, with the environmental enhancement that aquaculture offers as a very important opportunity. This feasibility study concentrated on Open Ocean Aquaculture (OOA) and its multi-use potential with offshore wind farms.

Most countries with access to the sea are engaged in aquaculture. Only rare cases, for example in developing nations, fundamentals of e.g. technical knowledge, resisting equipment and system design are lacking to successfully research and implement aquaculture. Although Germany has the above-mentioned factors in relative abundance, it can be counted with the few developing nations that have not adopted any form of aquaculture. The reasons for this stagnation are mainly as follows: Conflicts exist between interested parties on coastal land management issues. Such parties include commercial and pleasure boat traffic, gravel mining, marine and local fishing coops, and protected areas, such as national parks. Problems with regulation and assignment of areas in the North Sea and its near-shore waters arose due to these conflicts. In addition, complex local hydrodynamic conditions such as large wave heights and strong water currents have hindered the aquaculture development in Germany.

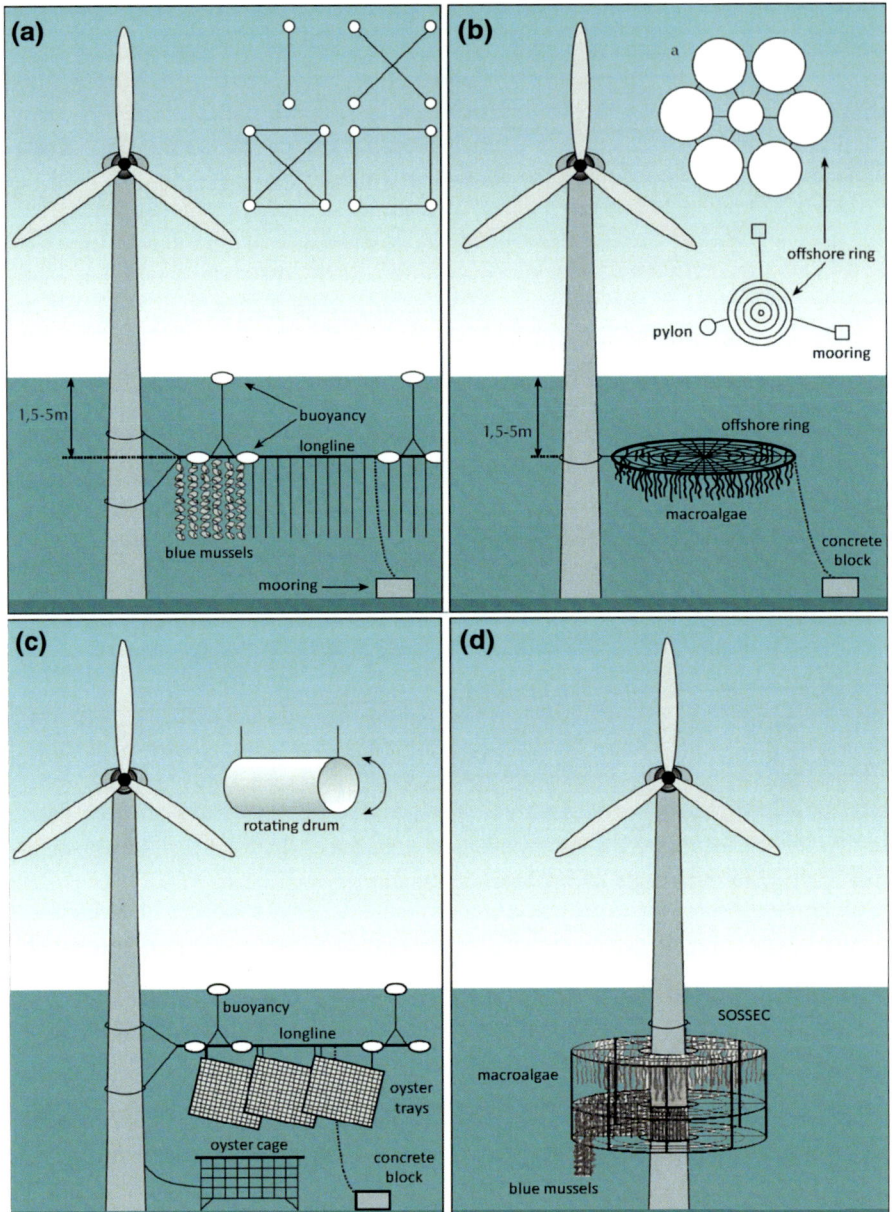

◄**Fig. 11.2 a–d** First drawings to show the multi-use concept combining aquaculture with wind farms. **a** *Top* a birds-eye view of longline constructions. Special pylon anchorages ensure access and maintenance of the wind energy plants by ship; *down* Side view of a submerged mussel and seaweed longline culture. The wind energy plant pylon is used as the anchorage for one end of the longline. The line can reach a length of 100–300 m. Longer lines may be mounted to the next pylon; **b** *Top* A bird's-eye view of a ring construction. The combined rings are fixed around the pylon as well as a pylon with a single offshore ring construction using two anchorages; *down* Side view of a pylon with offshore ring construction used for cultivation of algae and mussels. An anchor stone is placed at the outermost point of the ring construction; **c** Side view of a pylon combined with an 'on bottom' cage for oyster culture and oyster trays fixed to a longline as well as a rotating oyster cage (drum); **d** SOSSEC design (Submersible Offshore Shellfish and Seaweed Cage), which is during culture in a submerged mode and can be lifted to the surface for harvest. Modified after Buck (2000, 2001); *Images* AWI/Prof. Dr. Bela H. Buck

"*Offshore, Open Ocean, Far Out*"—currently unknown or seldom used terms among the German public, have become since the mid-1990s common "catch-phrases" in Asia and the USA. In an effort to cover the immense demand for seafood products, new regions beneath the coastal waters were sought. However, at that time only a few world-wide Open Ocean Aquaculture businesses are in operation and are still a new method of aquaculture internationally. Main research efforts in an international perspective were placed in the provision of sturdy off-shore technology, which resist the extreme environmental conditions and warrants safe utilisation of the installation. However, so far, little operational implementation of the results of extensive research has been undertaken.

Several types of cultivation in combination with the foundations of offshore wind turbines were suggested: (1) Predominately longline constructions for the cultivation of seaweed and mussels, by which submersible ropes provide the habitat on which the candidates can settle (Fig. 11.2a). (2) Other techniques, such as ring and cage constructions (Fig. 11.2b, c), which can be placed on the water surface and under water, as well as the construction SOSSEC (Submersible Offshore Shellfish and Seaweed Cage) (Fig. 11.2d) can be operated, too.

The advantage of submersible culture constructions is the avoidance of the impact of harsh weather conditions and strong wave mechanics. Thus, a combi-nation of these techniques with the main pillar of the wind turbine installation dispenses the need for a sophisticated mooring of the aquaculture system, which would impact strongly on the benthic ecosystem. Due to the solid construction of the wind turbine, such systems would be very safe.

The feasibility study suggested a variety of candidates, which can be cultivated under such hydrodynamic and environmental conditions within the North Sea, and hold large potentials on the German market. Additionally, the study pointed to the technical challenges and the importance of the socio-economic drivers. All users have to be addressed at the beginning of a new project in order to find an over-arching consensus of all parties thereto.

As predominately, already existing data sets were used in the feasibility study of offshore aquaculture potentials within the areas of wind farms in the German North Sea, this study was theoretical in nature. In order to determine the feasibility of aquaculture within the German North Sea, further experiments on a practical basis were conducted and led to the project development displayed in Fig. 11.1.

At the time when the study was published, we expected that the further development and modification of aquaculture systems would be directly dependent on the development of wind farms. Therefore, it was necessary that the development of future wind turbine techniques as well as aquaculture systems should not move ahead separately.

In summary, there was an ample need for practical research pertaining aquaculture development in the North Sea to overcome the current lack of knowledge in Germany in this sector.

11.3 Potential Species for Offshore Aquaculture

Suitable species for offshore aquaculture production in the German North Sea were identified during several scientific research projects. The selection was (1) based on literature research, (2) expert knowledge on candidate biology, (3) former projects as well as (4) confirmed by current practical research results. Accordingly, 21 aquaculture candidates of different trophic levels (extractive and fed species: seaweed, molluscs, crustacean and fish) are recommended for offshore cultivation in the German Bight (North Sea). In the following we present a collection of species tested starting with extractive species, followed by crustaceans and fish.

11.3.1 Seaweed Species

Seaweed aquaculture research in offshore sites focusses mainly on growth performance and culturing techniques. Various systems were installed in harsh offshore conditions, both at the wind farm foundation and at its vicinity (see subchapter on techniques and system design below), respectively. Following biological and ecological requirements, but also a regulatory and economic framework, five species of macroalgae were identified as potentially successful candidates for offshore cultivation in the German North Sea: Oarweed (*Laminaria digitata*), Sugar kelp (*Saccharina latissima*), Cuvie (*Laminaria hyperborea*), Dulse (*Palmaria palmata*), and Sea beech (*Delesseria sanguinea*).

11.3.1.1 Candidate: Laminarian Species

The genus *Laminaria* is one of the most important macroalgal genera in temperate coastal ecosystems, especially in the northern hemisphere. This is amongst other things reflected by its high species numbers, its considerable overall biomass, its dominance, and its economic significance (Bartsch et al. 2008). The utilisation of algal products plays an essential role in many fields of modern everyday life. *Laminariales* offer a variety of goods used in human and animal consumption, in industrial products or for bioremediation (Bartsch et al. 2008).

In the German North Sea, vital abundance of *L. hyperborea*, *L. digitata* and *S. latissima* is reduced to the rocky shore of the Island of Helgoland as the remaining coastline is characterized by soft-bottom substrate, not suitable for the fixation of *Laminaria* rhizoid holdfasts. For aquaculture production, *Laminaria* plants are "seeded" on ropes, which are subsequently fixed to various suspended or floating culture devices. Detaching *L. digitata* and *S. latissima* blade portions from the meristem induces them to become sporogenous far ahead of their natural reproductive season, making mature sporophytes available all year round (Buchholz and Lüning 1999; Lüning et al. 2000; Pang and Lüning 2004) (Fig. 11.3a–e).

In general, *Laminaria* in the German North Sea show a period of rapid growth from January to June and one of slow growth from July to September (Kain 1979).

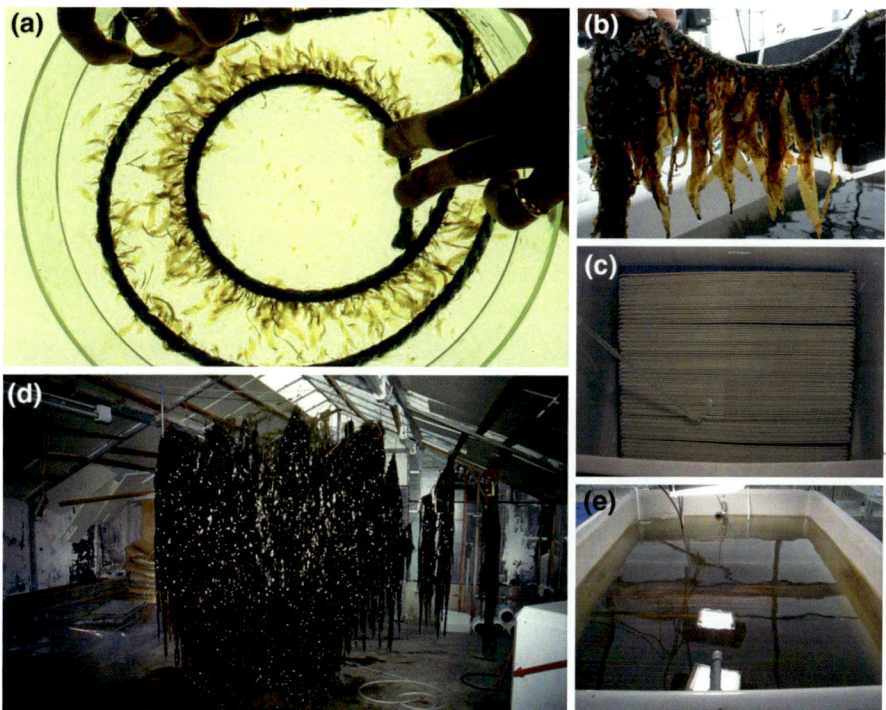

Fig. 11.3 a Rope with young *Saccharina latissima* 6 weeks after seeding; **b** Laminarian plants after 10 weeks cultivation in land based facilities; **c** Seeding rope used for offshore cultivation arranged in a "curtain"-mode before seeding; **d** Large *S. latissima* harvested from an offshore site after 5 months cultivation ready for the drying procedure; **e** Tank for seeding procedure and day-length adaptation. Photos **a** and **d** AWI/Dr. Cornelia Buchholz; **b–c** and **e** AWI/Prof. Dr. Bela H. Buck

Helgolandic *S. latissima* and *L. hyperborea* predominantly grow in winter and early spring similarly as in Norway (Bartsch et al. 2008). While growth of *L. hyperborea* totally stops in July, growth of *S. latissima* decreases substantially but does not cease. In contrast, the growth period of *L. digitata* from Helgoland extends from spring to summer and in September, it exhibits a growth rate of still 50% of the optimum (Lüning 1979). The seasonal variations in abiotic conditions affect growth performance, particularly in algae from high latitudes with a more pronounced seasonality of temperature, irradiance and photoperiod (Bartsch et al. 2008).

Exploring the feasibility of offshore kelp cultures, potential forces experienced by the attached algae were studied extensively. Sporophytes were seeded on ropes, which were designed for the use in offshore environments (Fig. 11.3a, c). Land-based cultivation took place in tank devices, where the day length could be adapted according to the development of sorus (Fig. 11.3b, e). When Laminarian thalli reached market size (1.5–2.0 m in length) plants were harvested and transferred to the lab to test its dry weight ratio (Fig. 11.3d).

The degree of exposure influenced morphology and shape of the algae and their resistance to environmental forcing considerably. *Laminaria* originating from sheltered conditions had wider blades with thick and undulate margins, while offshore sporophytes grown at exposed sites were thin and streamlined (Bartsch et al. 2008; Buck and Buchholz 2005). *S. latissima* sporophytes pre-cultivated onshore but transferred to the sea at very early stages developed a streamlined blade and resisted current velocities up to 2.5 m s^{-1}. If grown singly in currents of >1 mm s^{-1}, *S. latissima* can withstand the high energy environment experienced in offshore cultivation (Buck and Buchholz 2005). These experiments on *Laminaria* species show that adapted to strong currents as young individuals, they will grow well and produce large amounts of biomass at exposed sites of the German Bight (Buck and Buchholz 2005; Fig. 11.4a, b).

When combining *Laminaria* aquaculture with offshore wind farms, the foundations would provide a stable fixing structure for the seaweed cultivation systems (e.g. Buck 2002; Krause et al. 2003; Buck et al. 2004). These ideas led to several multi-use projects in the German Bight, such as No. 1–4, 9, 13, 15 (see Fig. 11.1) (Buck 2002; Michler-Cieluch et al. 2009a, b, Buck and Buchholz 2004a, 2005; Buck et al. 2012), where for the first time *Laminaria* species were tested in an IMTA approach with partners from the offshore wind industry.

Mass culture of *Laminaria sp.* and *Saccharina latissima* under the high energy offshore environment in the North Sea requires a rigid cultivation system for withstanding rough conditions, which can be handled while retaining the macroalgae. Buck and Buchholz (2004a) tested various carrier constructions and different mooring systems and their results led to a new patented ring carrier for macroalgae offshore cultivation (see subchapter on techniques and system design below) (Figs. 11.32 and 11.33).

For IMTA approaches in the German Bight, the three native *Laminaria* species are ideal extractive candidates: *Laminaria* cultivation brings clear advantages for the environment. Apart from their CO_2-consumption, kelps are also able to absorb large amounts of nitrogen and phosphate, thus helping to abate coastal

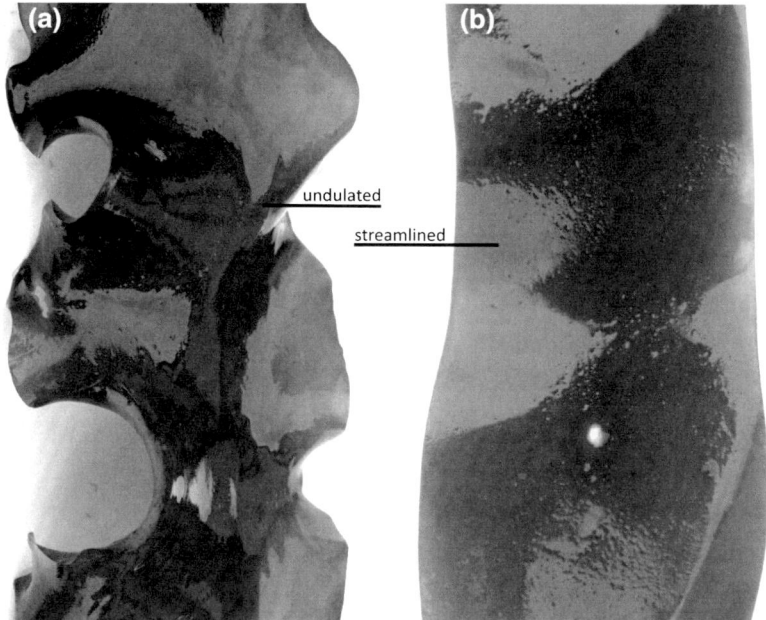

Fig. 11.4 a *Saccharina latissima* originating from a sheltered site with a wider blade and ruffled margins and **b** cultivated *S. latissima* from the offshore site with a streamlined shape (modified after Buck and Buchholz 2005). *Photos* AWI/Prof. Dr. Bela H. Buck

eutrophication (McHugh 2003; Lüning and Pang 2003; Fei 2004) (see Chapter Offshore Aquaculture with Extractive Species: Seaweed and Bivalves). This feature of *Laminaria* and other seaweeds qualifies them for use in sustainable maricultures. Modern integrated aquaculture systems have been developed and are constantly improved. The combination of seaweed culture with land-based fish culture or open marine cage culture has found great acceptance (Subandar et al. 1993; Petrell and Alie 1996; Ahn et al. 1998; Troell et al. 1999; Chopin et al. 2001; Neori et al. 2004). Macroalgae, including *Laminaria,* may act as biofilters for finfish aquaculture at offshore sites in the German Bight, removing dissolved excretions and surplus nutrients and providing extra oxygen and biomass.

11.3.1.2   Candidate: *Palmaria palmata*

Dulse is an edible intertidal or shallow subtidal red alga (Rhodophyta) of the Atlantic and North Pacific and its culture methods are well known (Browne 2001; Le Gall et al. 2004; Pang and Lüning 2004, 2006) (Fig. 11.5a).

   P. palmata has been proven to be a good candidate for integrated multi-trophic aquaculture (IMTA) in the projects *SEAPURA* (Lüning 2001) and "Offshore Site Selection" (Project No. 15; Matos et al. 2006; Grote 2016; Grote and Buck 2017)

Fig. 11.5 **a–c** Dulse (*Palmaria palmata*) grown in a land-based IMTA-system in co-culture with turbot for nutrient budget calculation for the multi-use of offshore wind farms. **a** *Palmaria palmata* (*Photo* AWI/Sina Löschke); **b** *Palmaria palmata* from IMTA tank culture (*Photo* AWI/Dr. Britta Grote); **c** IMTA tank system for cultivation of *Palmaria palmata* in project No. 15 "Offshore Site Selection". The *red* macroalgae were grown in tumble culture (*Photo* AWI/Dr. Britta Grote)

and is thought to be a suitable species for cultivation in multi-use approaches in offshore wind farms (Grote and Buck 2017) (Fig. 11.5b, c). The red alga can reach specific growth rates (SGR) between 1.1% d^{-1} at 14 °C, 13–14 h daylight and varying radiation in co-culture with halibut (Corey et al. 2014) and 6.18% d^{-1} at 14 °C, 16 h daylight and constant radiation in co-culture with turbot (Grote and Buck 2017). Latter results are similar to the maximum SGR of *P. palmata* reaching 7% d^{-1} at 6–14 °C and 16 h daylight with nutrients added (Morgan and Simpson 1981). In an IMTA-system with turbot, the excess N of 1 kg fish supported the growth of more than 6.5 kg of dulse, which removed a maximum of 0.76 mg N mg DW^{-1} d^{-1} at 14 °C (Grote and Buck 2017). Conservative estimates of yields of *P. palmata* at about 300 t $year^{-1}$ are expected to remove up to 30% of N excreted by 500 tonnes of salmon within two years (Sanderson et al. 2012).

Since *P. palmata* is one of the few seaweed in Europe used for food, its chemical composition has been investigated (Morgan et al. 1980a). In recent years, interest in health foods, food supplements, new protein sources and novel bioactive compounds led to further research of the chemicals found in seaweeds (Løvstad-Holdt and Kraan 2011; Indergaard and Minsass 1991). The red alga has also a relatively high protein content, which can be increased by IMTA cultivation, as the ammonium leads to increased N storage in the algal tissue (Grote 2016).

P. palmata requires a site with high water current for nutrient and CO_2 exchange across the surface of the fronds as well as for prevention of fouling (Werner and Dring 2011), thus making it a good candidate for offshore cultivation. Currents with flow rates of minimum 5–10 cm s^{-1} are needed for *Palmaria* cultivation. However, the exposure of the site to wave action should be moderate (Werner and Dring 2011). Therefore, for offshore cultivation of dulse wave exposure and current velocities need to be considered to reach optimal growth conditions and to reduce loss of biomass.

Another limiting factor for growth of *P. palmata* in the wind farms of the German Bight is the relatively high water temperatures in summer. Grote and Buck (2017) found reduced SGRs of 2.2% d^{-1} at 16.5 °C. Minimum winter temperatures are not considered to be damaging unless there is the unlikely threat of ice-formation which could cause abrasion of the cultures. Growth of *P. palmata* is optimal at temperatures between 6° and 12 °C, but the red alga will grow well to temperatures up to 15 °C (Morgan et al. 1980b; Morgan and Simpson 1981) or 17 °C (Grote and Buck 2017), with these differences probably resulting from different temperature ranges for different populations.

The size at which *P. palmata* should be harvested is a frond length of 30–40 cm (Werner and Dring 2011). It is crucial to monitor the algae during spring and especially during the summer months to ensure that the dulse is harvested before the fronds are overgrown by fouling organisms (Werner and Dring 2011). This is very important when *P. palmata* is grown for human consumption, which requires high quality harvests. As growth is expected to be reduced during the summer months due to higher temperatures in the German Bight, the optimal time for harvest would be in June (Grote and Buck 2017). As the growth of *P. palmata* was enhanced in the vicinity of fish farms at sea (Sanderson et al. 2012) and as it is an robust alga thought to withstand strong forces, *P. palmata* is thought to be an ideal extractive candidate for offshore IMTA (Grote and Buck 2017). However, the potential forces experienced by the attached algae offshore need to be studied in more detail.

11.3.1.3   Candidate: *Delesseria sanguinea*

Sea beech is an European endemic, sublittoral red alga with a distribution range from northern Spain and Portugal to northern Norway and Iceland (Lüning 1990). It is thought to be a candidate for offshore aquaculture systems due to its biological tolerance to environmental conditions (Lüning 1990); however, biological studies on multi-use and offshore aquaculture success of this species are still missing. The reproductive season of *D. sanguinea* lasts from October to February/April at Helgoland (Molenaar and Breeman 1997) and the red alga can tolerate temperatures between 13 and 23 °C, but temperatures for optimal growth lie between 10 and 15 °C (Lüning 1990). *D. sanguinea* is used in the cosmetics industry for its anti-coagulant properties and vitamin K content; the active principle being termed delesserine (Guiry and Blunden 1991).

11.3.2 Bivalve Species

Following mainly biological and technical requirements, but also a regulatory and economic framework, three filter-feeding bivalves were identified as potentially successful candidates for offshore cultivation in the German Bight: *Mytilus edulis,* *Crassostrea gigas* and *Ostrea edulis* (Buck et al. 2006a).

M. edulis and *O. edulis* are native species while *C. gigas* was introduced to German waters in the 1960s due to aquaculture activities in the Netherlands (Diederich et al. 2004; Smaal et al. 2009). By now *C. gigas* is distributed all over the German North Sea coastline and is commercially cultured since the 1980s. All three species are classified as extractive species in aquaculture and are therefore ideal candidates for offshore cultivation with IMTA concepts. They reduce nutrient emissions of higher trophic species by filter-feeding and require low maintenance as they do not require additionally feed (Pogoda et al. 2011).

During several studies, these candidates have been cultivated at various offshore sites and approved their suitability for offshore aquaculture operations in the future. Results of these studies clearly elucidated that all three bivalve species grow successfully under exposed conditions in offshore environments (e.g. Buck 2007; Brenner et al. 2012; Pogoda et al. 2011, 2013). As the "multi-use" of offshore wind farms for aquaculture installations will facilitate the further expansion of environmentally friendly and sustainable aquaculture, most test sites of the described cultivation experiments were conducted in North Sea areas of two offshore wind farms (OWF): "*Butendiek*" (in operation since 2015) and "*Nordergründe*"[1] (currently under construction in late 2016, Fig. 11.6).

11.3.2.1 Candidate: *Mytilus edulis*

Biological conditions: Blue mussels cultivated in offshore areas of the German North Sea, for the most part, show high growth rates compared to those grown in nearshore sites (e.g. Buck 2004, 2007). This is due to the fact that water quality (e.g. less urban sewage) and oxygen concentrations are more suitable and the infestation of parasites is low or non-existent (Buck et al. 2005; Pogoda 2012). However, in areas under estuarine influence and/or exposure to fluvial transport point to a comparable probability for high contamination loads similar to nearshore areas, thus potentially reducing fitness (Brenner et al. 2012). In Project No. 2 "Open Ocean Aquaculture" mooring devices were deployed to test the feasibility of mussel and seaweed aquaculture at offshore sites within areas of planned wind farms (Fig. 11.7a). Larval abundance of mussels tended to decrease with increasing distance from shore (Walter et al. 2002; Buck 2017). However, several offshore sites were identified, where larval concentration was sufficient enough to facilitate adequate natural seeding (Buck 2007) (Fig. 11.7b). Alternatively, limited spat

---

[1]*Nordergründe* is an area in the Weser estuary off the German Coast in the German Bight (Fig. 8).

Fig. 11.6 Transfer of offshore wind farm turbines to the OWF *Nordergründe* by tugboats. Nordergründe area, as part of the multi-use concept, was used for extractive species test sites, especially for bivalves and *Laminaria sp.* (*Photo* AWI/Dr. Bernadette Pogoda)

availability may be viewed as an advantage when moving offshore. The benefit for a low settlement can lead to a one-step cultivation technique (no thinning procedure) if collecting and grow-out sites are within the same region in the vicinity of offshore structures. The lower settlement success on one hand results in a limited commercial potential, but on the other hand eases handling and maintenance (Fig. 11.7c). During Project No. 3 "Roter Sand"[2] different types of submerged longline systems were developed and tested: a conventional polypropylene longline as well as a steel hawser longline, both with different versions of buoyancy modes. Spat collectors and grow-out ropes were suspended perpendicular from the horizontal longline for several months. These multi-use experiments have shown a massive settlement at offshore cultivation lines, making thinning essential (Buck 2007) (Fig. 11.8).

Resistance of mussels to severe conditions: The resistance of mussels to strong currents as well as high waves and swell depends on the degree and duration of these forces (Project No. 5 "MytiFit" and Project No. 10 "River Jade"). Information on hydrodynamic conditions is important due to the fact that e.g. mussels and oysters do adapt to harsh conditions but do not automatically grow fast. Even when flow rates are increasing and consequently deliver more food, which stimulates mussels and oysters to feed intensively, at a certain current velocity threshold growth is reduced due to a pressure differential between inhalant and exhalent siphons (Wildish and Kristmanson 1984, 1985; Rosenberg and Loo 1983). Further, exposure to high waves/swell also reduces production rates due to the loss of mussels through detachment (Scarratt 1993). Even if mussels cultivated in high energy environments are able to adapt to the permanent physical stress by increasing the strength and number of byssal-thread attachments, system design adapted from sheltered environments, such as collector devices, have to be modified to prevent detachment (Brenner and Buck 2010).

---

[2]*Roter Sand* is an offshore lighthouse located in an area called *Nordergründe* in the Weser estuary, German Bight/North Sea 17 nautical miles off the coast of the city of Bremerhaven (Fig. 8).

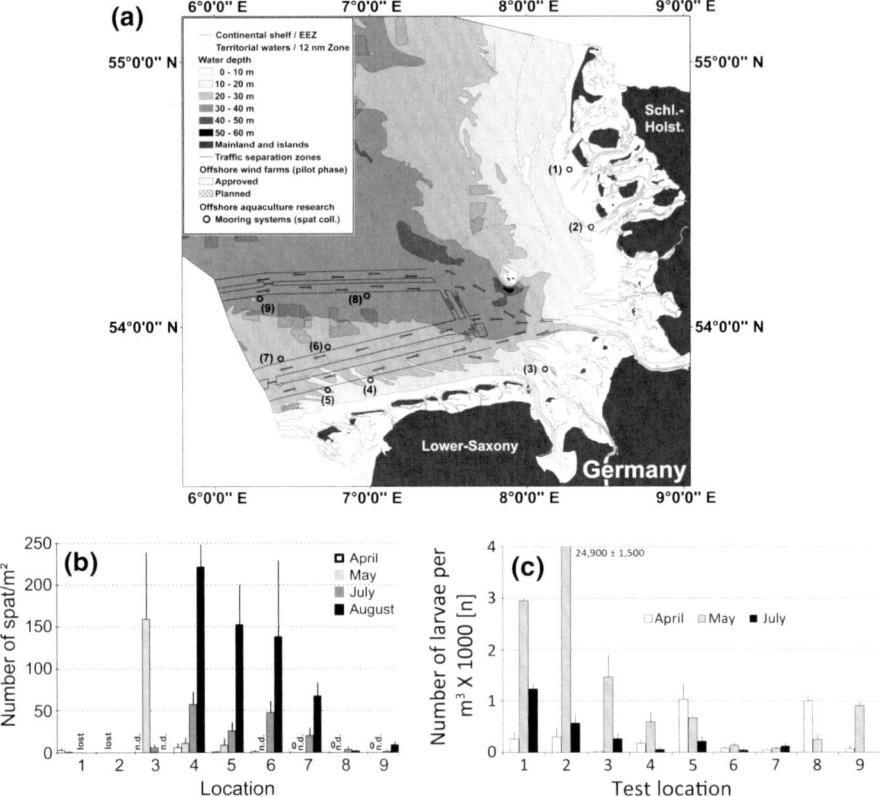

Fig. 11.7 **a** Shows the map indicating the stations where offshore test installations were moored; **b** displays the concentration of mussel larvae in the water column at the respective site and **c** represents the number of mussels settling on ropes (Buck 2017)

To follow the idea of multi-use of sites and support the co-existence of mussel cultivation off the coast in severe conditions where wind farms occur the attachment and detachment of mussels and their byssus threats from their holding devices were tested (Figs. 11.9a–c, 11.10 and 11.11a–m). Collector types for larval settlement attraction, good foothold and interweaving abilities, or collectors combining these properties should be developed and applied depending on the use for seed mussel collection or grow-out as well as for a one-step or a two-step (including thinning) cultivation method. Due to the harsh conditions in the open ocean collector surfaces should be precisely tailored to avoid loss of mussels. It may happen that not only one collector type for all purposes is the solution. Changing collector surfaces according to the prevailing conditions (current velocity, waves, size of the mussel, etc.) may be most beneficial.

Use of fouling organisms originating from foundations: In Project No. 14 "NutriMat" the use of harvested *Mytilus edulis*, originally settling in vast amounts on the foundations of offshore wind farms, was investigated (Weiss et al. 2012). The removal of fouling organisms at regular intervals due to regulated

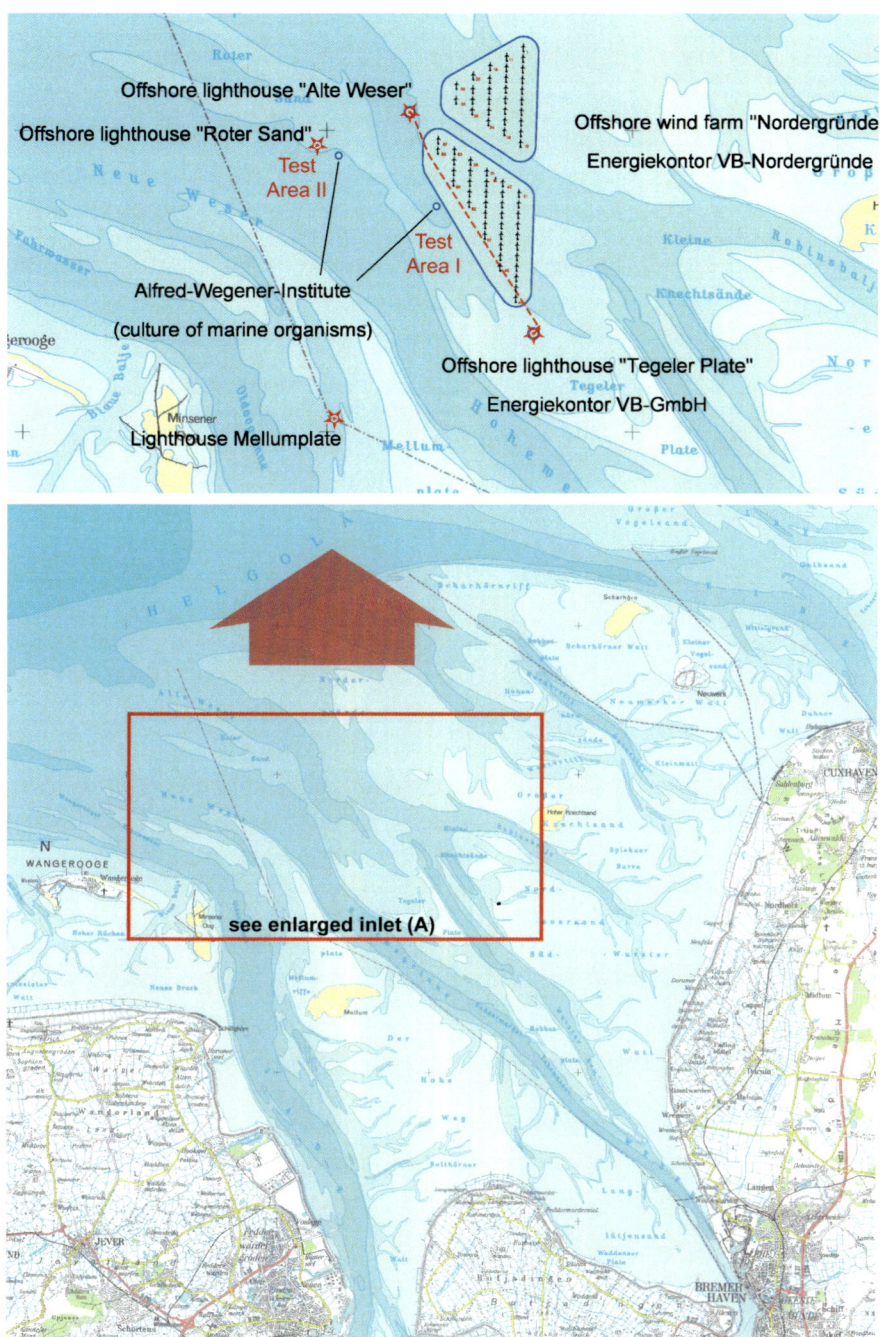

Fig. 11.8 Map of the southern German Bight. The enlarged *upper Inset* illustrates the test area No. I, *Nordergründe*, and the test area No. II at the offshore lighthouse *Roter Sand*. The wind turbines indicate the planned offshore wind farm "Energiekontor" (modified after Buck 2007)

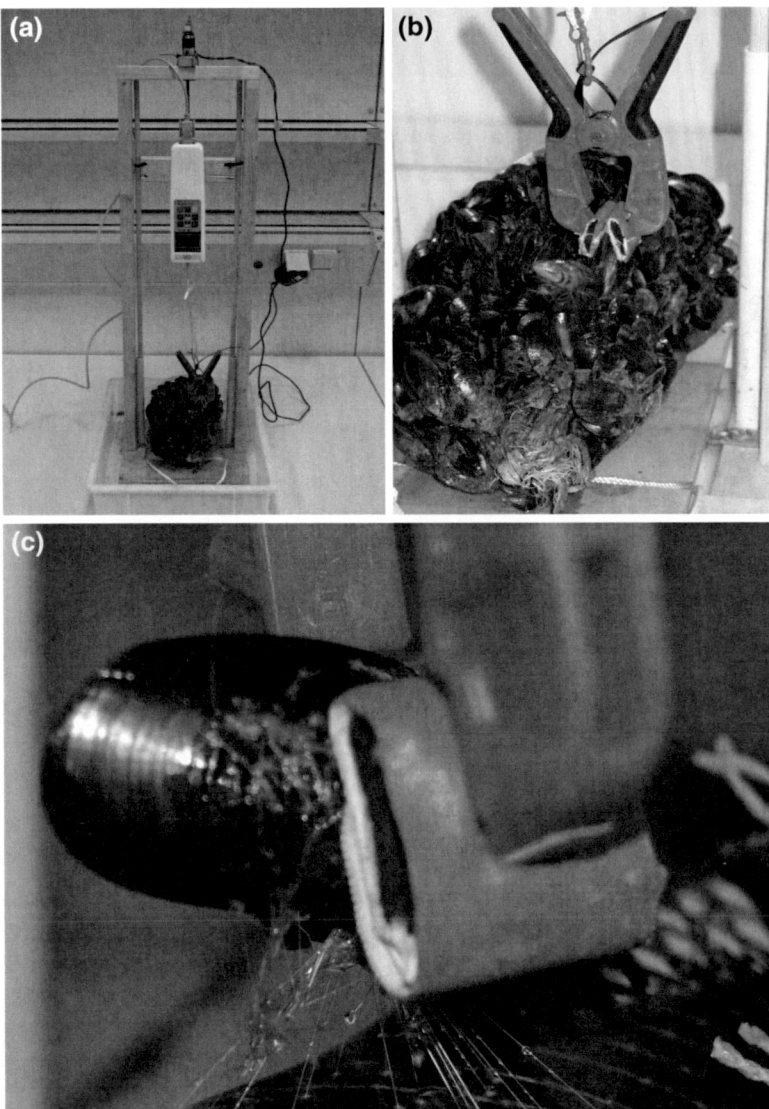

Fig. 11.9 **a–c** Investigation of the forces needed to detach mussels/byssus threads from substrates. **a** Dislodgement device with a force gauge; **b** sample and clamp to pick up a single mussel from a mussel conglomerate; **c** visible are various byssus threats when lifting up the clamp (modified after Brenner and Buck 2010)

construction surveys is mandatory to get an insight into corrosion, cuts or any other potential damage on the wind turbine foundation surface. However, the inspection and resulting removal opens an access to a high quality protein and lipid source. The meat of harvested Blue mussels as alternative protein source in aquaculture

Fig. 11.10 Collector types used for the detachment test. *ASW* Artificial Seaweed Collector (Size [*S*]: 10 mm; Material [*M*]: nylon rope as back bone with 10 cm long PP-leaves attached at both sites; Origin [*O*]: Japan); *LOC* Looped Christmas Tree (*M* Extruded polypropylene with a straight trim, strands of lead in the centre help sinking; *O* New Zealand); *LEC* Leaded Christmas Tree (*M* Extruded polypropylene with a looped trim, strands of lead in the centre help sinking; *O* New Zealand); *GAR* Galician Rope (*M* Rough surfaced nylon-PE ropes with strands; *O* Spain); *SSC* Self-Sinking Collector (*M* Polyester net formed as a tube, small stones help sinking; *O* Norway); *LAD* Ladder Collector (*S* 16 mm; *M* Parallel running PP-ropes connected every 35 cm by a plastic bar; *O* Norway); *AQU* Aquamats® (*M* Strands of PP-fleece material with ballast sleeve; *O* USA); *NFL* Naue® Fleece (*M* PP-fleece, cost-saving alternative to AQU; *O* Germany); *COC* Coconut Rope (*S* 24 mm; *M* rope of coconut fibres; *O* India); *REF* Reference Collector (*M* Bushy tufts of a unravelled 10 mm PP-rope; *O* Germany) (modified after Brenner and Buck 2010)

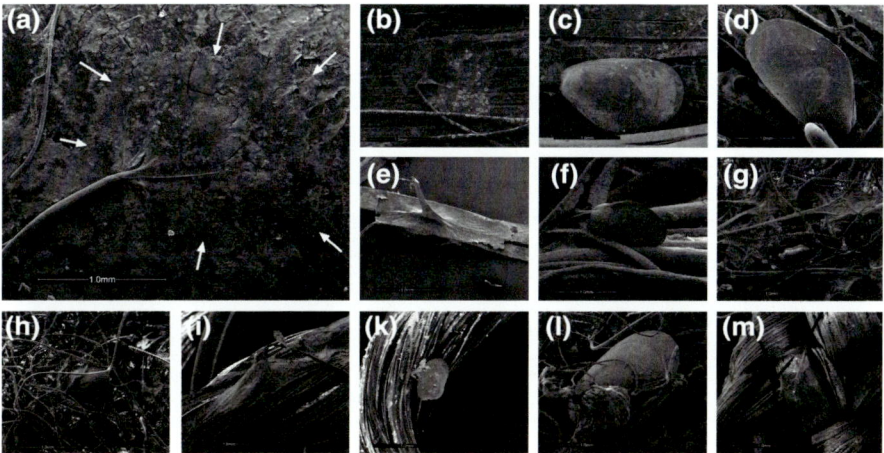

Fig. 11.11   a–m EM-photos (100-fold magnification) of mussel byssus threats/plaques attached to substrates. **a** Byssus plaque attached to a mussel shell. *White arrows* demonstrate the outstretched size of the plaque. **b** Byssus plaques on an ASW collector; **c** attached post-larvae on LEC collector and on **d** LAD collector; **e** byssus plaques on LEC collector; **f** attached post-larvae on COC collector; **g** byssus plaques and **h** attached post-larvae on NFL collector; **i** byssus plaques and **k** attached post-larvae on an SSC collector; **m** byssus plaques on an GAR collector (modified after Brenner and Buck 2010)

feeds was tested for the feeding of farmed turbot (*Scophthalmus maximus*) (Fig. 11.12a–e). For this purpose, mussels were scraped off the foundation and transferred to the land-based facilities to produce feed in various mixtures with

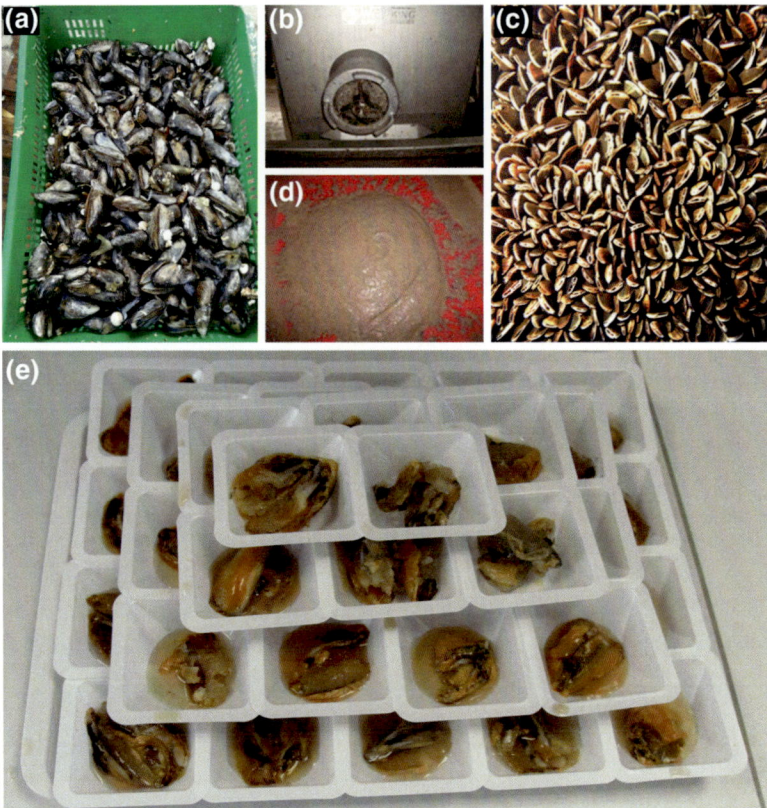

Fig. 11.12 a–e Mussels (*Mytilus edulis*) from offshore foundations to be used as fish feed. **a** Harvested mussels scrapped from the foundation; **b** grinder to chop the mussel meat; **c** Blue mussels attached to a foundation surface underwater; **d** mussel meat after grinding; **e** mussels to be tested for parasite infestation. Photos **a–b** and **d–e** AWI/Dr. Monika Weiß; **c** AWI/Prof. Dr. Bela H. Buck. Modified after Weiss et al. 2012

common fish feed and then fed to turbot, which was cultivated within a Recirculating Aquaculture System (RAS). These mixes of feed all containing different concentrations of mussel meal were tested and the resulting growth, acceptance, welfare and digestibility observed. The experiments revealed that a replacement of fish meal with mussel meal of 100% and 50% resulted in depressed growth; while a fish meal replacement with 10% and 25% of mussel meal did not reduce growth or negatively affect the health of the tested turbot. The outcome of this project was that mussel meal has a high potential to serve as supplement or fish meal replacement in feed for turbot raised in aquaculture systems and therefore reduces the impact of fisheries for fish meal production (Weiss and Buck 2017).

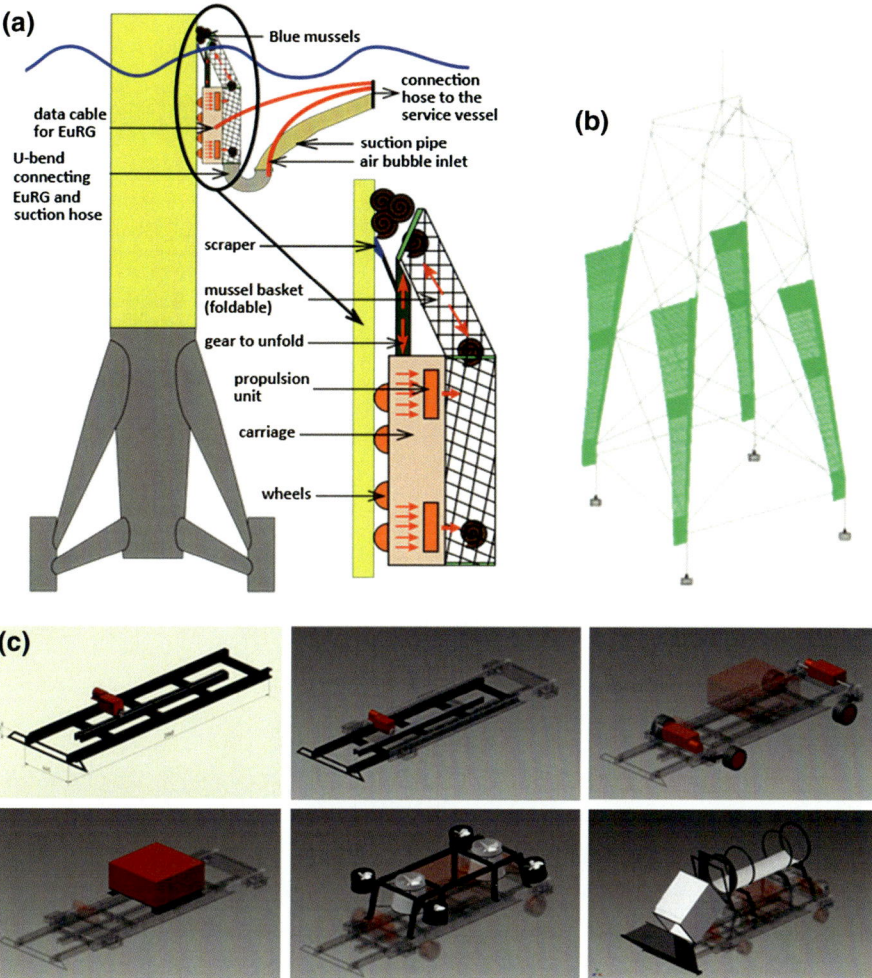

Fig. 11.13 a–c Harvest and cleaning device. **a** Side view of a foundation and an attached EuRG; the enlargement of the EuRG shows the structure in more detail; **b** is a computer model, which shows the load forces on the foundation and the carriage. **c** show some details of the development progress of the carriage and its superstructure. Modified after Weiss et al. 2012

Additionally, an automatic mussel harvesting technique by designing a remotely operating robot was developed (Fig. 11.13a–c), called EuRG (Ernte- und Reinigungsgerät, Harvest and cleaning device). It was used for harvesting mussels from the surface of the foundation. However, this vehicle was only designed on paper and never built.

11.3.2.2   Candidates: *Crassostrea gigas* **and** *Ostrea edulis*

<u>Health conditions:</u> In contrast to Blue mussel cultivation, which can rely on natural spatfall in offshore areas of the German Bight, offshore oyster cultivation depends on purchasing seed oysters and caging of animals. The biological performance of Pacific and European oysters was investigated at three different offshore sites in the German Bight during the first multi-use aquaculture project "Open Ocean Aquaculture" as well as in the project "Offshore Oyster Physiology" (OysterPhys) (Project No. 2 and No. 7, Fig. 11.1). Young oysters were cultivated in the open ocean 10–40 nautical miles off the coast in 2001–2003 and from April to October 2004 and 2007 in oyster lanterns (Fig. 11.14). Research focused on growth performance, condition and survival rates in these high-energy environments. For a higher resolution of overall condition, elemental and biochemical compositions as well as macroparasitic burden were analysed.

O. edulis and *C. gigas* obtained positive growth rates in terms of shell length (Fig. 11.15a) and dry mass outside their natural coastal habitat, which is normally

Fig. 11.14 Oyster lanterns filled with spat of *Ostrea edulis* and *Crassostrea gigas* for grow-out at offshore cultivation test sites (*Photo* AWI/Dr. Bernadette Pogoda)

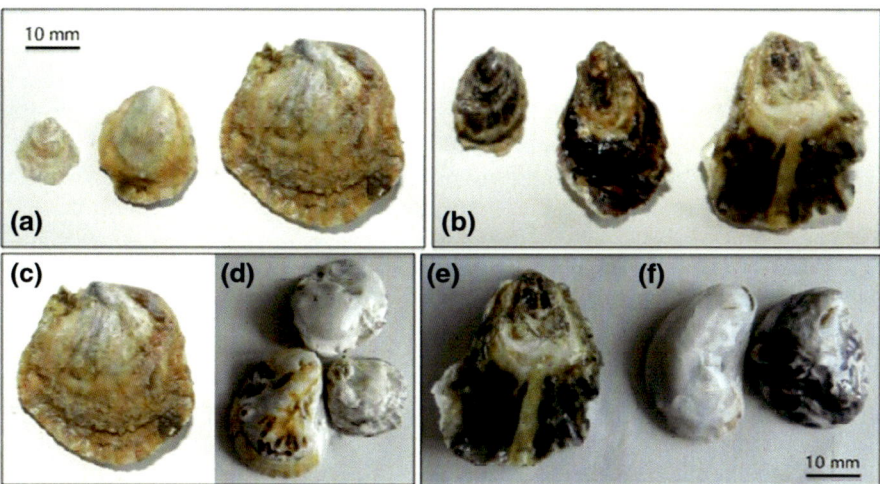

Fig. 11.15 a-f Shell growth of the European oyster *Ostrea edulis* **a** and the Pacific oyster *Crassostrea gigas* **b** during one offshore cultivation season (mid-April—mid-October); **c** *O. edulis* and **e** *C. gigas* showing normal shell growth from offshore sites in contrast to **d** and **f** with strong shell abrasion from an exposed cultivation site where tidal currents are high. Modified after Pogoda et al. 2012. *Photos* AWI/Dr. Bernadette Pogoda

located on the seabed of coastal seas. In general, growth rates were similar to those of oysters from nearshore coastal areas (Pogoda et al. 2011). This documents that offshore-cultivated oysters are able to achieve a "normal" or "natural" growth performance.

Furthermore, oysters developed their species-specific morphology and produced an "aesthetic" shell, which plays an important role for their market value (Matthiessen 2001) (Fig. 11.15a). Only oysters grown at sites where daily tidal currents and sediment loads are high developed a different shape (Fig. 11.15b). Thicker shells and a very compact appearance indicate that these animals strongly invested energy in shell growth to withstand the strong currents as well as to prevent shell abrasion (Newkirk et al. 1995; Pogoda et al. 2011). This emphasizes the importance of a detailed and thorough site selection prior to the start of offshore oyster cultivation or any aquaculture operation in general (see Pogoda et al. 2011).

According to typical size classes used in commercial aquaculture farms larger *O. edulis* spat from a Norwegian producer was used in the experiment of 2007 (Newkirk et al. 1995; Matthiessen 2001; Pogoda et al. 2011). These animals yielded significantly higher growth rates ($p < 0.0001$), even beyond those of Pacific oysters. The superior condition of these European oysters was reflected by an even higher condition index (Pogoda et al. 2011). Accordingly, oyster cultivation in harsh environments could benefit from using a bigger spat size class regarding physical constraints through high current velocities and wave action, as smaller animals, in a very sensitive growing phase, grow slower under severe conditions.

The European oyster showed constant increases in shell length and dry mass during the cultivation periods (Pogoda et al. 2011). This implies a high ability of dietary assimilation of the native European oyster, even when food availability is low in summer (Rick et al. 2006) and a good adaptation to this offshore environment (Newkirk et al. 1995; Matthiessen 2001; Laing et al. 2006). In contrast, seasonal variations with reduced growth rates in summer were observed for the Pacific oyster. Interestingly, this species-specific difference is also clearly reflected in parallel changes of the biochemical compositions.

The biochemical and elemental compositions were analysed to characterize the nutritional condition and for a better understanding of related energetic processes. Accumulation and depletion of metabolic energy reserves depend primarily on food quantity and quality, environmental effects on metabolic processes, and reproductive activities (Beninger and Lucas 1984; Whyte et al. 1990; Ruíz et al. 1992). Therefore, investigations focused on seasonal dynamics of the major energy storage products, namely carbohydrates, proteins and lipids, on the compositions of lipid classes and fatty acids as well as on carbon and nitrogen.

Both species utilized primarily glycogen as energy store during times of high food availability. After the phytoplankton spring bloom, when reduced growth rates were observed for *C. gigas*, glycogen contents were drastically depleted, whereas lipid contents increased. In contrast, *O. edulis* kept on growing and accumulating glycogen, while lipid contents remained relatively constant. These different strategies may be explained by the earlier maturity of Pacific oysters and the resulting conversion of carbohydrates to lipids to enhance the production of eggs, which are rich in lipid (Gallager and Mann 1986; Whyte et al. 1990; De la Parra et al. 2005).

In shellfish production, the condition index is commonly used to evaluate the effects of the surrounding environment on these organisms. It is an adequate parameter to describe the commercial quality, physiological state and health of bivalve molluscs. The most commonly applied condition index (CI) is the ratio of flesh mass to shell mass (e.g. Walne and Mann 1975; Davenport and Chen 1987). Condition indices for all three offshore-cultivated bivalve species support the positive results already observed for the growth performance at offshore sites. CI values for Blue mussels, European and Pacific oysters indicated excellent conditions (Pogoda et al. 2011).

Seasonal variations in lipid class compositions of offshore-cultivated oysters were essentially similar to those of nearshore-grown individuals (Pogoda et al. 2013). Triacylglycerols (TAG) are the main lipid stores in the investigated oysters and serve as short-term energy reserves. Together with glycogen, they accumulate during periods of high food availability and are depleted in periods of food paucity. Accordingly, the amount of TAGs is a sensitive indicator of the nutritional condition of an animal (Fraser et al. 1985). Expressed as the ratio of phospholipids to triacylglycerols (PL:TAG), values ≤ 1 indicate a good nutritional state (Abad et al. 1995; Caers et al. 2000) and European and Pacific oysters clearly improved condition during offshore-cultivation. PL and TAG levels reached equal proportions of

1 in summer, which indicate well-fed animals and excellent growing conditions at offshore cultivation sites in the German Bight (Pogoda et al. 2013).

Also the amount of essential fatty acids greatly affects growth and condition of oysters (Pazos et al. 1996). The fatty acid compositions of the European and the Pacific oysters were dominated by 16:0, 20:5(n − 3) and 22:6(n − 3), major components of phospholipids of typical marine organisms. Both species of offshore-cultivated oysters showed the accumulation of lipids as energy reserves during high food availability from spring to early summer. Diatom markers increased during spring and early summer in both oyster species and suggest a diet rich in diatoms (Pogoda et al. 2013). Increasingly high ratios of (n − 3)/(n − 6) during the cultivation experiment underline the excellent physiological condition of both offshore-cultivated oyster species (Pogoda et al. 2013; Pazos et al. 1996).

The combination of successful growth performance and obviously excellent overall condition of offshore-cultivated oysters resulted in insignificant mortalities. In contrast to commercial oyster production in nearshore environments, which often suffer from high mortalities, offshore survival rates for both oyster species were high (>96% in 2004, >99% in 2007) and encourage open ocean cultivation (Pogoda et al. 2011).

Parasite infestation: Studies on the macroparasite burden of offshore-cultivated European and Pacific oysters and Blue mussels reported a zero infestation at off-shore locations in the North Sea (Fig. 11.16a–c). In general, parasites can affect condition and health of host animals. Buck et al. (2005), Brenner (2009) and Pogoda et al. (2012) have shown that offshore grown mussels and oysters were free of macroparasites and that infestation rates increased with proximity of the sites to shore, respectively; intertidal mussels and oysters showed the highest numbers of parasites. The debate over the effects of parasites on the energy status and overall health of the host is still open as robust data to elucidate these issues is still lacking. Three major groups of macroparasites are known to infest North Sea mussels and oysters: shell-boring polychaetes, trematodes and mytilicolid copepods. Absence of trematodes at offshore locations can be explained by their complex life cycle: they often infest intertidal gastropods as first intermediate hosts (e.g. *Littorina littorea* and *Hydrobia ulva*). These are typical macroparasites of inshore bivalves, which, however, are completely absent in offshore cultivated oysters and mussels.

Due to the absence of these exclusively coastal organisms the parasite's life cycle cannot be completed in offshore regions (Buck et al. 2005). Mytilicolid copepods and shell-boring polychaetes (e.g. *Polydora ciliata*) are abundant in inshore waters (Thieltges et al. 2006). However, their short planktonic larval phase restricts successful dispersion to coastal waters. Larvae drifting away from the coast are bound to die due to predation or starvation in the absence of hosts, which are only available at very few selected offshore culture locations (Buck et al. 2005). These results present a commercial advantage of such offshore shellfish cultures.

All known micro- and macroparasites found in European coastal waters are harmless to consumers, but may have negative condition effects (macroparasites) and cause higher mortalities (microparasites) in infested hosts (Brenner et al. 2012).

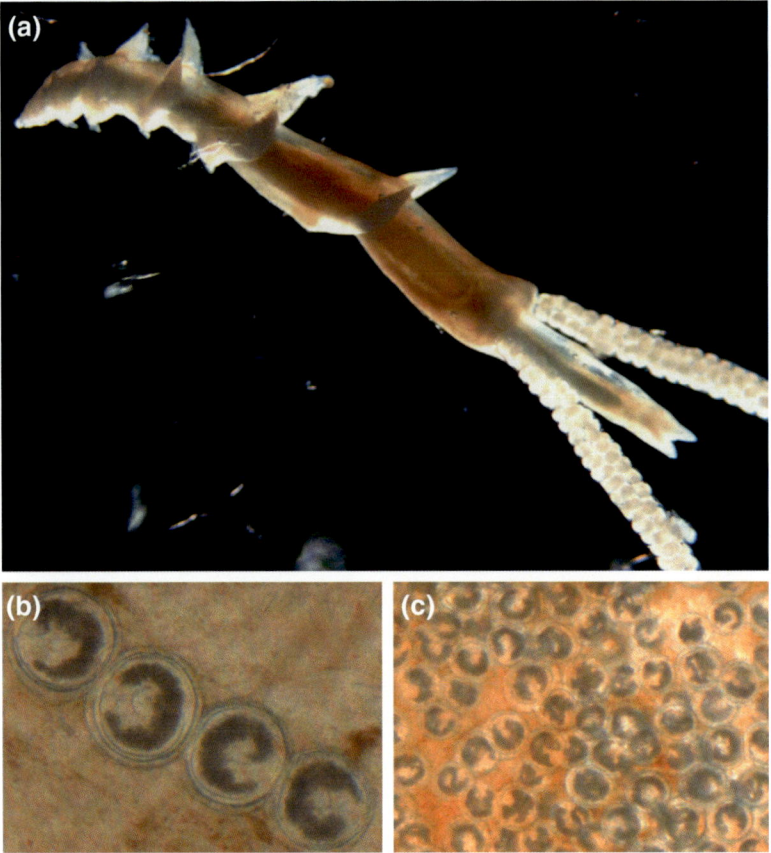

Fig. 11.16 **a** Mytilicolid copepod parasite *Mytilicola orientalis*; **b** trematode parasite *Renicola roscovita* located in the palps and **c** within the digestive gland (Photo **a**: AWI/Dr. Bernadette Pogoda, Photo **b–c**: AWI/Prof. Dr. Bela H. Buck)

Beside the potential harmful effect, some macroparasites cause aesthetic risks, since they are visible due to their colour (*Mytilicola intestinalis*) (Fig. 11.16a) or size (*Pinnotheres pisum*). Oysters are commonly eaten raw and consumers would not accept the appearance of e.g. parasitic copepods, as they are easy to recognize due to their bright red colour and size (up to 25 mm). This is an issue, as it would result in a serious commercial decrease of the shellfish value. Some macroparasites could also evoke a deteriorated morphological appearance, e.g. shell-boring polychaetes. As oysters and mussels represent high-value seafood products, an aesthetic appearance of the shell—especially on the oyster half-shell market—and meat is rather important. From an economic point of view the absence of macroparasites in shellfish products is certainly favourable. Furthermore, parasite infestations could reduce harvests and severely deplete local populations. Understanding the development of infestation patterns is therefore crucial for the successful site-selection in shellfish cultivation.

Restauration and ecosystem services: Because of their high ecological value oyster stocks are now in the focus of European conservation efforts. In the context of cooperation within the Oslo-Paris Commission (OSPAR), the native oyster was identified as a severely endangered habitat creating species and its protection in its area of distribution was concluded. According to the *Habitats Directive* for the protected habitat type "reef", a favourable conservation status has to be preserved or restored. In view of the ecosystem-related benefits of oyster reefs, especially the high biodiversity of species found in reefs, the Federal Agency for Nature Conservation (BfN) is engaged with the possibilities of a potential reintroduction of native European oysters in the North Sea. The recently launched Project No. 16 "Restore" aims at the development of strategies for a sustainable restoration of *Ostrea edulis* in the German Bight. Methods and procedures will be tested at experimental scales at different locations in the field. As fishery is completely prohibited within offshore wind farm areas in German waters, these are in the focus for first restoration sites and resemble yet another type of multi-use. Results of this investigation will support the future development and implementation of a German native oyster restoration program to re-establish a healthy population of this highly endangered oyster species in the German North Sea.

11.3.3   Crustacean Species

11.3.3.1   Candidate: *Homarus gammarus*

Krone and Schröder (2011) and Krone et al. (2013) investigated various artificial reefs (e.g. wind farms, wrecks) to proof if these new hard substrate structures, which are placed on the sand and silt dominated seabed of the German Bight, would provide a habitat for a hard bottom dweller like the European lobster. From these insights Krone (2012a) developed various habitat designs connected to foundations of offshore wind foundations, which allow mobile reef fauna, such as lobsters, to live in or hide. These new designs could be beneficial to reef fauna stocks.

The population of the European lobster *H. gammarus* in the German Bight is currently limited to the rocky area around the island of Helgoland. Despite management and reinforcement actions, the lobster population has still not recovered from a severe collapse in the 1950s and 1960s (Schmalenbach et al. 2011). A successful settlement of the animals in the stone fields that surround the individual wind turbines, could possibly contribute to the long-term stabilization of the population. Researchers are now exposing lobsters in this emerging habitat (Schmalenbach and Krone 2011). The project will investigate the feasibility and environmental consequences of such a lobster transaction. 2,400 animals were raised in a rearing facility on Helgoland in 2013/2014. Young lobsters, about five cm size, were then relocated by divers to the hard substrate surrounding wind

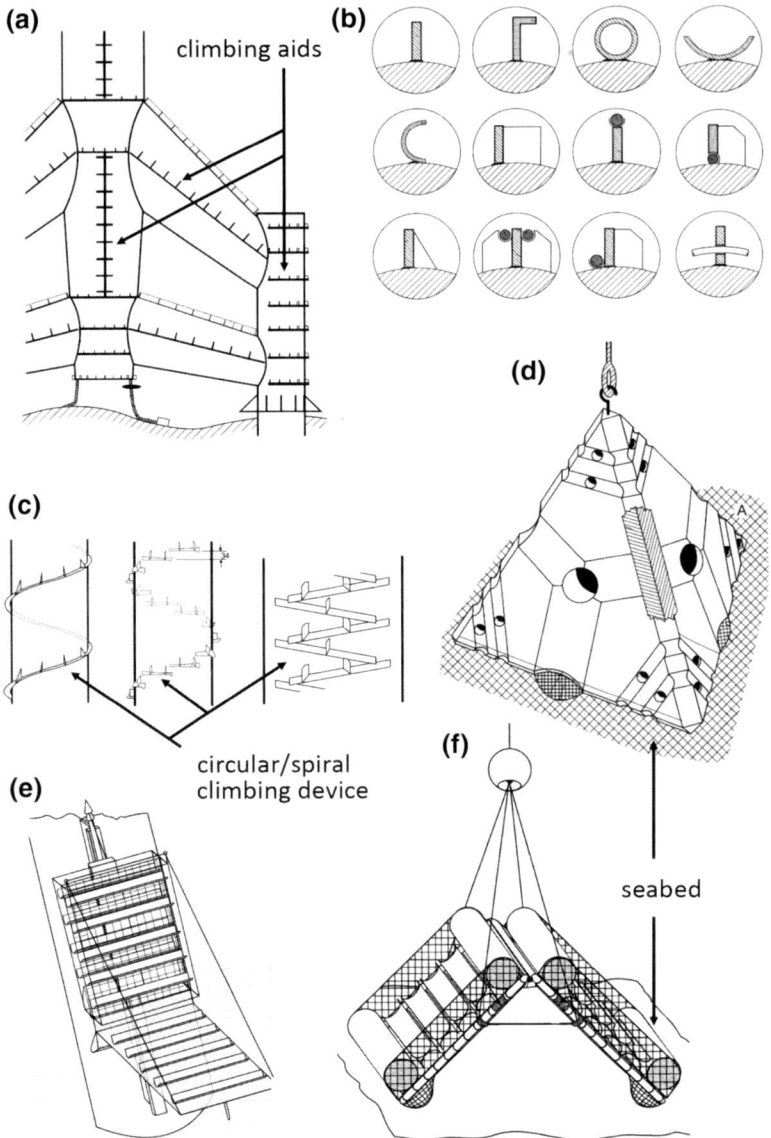

Fig. 11.17 **a–d** Wind farm foundations as artificial reefs. **a** shows a tripod foundation with climbing aids; **b** are potential guide profiles for the target species in a flat, round or even structured mode; **c** profiles attached to a foundation piece in a circular and spiral manner; **d** artificial habitat in polyhedron form for crustaceans; **e** scavenging carriage to harvest target species; **f** device for colonizing and harvesting marine hardground animals (all images modified after Krone (2012a); Krone et al. 2012; Krone and Krämer (2012) and Krone and Krämer (2011)

turbines in the Borkum Riffgat offshore wind farm 70 km off the German-Dutch coast. Researchers will survey the ecological habitat and how many of the juveniles settle successfully in the stone fields (Krone et al. 2013).

The shown foundation has climbing aids in the form of guide profiles, which are fastened in some sections to the offshore construction in a vertical, horizontal, and/or diagonal direction (Fig. 11.17a–c). These profiles are designed as rib-shaped flat or round profiles as well as structured by flat transverse profiles, which are also used for making breeding places (Krone et al. 2012). The pyramid-like structure (polyhedral form) relates to an artificial habitat to crustaceans on soft bottoms (Fig. 11.17d; Krone 2012b). Lobsters and other crustaceans as basically hard substrate animals also settle on soft soils when suitable habitats are available. This device can be placed next to wind turbine foundations. Another device is a scavenging carriage including a basic unit and a running unit that is configured to detachably couple the scavenging carriage with a rail device to be movable. The scavenging carriage is configured to detachably couple with the habitat carriage and the lifting device (Fig. 11.17e; Krone and Krämer (2011). Finally, one more device was invented to be used to establish and harvest a colony of benthic animals, in particular crabs. The device is hoisted and lowered in the folded form, and is laid out in the folded or fully open form. Tubular components improve drift and are particularly attractive as a habitat for crabs. This device can be used as an artificial reef to establish a colony and as a trap for effectively harvesting crabs in the immediate surroundings of wind turbines in offshore areas (Fig. 11.17f; Krone and Krämer (2012).

11.3.3.2   Candidate: *Cancer pagurus*

The edible crab or brown crab is a potential candidate for multi-use by passive gear fishery within offshore wind farms (OWF). As this species covers fairly the same niche as the European lobster and is also a high value seafood product on the market, it is another promising candidate that should be examined. Passive fishery on brown crab is very common along Northern European coasts and also in several offshore areas of the German Bight and reduces the ecological pressure on the endangered lobster population. Clear positive effects on the potential of OWFs to function as refuge for fish and marine animals were observed for crustacean species such as brown crab (*Cancer pagurus*) due to increased opportunities for shelter and food availability (Stelzenmüller et al. 2016).

11.3.4   Fish Species

Fish as a candidate for offshore multi-use concepts is new. Most of the studies on aquaculture in offshore wind farms concentrated on invertebrates and algae. In recent years, interest in offshore fish cultivation is increasing (Buck and Krause

2012). By now, only few studies focused on culture techniques, system design as well as on the commercial potential and the management of fish cultivation in wind farms (Buck et al. 2012). Existing studies on the biology of offshore fish cultivation, however, did not directly deal with fish cultivation in co-use with wind farms. Hundt et al. (2011) and Buck et al. (2012) provided a list of potential candidates for the cultivation in wind farms in the EEZ of the German Bight, such as European sea bass (*Dicentrarchus labrax*), cod (*Gadus morhua*), Atlantic halibut (*Hippoglossus hippoglossus*), turbot (*Scophthalmus maximus*), haddock (*Melanogrammus aeglefinus*), and Atlantic salmon (*Salmo salar*).

This selection of candidate species for offshore aquaculture in the German Bight was based on mainly four criteria: (1) natural occurrence in the North Sea (absolutely no introduction of non-native species), (2) physical requirements of candidates match the conditions in the German Bight, (3) current status of farming knowledge and available techniques, and (4) its economic feasibility (Buck et al. 2012). Generally, the harsh offshore conditions of the North Sea, namely strong tidal currents and wave heights, and the wide temperature difference in the shallow marginal sea limit the candidate list to few fish species. For the German Bight, the strong temperature difference lies between <6 °C in winter and >21 °C in summer and is probably the most limiting factor for cultivation of many fish species (Buck et al. 2012). Furthermore, Buck et al. (2012) conducted studies on the welfare of fish within net pens in RAS conditions that were similar to exposed conditions offshore. These results demonstrated, that a clear understanding of the dependence of fish fitness on strong hydrodynamic conditions is paramount.

11.3.4.1　Candidate: *Dicentrarchus labrax*

The European sea bass is a well-established species in aquaculture and with a production of over 60,000 tons per year one of Europe's most cultured marine fish species (FAO 2005). The on-growing usually takes place in coastal surface cages (FAO 2005), but the successful rearing under more exposed conditions was proven in flexible surface cages in the Mediterranean (Sturrock et al. 2008), which is the main site of production of European sea bass. Since sea bass is a physoclist species, in which the swim bladder is closed and pressure compensation is only possible by slow gas exchange, the ascent and decent of the cage has to include sufficient breaks to prevent a barotrauma of the fish's swim bladder (Korsøen 2011). However, the eurythermal species is capable to tolerate temperatures below 5 °C (FAO 2005), which would enable on-growing in the offshore regions of the German North Sea. Nonetheless, the average temperature of 10 °C in the North Sea (Wiltshire and Manly 2004) is below its optimum temperature range of 22–28 °C for growth (Lanari et al. 2002), which will result in an extended on-growing phase. A rough approximation of the actual length at the end of the on-growing can be based on the growth models by Lanari et al. (2002) and the annual temperature of the water in the German Bight (data BSH-German station bay, 10 m depth, daily mean values of 2009, www.bsh.de). The model suggests that an approximately 3 g fingerling

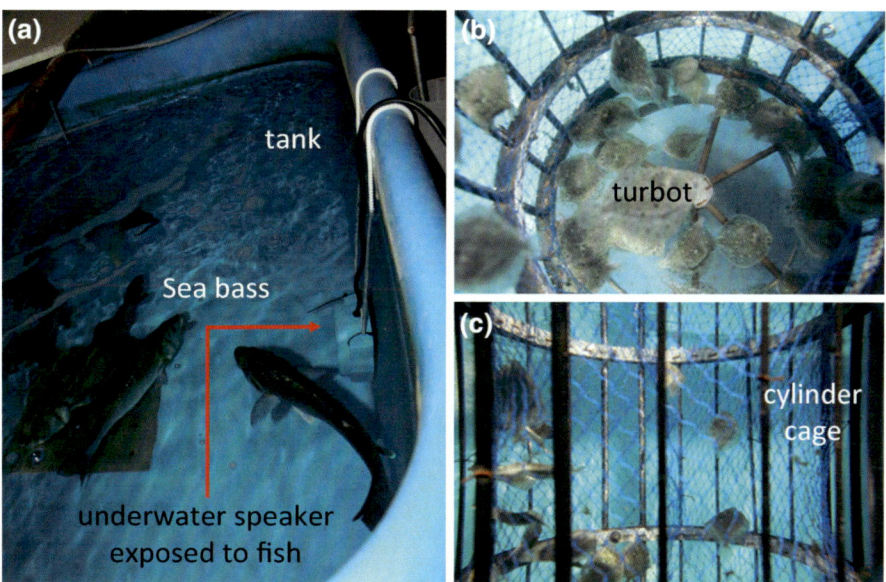

Fig. 11.18 a–c European seabass (*Dicentrarchus labrax*) and turbot (*Scophthalmus maximus*) for welfare experiments. **a** Sea bass in a tank of a recirculating system with speakers at the side wall of the tank to induce noises imitated of offshore wind turbines; **b** Juvenile turbots swimming in a cylinder-like cage model with a current velocity of 0.5 m/s to test welfare and behaviour and **c** in a current velocity of 0.9 m/s. *Photos* **a** University of Applied Sciences Bremerhaven/Gerrit Fiedler modified after Fiedler et al. 2011, **b–c** AWI/Jan Schmidt and Tim Heusinger modified after Buck et al. 2012

would need 31 months to reach the 300 g harvest weight (Buck et al. 2012). This exceeds the production of *D. labrax* in cages in the Mediterranean by 4–8 months (FAO 2005) making an offshore cultivation in the North Sea economically not feasible.

During a subproject of Project No. 13 "Open Ocean Multi-Use", the welfare of European seabass was tested by exposing fish in a tank device to loud noise using underwater speakers (Fig. 11.18a) (Fiedler et al. 2011). The noise was equivalent to the sound, which originated from offshore wind turbines in 2 m depth of the sur-rounding water column. Next to welfare observation and the fish behaviour regarding acceptance of feed, excitability, escape, aggression among fish of the stock as well as taking cholesterol samples to measure stress were conducted. Interestingly, fish did accept the loud noise in short time of 1–2 days. After adaption of the noise the fish showed no abnormal behaviour as well as cholesterol concentrations was within a threshold of fish not exposed to the noise.

However, at this stage it is important to understand that this test was just a preliminary work, which has to be conducted again in full scale, which means: more replicates, better infrastructure and experienced personal. The presented work was conducted during a student's project with limited access to specific infrastructure and financial support. Nevertheless, even if not planned under best conditions, the work shows a trend, which requires more research on this topic.

11.3.4.2 Candidate: *Gadus morhua*

Declining catches of Atlantic cod lead to a growing interest in aquaculture of this species. For more than thirty years, the intensive production of cod has been established (Moksness et al. 2004). The rearing of *G. morhua* requires the production of live food of different sizes and feeding protocols need to be followed strictly (Moksness et al. 2004). Of all the gadoid species, cod is thought to have the greatest potential for future development for aquaculture and some scientists believe that within the next 20 years similarly high production levels can be achieved as for the Atlantic salmon (Rosenlund and Skretting 2006). The suitability of *G. morhua* for offshore aquaculture has been successfully demonstrated in 3000 m^3 submersible cages including automatic feeding buoys by the company Sea Station™ off the coast of New Hampshire (Chambers and Howell 2006). As *G. morhua* is a physoclist species, necessary pauses need to be included during vertical movements of the cage (Korsøen 2011). One major problem for cod aquaculture is their ability to bite through the mesh or other materials, which makes them the fish species with the highest escapee rate (Jensen et al. 2010), which could have a negative impact on wild population by cross-breeding (Davies et al. 2008). The extreme temperature tolerance range from −1 to 23 °C and an optimum temperature range from 8 to 12 °C (Jobling 1988) make this species a candidate for aquaculture in the German Bight. However, aquaculture companies should be aware that the relatively high summer temperatures in the German Bight could enhance the risk of infections, as *G. morhua* is prone to bacterial fish pathogens such as *Franciscella spec.* or *Aeromonas salmonicidae* at prolonged temperatures of more than 12 °C (Buck et al. 2012).

11.3.4.3 Candidate: *Hippoglossus hippoglossus*

The Atlantic halibut is a much valued aquaculture candidate as the species is resistant to several common fish diseases, has high food conversion efficiency and a good meat quality due to its firm texture and long shelf life (Daniels and Watanabe 2010). However, the rearing of fingerlings is difficult as larvae need the right composition of live feed after the yolk sac stage and a precise temperature regime. The on-growing of *H. hippoglossus* takes place either in land-based tanks or troughs, or in net cages in the sea (Daniels and Watanabe 2010). The use of submersible cages was successfully demonstrated for halibut (Howell and Chambers 2005; Daniels and Watanabe 2010), offering significant advantages in contrast to surface culture, such as avoidance of the exposure to the seasonal high water temperatures and to the effect of UV radiation at the surface. The temperature tolerance range for the Atlantic halibut reaches from −1.3 to approximately 18–20 °C, depending on the oxygen saturation of the water (Daniels and Watanabe 2010). Although the oxygen saturation in the German Bight is very high below 10 m depth, temperatures in extreme years can reach values outside of the physiological capacity of *H. hippoglossus*. Furthermore, Atlantic halibut is sensitive to strong currents and turbidity (Buck et al. 2012). For offshore site selection, it is important to identify an area with a sufficient supply of oxygen, a moderate temperature regime as well as low current flow and turbidity.

11.3.4.4 Candidate: *Scophthalmus maximus*

The turbot is one of the most valuable food fish of the North-East Atlantic. Since the 1990s, intensive production methods are established for all stages of life of this species (Moksness et al. 2004). Today, the on-growing of turbot takes place mainly in land-based recirculating aquaculture systems. *S. maximus* can be held in very high stocking densities at optimum parameters (Daniels and Watanabe 2010). The culture in surface cages, or submersible cages, was successfully demonstrated at moderate velocities and a constant temperature regime (Daniels and Watanabe 2010). Since the growth of turbot rapidly decreases below 14 °C and above 20 °C and it stops feeding below 8 °C and above 22 °C, the temperature conditions in the German-Bight do not favour on-growing year-round (Person-Le Ruyet et al. 2006; Daniels and Watanabe 2010). In addition, this species is sensitive to the tide flow rates of up to 1.2 ms^{-1} (Buck 2002), which plays an important role for site selection in the German Bight (Buck et al. 2012).

Taking the results of current velocities and waves on cages and the nets (see Sect. 11.4 "Technologies" below) into account a further test was conducted to proof the behaviour of the culture candidates living in these high energy environments. These tests were conducted at the ZAF (*Zentrum für Aquakulturforschung/Center for Aquaculture Research*) under the umbrella of the AWI. A cage prototype (Fig. 11.18b, c) was manufactured and installed in a large raceway of a RAS (Recirculating Aquaculture System) including a water treatment device (drum filter, nitrifying and denitrifying filter, and protein skimmer) to guarantee best water quality during the experiment. This cylindrical cage (see section above) was mounted in a way that an adjustable pump (Hydor Koralia magnum 5, flow rate 6,500 l/h) induced a certain current velocity on the cage and its inner culture space. The cylinder cage in a size of 1:40 was taken from the current flume/wave tank investigations in Hanover.

The challenge in the subproject of Project No. 13 "Open Ocean Multi-Use" was to find fish species in small size to be used as model organism, such as juvenile turbots with a mean size of approx. 10 g (*Scophthalmus maximus*) (Buck et al. 2012). This fish size of the "model organism" would in terms of scaling correspond to a market sized adult turbot, which would be farmed in real offshore cages.

After acclimatizing the fish in the cage model after several hours the pump was switched on for a short term to induce a current velocity of 0.9 ms^{-1} at the exhaust pipe. The pump was adjustable in the tank so that the distance to the cage could be within a range of 25–150 cm. This distance to the cage and the resulting current velocity was calculated by using a corresponding situation as it occurs at the offshore site. After starting the experiment fish behaviour, escape response, welfare, and reaction at feeding were recorded using an underwater video camera.

The results of these investigations show a strong impact of strong current velocities on farmed turbot within the cage. Juvenile turbot are unharmed at velocities of 0–1.0 ms^{-1}. However, the fish are affected by stronger currents in a way that the fish will be transferred from their resting area on the bottom of the cage resulting in more activity of the fish swimming against the current trying to resettle on the bottom. At very strong current velocities of 1.5 ms^{-1} and higher it might happen that young turbots are pressed against the net of the cage leading to the risk to obtain skin injuries.

To summarize, these results show that turbot cultivation offshore is not suitable if current velocities exceed a certain value, making a site-selection-criteria process necessary. However, we have to take into account that we used small fish—the model organism—in the experiment, which allowed the upscaling to a size corresponding with real fish sizes to be farmed at offshore sites. As fish in a size we used in this investigation will never be transferred to offshore sites as it is described in this book we assume that larger fish, which normally will be cultured in offshore cages, will act differently.

11.3.4.5 Candidate: *Melanogrammus aeglefinus*

The haddock is a valued food fish on both sides of the North Atlantic, but its wild catches stagnate, so that haddock is an interesting candidate for aquaculture (Moksness et al. 2004). In recent years, the cooperation of a large commercial salmon production (Heritage Salmon Limited) with various Canadian research institutes lead to great advances in the management of broodstock, the feed of larvae, as well as the weaning and on-growing methods (Chambers and Howell 2006). Same as cod, haddock is a physoclist species and breaks in descents and ascents need to be included to prevent barotraumas due to rapid pressure changes by vertical movement of the cages. (Buck et al. 2012). Moreover, *M. aeglefinus* has the ability to penetrate even tight mesh (Özbilgina and Glass 2004), precaution needs to be taken to prevent escapes. The on-growing of *M. aeglefinus* in submersible offshore cages has proven to be very promising, reaching good growth rates in a 600 m^2 Sea Station at the temperature conditions off the coast of New Hampshire (Chambers and Howell 2006). However, the similarly high growth rates as cod in the first year decreased slightly and the growth potential over longer periods of time must be considered to be lower than in cod (Treasurer et al. 2006). Haddock can tolerate temperatures between 1 and 20 °C (Chambers and Howell 2006), making it a potential aquaculture species for the offshore area in the German Bight.

11.3.4.6 Candidate: *Salmo salar*

With an annual production of 1,244,637 t (Bilio 2008), the Atlantic salmon, *Salmo salar*, is by far the most produced marine species. Although *S. salar* is an anadromous species, which is spending a part of its live in freshwater (FAO 2004), it is an excellent candidate for aquaculture in marine waters and has its highest growth rate at salinities between 33 and 34‰. The temperature range for this species is 6–16 °C (FAO 2004), with temperatures optimal for growth between 8 and 12 °C (Boghen 1991). Although the temperature minima and maxima of this species can be exceeded in the surface waters of German North Sea, the Atlantic salmon is thought to be suitable for aquaculture in the German Bight (Buck et al. 2012).

11.3.5 Various Species in One Habitat—Managing Reef Effects

Following Krone (2012b), it is currently difficult to judge the ecological consequences associated with the introduction of the artificial reef type "wind power

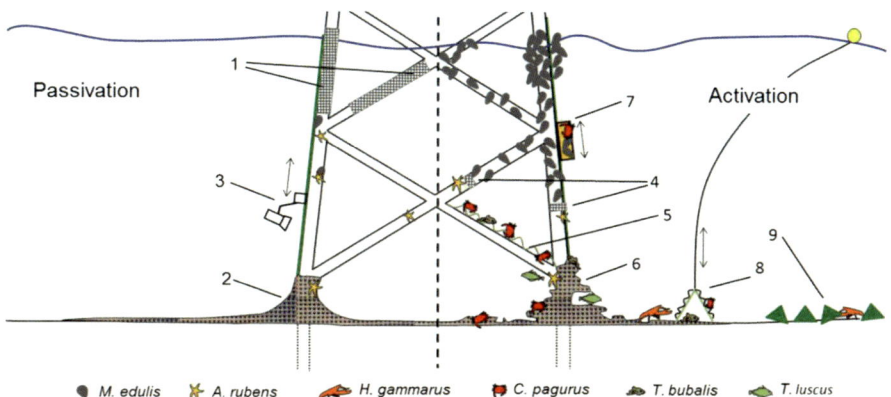

Fig. 11.19 Techniques for passivation and activation of wind power foundations and artificial reef fauna examples. 1 Electrochemical anti-fouling system for structures wetted by sea water; 2 Calcareous EAT-material designed to reduce structural diversity and as scour protection; 3 Device for using technical devices underwater, equipped with a camera; 4 The electrochemical anti-fouling system applied to prevent sea stars (A. rubens) to climb up to feed on the Bivalvia *M. edulis*; 5 Device for developing habitats in the underwater area of an offshore construction, paths and hideouts for crabs and demersal fish; 6 EAT-material used to create reef-like structures to enhance the reef species; 7 Device for colonizing and harvesting marine hard ground animalssuch as *M. edulis* and large vagile megafauna; 8 Transportable device for colonizing and harvesting invertebrates in the vicinity of offshore constructions; 9 Artificial habitat in polyhedron shape used as a fishing net barrier which is at the same time habitable for lobsters (modified after Krone 2012b)

foundation" and its reef effects on fish and benthos. The question whether the impact is positive, negative or negligible remains unanswered and depends on the emphasis one places on the different implications of the reef effect. The present findings, however, enable the designed structure to increase or to reduce biofouling and megafauna settlement at offshore wind power foundations (Fig. 11.19). Krone (2012b) defines it as "reef passivation". Alternatively, all potential reef functions could be activated if the development of a highly valuable artificial reef fauna is intended, such as "reef activation" (for more details see Krone 2012b).

11.4 Technology

11.4.1 Longline-Techniques and Combination to Wind Farm Turbines

In the coastal sea of Germany, there are some limitations to allow the expansion of mussel culture activities. First, nearly 98% of the coastal sea has a nature reserve status. Culture plot sizes are decreasing in order to follow the mussel management plans of Schleswig-Holstein and Lower-Saxony (e.g. CWSS 2002; Buck 2002) with no new licenses being approved in the future. Second, moving more seawards to abandon the preserved areas, environmental conditions in the North Sea are harsh and protected bays for safe mussel longline cultivation is inexistent. However, exposed locations off the coast are in the focus for many aquaculture activities worldwide (e.g. Polk 1996; Hesley 1997; Stickney 1998; Bridger and Costa-Pierce 2003) and generated a vast quantity of projects in high energy environments. With this in mind, a multiple use concept for another newcomer, the offshore wind farm operator, has been developed (Buck et al. 2004). The offshore wind farm turbines provide space and attachment devices for mariculture installations and therefore minimize the risks originating from high energy environments (Buck et al. 2006b). The viability of a mussel cultivation enterprise within offshore wind farm areas was intensively studied following various factors, such as (1) the technological and biological feasibility, (2) the legislative and regulatory constraints, (3) the environmental sustainability of farming aquatic organisms at all, and (4) the profitability of this potential commercial operation (for review see Buck et al. 2008a). However, without the solid foundations of the wind turbines as anchor or connection points, economic viable installations of equipment for an extensive mariculture would decrease in view of the high-energy environment in this part of the North Sea.

11.4.1.1 Mussel Longlines: Design

In Project No. 3 (Roter Sand) different anchored longlines were installed in order to test their suitability under open sea conditions in terms of material and functionality and to obtain insights on how to connect these systems to wind farm foundations. Various test devices of horizontal ropes anchored to the seafloor with buoys to provide flotation and vertical droppers for spat collection and/or mussel grow-out were deployed offshore in multi-use with the wind farm "Energiekontor" 17 nautical miles off Bremerhaven.

The first system had a polypropylene (PP)-based design (Fig. 11.20a) and was deployed by the Water and Shipping Agency (WSA) Bremerhaven in 2002. The 3-stranded polypropylene-based longline (ø = 32 mm) was made of two 35 m long segmental parts connected to each other. To keep the longline afloat each segment was equipped with one 80 kg buoyancy barrel in the centre and with 115 kg barrels at the coupling to the anchor-line. Further, a few 35 kg ball-like surface marker floats were fixed to a 5 m rope, which was connected to the longline. Three steel barrels filled with concrete (200 kg) tightened the longline to the sea bottom at the coupling to the anchor-line and in the centre. Another design was the steel hawser-based design (Fig. 11.20e), which was deployed a year later. The 3-stranded steel hawser-based longline (ø = 20 mm) had a similar set up as it was described for the polypropylene line, however, the longline consisted of more sections: seven 10 m long segments with each three 10 kg submersed floats in the centre and at each coupling elements 21 kg submersed floats. All floats were connected to a short wire element, which itself was spliced perpendicularly into the hawser and squeezed by one of the hawser's strands. An alternative attachment was the use of flat steel panels tightly fixed to the longline to which both, the floats and the collectors, were attached. Due to the weight of the steel wire, the longline had more floats than the polypropylene-based longline and no additional weights were attached to the longline. At all coupling elements pencil-like fenders (28 kg) were attached to mark the longline at the surface. The third system was based on the initial PP-design (Fig. 11.20f) but was equipped with V-shaped collector pairs.

All longline designs and moorings were based on the same fundamental setup having longlines operating in a submerged mode at a depth of about 5 m horizontally below the surface to avoid the destructive effects of surface waves. All longlines had a 10 m "undisturbed end" at both sides, which could not be retrieved when sampling or harvesting. Buoys were fixed along the longline in combination with suspended mussel spat collectors. The three-stranded collectors (Galician type, 2.5 m long), equipped with horizontal inserted pegs to prevent attached mussels to be shaken off through current and wave forces, were suspended every 2 m perpendicular from the longline in a parallel manner in the first two longline set-ups. In the third design V-shaped pairs consisted of two 2.5 m side pieces and were connected in series in a suspended manner and in perpendicular fashion from the longline. Each collector had an additional weight of approximately 3 kg at the bottom end and was on the top end hooked into the longline by shackles to ease the maintenance during harsh conditions. Tufts of unravelled polypropylene lines

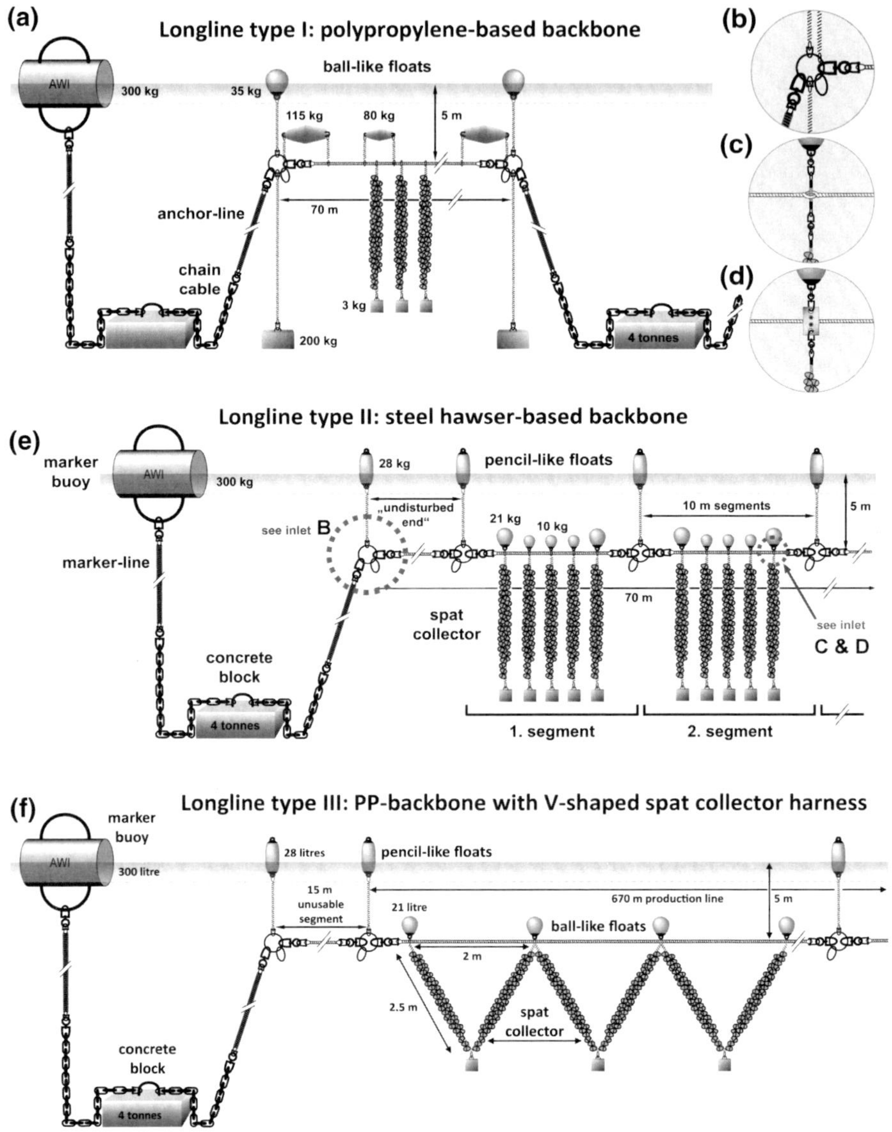

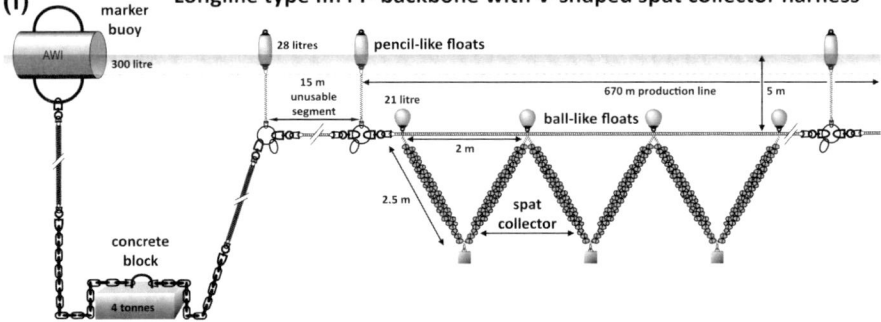

Fig. 11.20 **a–f** Submerged longline devices and the collector harnesses, which were developed at sea. One end of the backbone was attached to the wind turbine foundation. **a** Shows a polypropylene-based longline with coupling elements **b** and its equipment and connection with floats/collectors (**c–d**); **e** displays a steel hawser-based longline; **f** shows a longline with polypropylene (PP) as a backbone and V-shaped spat collectors. Images were modified after Buck 2007 and Buck et al. 2010. All images are not to scale

(Tortell 1976) were also connected to the longline to determine the settlement of mussel post-larvae. The longline was secured at each end with an anchor consisting of a 4 t concrete block. A chain cable attached to the concrete block was connected to the anchor-line, which itself held one end of the longline. Additionally, each concrete block was equipped with a marker-line and a marker buoy (300 kg). The marker line had a service load of several tonnes that it could be used to tauten the longline and to retrieve the concrete block at the end of the project. All mooring components were selected and sized for the use in muddy sea bottom and for long term durability. All longline segments had spliced eye loops with embedded galvanized thimbles at the ends and at all coupling elements shackles and swivels were used (Fig. 11.20b–d).

All longline systems and their mooring devices were first modelled and designed with the WSA. A local cable and wire manufacturer supported the engineering development for the culture design. All components were selected and procured to start the preparation and assemblage of the longline during wintertime.

After installation all longlines and their collectors were fully overgrown by mussels. Heavy mussel growth was found also on all mooring components and a lot of mussels were recorded on the undisturbed ends of the longline at each side. From the farmer's perspective, the offshore site was suitable in terms of settlement and grow-out. Problems encountered during the test trials included material failure, heavy fouling on the collector lines and other infrastructural units resulting in large ballast weights, while heavy predation on mussel seed by sea stars and Eider ducks also happened. Another difficulty was maintaining proper headline depth and its management with the right buoyancy due to mussel growth. This added drag forces to the entire longline, which in turn complicated the handling during hostile sea conditions.

The PP-longline design was resistant to the high energy environment of the North Sea, but chafed within the inner parts of the eye loops and therefore some of the thimbles were lost. These stress forces affect the lifetime greatly. The segmentation set-up in the second longline design allows the replacement of spare parts. The weight of the suspended collector harness increased significantly with time, and extra floatation had to be added to support the buoyancy of the longline. However, the time intervals chosen for service were too long (particularly during the summer growing season) so that parts of the collector harness including the floats submersed fully and partly touched the seafloor. Additionally, this process was exacerbated by the fact that the loss of buoyancy resulted in a further lowering of the total construction. When the longline was retrieved the concrete blocks were broken easily out of the clay at the site and raised by the anchor chain. Most of the mooring system, in particular the longline and its harness, was later inspected onshore. On a few spots near the eye loops some slight damage of the material was found, but, after removing the fouling, most of the line looked intact and almost new.

11.4.1.2   Mussel Longlines: Forces Induced by Collectors

As the distance between wind turbine foundations within an offshore wind farm are expected to be between 700 and 2500 m, the forces induced by the longline can be estimated to be very high. Additionally to the investigation on forces impinging on the foundations by the attached longlines, the forces induced by the hundreds of fully grown, market-sized Blue mussel collectors was calculated as well. To get more insight into these forces the drag coefficient of this type of collector had to be investigated. First tests were conducted in a towing tank of the *Hamburg Ship Model Basin* (*HSVA*, Hamburg) to determine the drag coefficient (C_D) for the collectors (Fig. 11.21a–i). The length of the tank allowed sufficiently long measuring times (length 80 m, width 3.8 m, and depth 3 m). Mussel collectors, with mussels and without mussels (naked) were prepared for measurements of drag by attaching the proximal end of the collector to a string. This string was directed over a guide reel fixed to the bottom of a beam. A cavity within the beam allowed the thread to be hooked onto the digital gauge, which was placed on the control panel of the carriage. The beam was fastened to the base frame of the towing tank's carriage. It was lowered 1 m deep into the tank filled with water. Collectors tied to

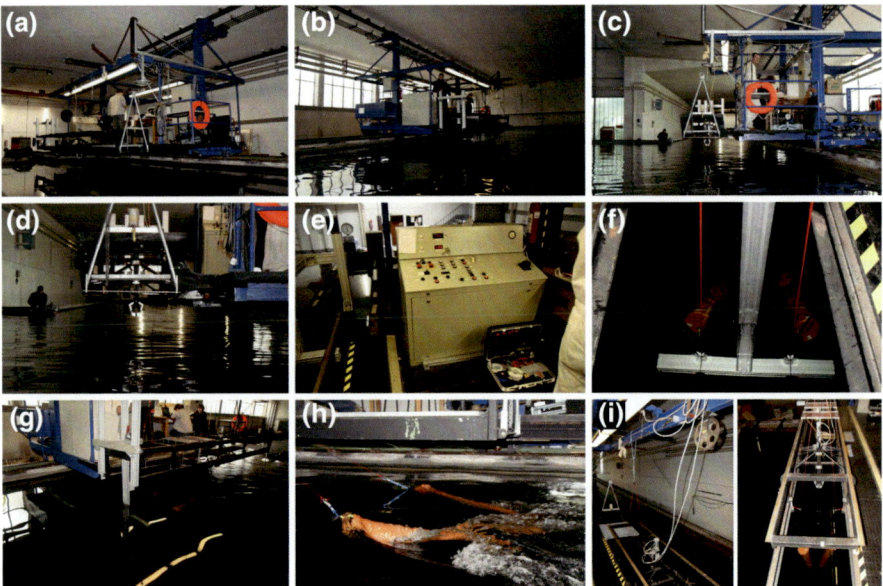

Fig. 11.21 a–i Towing tank at the HSVA (Hamburg Ship Model Basin). **a** Main carriage with the operation desk and the equipment holding the collectors; **b** carriage under way; **c** carriage displaying the Bosch-profile, which holds the experimentals; **d** single holdings device to which different collector types (artificial, mussel) in various modes (single, double, V-shape) can be attached; **e** operation desk controlling the speed of the carriage; **f** V-shape artificial collector pair to be tested; **g** carriage with V-shape artificial collector pair under way; **h** V-shape artificial collector pair during high speed under way; **i** walkable platform to manage the experimentals. *Images* AWI/Prof. Dr. Bela H. Buck and Björn Hendel

the thread of the gauge were towed through the tank by the carriage at velocities of 0.5, 1.0, 1.5, 2.0, and 2.5 m s^{-1}, respectively.

The drag measurements of each collector were used to calculate the dimensionless coefficient of drag (C_D) by applying the equation:

$$C_D = \frac{2F_{Drg}}{\rho \cdot A \cdot v^2} \tag{11.1}$$

F_{DRG} denotes the measured drag, ρ the density of water, A is the surface area exposed to the current of velocity v. While for relatively rigid or bluff organisms the projected area of the organism across the flow is generally used, we decided to test an artificial test body (collector) and a fully grown mussel collector as well. For solid objects being accelerated in fluid in addition to drag a force occurs which is, commonly described as acceleration reaction (e.g. Daniel 1984; Denny et al. 1985; Denny 1988). In order to take orbital motions into account the drag F_{Drg} from Eq. (11.1) has to be written in vector notation, and an acceleration term is added (Morison et al. 1950):

$$\overrightarrow{F}_{Drg} = C_D \cdot A \cdot \frac{1}{2} \cdot \rho \cdot \overrightarrow{|v|} \cdot \overrightarrow{v} + C_M \cdot \rho \cdot Q \cdot \frac{d\overrightarrow{v}}{dt} \tag{11.2}$$

where C_M is the dynamic drag coefficient and Q the volume of water displaced by the object. In order to initially avoid uncertainties with C_D and the area A for flexible organisms it is useful to write the first term in Eq. (11.2) in terms of dynamic pressure ($\overrightarrow{F^*}$), which can be exactly computed from current measurements:

$$\overrightarrow{F^*} = \frac{\overrightarrow{F}_{Drg}}{C_D \cdot A} \tag{11.3}$$

i.e.

$$\overrightarrow{F^*} = \frac{1}{2} \cdot \rho \cdot \overrightarrow{|v|} \cdot \overrightarrow{v} \tag{11.4}$$

For a detailed demonstration of the time history of drag forces occurring under the action of tidal currents, wind wave and swell model calculations were carried out using the software "WaveLoads" developed by Mittendorf et al. (2001), which solves Eq. (11.2) for offshore structures. The forces were computed for a hypothetical (artificial) test body (cylinder) of 2.5 m in length and such a diameter that the area exposed to the current corresponded with that of the plan area of a typical fully grown mussel collector as determined before. The drag coefficients measured in the towing tank experiments were used instead of the ones determined for cylinders (Fig. 11.22a–f). The cylinder was exposed perpendicular to the flow direction. The results render information on the distribution of horizontal and vertical forces which act on the test bodies and mussel collectors.

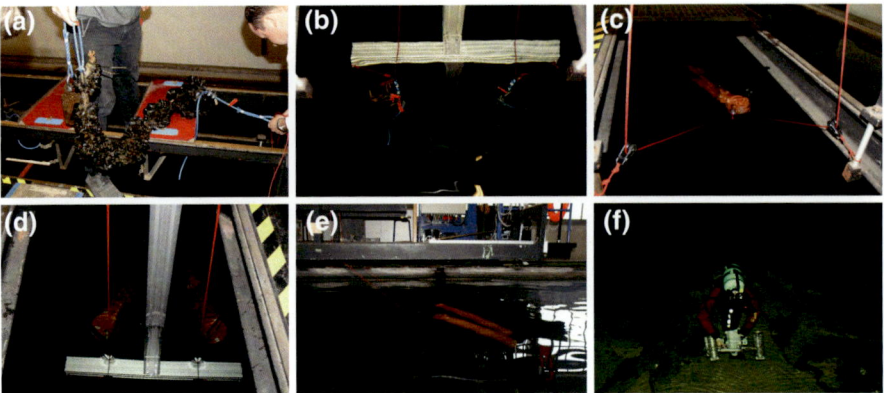

Fig. 11.22 **a–f** Mussel collectors and artificial test bodies in the towing tank. **a** Mussel (*Mytilus edulis*) collector in preparation for the towing experiment; **b** crossbeam with two mussel collectors in a V-shape; **c** single test body in test mode; two test bodies (V-shape) in test mode at the crossbeam (**d**) and from side view (**e**); **f** scuba diver in the towing tank to monitor the behaviour of test bodies during towing. *Images* AWI/Prof. Dr. Bela H. Buck and Björn Hendel

The drag coefficient C_D of the artificial test body as well as the fully grown mussel collector was calculated at velocities of 0.5, 1.0, 1.5, 2.0, and 2.5 m s^{-1} to range between 0.2 and 1.15. The graphs for moth collector types show a pronounced bend at a velocity of 1 m s^{-1}. This appears to approximately mark the minimum speed required for orientation of the collectors within the water column into the downstream direction. At higher velocities ≥ 1.5 m s^{-1}, $\underline{C_D}$ was almost a constant.

11.4.1.3   Mussel Longlines: Attachment to Foundations

The development and the conceptual design of offshore foundation structures are complex and require an interdisciplinary approach. One of the most important questions pertains whether it is technically possible and economic feasible to use offshore foundation structures as fixation device for aquaculture operations, such as a longline construction as one possible culture design. Therefore, wind energy converters in offshore sites do need an exact description of the loading on the plant. In general, these loads are caused and influenced by the environmental conditions such as wind, wave, currents and the soil properties (GL 2005). From the engineers point of view it is indispensable to have an estimation of the additional loadings caused by attached longline constructions.

In Project No. 6 "AquaLast" supplemental loads on the support structures of offshore wind energy converters (OWEC) caused by attached longlines were investigated. Several alternative connection points were tested. As monopile and tripod foundations are the most common foundation structures to date, these both were considered within the modelling approach calculating the respective loads from wind and waves. Both foundation structures are in the dimension of 4–5 MW wind turbines. While the monopile design was calculated for a water depth of 10 m the tripod design was computed for a water depth of 30 m.

The monopile was calculated and designed with the representative environmental loads from the area of the planed wind farm *Nordergründe* (see Fig. 11.8). For the static calculation of the monopile we use the structural analysis program *RSTAB* in the version 5.14.191 and the add-on module *STEEL* from *Ingenieur-Software Dlubal GmbH*. The 2nd order theory is used to get the results. In Fig. 11.23a, b the static system, the bending moments and the transverse loads based on the maximum loads from the wind energy converter are plotted.

(a)   **(b)**

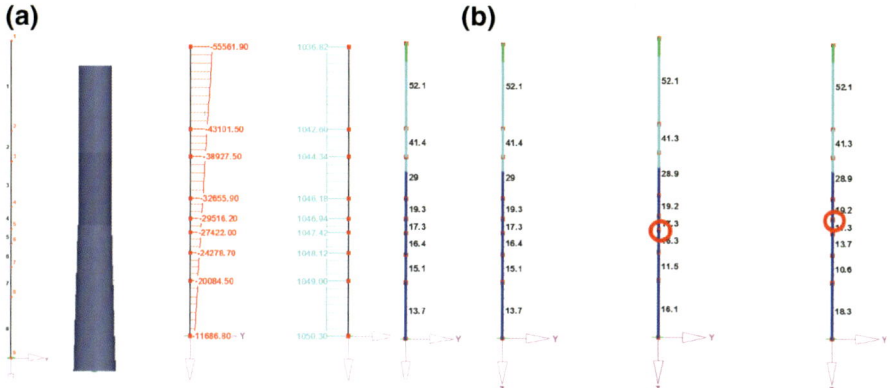

Fig. 11.23   a Static system of a monopile, moments and transverse loads including the distribution of stress; **b** diagram of the maximum tension at the monopole at different potential longline connection points. Modified after Buck and Wiemann (unpublished data)

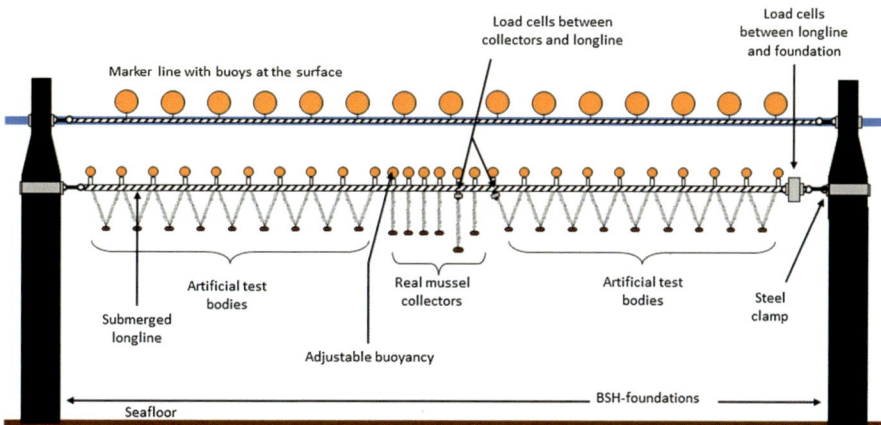

Fig. 11.24 First submerged test longlines in a multi-use-concept by using former ODAS-foundations, which are located in the vicinity of the offshore wind farm *Butendiek*. Modified after Buck et al. 2006b

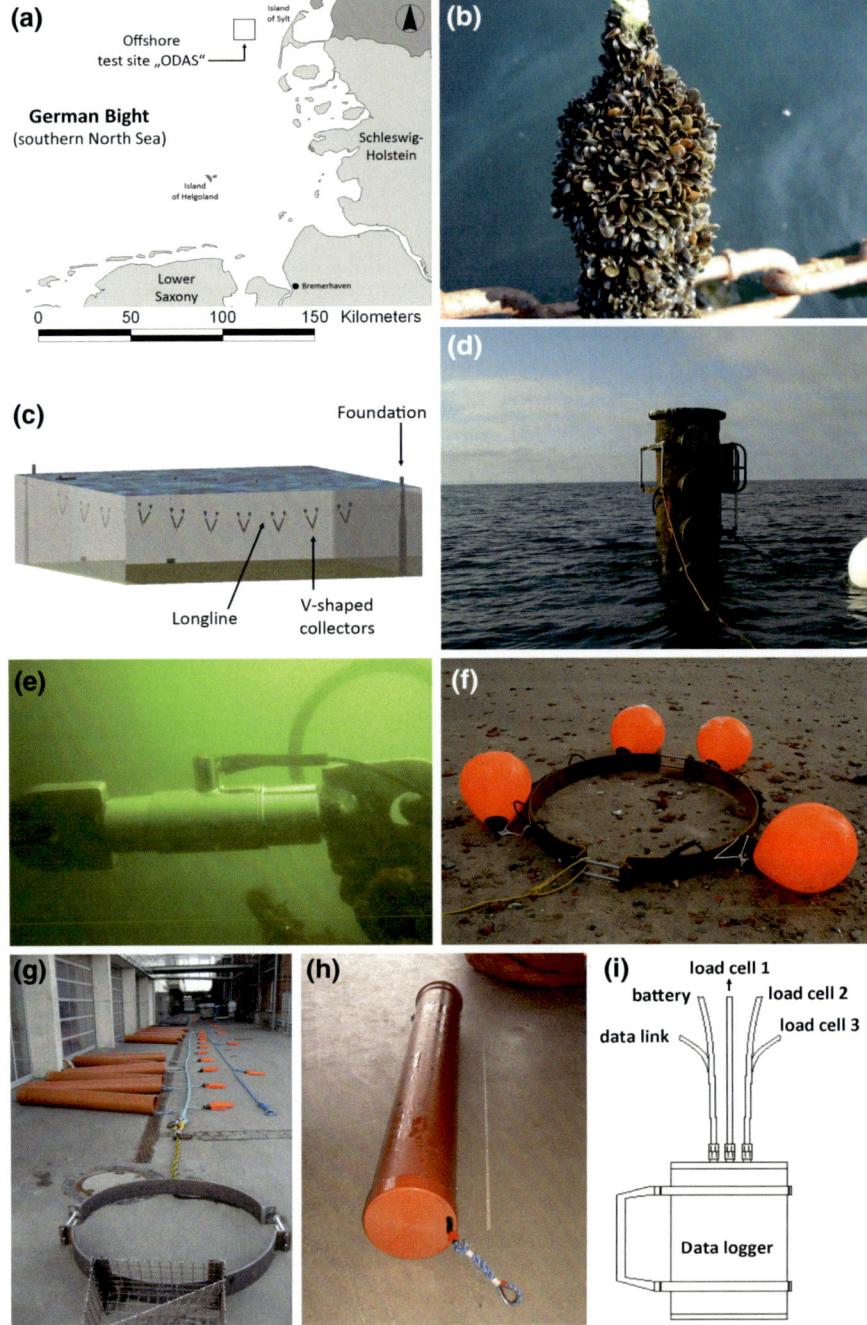

(a) Offshore test site „ODAS"

German Bight (southern North Sea)

Island of Sylt

Schleswig-Holstein

Island of Helgoland

Lower Saxony

Bremerhaven

0   50   100   150   Kilometers

(c) Foundation

Longline   V-shaped collectors

(i) load cell 1

battery   load cell 2

data link   load cell 3

Data logger

◄**Fig. 11.25 a** Location of the test site 16 nautical miles off the island of Sylt (ODAS-foundation); **b** fully overgrown collector with mussels (*Mytilus edulis*); **c** Drawing of the longline device connected in between two foundations showing the V-shaped collector types; **d** former ODAS-foundation simulating the wind turbine monopile foundation, to which one end of the longline was connected; **e** underwater load cell; **f** steel clamp to connect one end of the longline to the foundation (the buoys were only fixed to the clamp during transfer from the vessel to the foundation); **g** complete longline harness on land; **h** artificial test body; **i** drawing of the underwater data logger. *Images* AWI/Prof. Dr. Bela H. Buck

11.4.1.4 Mussel Longlines: Attachment to a Monopile

For the connection of a longline and a monopile foundation a numerical model was built up to provide information of the parameters influencing the reactions at the longline fixation. The analysed load cases were quasi-static calculations for extreme loads. In support of the model approach, an entire monopile-longline-monopile construction was set up in the offshore waters 16 nautical miles off the Island of Sylt at a water depth of about 17 m (Figs. 11.24 and 11.25a–i). Loads at one junction of the longline with the foundation as well as in between various collector types and the longline were measured and environmental conditions were assessed by sensors installed on site over a period of 5 months (May-September 2006).

For the test period a submerged longline design was chosen due to its ability to withstand strong oceanographic conditions (Buck 2007, Langan and Horton 2003). The submerged construction consisted of a buoyed, horizontal running backline, which is commonly described as a back-bone of the entire harness (Figs. 11.24 and 11.25c). This back-bone (polypropylene, 32 mm Ø) was fixed via spliced ends (including thimbles) to a 1.5 m chain with shackles and swivels (service loads above 20 tons) at both ends to the test piles in a depth of about 5 m (low tide level). In this project the investigation of the settlement success of mussel larvae on artificial collector substrates was not the main objective. Moreover, the concentration of mussel larvae in the water column even in June to July, which is the main peak season for drifting mussel larvae (Buck 2017), was already determined to be very low at the study site. Thus, artificial test bodies were introduced, which had a similar diameter, shape and weight to simulate a fully colonized market-sized mussel collector. For this purpose, 22 V-shaped test bodies were fastened perpendicular to the longline and served as "non-collecting" collector harness.

The mechanical loads were measured at three strategically important positions along the line using submersible force sensors (*Althen TCA-256*). One sensor acquired the tension of the main leash rated for a load of 200 kN was located directly at the attachment of the long line. Two additional force sensors each ruggedized for 10 kN of load were placed above the V-shaped mussel collectors in order to confirm the charge on a single mussel collector.

Data from the force sensors was recorded on a custom made submersible data logger (*iSiTEC*) located in a cage fixed to one pile in 5 m water depth. The measurement was accomplished in a 1 h interval, whereby 180 measurements were noted within a 3 min burst. This cage also held a battery supplying energy for the sensors and the logger for roughly 3 month together with an autonomous underwater tide and wave recorder (*RBR TWR-2050*). The tide and wave recorder was used, to meet exact statements about the arising wave spectra during the test period. Data noted during the trial period were acquired all 30 min in one burst with a 512 sample measurement length and were averaged afterwards. This power saving strategy allowed for a time span of approx. 3 months before the inbuilt batteries were exhausted.

Measurements of currents in the plane of the collectors, approx. 7 m below water surface was achieved by applying a current meter (*AADI RCM-9LW*) at a mooring site in approximately 30 m distance from the longline demonstrator. The current meter was additionally equipped with pressure, temperature and conductivity sensors to address the oceanographic conditions as well as a turbidity probe and a shallow water oxygen sensor, both to account for growth parameters of the biological setup. Due to the high capacity of the *RCM-9LW* batteries no power saving precautions had to been taken and a sample interval of 10 min was realized. Table 11.1 summarizes the applied sensors and acquired parameters during the test construction.

The longline (LL) was represented by finite elements (FE) and the calculation was done by commercial FE software. The FE were realised by 3D tension-only spar elements. These elements only considered axial tension forces and had three translational and no rotational degrees of freedom. No bending moments could be transferred. The ends of the longline model were fixed in all three translational degrees of freedom. There was no mass for the longline added because it was small compared to external loading and the mussel collectors were assumed to be floating, thus only hydrodynamic forces were applied on the collectors. The mussel collectors were represented by points, which were in this case FE nodes between the 3D spar elements.

The complex dynamic behaviour of a floating longline (Raman-Nair and Colbourne 2003) and the load calculation based on statistic wave and current fields

Table 11.1 Overview of the applied sensors used in the longline demonstrator setup

Applied sensor and the acquired parameters		
Device	Parameter	Unit
Aanderaa RCM9-LW	Current speed	cm/s
	Current direction	°deg
	Temperature	°C
	Conductivity	mS/cm
	Depth	kPa
	Turbidity	NTU
	Oxygen	μM
RBR TWR 2050	Tide spectrum, $T^{\frac{1}{3}}$	s
	Wave spectrum, $H^{\frac{1}{3}}$	m
	Temperature	°C
Althen TCA-256	Stress loads	kN from mA

(GL 2005) are reduced to a static worst case consideration. This was done to give a first approximation of the additional loading on OWEC support structures. In the OWEC dimensioning process extreme load cases like the 50 year wave or emergency stops were taken into account. Thus, the extreme load case of the 50 year wave was simulated as a quasi-static simulation. No dynamics and thus no inertia forces were taken into account. The very time consuming process of dynamic modelling would be necessary for fatigue analysis. In that case, the varying loads caused by the longline were of interest. Modal analysis of these loads was also important on the support structure to avoid excitation in or near the eigenmodes of the plant. In any case, the dynamics were highly depending on the sea state. This was modelled by wave theories and current assumptions. These were highly statistic inputs and could only be a rough approximation of reality.

For these reasons, a static analysis of the problem was a reasonable first approach to the complex problem. The forces of the longline were depending on the single collector forces. These were considered to be concentrated loads, which were all applied in the same angle of attack relative to the longline and with similar force value. In this case, the reactions were maximal because there was no neutralization of the forces taking place. The longline was oriented in one plane. In consequence the LL could be approximated by a 2D model as shown in Fig. 11.26a.

The schematic 2D LL picture in Fig. 11.26a shows the pre-deformed model as a starting point of simulation. This deformation was the approximate form of a catenary curve. A curve like that would be obtained for a cable under self-weight or in a uniform loading caused by even very low currents perpendicular to the LL. The pre-deformation was needed to define the maximum sag without loading. It was also needed for better convergence of the numerical calculation.

The following parametric studies for this model were conducted: First, a single load was applied vertically in the middle of the LL. The maximum sag was variable. For this case an analytical calculation was carried out to validate the model itself and to understand the main parameters of interest in the problem.

The second parametric study was made for forces on each FE node that represents a single or V-mussel collector under varying angles of attack. The loads in both parametric studies were applied as ramped loads to ensure convergence. The reaction forces in both supports were calculated. All simulations were quasi-statically calculations for the worst case loading investigated.

According to theoretical analysis, a single load in the middle of the longline led to a V-like deformation (Fig. 11.26b). The relative force in a single loaded cable in equilibrium without consideration of strain and mass is defined as

$$\frac{F_s}{F} = \frac{1}{2\sin(\beta)} \qquad (5)$$

where F_s is the force in the cable and F is the external load. β is the angle between cable and horizontal axis. F_s equals the absolute value of the reactions. This formula shows the fundamental difference between cables and beams. Beams could carry bending moments and cables could not. This was due to the fact that at angle β of $0°$

Fig. 11.26 a Schematic longline with nodes representing the mussel collectors; **b** free body diagram of a single loaded cable

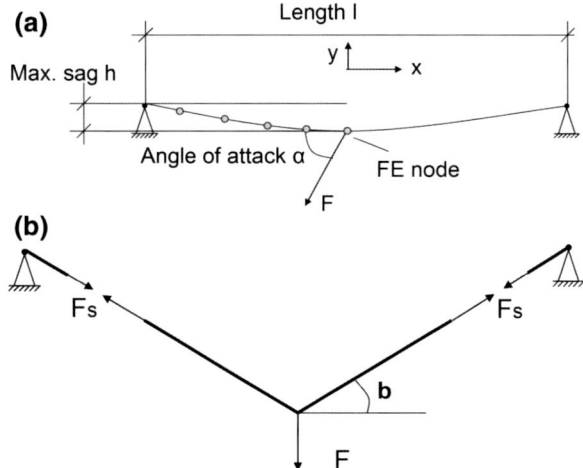

the vertical reactions in a beam were a half of the applied load and for a theoretic cable the reactions were infinite.

In the following graph, the theoretical and simulated values for the relative force were shown as a function of the relative maximum sag h/l. The relative maximum sag h/l was the ratio of the maximum sag to total distance between the supports (Fig. 11.27a). The theoretical values for h were based on curve length of the pre-deformed model. The curve length was then used to get the maximum sag of the foreseen V-form after loading. That value of h was then used with the formula described above. The theoretical and simulated values for the relative force are shown in Fig. 11.27a, b.

With values for h/l bigger than 0.05 the relative force was smaller than 4 and was approaching 1 with increasing h/l ratio at nearly 0.25. The more the longline was tightened and the h/l ratio was decreasing the numerical model of the longline shows an increasing relative force in the supports. In Fig. 11.28a the pre-deformed model is shown and the deformed model under single load depicted as an arrow is also sketched. The deformed model had a V-form similarly to the theoretical cable under the same loading. The second parametric study for different angles of attack and different maximal sags is shown (Fig. 11.28b). As an example the deformed and not deformed model of a longline loaded at equally spaced points with a loading angle of 60° is shown in Fig. 11.28b. The forces are depicted separately in their coordinate direction components. In the middle part smaller loads represent the single mussel collectors whereas the rest represents V-collectors. A deformation of the LL model in force direction was observed. Figure 11.28c shows the results for the second parametric study. The relative force in the right and left support was calculated for different values of angle of attack and varying maximal sags. In this case the relative force was the reaction at the end of the longline divided by the sum of total collector forces that were applied. In general the reactions in the left support were bigger than in the right except for angles of 90° where the loading was

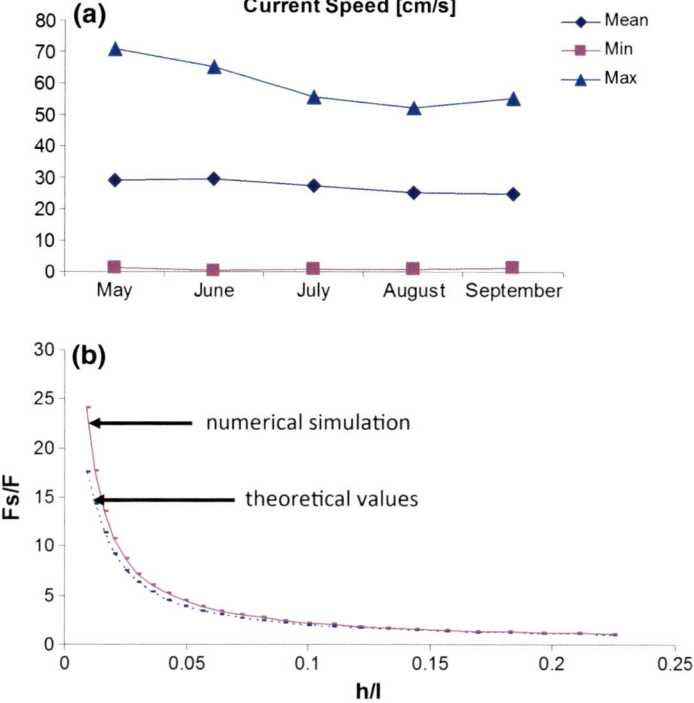

Fig. 11.27   a Results of the current-meter at site; **b** free body diagram of a single loaded cable. Buck and Wiemann (unpublished data)

symmetric thus the relative forces are equal. Depending on the angle the loads in the right support increased with decreasing angle to a certain point and then decrease again. In the right support the reaction decreased for angle smaller than $90°$. With increasing h/l ratio the reactions decreased as well.

11.4.1.5   Mussel Longlines: Attachment to a Tripod

In a final approach, a tripod construction was chosen as a potential connection point for aquaculture devices. A tripod is an offshore foundation construction for water depths of more than 25 m. The weight of a tripod for a 4–5 MW turbine is up to 1000 tons of offshore steel S355 NL/ML. The tripod considered within this project is constructed and calculated for the area "Borkum Riffgrund", where the German research platform *FINO 1* (N54° 0.86′ E 6° 35.26′) is located. We employed the common wind loads and the waves for this position to create a tripod model, which was used as basis for the following investigations. Figure 11.29a–f shows the construction and the static model for the calculated tripod as well as loads on the tripod and the tension of the steel induced by a connected longline device. A tripod

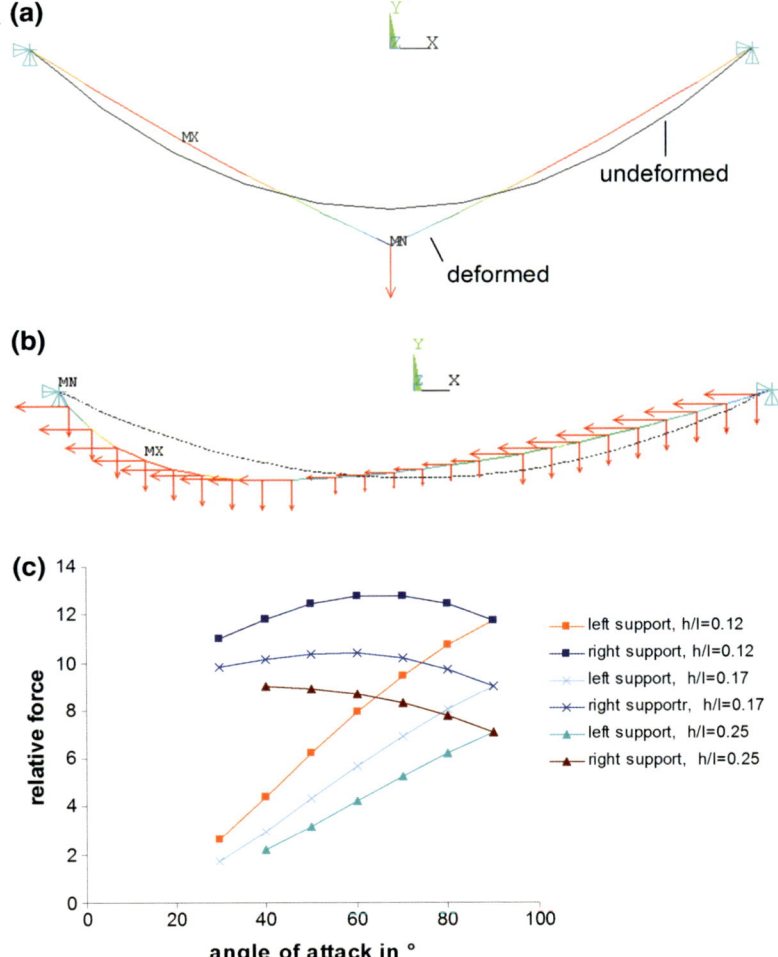

Fig. 11.28 a Preformed and deformed numerical model of the single loaded longline; **b** Preformed and deformed numerical model of the single loaded longline; **c** Relative forces for different angles of attack. Buck and Wiemann (unpublished date)

has a main joint, which is the technical limit for the depth of the connection point. This is located about 5 m below the water line (Fig. 11.30).

11.4.1.6 Additional Forces Measurement

To estimate the maximal forces in order to qualify the load sensor class some previous calculations were conducted. Figure 11.31a shows a schematic depiction of a single collector load system. Further, the load forces of the attached buoys were

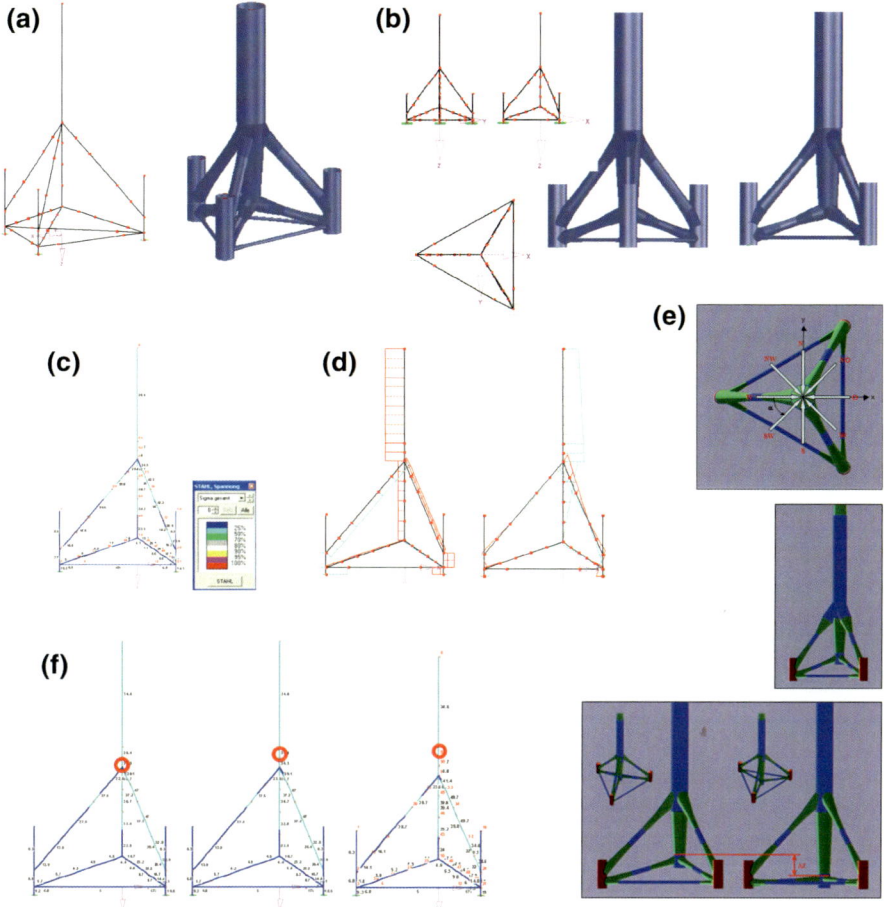

Fig. 11.29 a Representation of a beam model (*left*) and of a tube geometry (*right*) of a tripod foundation for offshore wind generators (4–5 MW class); **b** different views of a tripod foundation; **c** maximum tension utilization of steel at the various single components; **d** Representation of the progress of the transverse force (*left*) and of the progress of the momentum (*right*); **e** static system of a tripod; **f** representation of longline connection points at the tripod as well as the utilization of the maximum acceptable tension of the steel. Modified after Buck et al. 2006b, 2008a, b; Buck and Krause 2012

calculated similar to the collector (Fig. 11.31b). As the vertical forces were caused by the mass of the collector and the buoyancy the mass force was defined as

$$F_m = m \cdot g \tag{11.6}$$

with

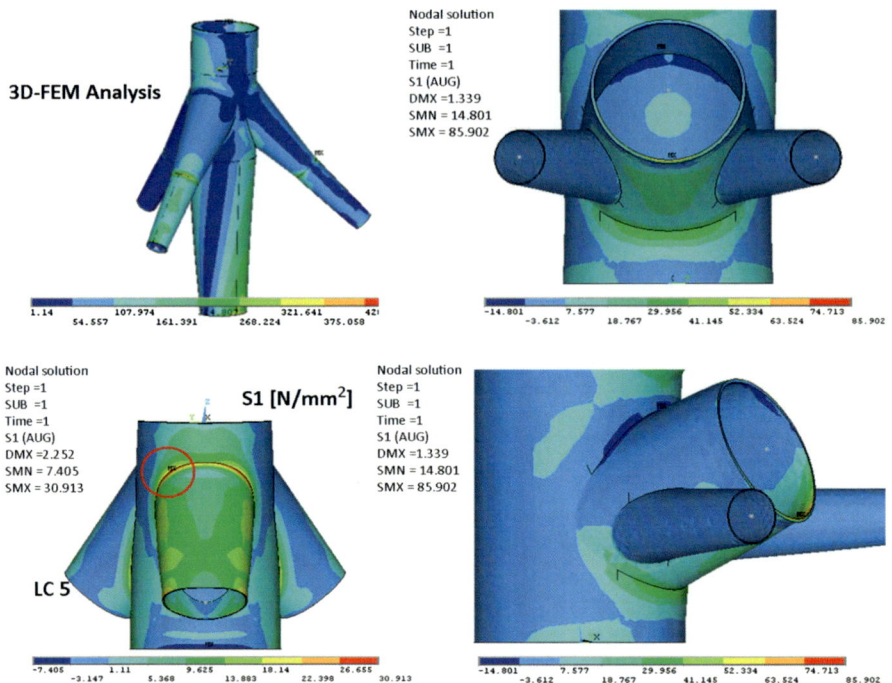

Fig. 11.30 Suggested connection points for the tripod construction including the tensions and load forces. Modified after Buck and Krause 2012

$$m = m_c + m_l + m_a \qquad (11.7)$$

where m_c is the collector mass, m_l the line mass and m_a the additional mass. The buoyancy was calculated applying

$$F_b = V \cdot \rho \cdot g \qquad (11.8)$$

where V is the volume of the buoy, ρ is the density of seawater and g the gravity constant.

Given the resulting forces from the single collector and the buoy it was possible estimate the total load from the collectors and finally the longline system itself.

11.4.1.7   Conclusion and Outlook

The results of the first parametric study showed good correlation to theoretical results. However, for very small values of h/l the difference between analytical and numerical values increases. This was mainly caused by length variation due to increasing strain for a tightened LL. The strain was not taken into account in the

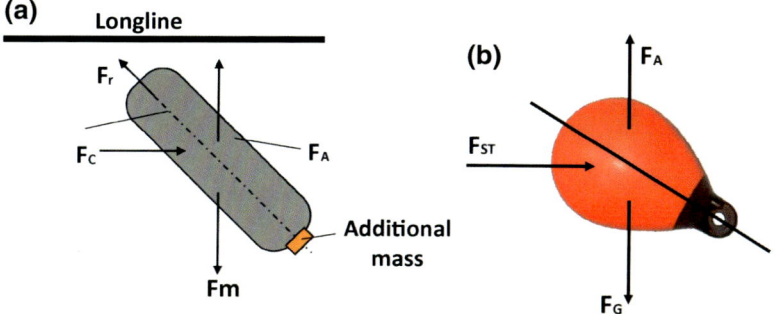

Fig. 11.31 a Model of collector loads for a single collector attached to a longline system; **b** Schematic diagram of a buoy and its forces. Modified after Buck et al. 2006b

analytical calculation. Hence, in the numerical calculation the calculated forces were smaller because the strain in the LL results in larger effective maximal effective sag. The rapprochement to relative force of one for β approaching 30° is suitable to the theoretic values.

The results present a numerical model with different parametric studies in order to assess the loads at the fixation of a longline for a worst case scenario. The model has been verified for the case of a single load point by analytical calculations. The parametric studies give an overview over the influence of the force angle and h/l ratio parameters on the forces in the ends of the longline. This study also shows that in the case of high loads a load reduction could be achieved by an extension of the LL. Thus, large maximal sag leads to load reduction.

To provide detailed information for the additional loading on the support structure of offshore wind energy converters, a detailed examination of the sea states should be taken into account. Factors like wave height and direction as well as current speed and direction will influence the forces. To estimate the occurrence of worst-case scenarios, a dynamic analysis should be performed and would be useful for providing information about inertia forces.

11.4.2   Seaweed Cultivation Devices

In the German North Sea, macroalgal farming of brown algae in offshore environments was launched by the German Government in 1993 and conducted at the marine Station of the *Biologische Anstalt Helgoland* (BAH) off the island of Helgoland, North Sea (see also Sect. 3.1 "Offshore Aquaculture with Extractive Species: Seaweed and Bivalves"). One major component of this study was to develop an appropriate technical device to grow macroalgae. The system had to withstand the harsh environmental conditions of the German North Sea shelf, where maximum wind speeds can be 150–180 km/h and wave amplitudes commonly reach 5–8 m during storms. Investigations were resumed in 2002–2004 aiming at using this new aquaculture technology in conjunction with offshore wind farms (Buck 2004; Buck and Buchholz 2004a).

To get more insight into the cultivation of plants under hostile conditions in multi-use with offshore wind farms, several known carrier designs for algal culture were built and deployed, subsequently resulting in the final modular construction named the "Offshore-Ring" (Fig. 11.32a–l; Fig. 11.33a–p; see also see also Sect. 3.1 "Offshore Aquaculture with Extractive Species: Seaweed and Bivalves"). The performance of the various test designs under offshore conditions, length changes and, where possible, the biomass yield of Sugar kelp Saccharina latissima (Laminaria saccharina) on these constructions were investigated at different locations. The results are of high importance for the future utilisation of exposed offshore locations in combination with offshore platforms, especially when considering multi-user concepts in offshore areas combining wind farm installations (e.g. Buck 2002; Krause et al. 2003; Buck et al. 2003, 2004).

Experimental offshore farming of Sugar kelp was first conducted in 1994 and 1995 at Helgoland (North Sea, Germany) as well as in 2001 and 2002 near the island of Sylt and in the outer estuary of the river Weser (17 nautical miles off the coast of the City of Bremerhaven). The study sites are characterized by various hydrographic features. Peak wind velocities of ≥ 6 Beaufort and ≥ 8 Beaufort were noted down during years of the experimental studies.

Four different cultivation systems were designed and deployed in the study area in order to find the most suitable design for offshore use (Buck and Buchholz 2004a). These included longline (Fig. 11.32e), ladder ("tandem longline") (Fig. 11.32f), grid (Fig. 11.32g) and a ring-shaped design (Figs. 11.32a and 11.33a–p) for attachment of algae seeded culture lines. Each of these different constructions varied in mooring design, floatation and culture units. Concrete blocks of 2.5, 4 and 4.5 t were employed in a single, twin or radial mooring geometry in order to securely moor the carrier constructions. The ladder and grid constructions were oriented parallel to the main direction of the tidal current. Starting from the anchor stones chains with a service load (SL) of at least 8 t were used to connect the concrete with the mooring line (SL 12 t). The service loads corresponded to a threefold collapse load. The mooring line itself held the culture unit, which was designed to float at or 1–1.5 m below the water surface. The floating system consisted of ball-like floats or pencil-like fenders, which were connected by ropes to the culture unit to provide sufficient buoyancy. All connections between ropes, chains, floats and concrete blocks contained triple rings (SL 6 t), shackles (SL 6.5 t), warbles (SL 6.5 t) and thimbles, in case of eyes at rope ends. The longline design consisted of a 50 m long, horizontal carrier rope anchored by a 4 t twin mooring system. It served to fasten culture lines perpendicular to the water surface, each kept straight by a concrete weight (2.5 kg) (Fig. 11.32e). This method had been successfully employed by Kain and Dawes (1987) and Perez et al. (1992).

Between December 1994 and April 1995, a total of 140 × 5 m long culture lines with young Saccharina sporophytes were transferred from the laboratory and

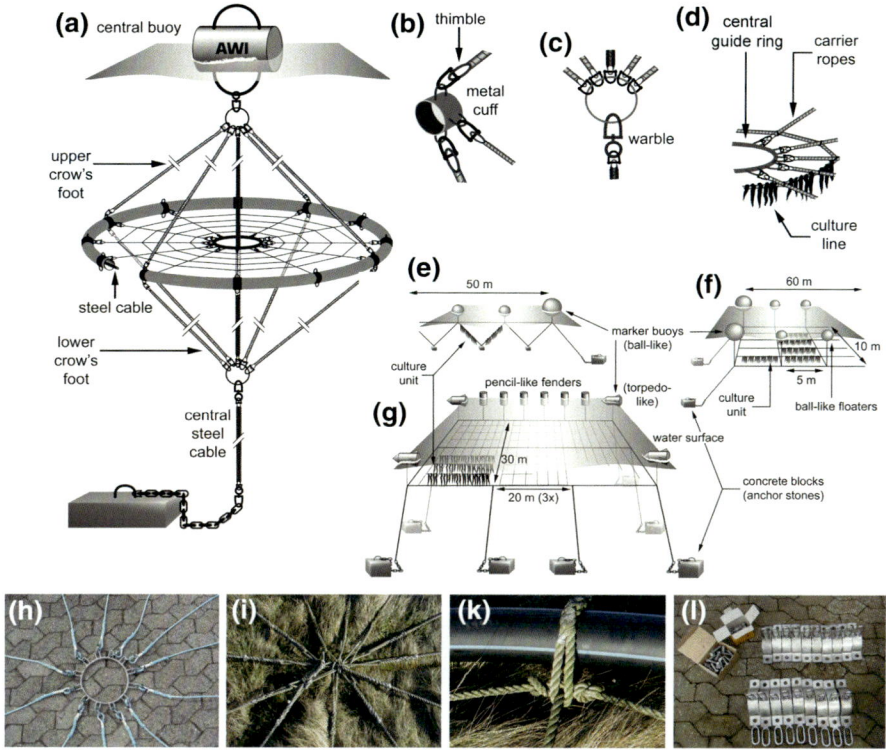

Fig. 11.32 **a–i** Development of system design to culture seaweed and invertebrates in offshore wind farms. **a** The successful ring design for the culture of *Laminaria* at offshore locations; major elements of the system design are magnified in **b** metal cuffs, to which the crow's feet and the ring tube are attached; **c** the transition between central steel cable of the mooring and that of the lower crow's foot, and **d** the central guide ring with attached carrier rope and culture line; **e** longline construction with perpendicular culture unit; **f** ladder construction, with culture lines knotted between the "steps"; **g** grid design with rectangular culture units; **h** central guide ring of the "Offshore-Ring" after modification; **i** central junction point for the carrier ropes in the first "Offshore-Ring" version; **k** interloops at the ring frame to connect the carrier ropes (first version) and **l** metal cuffs to connect carrier ropes (after modification): **a–g** Modified after Buck and Buchholz 2004a, **h–l** AWI/Prof. Dr. Bela H. Buck

knotted at 3 m intervals to the horizontal carrier rope. Later, two adjacent 3 m long culture lines were connected at their lower ends (V-shape). *S. latissima* on these lines were 2–3 mm at transplantation. The ladder design was 60 m × 10 m in size and was positioned horizontally 1 m below the sea surface by 24 concrete weights (each 1 kg under water) and air-filled buoys on the surface (Fig. 11.32f). 110 l ball-shaped buoys at the corners of the ladder construction were meant to keep it afloat. It was suspended between four anchor stones (4.5 t) in a double twin mooring shape. 5 m long culture lines were knotted in between the "steps". Experience from the "ladder" went into the construction of a grid system. The grid

Fig. 11.33 **a–p** First offshore seaweed cultivation project in the mid-1990s. **a** Deployment of concrete blocks as mooring device using an offshore buoy layer; **b** buoy layer with mooring equipment; **c** central part of the first ring construction showing carrier ropes; **d** set-up of mooring devices for the first deployment of the "Offshore-Ring"; **e** first floating tests with ring devices and its potential combination; **f** small ring versions at the pier; **g** deployment of various ring devices in combination; **h** entanglement after a storm event; first "Offshore-Ring" several nautical miles off the Island of Helgoland with young Laminarian sporophytes in a **i** submerged and a **k** floating mode; **l** flow meter for the drag measurements offshore; **m** drying plants after harvest; **n** harvest of the first "Offshore-Ring" in the harbour of Helgoland; **o** preparation of *Saccharina/Laminaria* harvest of the ring formerly located at Helgoland Roads in the harbour of Helgoland. The ring was lifted from the water by a land-based crane; **p** control of the ring construction and length measurements of farmed plants. *Images* AWI/Dr. Cornelia Buchholz

construction had been in use off the Isle of Man (Kain 1991) and in Brittany (Perez et al. 1992). Based on the above and our own experiences, a grid system depicted in Fig. 11.32g was set up. The grid measured 60 m × 30 m and was submerged at a depth of 1.2 m. The grid was designed to hold 1400 m of culture line in an area of 0.18 ha. A radial mooring system was used with 10 concrete blocks (2.5–4.5 t). The frame material employed tube was a "Herkules" rope, which is commonly used in commercial fisheries. This rope contains in its core several subcores, each made of six strands of steel. This way the rope was heavier than the surrounding seawater, which reduced the risk of potential damage at the weight attachment points. The inner supporting ropes of the construction were made of *Polystar*, a mixture of polypropylene and polyethylene, a material with excellent references in steel grades. Four metal torpedoes served as buoyancy devices at the corners, another 72 floats were pencil-like fenders with 23 kg buoyancy each. The final ring device (Patent No. PCT/DE2005/000234; Buck and Buchholz 2004b) had a total diameter of 5 m and consisted of polyethylene tubes with a 10 mm thick wall and a diameter of 110 mm that were welded to rings (Figs. 11.32a–d, h–l and 11.33a–p). The rings were weighed down by a steel cable (30 mm in diameter) inserted into the tube and obtained their buoyancy through eight elongated fenders (23 kg buoyancy each). They consequently floated at a depth of 1.2–1.5 m. Carrier ropes were suspended radially and 80 m of culture line could be fastened like cobwebs on each ring. A crow's foot was used to fasten the ring on a common mooring system. Due to permanent chafing of the carrier ropes with the fender ropes and because the fenders themselves got entangled with each other a modified system was developed. This consisted of one centre buoy (300 kg buoyancy) with a connected reverse crow's foot and a centre guide ring to prevent chafing of the mooring line with the carrier ropes (Fig. 11.32a, d). Furthermore, all radial splices, which connected the carrier ropes to the polyethylene tubes, were replaced with metal cuffs (Fig. 11.32b). Three loops were welded to these cuffs, one to the centre to fix the carrier ropes and the other two to the bottom and the top of the cuff, to connect both crow's feet. An important feature common to all constructions was their ability to adjust the depth of culture lines to 1–1.5 m as this appeared to prevent possible PAR and UV damage to the young algae while providing enough light for successful photosynthesis. Moreover, the most turbulent upper meter of the water column could thus be avoided.

In the following, an overview of technical experience with the different culture devices in the offshore environments of the German North Sea is given. With the longline design, only 65 of the 140 culture lines, which were fastened on the longline system, have been retrieved and 20 of these were evaluated for further investigation. Due to very stormy weather the farm had not been visited very often. Every chance during calm weather was used to change the horizontal carrier rope and supply new culture lines. The study, however, revealed that the weights on the culture lines were not sufficiently heavy so that they were frequently tossed across the carrier line resulting in the removal of the young plants by friction and causing them to become entangled. Other culture lines, consisting of three twisted strands, were untwisted by the current and turbulences and consequently the individual strands were torn. Some improved performance of the longline was obtained by connecting pairs of only 3 m long culture lines at the lower end like a V-shape. Length data used here were taken from these lines. The ladder construction revealed problems pertaining to the durability of the frame material, the attached weights being potential breaking points. Moreover, the 110 L buoys at the corners were very instable and had to be exchanged several times. These drawbacks were taken into account during the development of the grid system. The grid system proved much more stable compared to the "ladder" even though the mooring ropes could not be adjusted to their optimal length to accommodate the full tidal differences. The use of elongated fenders instead of ball floats protected the construction by better riding the swell, which resulted in continuous vertical movement, while the balls used previously had resulted in jerking behaviour that created substantially more stress on all the materials. Culture lines were knotted into the grid from a small rowing boat, a procedure that needed smooth sea and calm weather and could only be managed during the period of slack tide (maximum 30 min). 880 m of culture line were transferred to the grid and later harvested. Finally, the "Offshore-Ring" was the best design. So far, individual rings of 5 m diameter showed a superior performance in comparison to the other tested carrier constructions. They remained stable and in place during all weather conditions, provided their moorings were tended regularly, at least after storms, which imposed some wear on them. In addition, they allowed equipment with culture lines to be performed onshore, the rings subsequently being towed to their mooring locations and fastened relatively quickly during slack tide. With the ring construction the harvesting period could be prolonged by moving complete rings onshore. Moreover, sampling of the seaweed culture was more easily done due to the possibility of heaving up the ring construction with a ship's crane.

To conclude, our experience is that the major key conditions for offshore culture were fulfilled such as the pre-cultivation of healthy plants that were well attached to the culture lines. Another key factor, i.e., reduction of mechanical abrasion, was a major problem on the longline system, because of high turbulences. Longline systems are hence considered unsuitable for macroalgal culture under offshore North Sea conditions. The ladder system was more apt to damage than the improved grid system, e.g., at the fastening points of weights, and should therefore also be rejected in future considerations. A further problem of all carrier constructions except the rings was the necessity to fix them at permanent offshore sites. This led to the logistic and cost problems of efficient transfer of sporelings from the laboratory (or hatchery facility) to the grow-out location as well as appropriate tending of the carrier system under the prevailing rough weather conditions. Labour requirements were also enormous. Every single culture line had to be fastened to the carrier system from a small rowing boat, and this was only suitably done during slack tide. Work was seriously impaired by the difficulty of getting ship time and divers, while also waiting for calm seas and all of this at the 30 min of slack tide and during working hours. The ring construction using its present dimensions has proven stable in offshore conditions (*Helgoland Farm*, *Helgoland Roads*, and *Roter Sand*). The new ring construction, with a central steel cable and central buoy, reduced tractive power and tension in high velocity currents and when being moved for sampling or harvest (Fig. 11.32a–d, h–l). The two crow's feet with the metal cuffs greatly prevented torsion of the ring when lifted. The depth of a ring could be adjusted by insertion of steel ropes into its cavity and the buoyancy of the central buoy could also be adjusted by changing its size. This way the ring could be kept at an appropriate depth to avoid exposure to stressful surface turbulence and admit sufficient light for algal photosynthesis even with increasing weight of algae.

A major advantage of the ring system compared to the other systems was that the ring could be equipped onshore with 80–100 m of culture line and subsequently towed to the mooring site, where it could easily be moored by the ship's crew. The reverse took place at harvest time and was also most advantageous. The ring diameter of 5 m could be managed by cranes from relatively small vessels. This way the algae on the ring construction could be examined at most dates and at harvest while in the worst case it was at least possible to tow the ring into the harbour, where a larger crane could lift it onshore. The described characteristics and the modular nature of the ring construction promise to make it a sensible and effective choice to be used in aquaculture situations where offshore wind farms are located (Buck 2002; Krause et al. 2003; Buck et al. 2003, 2004; Buck and Buchholz 2004a). Moreover, one could transfer the technique to less developed countries using suitable materials, e.g. bamboo or rattan, and the craftsmanship of local people. However, lifetime and stability of these systems using materials in these countries will have to be tested before large-scale employment.

Concerning the most favourable location for aquaculture of macroalgae, our experiments suggest that fairly exposed sites with rough conditions are suitable, however, only if the carrying support structure is sufficiently rigid to withstand the rough to extreme conditions encountered in most of the trials. Aquaculture in sheltered waters must avoid shallow areas, like in the Sylt backwaters, because of possible contact with the seabed and the high siltation and suspended solid load which creates low light conditions. Any location selected for seaweed culture should have a minimum depth of 5–8 m. Offshore areas, such as "Roter Sand" and *Helgoland Roads*, seem to be well-suited for future commercial-scale seaweed culture.

11.4.3 Fish Cages in Multi-use with Offshore Wind Turbine Foundations

11.4.3.1 Integrating a Fish Cage into a Tripod Foundation

So far, no findings exist on the question how fish cages interact with offshore wind energy converters and how large additional forces could grow in the presence of waves (Goseberg et al. 2012). To investigate the potential multi-use of offshore wind turbine foundations with offshore submersible fish cages the offshore wind farm "*Veja Mate*" was taken as a case study site. Next to different foundation types, a triple foundation of "Bard Engineering" was used to erect the 5 MW turbines and towers (Fig. 11.34a–g). The advantage of the triple foundation in direct comparison to other foundation types is that due to its three-monopile character with a rigid platform on top it creates an open space under the platform above the water surface down to the seabed marked-off by the three foundation piles (Fig. 11.34d–g). The initial idea was to design different cages types to be installed exactly in this open space allowing submerging it towards the seafloor while also enabling it to be in a floating mode during maintenance and harvest procedures (Figs. 11.35a–d and 11.38a, b, e, f).

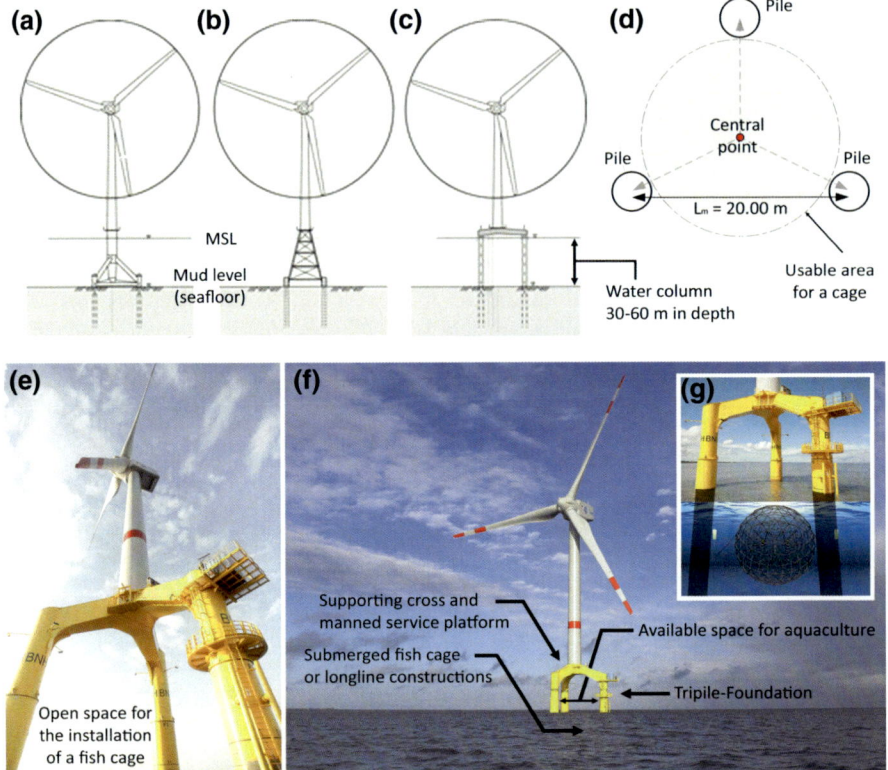

Fig. 11.34 **a–e** Support structures (foundations of offshore wind energy installations) and tripile constructions for the co-use for fish cages: **a** tripod; **b** jacket; **c** tripile; **d** usable horizontal section of the area under the tripile foundation; **e** shows the open space within a tripile foundation to be used for aquaculture purposes; **f** displays a lateral view of the "BARD-Wind-Turbine" and the access to the fish cage; **g** displays a photo animation and gives an idea how a fish farm, such as an "Aquapod", could be moored below. Modified after Buck and Krause 2012; Buck et al. 2012. *Images* (**a–d**) Dr. Nils Goseberg, (G) OFT 2010

In close cooperation with partners from wind industry, fishery and science it was investigated to which extend multi-use synergies among different stakeholders (wind, aquaculture) can be realized at a supporting structure of an offshore wind energy installation (Hundt et al. 2011). In this context, the technical design and the integration of cages in the support structure was put forward. Alternatively, if stand-alone and self-supporting structures can be set-up in the open field between wind turbine foundations moored on the seafloor and eventually partly to a foundation.

The investigations concentrated on the following aspects: (1) Fundamental development of a cage basis, (2) pre-design of the suggested fish cage(s) construction, (3) investigations on the designed cages in a current flume and a wave tank, and (4) design reiteration of the pre-designed cages following the current

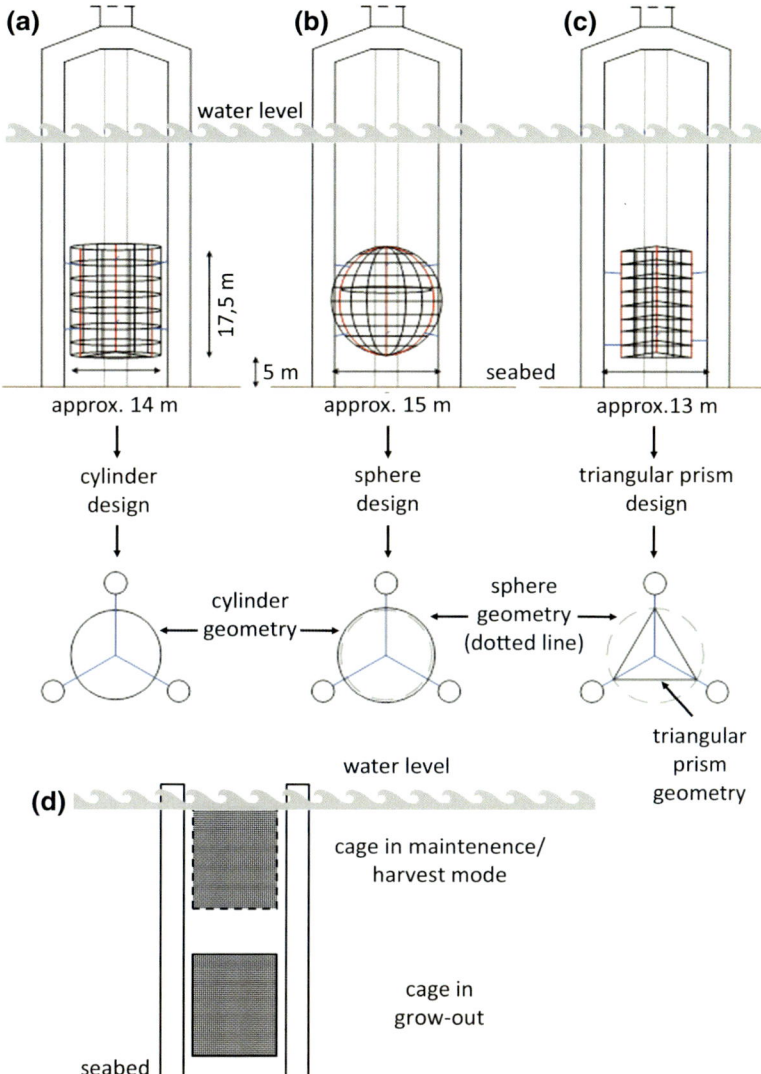

Fig. 11.35 a–b Support structures (tripile foundations of offshore wind energy installations), the integration of different cage designs and its geometric views. **a** Cylinder cage design; **b** sphere cage design; **c** triangular prism cage design; **d** operation modes (floating, submerged) of the cage in the tripile. Modified after Buck et al. (2012). *Image* **a–c** Jan Dubois

flume and wave tank results to optimise the final cage design. However, the evaluation of the potential impact of the additional loads due to the attached cage structures on the dimensioning of the support structure was not focused on. The basic data of the test site at "*Veja Mate*" are shown in the following table (Table 11.2).

Table 11.2 Site specific data at the case study site "*Veja Mate*"

Site specific data at the offshore wind farm "*Veja Mate*"							
Position	(1) 54° 20′ 30.07″N 05° 49′43.30″E (2) 54° 22′ 36.71′ N 05° 54′ 34.61″E (3) 54° 16′ 29.11″ N 05° 49′38.99″E (4) 54° 16′ 29.44″ N 05° 54′ 37.32″E						
Water depth (m)	39–42						
Significant wave heights divided in three trimester (%) (m)	1	2	3	4	5	6	>6
November–February	17.4	48.3	31.7	15.4	0.8	0.4	<0.1
March–June	36.8	44.4	13.8	4.6	0.3	0.2	<0.1
July–October	39.2	40.7	13.6	4.5	1.5	0.6	<0.1
Extreme events (m)	10.81 m within 50 years						
Current velocity (m s^{-1})	Max. tidal current 0.98						

To proceed with the development of different cage models the area between the three piles of the tripile geometrically usable was taken into account for the maximum size of the cage (Figs. 11.34d and 11.35a–c). Additionally, the cage had to be planned in a size that would fit between the piles to get it in and out for installation, maintenance and repair. Therefore, next to the size of the cage the coupling between cage and piles had to be planned as well to get the maximum dimension of the entire cage construction. The following additional parameters and characteristics were taken into account during the fundamental development phase:

(1) simple cage design with only a few crossbars to ease construction and installation,
(2) the height of the cage depends on the depth of the water to allow fully submergence and also the draught (up to 8 m) of the maintenance vessels when accessing the wind turbine (Figs. 11.35d 11.38a, b, e, f),
(3) the attachment mechanism (e.g. rails) mounted exteriorly of the cage to avoid any interference with the net,
(4) optimizing the coupling design symptoms of fatigue that were not primarily considered as the malfunction of the maximum load on it,
(5) the attachment device should not interfere with the sacrificial anode at the piles,
(6) the cage should be additionally fixed to the support cross and to an anchor at the seafloor,
(7) the coupling of the cage with the pile can be realized via rigid or flexible attachments or independently via an external mooring. The various options were reduced to a rigid attachment to the piles of the offshore wind energy foundation, as flexible couplings via ropes or cables could lead to an entanglement,
(8) to allow a preferably large surface area for flatfish to settle. Therefore, additional plane levels were integrated into the inner parts of the cage, which also supported the bracing of the cage and therefore supported structural stability, even if the conditions of production were more complicated as well as the loads on the entire cage induced through currents and waves,

(9) to identify the overall loads the load forces of the cage structure and the load forces of the entire net harness (including the mesh size) were superposed.

Cylinder design: A cylinder cage design perfectly fits into the inner part of a triple and the connection points from the cage to the piles are adjustable in height. The distance from the cage top to the cage bottom was calculated with 17.5 m. Further, to avoid an additional scouring in the vicinity of the piles at the seafloor a distance of 5 m from the cage bottom to the seafloor in the submerged mode was provided. The resulting cage diameter was calculated being 14 m leading to a cage volume of about 2650 m^3 (Figs. 11.35a and 11.36a–e).

The cylinder design was calculated with and without internal levels. The current regime within the entire foundation-cage construction would be changed

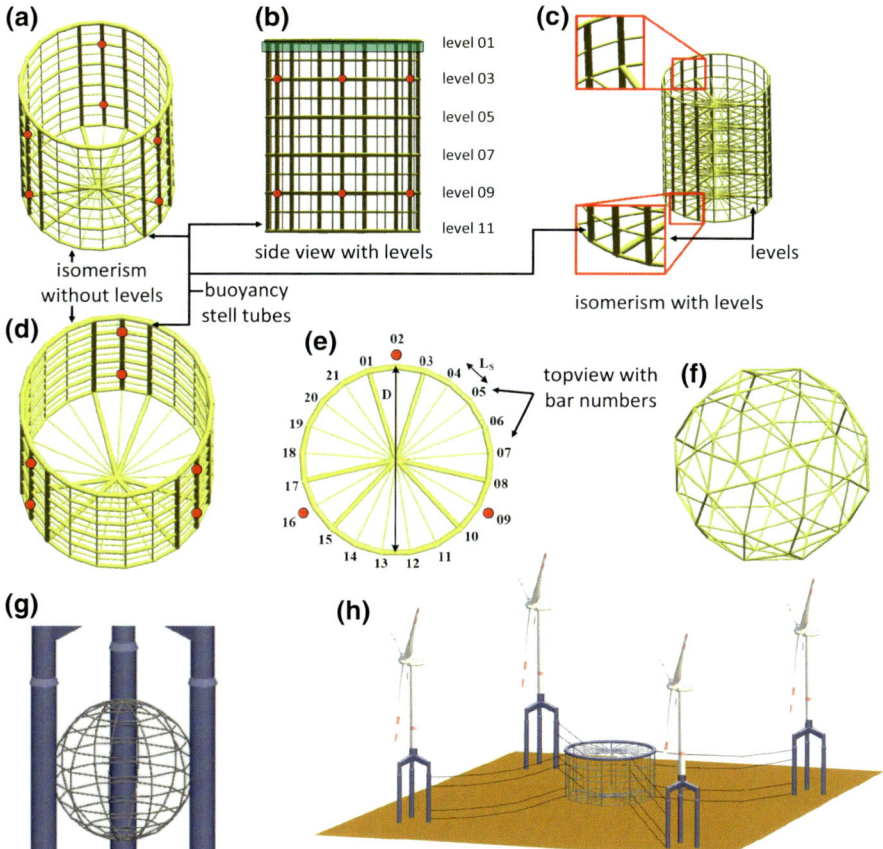

Fig. 11.36   a–h Cage designs integrated into the tripile foundation as well as outside of the tripile. **a–e** Visualisation of cylinder-like cages with and without horizontal levels, support (bearing) points are displayed in *red* and the buoyancy device is shown in *dark green*; **f** cage design with a triangular bar interconnectedness of the spheres' surface; **g** cage design with meridional bar direction; **h** cage concept not integrated into a tripile foundation but installed in the centre of various wind turbines and connected to the foundations. Modified after Buck et al. 2012. *Images* Jan Dubois

significantly which in turn could lead to more stress on the installation and the farmed candidates. The cage structure has additionally vertical steel pipes, which also act as buoyancy device.

Spherical design: The coordination of crossbars and the entire skeleton to design the spherical structure and its attachment device for the foundation piles simply depends on the height of the cage. We planned to connect the cage with the piles at least at six points to absorb axial forces and to avoid the generation of a kinematic positioning. The connections to the support structure were installed in the upper and lower third of the sphere (Figs. 11.35b and 11.36f, g). The cage volume of the sphere is approx. 1,700 m^3.

As levels in the internal space would create wedge-shaped units in the upper and the lower third of the sphere, further developments would not include additional levels. Further, these levels would not explicitly improve the stability of the cage. The cage design with meridional bar direction (Fig. 11.36g) was rejected due to the sharp angles on the top and the bottom of the age as well as due to its poor load-bearing behaviour. Another design of the skeleton of the cage was the use of a triangular bar interconnectedness of the spheres' surface (Fig. 11.36f).

Triangular Prism: The height was calculated to be 17.5 m and an edge length of 13 m leading to a volume of about 1,250 m^3 (Fig. 11.35b). This is a comparably small volume compared to the other cages, however, the advantage of this design is the reduced tensile loads on the horizontal crossbars through the prevailing bearing forces as the three edge cage piles spread the forces over the entire cage height. Another disadvantage of this cage design is the existence of the three edges along the entire cage height. In worst case scenarios with high waves and strong current velocities these edges could function as energy focussing zones leading to the fact that the fish could be pushed against the net, which in turn could lead to stress and interference of the fish.

Tripile cage designs: This cage design relies on the structure of the tripile as the foundation piles also function as the outer piles of the cage holding the net. In comparison to the cylinder cage the height of the tripile cage would be only 15 m as the distance from the cage bottom to the seafloor will be increased by 2.5–7.5 m to avoid any scour effects. The volume of the cage would be 3,000 m^3.

Large-scale cage design: These large cages are designed with a volume of approx. 20,000 m^3 and could be installed in the vicinity of the tripile and connected between 3 and 4 foundations (Fig. 11.36h).

11.4.3.2 Impact of the Integrated Cage on the Support Structure

To integrate a cage into the internal space of tripod foundation it is of the utmost importance to get more insight about the impact of the fish cage on the tripile-dynamics, such as an oscillatory instability of the foundation piles between the water surface and the seabed.

To start the laboratory experiments *BARD Engineering* constructed a model of the tripile, which was identical to the tripile used at the offshore wind farm

"*Veja Mata*". This model was scaled down to a size of 1:40 and was used to integrate the new cage designs. After the first experiments were conducted in the wave flume of the *Ludwig-Franzius-Institute for Hydraulic, Estuarine and Coastal Engineering* we identified that the piles could under extreme conditions tend to local vibrations (Fig. 11.37a). This effect was even intensified on the foundation piles when integrating the cage models (Fig. 11.37b–d). Additionally, we found that rigid constructions would lead to a static fatigue of the connection devices between the cage and the piles due to the constraining force resulting from the differential deformation of the piles of the tripile. A solution would be a more flexible coupling element (e.g. spring bars) as well as a certain clearance between the different installations to avoid that all three piles at the same time transfer the forces through their vibration on the cage (Figs. 11.37e–g and 11.38c, d).

11.4.4 Supporting Devices

11.4.4.1 Offshore Seed Collector and Grow-Out Devices

In order to find suitable offshore locations, where the biological background conditions were suitable for both wind and aquaculture farms, in Project No. 2 "*Open Ocean Aquaculture*" and 3 "Roter Sand" a number of offshore wind farm sites within the German Bight were surveyed in 2002. Subsequently, nine locations 10–40 nautical miles off the German North Sea coast were selected for further investigation (Fig. 11.7a). Site criteria for this survey were the vicinity to a planned wind farm, the distance from the coast, the water depth, the water quality, and the substratum. In January 2003, the selected offshore locations were equipped with test moorings to investigate the potential of offshore seaweed, mussel and oyster farming. For this purpose an offshore seed collector as well as grow out device for extractive species was developed (Fig. 11.39a, b).

The mooring's marker buoy had a buoyancy of 300 kg and was connected to a 2 ton concrete block with a 22 mm steel wire and a heavy buoy chain (Fig. 11.39a). At 3 m below the surface a 1 x 1 m metal frame was fixed to the wire, providing a holding unit for two spat collectors clamped into the frame. The depth was chosen because of multi-annual data (Walter and Liebezeit 2001; Joschko et al. 2008) indicating little settlement in depths < 4 m. Each collector consisted of a polypropylene carrier rope (10 mm) with four inserted transverse elements to enlarge the surface area. The elements were made of 15 cm long pieces of the same polypropylene rope, which were frayed manually in 1100 single fibres to produce a bow-tie-like bundle. This type of collector is equivalent to the type used by Tortell (1976) and Dare et al. (1983) and has proven successful in its ability to attract mussel spat in tens of thousands of individuals per meter (Walter and Liebezeit 2003). Once a month if possible (February, March, April, May, July, September), samples were collected on a 5-day cruise using the research vessels *RV Heincke*, *RV Uthörn*, and *RB Remzy*. All

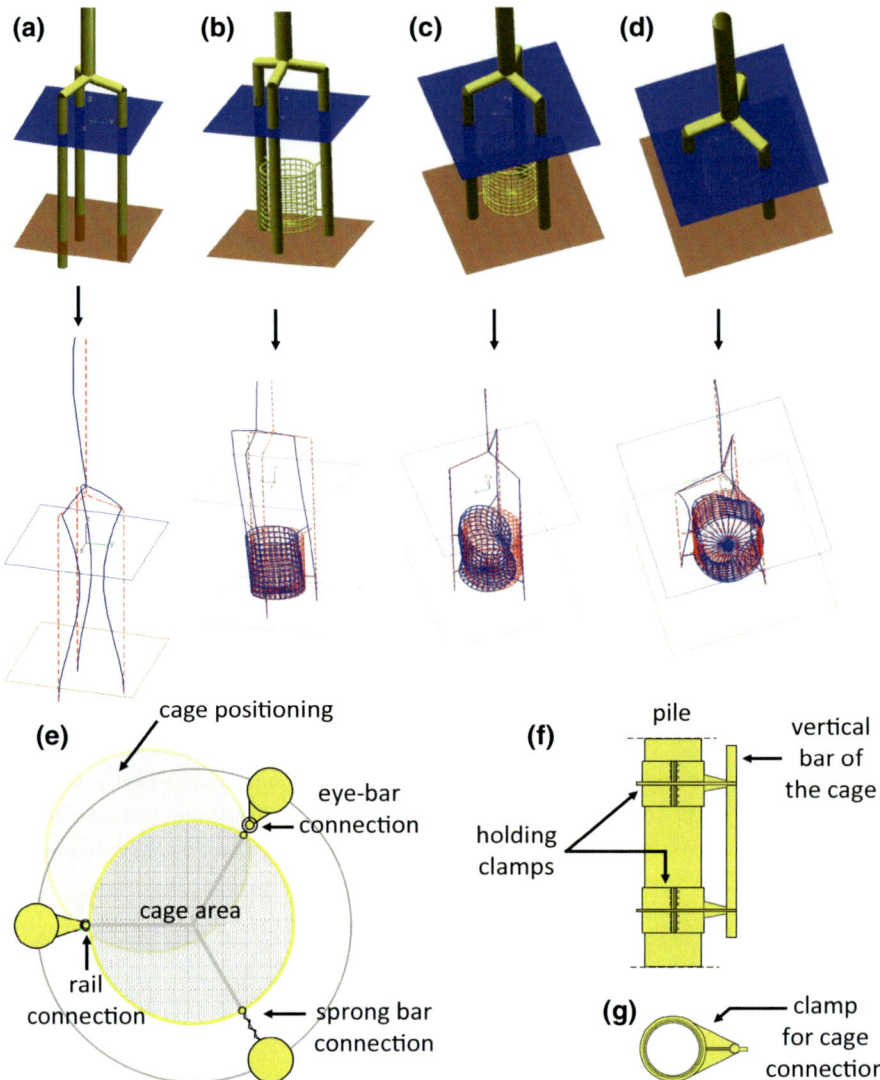

Fig. 11.37 a–g Vibration of the tripile and the connected cylinder design cage and connection points. **a** bending with local vibration (tripile without cage); **b–d** bending with local vibration (tripile with cylinder cage) showing cumulative effects; **e** top view including different connections from cage to tripile; **f** pile with connecting clamps; **g** top view of a clamp. Modified after Buck et al. 2012. *Images* Jan Dubois

moorings were deployed adjacent to proposed wind farms. For safety reasons they were placed at least 1 nautical mile from the wind farm areas.

At a later stage a second version of an offshore test device for mussel seed collection was developed and tested during the Project No. 5 "MytiFit". Figure 11.39c–e shows the new version of the test device.

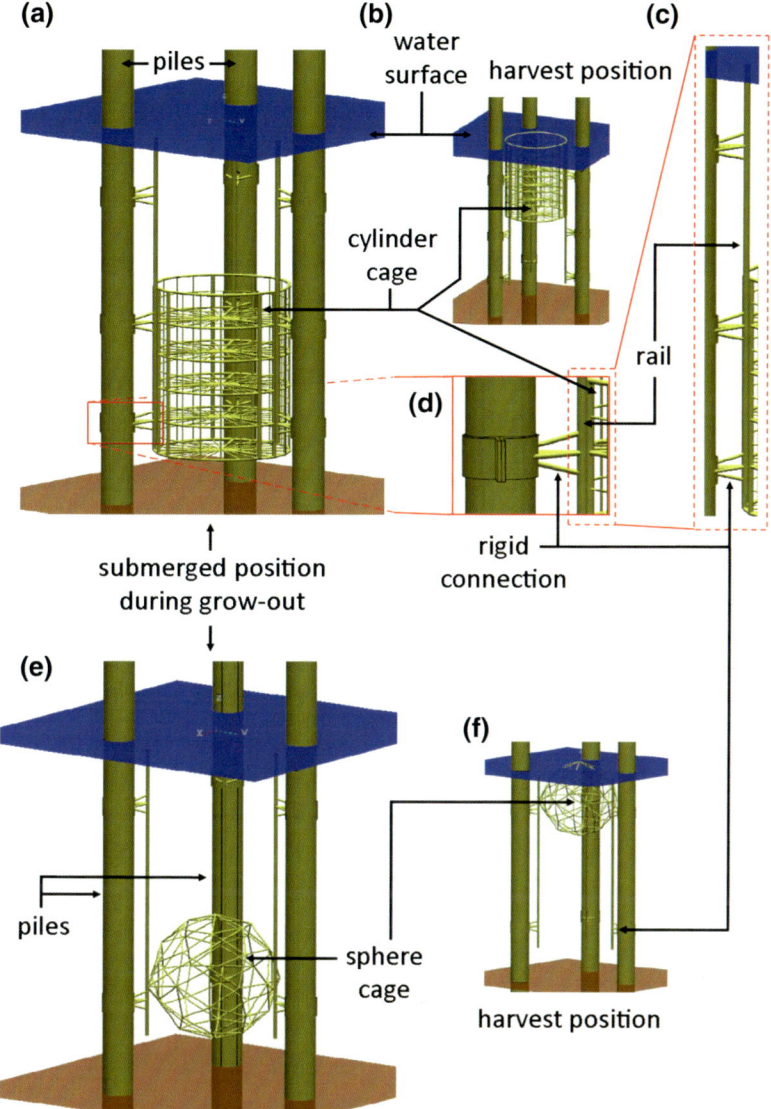

Fig. 11.38 a–f Cage connections and cage positions during maintenance/harvest and grow-out. **a** cylinder cage in grow-out position; **b** cylinder cage in maintenance and harvest position; **c** and **d** enlargements of the cage-rail-pile connection; **e** sphere cage in grow-out position; **f** sphere cage in maintenance and harvest position; Modified after Buck et al. 2012. *Images* Jan Dubois

11.4.4.2 Underwater Inspection Device

The underwater inspection device is a camera system, which helps the farmer to inspect his longline or tube system and being independent in this need to order a

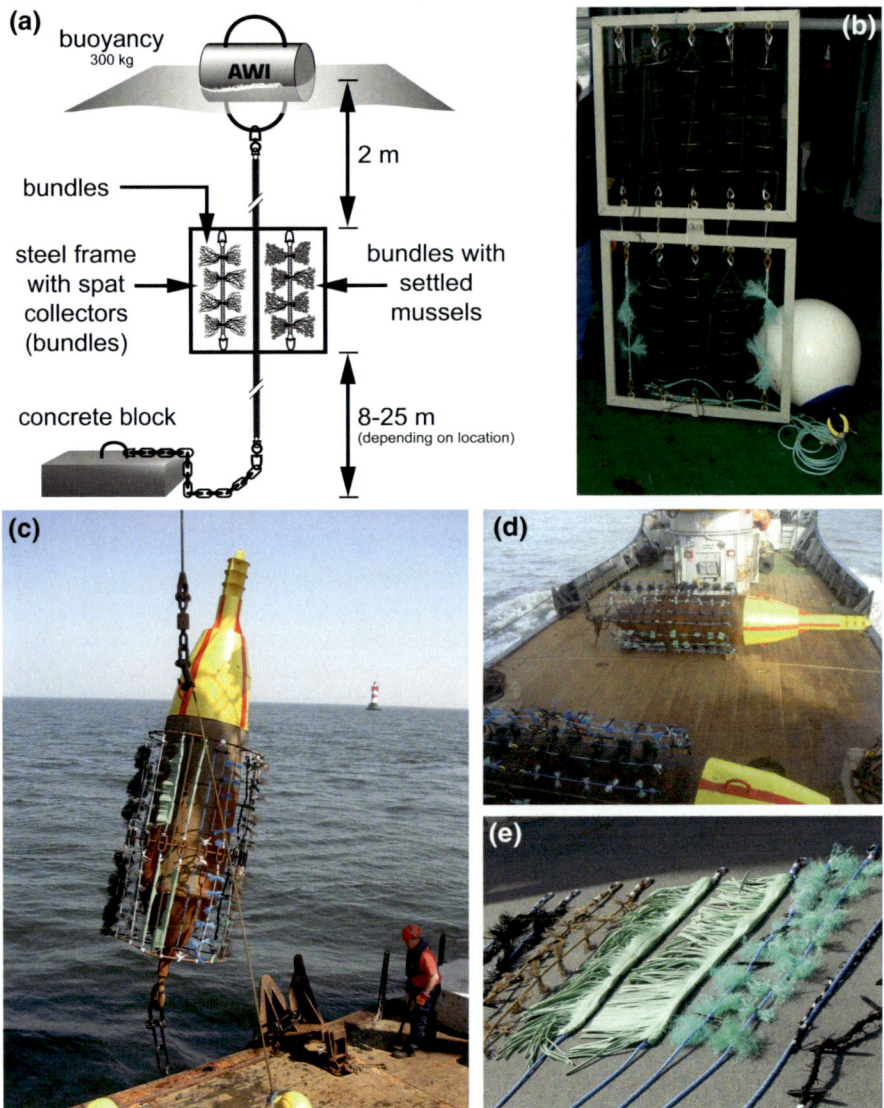

Fig. 11.39 a–e Spat collecting and grow-out devices used for offshore multi-use projects. **a** Offshore spat collector "First Generation" with holding supports for seed collectors and mussel spat (modified after Buck 2017); **b** Same mooring device for offshore oyster tests (*Photo* AWI/Prof. Dr. Bela H. Buck); **c** Offshore spat collector "Second Generation" (modified after Brenner et al. 2007); **d** New collector buoys during transfer at sea (Photo: AWI/Thomas Manefeld); **e** Collector types tested at the offshore wind farm site (Photo: AWI/Thomas Manefeld)

scientific diving team (Buck and Wunsch 2005; Patent: DE 10 2005 020 070). The camera device was developed to allow the farmer to inspect the culture candidate suspended in the water column, while at the same time taking samples

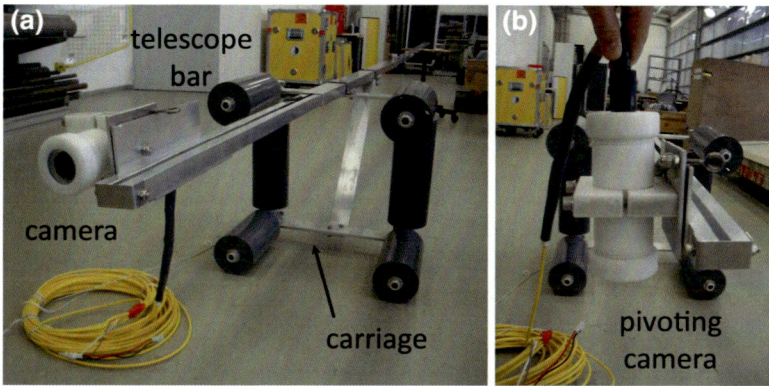

Fig. 11.40 a–b Inspection device for under water constructions including a positioning system. **a** Shows the complete device with the camera and the carriage, which wheels over the offshore mussel longline and tube system for inspection; **b** displays the pivoting equipment for the digital camera (*Photos* AWI/Prof. Dr. Bela H. Buck, modified after Buck and Wunsch 2005; Patent: DE 10 2005 020 070)

(Fig. 11.40a, b). The camera was mounted to a pivoting arm to enable 360° view and was adjustable to the depth of the backbone of the longline. The complete inspection system was used to wheel over the longline or to move a small carriage adapted to the width of a long tube. The development of the device was conducted during the Project No. 3 "Roter Sand" and was modified in Project No. 6 "AquaLast".

11.4.4.3   Automated *Saccharina* Seeding Tank

To ease the cultivation of *Saccharina* during lab-phase and to allow a several 100 m long seeding rope as substrate for the sporophyte various "curtain"-systems were developed (see also Fig. 11.3c). Here, a tank with rotating drums was set-up to ease handling, allow a constant seeding while at the same time enable the use of long ropes, thus avoiding the interconnecting of several short pieces of substrate (Fig. 11.41a, b). However, the main purpose of this system is that the plants experience a certain current due to the rotation and therefore adapt to a minimum current. The advantage is that due to this current, which can be increased by accelerating the rotation velocity of the drums, leading to an adaptation of the holdfast of the plants to more hostile environments. When transferring the plants at sea in high energy environments plants will not detach from its substrate (Buck and Buchholz 2005).

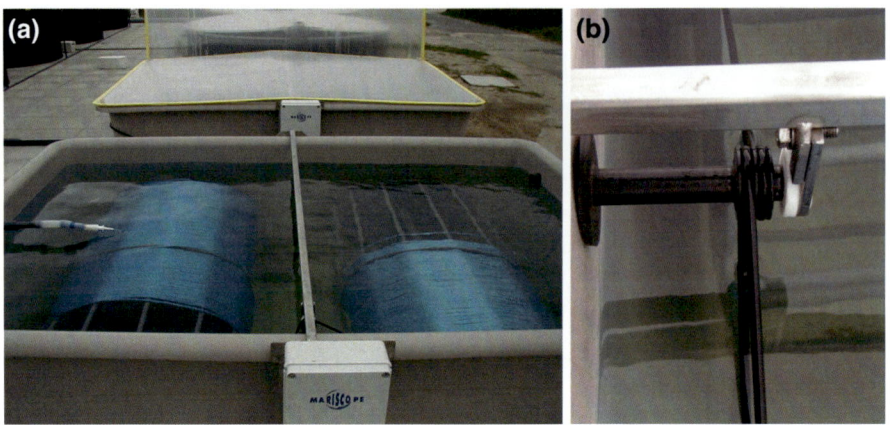

Fig. 11.41 **a–b** Automated seeding and on-growing device for *Laminaria sp.* culture. **a** Tank with rotating drums equipped with the seeding line. **b** Transmission shaft for the rotation with a power unit aside of the tank (*Photos* AWI/Prof. Dr. Bela H. Buck)

11.5 Stakeholder Attitudes, Perceptions and Concerns

While early studies mostly focussed on technological feasibility, environmental impacts, and profitability, the perceptions of stakeholders in the offshore realm have experienced increasing attention in the North Sea context. A number of studies have collected and analysed stakeholder attitudes, their interests and concerns over the past decade (Michler-Cieluch et al. 2009a, b; Michler-Cieluch and Kodeih 2008; Michler-Cieluch and Krause 2008; Vollstedt 2011; Wever et al. 2015; Michler-Cieluch 2009) (Projects No. 4 "Coastal Futures", Project No. 13 "Open Ocean Multi-Use"). Many of the stakeholders from the fishery and offshore wind industries, public administration, environmental groups and the research community have been part of the ongoing offshore research process since its very beginnings and still remain to date. Table 11.3 gives an overview of the most relevant actor groups in this context.

In the early process of identifying key stakeholders in the offshore realm, it became apparent that the types of actors involved in, or affected by offshore ventures clearly differ from stakeholders in coastal areas, and so do their interests and concerns with respect to offshore uses (Krause et al. 2003; Michler-Cieluch 2009).

Coastal areas have a long history of a great variety of uses, so user patterns and stakeholder networks have naturally grown over a long time period. Offshore areas to the contrary have only recently experienced intense utilisation due to technological advancements. This resulted in conflicts, as new types of use emerged or intensified, while other uses are marginalized or even banned. The offshore wind energy sector has turned into a powerful, international player that has benefitted immensely from the current political agenda in Germany. Other users, such as fisheries, are pushed out of large areas of the ocean. Within this highly contested

Table 11.3 Overview of stakeholder groups and corresponding institutions (modified after Vollstedt 2011)

Stakeholder group	Detailed description of stakeholder group
Fisheries	• Fisheries association • Fishing companies • Mussel and aquaculture farms and producer
Public administration	• Regional and national public authorities concerned with nature conservation, agriculture, fisheries, renewable energy and environmental protection • Water and shipping directorates • Public authorities involved in maritime Spatial planning, ICZM and authorization of offshore wind farms
Offshore wind energy	• Regional and national wind farm companies concerned with planning, construction, maintenance and service of offshore wind farms • Wind farm associations
Environmental organisations	• Regional, national and international organizations engaged with nature conservation, protection of the marine environment and environmental protection
Fish industry	• Producer of frozen fish products for consumption • Associations concerning fish processing, fish wholesale and fish importers
Marine technique	• Companies and institutes involved in planning and construction of maritime technology
Promotion of economic development	• Regional companies involved in business development and project investment
Research institutes	• Research institutes related to aquaculture and fishery

socio-political landscape, it is of overriding importance to understand the mind-sets of the actors involved, appreciate potential (beneficiary as well as harmful) impacts of their activities and interests, and identify, if possible, "win-win" solutions.

Recent studies point to a generally supportive attitude of the majority of stakeholders towards spatial and/or operational integration of marine aquaculture and offshore wind energy (Wever et al. 2015; Michler-Cieluch and Kodeih 2008; Vollstedt 2011). Many stakeholders that were consulted believe that the combination of a limited number of sustainable marine uses—such as offshore wind energy and fish farming - appears as an attractive solution to increasing, and competing demands for limited ocean space. Figure 11.42 displays the general acceptance of co-use of different interest groups as compiled by Michler-Cieluch and Kodeih (2008).

However, when it comes to the details of a hypothetical co-management scenario highly controversial attitudes, perceptions, concerns, and interests surface. Of overriding concern to many of the stakeholders are potentially harmful impacts of offshore aquaculture systems to the marine environment. In particular environmental agencies and organisations, but also researchers from a range of disciplines, are highly concerned especially about the impacts of nutrients on the benthic

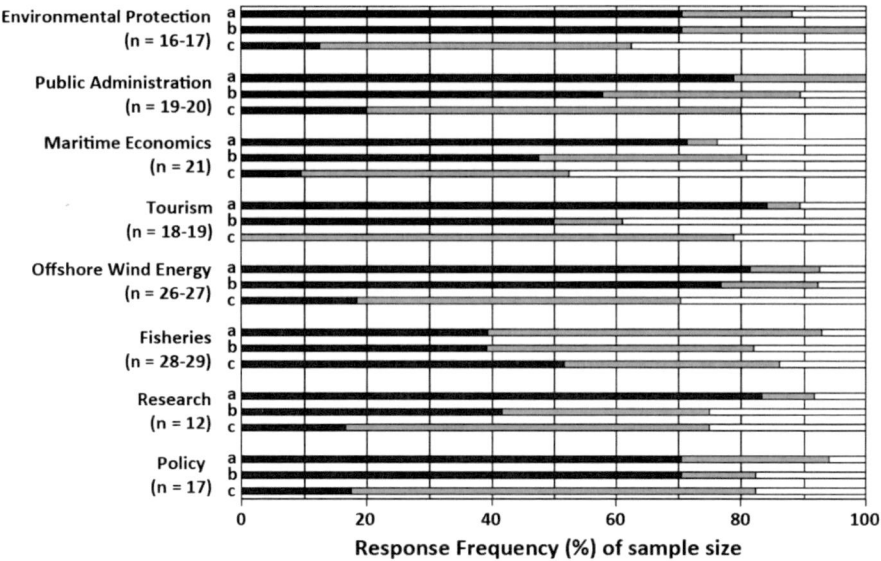

Fig. 11.42 Responses of various stakeholders to the statements: "A concurrent cultivation of mussels and seaweed in offshore wind farms is...: (**a**) "... a reasonable co-use of those marine areas occupied by wind farms." (Cramer's V = 0.33); **b** "... reduces conflicts between interest groups due to the twofold use of the same marine area." (Cramer's V = 0.28); **c** "... negatively affects security in offshore wind farms." (Cramer's V = 0.32). *n* sample size; *black bars* agree, *grey bars* disagree, *white bars* do not know. Modified after Michler & Kodeih (2008)

environment and pelagic fauna. The need for locale-specific nutrient budgeting and carrying capacity analyses, and further research into the possible role of IMTA systems were clearly articulated throughout the process (Wever et al. 2015). Other concerns relate to transmission of diseases and genetic change of wild populations by fish escapes (Vollstedt 2011). Environmental impacts of open offshore systems clearly need to be evaluated and appropriately quantified as part of the site selection process (Buck et al. 2012; Pogoda et al. 2015). However, while the specific environmental concerns may be alleviated as more precise, on the ground information on impacts becomes available, to some stakeholders the mere accumulation of ocean uses is a threat from the environmental protection point of view. In fact, while offshore wind farms themselves are subject to environmental concerns (such as increased noise levels, risk of collisions, changes to benthic and pelagic habitats, alterations to food webs, and pollution from increased vessel traffic, see e.g. Bailey et al. 2014 and Köller et al. 2006), some environmentalists believe in beneficial effects of single-use wind farms, in which fishing is prohibited, such as reduced pressure on fish stocks and recuperation of the benthic environment. Additional uses such as aquaculture would rescind such effects (Wever et al. 2015).

Turning to the potentially most involved actor groups—the offshore wind farm developers and operators, and the fisheries sector—a critical attitude is prevailing. Both sectors are highly sceptical with respect to the economic, technological,

Table 11.4 Compilation of the strengths, weaknesses, opportunities, and threats (SWOT-analysis), which identify and assemble internal and external favouring or inhibitory factors (potentialities and restrictions) of interrelated O&M *activities of offshore wind farms and mariculture installations* (modified after Michler-Cieluch et al. 2009a)

	Potentialities	Restrictions
Internal	*Strength*	*Weaknesses*
	• Development of a flexible, collective transportation scheme	• Little to no interest in joint planning process
	• Sharing of high-priced facilities	• Little willingness to engage into new fields of activity
	• Rationalization of operating processes	• Ambiguous assignment of rights and duties
	• Shortening of adaptive learning process for any offshore works by making use of available experience and knowledge	• Problems of interfering operations
		• Lack of motivating force due to doubtful mutual cost benefit
External	*Opportunities*	*Threats*
	• Available working days coincide	• Unfavourable accessibility of wind farm location inhibits joint O&M
	• Transportation and lifting devices are indispensable	• Lack of regulatory framework supporting co-management arrangements
	• Availability of a wide range of expertise (hard and soft skills)	• No access rights within wind farm area for second party
	• Lack of legislation in EEZ favours implementation of innovative concepts	• Unsolvable problems of liability
		• Dissimilar lease tenures

operational and biological feasibility of such an endeavour (Wever et al. 2015; Vollstedt 2011). Michler-Cieluch et al. (2009a) identified a number of potential synergies and benefits relating to functional aspects such as joint use of transportation infrastructure, and organizational features such as the prospect to combine offshore working pattern, both of which hold the potential to reduce costs (e.g. for security systems or by vessel sharing) for both participating parties. Table 11.4 compiles potentials and restrictions related to interrelated operation and maintenance activities of offshore wind farms and aquaculture installations. However, considering the variety of technological designs in offshore wind energy installations currently on the market, and the range of potential, yet to be developed technical solutions for offshore aquaculture installations as described in the Chap. 4 "Technologies", a reliable estimation of financial benefits and overall economic viability needs to be done on a case-to-case basis. Even though the offshore wind energy sector is a highly innovative and dynamic sector that has been described as willing to take risks (Byzio et al. 2005), at this point in time the technical, legal, actuarial and operational concerns still appear to prevail (Michler-Cieluch and Kodeih 2008; Michler-Cieluch and Krause 2008).

The fisheries sector from the outset faces a different situation: while the offshore wind energy sector is privileged with stable and outspoken political commitments and to-date exclusive user rights for vast ocean areas, the fisheries sector is losing ground. This development however has not resulted in any noticeable efforts of the fisheries sector to diversify into marine farming so far. In fact, local fishermen appear rather reluctant to the idea of engaging with marine farming (Wever et al. 2015; Michler-Cieluch 2009). Neither do they possess the specific knowledge of marine farming, nor the sufficient investment capital, nor do they seem to be willing to move away from their original profession. By some fisheries representatives, the efforts to engage fishermen into alternative occupations are perceived as lip service to avert from the fact that the expansion of offshore wind energy goes hand in hand with the closing of fishing grounds, thus threatening the very existence of the fisheries sector.

Earlier stakeholder efforts focused on the potential role of nearshore aquaculturists, in particular mussel farmers, in offshore mariculture operations (Michler-Cieluch and Kodeih 2008). Similarly to the fishermen, the local mussel farmers also generally appeared reluctant to the idea of expanding their businesses into offshore waters. Here again the disbeliefs in economic efficiency and technical feasibility, and a generally low readiness to assume risks were identified as the main causes for a sceptical attitude towards offshore mariculture. Moreover, offshore mariculture was found to be perceived as "intruder" that could displace traditional ways of fishing from the Wadden Sea. The fisheries sector thus felt as a "two-fold loser" that is threatened not only by the loss of fishing grounds due to the expansion of offshore wind farms, but also the loss of their occupational identity if forced into offshore farming in order to secure income. Table 11.5 displays selected supportive, as well as opposing statements from fisheries and offshore wind energy representatives.

These results seem surprising, as both fisheries and nearshore aquaculture sectors possess developable equipment as well as valuable skills and knowledge to work in open waters, and both sectors, for different reasons, are confronted with the need to extend and diversify their businesses. They reveal the Achilles heel of socioeconomic studies so far: the lack of a clearly identified target group. It is not yet clear who would actually be willing and capable to develop and operate a marine aquaculture facility in offshore waters. In fact, due to the remoteness of offshore installations, operations would have to rely heavily on automated processes. Stakeholders have repeatedly raised their worries that the technology would be attractive only to large, possibly foreign investors and would hardly generate any income or employment effects to the region (Wever et al. 2015). It is also not clear under what kind of arrangement the two very heterogeneous actor groups (wind farm operators and aquaculture operators) with substantially different interests and concerns could be brought together in a co-management scheme. Financial and operational benefits from sharing a common facility and space may only be attained when a high degree of cooperation between the co-users is achieved, not only

Table 11.5 Statements made by representatives of the offshore wind farm and fisheries actor group towards potential 'Wind Farm-Mariculture Integration' (modified after Michler-Cieluch and Kodeih 2008)

Actor Group	Statements
Fisheries	
Statements in support	• "It is no alternative for the mussel fishery sector but an additional possibility. Synergetic effects are always desirable" • "Could open up interesting possibilities and have advantages for both wind farmers and fishermen"
Critical statements	• "There are no reliable original data" • "Loss of fishing grounds remains unchanged for the fisheries sector. Mariculture of this kind is not economical and no alternative to sea fishing" • "Too little mussel spat available in the region, I think it will stay a fantasy" • "It only serves to show wind farms in a better light and to enhance public acceptance"
Offshore wind energy	
Statements in support	• "New innovative idea. For sure there are many open questions to be resolved such as technical feasibility or cross-linking of branches" • "Additional uses are welcome, also for the purpose of increasing acceptance among fishermen" • "Maximization of the economic value of ocean territory from an ecological point of view"
Critical statements	• "I don't see any added value to the manufacturers/suppliers of wind turbines or to the operators. The use of offshore wind energy has nothing to do with mariculture" • "Attaching [mariculture devices] to turbine foundations could be a problem and requires third party certification" • "The problem will be to make sure that operation and maintenance of the wind turbines is not impeded"

during operation and maintenance, but already in the early stages of technological development. Lack of social ties and common identity make collective action more difficult to organize. Appropriate institutional design of co-managed facilities must compensate for the absence of tight social networks and long traditions of collaboration (Michler-Cieluch 2009). Figure 11.43 displays some of the key framework requirements for ensuring sound cooperative work between the two actors. Further research is clearly needed to identify the target group and their specific requirements for operation, and infer common ground and modes of operation between the co-users.

The methodology used to obtain the presented information shown is thoroughly described in Michler-Cieluch and Kodeih (2008).

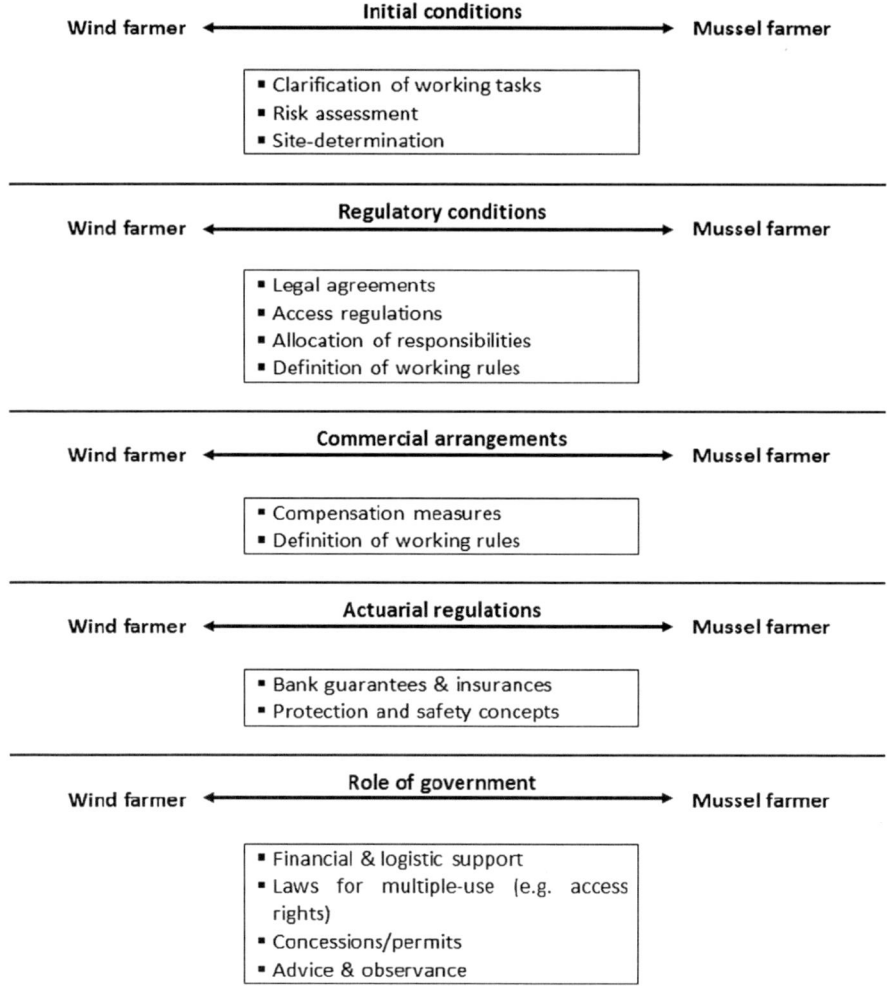

Fig. 11.43 Framework requirements for managing 'wind farm–mariculture integration' (modified after Michler-Cieluch and Krause 2008)

11.6 Economic Calculations

For the last 25 years, total mussel production in the German North Sea varied from 5000 to 50,000 annual tons. As these mussels were produced by using the on bottom-culture technique, the farmer depends on the bioavailability of seed mussels obtained from wild beds in the coastal sea. Walter and Liebezeit (2001, 2003) commenced research into whether or not suspended culture techniques could be used to obtain seed mussels in a nearshore areas of the German Bight. They found that spat can be obtained even in years with a low spat fall when using the floating longline technique, commonly described as off-bottom culture (Hickman 1992). Due to conservation measures applied to nearly 98% of the German North Sea coast, the development and expansion of the mussel aquaculture sector is limited insofar as the current area of mussel culture plots will not be enlarged (CWSS 2002). As already described previously in this book, moving off coastal areas to the open ocean, where offshore wind farms are planned, is a potential solution (Buck 2002, 2007). Advantages of performing mussel cultivation activities within offshore wind farm territories are manifold, such as the placement of mariculture devices in defined corridors between wind farm turbines as well as sharing infrastructure for regular servicing and joined multi-use sources of transportation. This provides an opportunity for both enterprises to share these high-priced facilities (Michler-Cieluch et al. 2009b). Further, charter contracts for specially designed mussel harvesting vessels could be aimed as a solution for transporting wind farm technicians to the offshore location at times of planned, preventive operation and maintenance activities (Michler-Cieluch et al. 2009b). Altogether, the viability of a mussel cultivation enterprise within offshore wind farming areas depends on various factors such as (1) the technological and biological feasibility, (2) the legislative and regulatory constraints, (3) the environmental sustainability of farming aquatic organisms, and (4) the profitability of this potential commercial operation (see review by Buck et al. 2008b). As for some offshore aquaculture projects on pilot scale, the economic part is often unrepresented or even ignored. Therefore, this sub-chapter, which is the summary of the Project No. 8 "MytiMoney" (Fig. 11.1), provides a first insight into financial considerations associated with moving mussel cultivation close to German offshore wind farms, aiming to demonstrate the commercial potential from an economic perspective of a new enterprise that has not yet become established even on a pilot scale (Fig. 11.44a).

In the following the most relevant parameters that have an impact on potential commercial exploitation, an investment appraisal, an enterprise budget analysis, a break-even analysis, and a sensitivity analysis of various scenarios to evaluate economic profitability of mussel cultivation offshore are outlined.

11.6.1   Basic Data

The offshore wind farm, which acts as a case study site, is called *Nordergründe* (see Fig. 11.8) and is located close to the offshore lighthouse "Roter Sand". Site-specific data are shown in Table 11.6 and Fig. 11.44a–c.

The economic analysis consists of an investment appraisal by calculating the following: (1) net present values (NPV), (2) the internal rate of return (IRR), (3) an enterprise budget analysis, (4) a break-even analysis, (5) and a sensitivity analysis of changes of the most important parameters (all numbers are in real terms, taxes were not considered; see also e.g. Hatch and Tai 1997; D'Souza et al. 2004; Engle et al. 2005; Pomeroy et al. 2006; Whitmarsh et al. 2006; Liu and Sumaila 2007). According to the operating life expectancy of main components of the complete

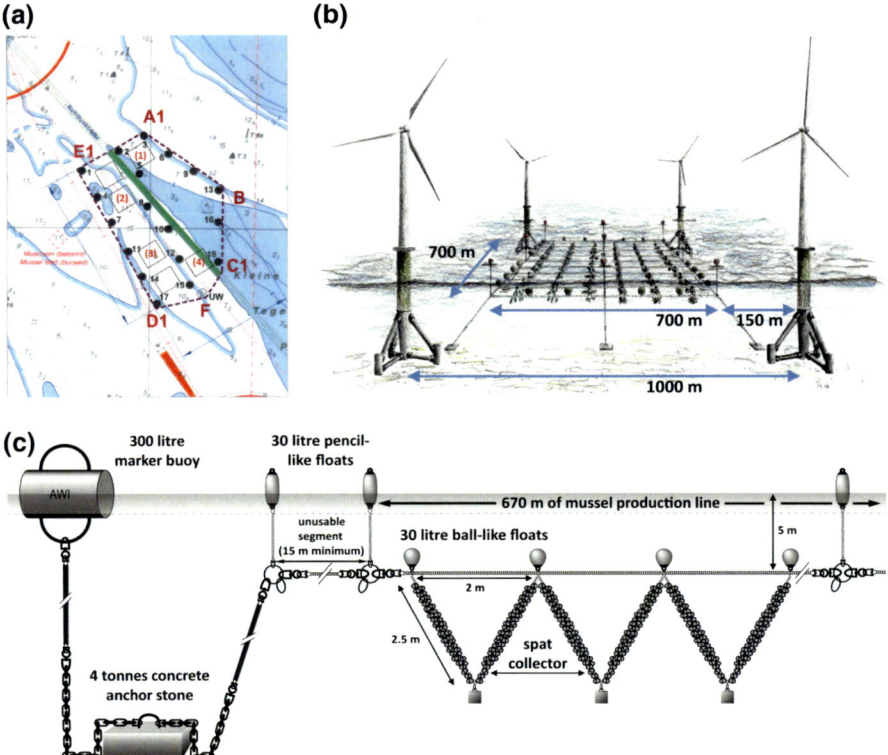

Fig. 11.44 a Map of the planned wind farm *Nordergründe* displaying 18 offshore wind turbines (numbers without brackets) and six single mussel plots designated by the wind farm company (bird's eye view). Four of these six designated plots were calculated according to our mussel cultivation projections (numbers in brackets). **b** Presents a design of a single mussel plot within a group of four wind turbines (not to scale) (modified after Buck et al. 2010). **c** Example of a submerged longline system design with a V-shaped spat collector harness. In this image only a part of the 700 m long longline is presented (not to scale). Modified after Buck et al. 2010

longline device, the enterprise budget for a four-year life cycle was calculated. Additionally, a new vessel as well as the possibility of using existing capacities of the mussel farmer community in Lower Saxony will be taken into consideration. This led to four different scenarios. First, a basic scenario for the farming of consumption mussels with a new appropriate vessel as well as a new land facility was assumed. In a further scenario, the above-mentioned existing capacities of mussel farmers were addressed. The economic analysis was organized as follows: (1) basic parameters for farm size, culture technology and biomass gain are described (Table 11.6); (2) time schedule of the farm set-up and harvest operations were presented; (3) basic data on costs and investment were specified; and (4) finally a sensitivity analysis was outlined.

Table 11.6 Site-specific data of the offshore wind farm-aquaculture multi-use concept at the offshore lighthouse "Roter Sand" (modified after Buck et al. 2010, based on Buck 2004 and Buck et al. 2008a, b)

Basic data	
Distance to shore (City of Bremerhaven)	17 nautical miles
Number of planned wind turbines/power	18/5 MW-class
Salinity	20–33‰ (due to tidal current and the influence of the Weser estuary)
Depth	10–15 m
Condition of the sea bottom	Soft bottom (Wadden Sea)
Turbidity and light	High sediment load (due to the Wadden Sea)
Wave climate	exposed
Current velocity	0–1.2 m s^{-1} (depending on the tide)
Significant wave heights	0–6 m
Nutrients	Eutrophic situation
Water temperature	1.5–18 °C
Wind velocities	Up to 8 Beaufort
Distance between turbines	Approx. 1,000 m
Minimum spacing between turbines and any aquaculture co-use	150 m
Size of aquacultural area (single mussel plot)	700 × 700 m (490,000 m^2 = 0.49 km^2 = 49 ha = 121 acre)
Number of single mussel plots	4 (196 ha = 484 acre)
Total length of the collector harness per longline	1675 m (335 V-shaped collector pairs having each a length of 2 · 2.5 m = 5 m, 71 · 1675 m = 118,925 m per single mussel plot
Biomass of mussels per meter of collector	10–15 kg (16.75 metric tons · longline^{-1}, 1190 metric tons · plot^{-1})

11.6.2   Data for the Economic Analysis

(1) See Table 11.6
(2) As the best and safest working conditions are to maximize onshore activities (Sørensen et al. 2001), the setup of all longline devices should take place on land-based facilities and transferred to the aquaculture site during spring to allow settlement in May of the same year. It was planned to install two full mussel plots, which were scaled up in the following year, which would then be equivalent to four mussel plots in operation. Exchange of longline devices after its operating life expectancy can automatically be done during or after maintenance or harvest procedures. Consumption mussels reach market size after approximately 1.5 years; therefore, the farm will operate at full scale in the second year, which results in a form of shifting cultivation (Bartlett 1956). While only in the first year of the enterprise no mussels can be harvested due to their growth period to reach market size, in the following years two plots can be

harvested biennially each year (6 harvests in four years). Once the longline is transferred at sea, deployed and ready for cultivation the production undergoes two cycles: Spat collection (April–June) in year one and maintenance of longlines to remove fouling organisms and modify buoyancy (August–May) in year one as well and grow-out to consumption size within 15–18 months (market size: <5.5 cm) and harvest in August–November in the second year.

(3) All costs were itemized by scenarios of production for consumption mussels (Buck et al. 2010 also calculate seed mussel production). Cost calculations were based on data gathered from existing traditional nearshore mussel culti- vation activities. Some nearshore cultivation plots have a distance of 10–55 nautical miles to the port of trans-shipment, which is within the scale of the planned offshore site (17 nautical miles) (BSH 2016). Offshore operations are more labour and time intensive than nearshore sites. Much of the labour is for maintenance that includes deploying or retrieving of moorings or other parts of the construction harness, which may lead to generally higher operation and production costs. However, some production steps necessary for nearshore operations cease to exist offshore, which in turn leads to cost reduction. The annual fixed costs consist of depreciation, licenses, motor overhaul, interest on fixed capital and miscellaneous costs like insurance premium and administra- tive cost. Interest rate was assumed to be 7%. Variable costs are fuel expenses, wages, repairs and maintenance, miscellaneous costs and interest on variable costs. Fuel was assumed to cost 0.55 € liter^{-1}, wages are calculated with 3,333 € month^{-1}. When using existing capacities of mussel farming in Lower Saxony, an investment for retrofitting at the beginning of the enterprise will be required. In the scenarios where new capacities have to be established, investment into a new appropriate vessel as well as into a new land facility was considered. All other costs are assumed to be similar to those used for the basic scenario.

(4) A sensitivity analysis was calculated to explore the effects of changes in the key parameters that reflect uncertainty, such as the biomass gain and/or the development of costs and prices. Here, NPV and IRR for different mussel prices, different biomass gain, different developments of single cost compo- nents as well as an increase in overall costs were calculated as well as different discount rates on NPV.

11.6.3   Calculation and Results

Following the data of the Federal Agency for Agriculture and Food (2007) and the State Fisheries Agency, Bremerhaven, Germany (SFA) (2008), the average market price per kg of consumer mussels has been relatively stable until 1975 (below 0.2 € kg^{-1}). Afterwards, the price has been subject to fluctuations ranging between

0.50 and 1.96 € kg^{-1}. According to the development of the market price of Blue mussels, a price of 1.0 € kg^{-1} of mussels was used in the base scenario (SFA 2008). Thus, a single longline could have a production value of 16,750 € (1,000 € × 16.75 metric tons) and a single mussel plot of approx. 1,190,000 € (1,000 € × 1190 metric tons).

The cost of longlines including the complete harness is the sum of various individual costs and levels around 15.80 € meter^{-1} of longline (including collectors, mooring constructions, connecting pieces for the entire longline device, such as shackles, swivels, rings as well as the complete buoyancy). Costs were calculated by Sahr (2006) using the equations for the definition of key cost data published in Pelz (1974). This leads to an overall investment cost of approximately 835,500 € single mussel plot^{-1} every four years. In line with the estimates of Whitmarsh et al. (2006) the operating life expectancy is assumed to be four years for longlines and collectors, six years for buoyancy, and 10 years for anchors. A vessel adapted for performing offshore operations is needed. In the base scenario, an investment in a new vessel (45 m class, 430 GRT, 500 kW) for around 4 million € (Sahr 2006), including all necessary equipment for longline cultivation, was assumed. This case also includes a complete motor overhaul after 10 years with 385,000 € (assuming motor costs to be 17.5% of total vessel investment and retrofitting to be 55% of the amount of 17.5%; Sahr 2006). Because the mussel farmer community already disposes of mussel farming cutters used for bottom culture, we also calculated the NPV with the assumption of using existing capacities of mussel farmers. Investment will then be reduced to the retrofitting of the vessel only, which was calculated with costs of about 750,000 € (Sahr 2006). Capital investment costs include the costs of a land facility for the purpose of equipment storage and for carrying out land-based activities, such as tying and repairing collectors and other equipment. Investment costs for a land facility are assumed to be 1,500,000 €. Licensing costs for a single mussel plot at the offshore site Nordergründe is based upon the scale of charges and fees of the State of Lower Saxony (NKüFischO 2006). Following the fees for mussel license areas, only the bureaucratic work load will be charged, which was calculated by the *State Fisheries Agency* (SFA) in Bremerhaven with a nonrecurring charge of approximately 1,000 € (personal communication with Brandt from SFA). Miscellaneous fixed costs (e.g., insurance premiums) are assumed to be 5% of depreciation leading to a total sum of 151,127 € in four years. Interest on fixed capital is 232,951 € for a four-year period. Total fixed costs were 3,560,817 €.

11.6.3.1 Operation Costs

The experience of the bottom-culture aquaculturists indicates that approximately 70 days per year are needed for labour at four culture plots, amounting altogether to 280 offshore working days in four years. Taking into account 61.8% of full load engine performance in a 24 h day, fuel costs per day at sea are estimated to be 1200 € (Gloy 2006; Sahr 2006). This totals 84,000 € per year or 336,000 € in four years. Two full positions and two seasonal employees are required per year. The latter are employed only in times of the heaviest workload in the 6 months from spring to autumn. Labour costs total 479,952 € in a four-year period. Costs of maintenance and repairs, estimated as 10% of the yearly depreciation, are 302,254 €. Miscellaneous variable costs are estimated to be 5% of depreciation, total 151,127 € in four years. Interest on operating capital sums to 88,853 € in four years. Total variable costs were 1,358,186 €.

11.6.3.2 Enterprise Budget Analysis

Costs and receipts of two case-scenarios were calculated for consumption mussels. Scenario 1: Production of consumption mussels with Investment into a new vessel. This is the base scenario assuming a four million € investment into a new vessel for farming of mussels for consumption. A general overhaul of the motor is necessary after 10 years and is calculated with 385,000 €. Net returns for an average four year period sum to 4,594,996 €. Scenario 2: Production of consumption mussels using free capacities of existing mussel farmers. For this scenario, retrofitting costs for the vessel are about 750,000 €. No land-based facility is included. This leads to net returns of approximately 6,022,000 € in four years, which is 1.3 times higher than in the base scenario.

11.6.3.3 Productivity Measures

Break-even yield and break-even price were calculated to estimate the minimum level of biomass production and the minimum price per kg mussel to enable the enterprise to cover cost. Assuming a biomass of 10 kg meter^{-1} (consumer mussels) the break-even price is 0.52 € when a new vessel and land facility is taken into calculation. Using existing equipment, a break-even price of 0.37 € results. In the case of seed mussels the break-even price varies between 0.34 and 0.49 €. Break-even yield for the consumer mussel scenarios lies between 3.67 kg and 5.17 kg per meter longline, respectively, assuming a mussel price of 1 € kg^{-1}. In the seed mussel scenario the break-even yields range from 3.42 to 4.92 kg. Actual prices and yields observed at field experiments are higher than the break-even values. This indicates the profitability of both practices, while the consumer mussel production is clearly more above those criterions for economic viability.

11.6.3.4 Investment Appraisal

Assuming the operating life expectancy of a new vessel to be 20 years, we calculate the NPV of cash flows over 20 years with a discount rate of 7% in the basic model. This rate is chosen according to Liu and Sumaila (2007), who argue that the most frequently used discount rate by Nature Resources Canada is within a range of

5–10%. D'Souza et al. (2004) used 7, 9 and 11%, while Whitmarsh et al. (2006) limits the discount rate to 8%. Due to the sensitivity of the NPV to the discount rate, values ranging from 6 to 9% were used. In the base scenario, the price for one kg of mussels was assumed to be 1.0 €. Net present value amounts to 5,667,073 €, with an IRR of 14.73%. When using existing capacities of mussel farming in Lower Saxony, an investment of about 750,000 € for retrofitting of the vessel at the beginning of the enterprise will be required. All other costs are assumed to be similar to those from the basic scenario. NPV levels around 9,622,937 € and an IRR of 28.11%. Economically, the most promising enterprise is the production of consumer mussels if existing equipment can be used. But also in the case of a new vessel and a new land facility profits are likely, since the IRR levels at 14.73%. This should be in most cases higher than the costs of capital.

11.6.3.5   Sensitivity Analysis

A sensitivity analysis was carried out to assess the economic feasibility, if key parameters of the economic analysis are changing. As the biomass harvested was assumed to be at a low level, the positive impact of a 25 and 50% biomass increase was estimated for consumer mussels as well as an increase of 10% for seed mussel yield. Fuel costs were increased by 10 and 20% per year, wages by 3% per year, longline costs by 5% per year and total costs by 5% per year. Discount rates were varied from 5% over 6 to 8%. The mussel price was changed by 10% in case of consumer mussels and by 20% in case of seed mussels. The results are shown in Table 8. The overall result shows the capacity of the production of consumer mussels with existing equipment to withstand cost increases quite well. In case of a new vessel and new land facility NPV remains positive except for an overall cost increase of 5% per year. All calculated discount rates leave NPV to be positive.

11.6.4   Final Conclusion

Assuming a baseline production of 2380 tons of consumption mussels per year (2 plots) the results of the economic study show that the base scenario is clearly beyond the break-even point. Varying parameter values, such as investment costs concerning longlines, new vessels or retrofitting, operating costs like wages and fuel, biomass yield, market price, total cost increases, and different discount rates, show different levels of feasibility. Offshore mussel production for consumption is profitable, but profits are less with a new vessel and a new land facility and higher in the scenarios without a new vessel and a new land facility, respectively. The NPV and IRR are large enough that this business can be recommended as long as there are existing capacities. Of course, all businesses can become profitable and respectively more profitable if costs can be reduced and revenues increased.

The lack of practical experience of culturing mussels in exposed environments precludes estimating effects of economic risks.

11.7   Ownership Issues

11.7.1   Modes of Cooperation

The relevant government policy as a central framing condition must be part of any analysis of likely management scenarios for an integrated wind energy-mariculture facility. In the North Sea area, the here presented concept is ahead of the current institutionalized regulatory system. Indeed, a systematic regulation addressing this multi-use concept in the context of industry support is yet lacking. While current legislation may preclude concurrent economic activity within offshore wind farms, that likely can be interpreted as a de facto law absent any regulatory consideration on this multi-use issue. However, given the strong push for spatial efficiency and multi-use concepts in the maritime waters in the EU and elsewhere (Krause et al. 2003; Lutges and Holzfuss 2006), it can be expected that more comprehensive regulatory frameworks will develop in due time. Krause et al. (2011) identified three likely avenues under which an integrated mariculture-wind energy facility may be organized. These are not exhaustive, or mutually exclusive from each other, but rather provide a straightforward method for categorizing potential outcomes. In the following, a brief synopsis of these three different ownership scenarios of multi-use in the offshore realm is provided.

11.7.1.1   Sole Owner

At the extreme end, a sole owner situation could be envisaged, in which a multiple use business plan could be enacted by a sole company without any cooperation. This set-up may accommodate especially the interests of the wind energy producers, who would have easier access to the financial resources needed. In view of the current complexity of drafting and following a contract with an outside firm makes this sole owner approach highly appealing. Indeed, from an economic point of view, governance structures that have better transaction cost economizing properties are preferable. Additionally, transaction cost economics suggests that full vertical integration completely resolves issues related to hold-ups and misaligned incentives (Williamson 1979, 1981; Johnson and Houston 2000).

Undeniably, the potential for further net revenue via mariculture may be alluring to a wind energy firm, especially in the light that the area occupied by wind turbines is roughly 1–3% of the total area of an offshore wind farm (Mee 2006). Economies of scope may act thus as a financial catalyst, i.e. simultaneously producing two products with a lower average cost than if undertaken separately.

Undertaking this economic decision as a sole firm partly however rests on the ability of the wind energy producer to culture products at a similar or lower average cost than if they had negotiated a contract or formed a joint venture with a firm who specializes in mariculture. The lack of knowledge on the different modes of conduct in the mariculture section could act however, as a major impediment to such sole ownership scenario of a wind farm enterprise. Thus, while a sole ownership approach may initially appear promising, the degree of risk involved in operating two very different businesses at the same location is high. The relative risk of internalizing both productive activities can be somewhat combatted by the degree to which personnel with specialized knowledge could be brought into oversee and conduct these operations. Current lines of research are assessing the economic merits of a joint mariculture-wind energy facility and will help illuminate the viability of such a venture from multiple perspectives (Griffin and Krause 2010).

11.7.1.2   Negotiated Contract

Forming alliances is a common commercial strategy that is employed to organize and mitigate activities that are riskier than a firm's average inside project. These alliances occur more in riskier industries (Robinson 2008). Expanding to an industry-level analysis, Robinson (2008) found that alliance intensity across industries is positively associated with the risk difference between the two industries. This dynamic could play an important role in alliance formation versus single firm management of a multi-use facility. Therefore negotiated contracts are a alternative path to mitigate and manage the risk associated to an integrated facility. Such categories of agreement may cover a multitude of different settings, such as a joint venture or a consortium or any form of subcontracts. Central hereby is the fact that the outlined interdependence between firms must provide benefit to each party (Pareto-improving) and be perceived as fair by the participating entities. Continued cooperation between parties must be sustainable by the underlying game structure (Grandori and Soda 1995). Alliance between firms that both hold unique capabilities that neither partner could efficiently provide alone, have the highest potential for coordination. Michler-Cieluch and Krause (2008) showed that under such an umbrella there is sufficient scope for such wind farm-mariculture cooperation in terms of operation and maintenance activities.

However, the process of drawing up a contract that delineates the lines of cooperation between firms is fraught with challenges. Hold-up hazards increase when complexity and uncertainty make writing and enforcing contracts difficult (Williamson 1979), and when products require asset-specific investments, two conditions that hold in this case. Only when there are offsetting economic benefits and sufficient efficiency scope to doing so firms are compelled to engage in integrated organisational structures over simple contracts or sole ownership (Johnson and Houston 2000). Such economic benefits pertain to any of the previously outlined benefits from cooperation, such as reduced production costs, organisational efficiencies, or pooling risk—but these benefits are not guaranteed. Because of the parties' inability to write an a priori comprehensive agreement that covers all future contingencies, Nielsen (2010) argued that all alliance contracts are necessarily incomplete. Thus, these contracts may enhance or prohibit desired outcomes. In order to be successful, all stakeholders involved in such joint cooperation agreements must be informed and clear about their expectations, rights and duties involved from the onset if being engaged in such multiple use setting.

Regarding the predictors of success in joint ventures and other alliances, an amount of considerable research exists. Johnson and Houston (2000) find that only joint ventures between firms in related businesses are likely to generate operating synergies. Thus, combinations of dissimilar firms can reduce value by contributing to bureaucracy and lack-of-focus. Beamish (1994) finds that the good intentions and rational motives behind alliances are often not congruent with the strategic direction of either firm on its own. This inconsistency can lead to poor performance and instability. Indeed, Kumar (2007) finds that in incidences where firms with asymmetric resource endowments enter into a joint venture, asymmetric wealth gains arise via the negative wealth transfer effects of resource appropriation by the firm with more valuable resources. Therefore, initial collaborative research between sectors prior to the design and execution of a commercial agreement by act as avenue to overcome this potential pitfall Michler-Cieluch et al. (2009a). In the German case study, our interviews and survey work suggests that the stakeholders in a potential mariculture-wind energy facility may be amenable to some type of contracted agreement. Respondents have suggested that they would be open to the idea of contracting out culturing activities at the site of an offshore wind farm as a prior joint research initiative and feasibility study. It does seem unlikely though at this point that a contracted solution could occur in the absence of some intervening third body (Michler-Cieluch et al. 2009a). However, an advisory or some other external independent mediating group or entity that helps to coordinate and facilitate the entire process generally improves the chances of reaching a successful agreement (Noble 2000).

11.7.1.3 Legislated

In the third ownership option, a legislative prescription could attain desired policy goals of spatial efficiency in the ocean area. The use of mandates, subsidies, tariffs, and other policy tools can change the incentives of the current economic environment to make the multi-use concept economically viable in cases where it may not be to date very appealing or in places where finding a direct commercial market solution for multi-use in the offshore setting is not possible. Increasingly, policy makers may find policy instruments as a palatable solution for achieving policy goals, especially as there is a growing focus on coastal zone management and the efficient and equitable use of coastal resources in the EU, US, and elsewhere (Krause et al. 2003, Lutges and Holzfuss 2006; Rhode Island Coastal Resources Management Council 2010). Mariculture can offer expanded employment opportunities to rural peripheral regions and displaced fishermen in the area of a wind energy facility. Furthermore, it has the potential to make wild harvest fisheries more productive if mariculture areas act as nurseries for wild fish (Mee 2006). Indeed, multi-use layering of economic activities can maximize the value of offshore resources while reducing conflict between stakeholder groups. Since regulators have already shown they are comfortable with using legislation to spur growth in the offshore wind energy industry, such type of legislative promoting a multi-use concept would not be an uncommon step. Indeed, a clear, coherent, and stable regulatory framework is a bare minimum when firms make financial decisions in the inherently risky offshore marine environment. Managers need to be able to predict with some certainty the expected outcomes of changes in strategy, be it an internal decision or the decision to form an external alliance. To the contrary, fragmentation in the structure of State decision making is shown to lead to more elaborate and costly inter-organisational networks (Carroll et al. 1988). However, the decision to actively foster cooperation on a multi-use concept should largely be dependent on market conditions, which must be somewhat accommodated for by the legislation. The latter may be driven by the recognition of the potential social benefits available from multi-use facilities. Williamson (1981) stated that "there are so many different types of organisations because transactions differ so greatly and efficiency is only realized if governance structures are tailored to the specific needs of each type of transaction." The legislative ownership discussion this far have attempted to frame the potential cooperation in a multi-use setting in the context of the broader social, political, and economic spheres. Next, the discussions also acknowledge and illuminate the perceptions and characteristics of the particular industries themselves. It appears clear that uncertainty and risk are large components of this discussion. These were reinforced and frequently voiced by the survey respondents in the German case study. It can be expected that the likelihood and form of collaboration in the near future will be shaped by how well this risk and uncertainty is addressed.

11.8   Regulation of Aquaculture Within the German Bight

The EEZ is a special area—it is not a state territory even if a coastal state has there sovereign rights and jurisdiction. This is an area where three legal systems come together: international law, law of the European Union and national law (Buck et al. 2003). There are no commercial aquaculture operations in the German EEZ and no approval procedure has been completed so far. Additionally, some sites are not suitable for aquaculture, especially because of nature conservation and shipping.

11.8.1   Mariculture in the German Exclusive Economic Zone

Mariculture in the Exclusive Economic Zone (EEZ) could be in the future an interesting completion of fishery, not least because of overfishing of natural stocks. Nevertheless, its legal questions are still not fully answered (for an overview of legal frame of mariculture see Kersandt 2012 as well as in BLFRG 2012; ECR 2006, 2007, 2008; EC 2000, 2008, 2011; SAG 1965; VüAs 1997).

11.8.1.1   The System of Law in the EEZ

The EEZ cannot be considered as a part of the state territory. It is a *sui generis* zone and the coastal state can only use it and regulate its use as outlined in the relevant provisions of the United Nations Convention on the Law of the Sea (Dahm et al. 1989; Gündling 1983). This effects the first restriction for the German law-maker. Furthermore, Germany as a member of the European Union, has to respect the European law. Although the EEZ is no sovereign state territory, the EU is allowed to issue the law: The EU has legislative authority in the area outside the members' territory, where the international public law gives such authority to the state, e.g. fisheries on the High Sea (Grabitz et al. 2010). Some directives and regulations refer to aquaculture, e.g. regulation concerning use of alien and locally absent species in aquaculture, directive on animal health requirements for aquaculture animals, products thereof and on the prevention of diseases or regulation on conditions for market placing and the import of aquaculture animals. Some general (i.e., not specific to aquaculture) rules are relevant to aquaculture, e.g. Marine Strategy Framework Directive, Water Framework Directive or EIA Directive.

   The legislative authority usually belongs to the Federal Republic of Germany and not to the German States (*Länder*). It depends on the jurisdictional provisions of Article 74 GG, e.g. the law relating to economic matters (no. 11), deep-sea and coastal fishing (no. 17), and protection of nature and landscape management (no. 29) or management of water resources (no. 32).

11.8.1.2   The Approval Procedure

The basis for the approval procedure of aquaculture in the EEZ is Marine Facilities Ordinance (SeeAnlV), which is based on Federal Maritime Responsibilities Act. The agency deciding on the approval of mariculture in the EEZ is Federal Maritime and Hydrographic Agency (*Bundesamt für Seeschifffahrt und Hydrographie, BSH*). Aquaculture, as any other facility that is not purposed to produce or transmit energy, needs an approval. For energy projects, a plan approval procedure is mandatory. The approval, different than the plan approval, doesn't have the "concentration effect": the applicant needs to get all necessary permits on his own

(Dahlke 2002). Especially provisions of the German Federal Nature Conservation Act (Bundesnaturschutzgesetz - BNatSchG) are here determined, e.g. those about protected species (Art. 44 ff. BNatSchG), protected habitats (Art. 30 ff. BNatSchG) and Natura 2000 areas (Art. 31 ff. BNatSchG) (BNatSchG 2009).

An approval is a non-discretionary administrative act, i.e., it has to be granted when there are no reasons for denying approval (Dahlke 2002). According to Sect. 7 SeeAnlV the approval has to be denied when

- it is likely to impair the safety and efficiency of navigation or poses a threat to the marine environment or
- the requirements of regional planning or overriding military or other public or private predominant concerns are against the approval.

Before the approval has been granted the Federal Waterways and Shipping Agency (former Waterways and Shipping Directorate) has to give a declaration of agreement. It can only be denied if the facility impairs the safety and efficiency of navigation and it is not possible to prevent or compensate the detrimental effects through conditions or requirements, Sect. 8 SeeAnlV.

The term "marine environment" is explained in Sect. 5 subsection 6 no. 2 SeeAnlV. It refers to the definition of "pollution of the marine environment" in Article 1 (1) no. 4 UNCLOS. The approval is to be denied when the pollution gives rise to concern. Aquaculture projects in the EEZ can require environmental impact assessment. The Environmental Impact Assessment Act, which implements the EIA Directive, intends the examination only for intensive fish farming and not for other candidates like mussels or seaweeds. Whether the assessment is mandatory or not depends on the yearly fish amount: it is mandatory if this amount is larger than 2500 t. Otherwise, there is a general (500 t–2500 t) or site-related (250 t–500 t) preliminary examination of the environmental compatibility to undertake. Depending on the result, the assessment is to the end or the whole EIA has to be done.

11.8.1.3   The Sites for Mariculture Facilities

Spatial Planning Ordinance for the German EEZ in the North Sea (also in the Baltic Sea) regulates targets and principles of spatial planning. Marine aquaculture is identified as a meaningful economic sector in the future. Negative impacts of the mariculture on the marine environment shall be avoided. Facilities for mariculture shall be established preferably in combination with existing installations, Sect. 3.6.1 (1) and (2) Spatial Plan North Sea. Spatial planning has an effect on the site for mariculture facilities. In some areas, e.g. in shipping priority areas, measures and projects which are not compatible with the character of this area are not permitted, as stated in Sect. 3.1.1 (1) Marine Spatial Plan North Sea. In the reservation areas, other uses are also allowed, but the main use is given special consideration. This needs to be taken into account in a comparative evaluation assessment with other spatially significant planning tasks, measures and projects. Corresponding rules are provided for the Baltic Sea.

There are also other sites where mariculture is not allowed. In Natura 2000 areas, which are protected areas under national law, mariculture facilities are prohibited. 2005 the Eastern German Bight became a nature conservation area under German law, other sites—*Dogger Bank*, *Sylt Outer Reef* and *Borkum Reef Ground*—were nominated to the European Council. In Sect. 4 subsection 2 no. 1 of designating ordinance mariculture facilities are prohibited. It is likely that other ordinances will have the same regulation.

11.8.2   Approval procedure of offshore multi-use installations

Changing marine utilization patterns represent considerable challenge to society and governments. Indeed, the ongoing intense competition for ocean space leads and has led inevitably to disputes. As a case in point, the somewhat private ownership of ocean space, which usually is assigned via central government authorities, has fostered the raise of "client mentalities" of the ocean pioneers. This has resulted in a complex mix of ownership, associated commons and private property. Thus, the central question remains how to operationalise the multi-use dimension of offshore installations within marine spatial planning?

Centralized authority planning, such as showcased by the existing permitting procedure in the offshore realm in Germany that is charged by one lead agency (Federal Maritime and Hydrographic Agency) has given way to new forms of governing—tendency to define issues at only one scale and only one installation at time. Indeed, in the case of Germany, a highly comprehensive regulatory framework for offshore wind energy, but only a weak and uncertain framework for offshore aquaculture installations is in place. For the latter, technological as well as ecological standards are yet needed. In Fig. 11.45 we present a permitting procedure that possibly integrate multi-use installations within the same suite of permitting process. This is showcased by the integration of offshore wind farm installations and open ocean aquaculture. By acknowledging throughout the process the different demands, potential impacts and outcomes of each partner within a multi-use system, the permitting process could be streamlined and efficient. By this, the permit process would become more specific articulating societal choices about goals and their related social values in a multi-use context. Open issues pertain to e.g. who should be included in a balanced jurisdiction and what should that jurisdiction do? What criteria are relevant to opt for open ocean aquaculture and what are the implications of such criteria and how to include cross-border users? E. g. regions, where offshore wind farms are planned, are also important fishing regions of German, Belgian, Danish and Dutch fishermen.

In summary, the current gap between oceans as commons and ocean as private property as well as diverging views and pictures leads to a contested sea space.

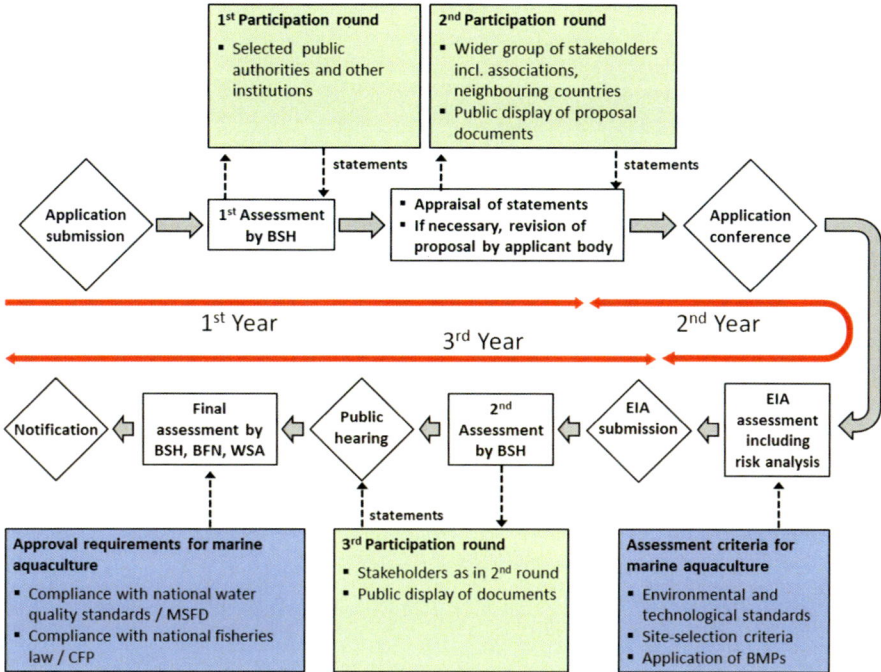

Fig. 11.45 Potential approval procedure of multi-use offshore installations (here exemplified by offshore wind farm and offshore aquaculture) in the German EEZ according to the Marine Facilities Ordinance. BFN = Bundesamt für Naturschutz (Federal Agency for Nature Conservation), BMP = Best Management Practice, BSH = Bundesamt für Seeschifffahrt und Hydrographie (Federal Maritime and Hydrographic Agency of Germany), CFP = Common Fisheries Policy, EIA = Environmental Impact Assessment, MSFD = Marine Strategy Framework Directive, WSA = Wasser- und Schifffahrtsdirektion (Waterways and Shipping Office) (modified after Wever 2011)

A more holistic, integrated approach to ocean management that acknowledges the interconnectedness of human and natural systems is timely. However, there is a high risk of failing in the current window-of-opportunity to integrate open ocean aquaculture within the emerging management of the marine realm. What is at odds is the management discourse of the politically powerful vs. newcomers, reaffirming the socially constructed nature of knowledge.

References

Abad, M., Ruiz, C., Martinez, D., Mosquera, G., & Sánchez, J. L. (1995). Seasonal variations of lipid classes and fatty acids in flat oyster, *Ostrea edulis*, from San Cibran (Galicia, Spain). *Comparative Biochemistry and Physiology, 110,* 109–118.

Ahn, O., Petrell, R. J., & Harrison, P. J. (1998). Ammonium and nitrate uptake by *Laminaria saccharina* and *Nereocystis luetkeana* originating from a salmon sea cage farm. *Journal of Applied Phycology, 10,* 333–340.

Bailey, H., Brookes, K. L., & Thompson, P. M. (2014). Assessing environmental impacts of offshore wind farms: Lessons learned and recommendations for the future. *Aquatic Biosystems, 2014*(10), 8.

Bartlett, H. H. (1956). Fire, primitive agriculture, and grazing in the tropics. In W. L. Thomas (Ed.), *Man's role in changing the face of the earth* (pp. 692–720). Chicago and London: The University of Chicago Press.

Bartsch, I., Wiencke, C., Bischof, K., Buchholz, C., Buck, B. H., Eggert, A., et al. (2008). The genus Laminaria sensu lato: Recent insights and developments. *European Journal of Phycology, 43,* 1–86.

BLFRG. (2012). Basic law for the federal Republic of Germany in the revised version published in the Federal Law Gazette Part III, classification number 100-1, as last amended by the Act of 11.7.2012, Federal Law Gazette I, 1478.

Beamish, P. (1994). Joint vetures in LDCs: Partner selection and performance. *Management International Review, 34,* 60–74.

Beninger, P. G., & Lucas, A. (1984). Seasonal variations in condition, reproductive activity, and gross biochemical composition of two species of adult clam reared in a common habitat: *Tapes decussates* L. (Jeffreys) and *Tapes philippinarum* (Adams & Reeve). *Journal of Experimental Marine Biology and Ecology, 79,* 19–37.

Bilio, M. (2008). Controlled reproduction and domestication in aquaculture. The current state of the art. *Aquaculture Europe Part I-IV, 33,* 1–24.

Boghen, A. (1991). *Cold-Water aquaculture in Atlantic Canada.* New Brunswick: Canadian Institute for Research on Regional Development.

Brenner, M. (2009). *Site selection criteria and technical requirements for the offshore cultivation of Blue mussels (Mytilus edulis L.).* PhD Thesis. Jacobs University Bremen, Germany.

Brenner, M., & Buck, B. H. (2010). Attachment properties of blue mussels (*Mytilus edulis* L.) byssus threads on culture-based artificial collector substrates. *Aquacultural Engineering, 42,* 128–139.

Brenner, M., Buck, B. H., & Köhler, A. (2007). New concepts for the multi-use of offshore wind farms and high quality mussel cultivation. *Global Aquaculture Advocate, 10,* 79–81.

Brenner, M., Buchholz, C., Heemken, O., Buck, B. H., & Köhler, A. (2012). Health and growth performance of the blue mussel (*Mytilus edulis* L.) from two hanging cultivation sites in the German Bight: a nearshore—offshore comparison. *Aquaculture International, 20,* 751–778.

Bridger, C. J., & Costa-Pierce, B. A. (2003). *Open ocean aquaculture: From research to commercial reality.* Baton Rouge, Louisiana, USA: The World Aquaculture Society.

BSH. (2016). *Bundesamt für Seeschifffahrt und Hydrographie (Federal Maritime and Hydrographic Agency), Maps of North Sea Stakeholders, CONTIS-Geodata (Continental Shelf Information System).* Germany: Hamburg and Rostock.

Browne, L. (2001). Mariculture of the edible red alga *Palmaria palmata.* PhD thesis, Queen's University Belfast.

Buchholz, C., & Lüning, K. (1999). Isolated, distal blade discs of the brown alga Laminaria digitata form sorus, but not discs, near to the meristematic transition zone. *Journal of Applied Phycology, 11,* 579–584.

Buck, B. H. (2000). *Open Ocean Aquaculture und Offshore-Windparks: Eine Machbarkeitsstudie über die multifunktionale Nutzung von Offshore-Windparks und Offshore-Marikultur im Raum Nordsee.* Reports on Polar and Marine Research, 412. Alfred Wegener Institute for Polar and Marine Research, Bremerhaven, Germany.

Buck, B. H. (2001). *Combined utilisation of wind farming and mariculture in the North Sea.* Biennial Report of Science Highlights at the Alfred Wegener Institute Helmholtz Centre for Polar and Marine Research, 33–39.

Buck, B. H. (2002). *Open Ocean Aquaculture und Offshore-Windparks: Eine Machbarkeitsstudie über die multifunktionale Nutzung von Offshore-Windparks und Offshore-Marikultur im Raum*

Nordsee. Reports on Polar and Marine Research, 412. Alfred Wegener Institute for Polar and Marine Research, Bremerhaven, Germany.

Buck, B. H. (2004). *Farming in a high energy environment: potentials and constraints of sustainable offshore aquaculture in the German Bight (North Sea)*. PhD thesis, University of Bremen, Germany.

Buck, B. H. (2007). Experimental trials on the feasibility of offshore seed production of the mussel Mytilus edulis in the German Bight: Installation, technical requirements and environmental conditions. *Helgoland Marine Research, 61*, 87–101.

Buck, B. H. (2017). Mytilid larval appearance and settlement at offshore wind farm sites in the German Bight—a trial to estimate multi-use potentials for bivalve seed collection and cultivation. Helgoland Marine Research.

Buck, B. H., & Buchholz, C. M. (2004a). The offshore-ring: A new system design for the open ocean aquaculture of macroalgae. *Journal of Applied Phycology, 16*, 355–368.

Buck, B. H., & Buchholz, C. M. (2004b). Support device for the cultivation of macro organisms in marine waters. Patents: PCT/DE2005/000234.

Buck, B. H., & Buchholz, C. M. (2005). Response of offshore cultivated *Laminaria saccharina* to hydrodynamic forcing in the North Sea. *Aquaculture, 250*, 674–691.

Buck, B. H., & Krause, G. (2012). Integration of aquaculture and renewable energy systems. In R. A. Meyers (Ed.), *Encyclopaedia of sustainability science and technology* (Vol. 1, pp. 511–533). Springer Science + Business Media LLC.

Buck, B. H., & Wunsch, M. (2005). Inspection device for under water constructions including a positioning system. Patent: DE 10 2005 020 070.

Buck, B. H., Ebeling, M., & Michler-Cieluch, T. (2010). Mussel cultivation as a co-use in offshore wind farms: Potentials and economic feasibility. *Aquaculture Economics and Management, 14*, 255–281.

Buck, B. H., Krause, G., & Rosenthal, H. (2004). Extensive open ocean aquaculture development within wind farms in Germany: The prospect of offshore co-management and legal constraints. *Ocean Coastal Management, 47*, 95–122.

Buck, B. H., Krause, G., Rosenthal, H., & Smetacek, V. (2003). Aquaculture and environmental regulations: The German situation within the North Sea. In A. Kirchner (Ed.), *International marine environmental law: Institutions, implementation and innovation. International environmental law and policies series of Kluwer Law international 64* (pp. 211–229). The Hague: Kluwer.

Buck, B. H., Walter, U., Rosenthal, H., & Neudecker, T. (2006a). The development of Mollusc farming in Germany: Past, present and future. *World Aquaculture Magazine, 6–11*, 66–69.

Buck, B. H., Berg-Pollack, A., Assheuer, J., Zielinski, O., & Kassen, D. (2006b). Technical realization of extensive aquaculture constructions in offshore wind farms: consideration of the mechanical loads. In *Proceedings of the 25th International Conference on Offshore Mechanics and Arctic Engineering, OMAE 2006: Presented at the 25th International Conference on Offshore Mechanics and Arctic Engineering*, 4–9 June 2006, Hamburg, Germany.

Buck, B. H., Thieltges, D., Walter, U., Nehls, G., & Rosenthal, H. (2005). Inshore-offshore comparison of parasite infestation in *Mytilus* edulis: Implications for open ocean aquaculture. *Journal of Applied Ichthyology, 21*, 107–113.

Buck, B. H., Krause, G., Michler-Cieluch, T., Brenner, M., Buchholz, C. M., & Busch, J. A. (2008a). Meeting the quest for spatial efficiency: Progress and prospects of extensive aquaculture within offshore wind farms. *Helgoland Marine Research, 62*, 269–281.

Buck, B. H., Zielinski, O., Assheuer, J., Wiemann, K., Hamm, C., & Kassen, D. (2008b). Technische Umsetzung von extensiven Marikulturanlagen in Windparks: Betrachtung der mechanischen Lasten (Technical realisation of extensive aquaculture in offshore wind farms: Calculation of mechanical loads). Project Report FV 174.

Buck, B. H., Dubois, J., Ebeling, M., Franz, B., Goseberg, N., Krause, G., et al. (2012). Projektbericht multiple nutzung und co-management von offshore-strukturen: Marine aquakultur und offshore windparks—Open Ocean Multi-Use (OOMU). Gefördert durch das BMU—Förderkennzeichen 325206. Projektleitung: IMARE GMBH Bremerhaven—Germany.

BNatSchG. (2009). Bundesnaturschutzgesetz, Federal Law Gazette I, 2542/2009, last amended by Federal Law Gazette I, 3154.

Byzio, A., Mautz, R., & Rosenbaum, W. (2005). *Energiewende in schwerer See? Konflikte um die Offshore-Windkraftnutzung*. München: Oekom.

Caers, M., Coutteau, P., & Sorgeloos, P. (2000). Impact of starvation and of feeding algal and artificial diets on the lipid content and composition of juvenile oyster (*Crassostrea gigas*) and clams (*Tapes philippinarum*). *Marine Biology, 136*, 891–899.

Carroll, G., Goodstein, J., & Gyenes, A. (1988). Organisations and the state: Effects of the institutional environment on agricultural cooperatives in hungary. *Administrative Science Quarterly, 33*, 233–256.

Chambers, M. D., & Howell, W. H. (2006). Preliminary information on cod and haddock production in submerged cages off the coast of New Hampshire, USA. *ICES Journal of Marine Science, 63*, 385–392.

Chopin, T., Buschmann, A. H., Halling, C., Troell, M., Kautsky, N., Neori, A., et al. (2001). Integrating seaweeds into marine aquaculture systems: A key towards sustainability. *Journal of Phycology, 37*, 975–986.

Corey, P., Kim, J. K., Duston, J., & Garbary, D. J. (2014). Growth and nutrient uptake by *Palmaria palmata* integrated with Atlantic halibut in a land-based aquaculture system. *Algae, 29*, 35–45.

CWSS. (2002). *Shellfish fisheries. An overview of policies for shellfish fishing in the Wadden Sea*. Wilhelmshaven (Germany): Common Wadden Sea Secretariat.

Dahlke, C. (2002). Genehmigungsverfahren von Offshore-Windenergieanlagen nach der Seeanlagenverordnung. *NuR, 8*, 472–479.

Dahm, G., Delbrück, J., & Wolfrum, R. (1989). *Völkerrecht, Band I/1: Die Grundlagen. Die Völkerrechtssubjekte*. Berlin: de Gruyter.

Daniel, T. L. (1984). Unsteady aspects of aquatic locomotion. *American Zoology, 24*, 121–134.

Daniels, H. V., & Watanabe, W. O. (2010). *Practical flatfish culture and stock enhancement*. New Jersey: Wiley Blackwell.

Dare, P. J., Edwards, D. B., & Davies, G. (1983). *Experimental collection and handling of spat mussels (Mytilus edulis L.) on ropes for intertidal cultivation*. London: Ministry of Agriculture Fisheries and Food, Directorate of Fisheries Research.

Davenport, J., & Chen, X. (1987). A comparison of methods for the assessment of condition in the mussel (*Mytilus edulis* L.). *Journal of Molluscan Studies, 53*, 293–297.

Davies, I. M., Greathead, C., & Black, E. A. (2008). Risk analysis of the potential interbreeding of wild and escaped farmed cod (*Gadus morhua* Linnaeus). Reports and studies—Joint group of experts on the scientific aspects of marine. *Environmental Protection, 76*, 112–132.

De la Parra, A. M., García, O., & Fuencisla, S. J. (2005). Seasonal variation on the biochemical composition and lipid classes of the gonadal and storage tissues of *Crassostrea gigas* (Thunberg, 1794) in relation to the gametogenic cycle. *Journal of Shellfish Research, 24*, 457–467.

Denny, M. W. (1988). *Biology and the mechanics of the wave-swept environment*. New York: Princeton University Press.

Denny, M. W., Daniel, T. L., & Koehl, M. A. R. (1985). Mechanical limits to size in wave-swept organisms. *Ecological Monographs, 55*, 69–102.

Diederich, S., Nehls, G., van Beusekom, J. E. E., & Reise, K. (2004). Introduced Pacific oysters (Crassostrea gigas) in the northern Wadden Sea: Invasion accelerated by warm summers? *Helgoland Marine Research, 59*, 97–106.

D'Souza, G., Miller, D., Semmens, K., & Smith, D. (2004). Mine water aquaculture as an economic development strategy: Linking coal mining, fish farming, water conservation and recreation. *Journal of Applied Aquaculture, 15*, 159–172.

EC. (2000). Directive 2000/60/EC of the European Parliament and of the Council of 23 October 2000 establishing a framework for Community action in the field of water policy, OJ L 327, December 22 2000, 1–73.

EC. (2008). Directive 2008/56/EC of the European Parliament and of the Council of 17 June 2008 establishing a framework for community action in the field of marine environmental policy, OJ L 164, June 25 2008, 19–40.

EC. (2011). Directive 2011/92/EU of the European Parliament and of the Council of 13 December 2011 on the assessment of the effects of certain public and private projects on the environment, OJ L 335, 17.12.2011, 1–14; see also Directive 2014/52/ of the European Parliament and of the Council of 16 April 2014 amending Directive 2011/92/EU on the assessment of the effects of certain public and private projects on the environment, OJ L 124, April 25 2014, 1–18.

ECR. (2006). European Council Directive 2006/88/EC of 24 October 2006 on animal health requirements for aquaculture animals and products thereof, and on the prevention and control of certain diseases in aquatic animals, OJ L 328, November 24 2006, 14–56.

ECR. (2007). European Council Regulation No 708/2007 of 11 June 2007 concerning use of alien and locally absent species in aquaculture, OJ L 168, June 28 2007, 1–17.

ECR. (2008). European Commission Regulation No 1251/2008 of 12 December 2008 implementing Council Directive 2006/88/EC as regards conditions and certification requirements for the placing on the market and the import into the Community of aquaculture animals and products thereof and laying down a list of vector species, OJ L 337, December 16 2008, 41–75.

Engle, C. R., Pomerleau, S., Fornshell, G., Hinshaw, J. M., Sloan, D., & Thompson, S. (2005). The economic impact of proposed effluent treatment options for production of trout *Oncorhynchus mykiss* in flow-through systems. *Aquacultural Engineering, 32,* 303–323.

FAO. (2004). *Food and Agriculture Organization of the United Nations. Cultured aquatic species information programme. Salmo salar.* FAO Fisheries and Aquaculture Department. Retrieved October 12, 2016, from http://www.fao.org/fishery/culturedspecies/Salmo_salar/en

FAO. (2005). *Food and Agriculture Organization of the United Nations. Cultured Aquatic Species Information Programme. Dicentrarchus labrax.* FAO Fisheries and Aquaculture Department. Retrieved October 12, 2016, from http://www.fao.org/fishery/culturedspecies/Dicentrarchus_labrax/en

Federal Agency for Agriculture and Food. (2007). Der Markt für Fischereierzeugnisse in der Bundesrepublik Deutschland im Jahre 2006: Bericht über die Versorgung der Bundesrepublik Deutschland mit Fischereiprodukten aus Eigenproduktion und Importen sowie die Exportsituation. Hamburg, Germany: Bundesanstalt für Landwirtschaft und Ernährung Referat 523 – Fischwirtschaft.

Fei, X. (2004). Solving the coastal eutrophication problem by large scale seaweed cultivation. *Hydrobiologia, 512,* 145–151.

Fiedler, G., Heuer, R., Koch, L., & Steinberg, R. (2011). *Windenergieanlagengeräusche als Stressor. Projektierung Biotechnologie WS 2010/11.* University of Applied Sciences Bremerhaven (Germany).

Fraser, A. J., Tocher, D. R., & Sargent, J. R. (1985). Thin-layer chromatography—Flame ionization detection and the quantification of marine neutral lipids and phospholipids. *Journal of Experimental Marine Biology and Ecology, 88,* 91–99.

Gallager, S. M., & Mann, R. (1986). Growth and survival of larvae of *Mercenaria mercenaria* (L.) and *Crassostrea virginica* (Gmelin) relative to broodstock conditioning and lipid content of eggs. *Aquaculture, 56,* 105–121.

Gloy, C. (2006). *Wirtschaftliche Bewertung einer kombinierten Windenergiegewinnung–Saatmuschel-Konsummuschelproduktion im Offshore-Bereich (Abschätzung von Kosten-Nutzen).* Bremerhaven: CG Gloy.

Goseberg, N., Franz, B., & Schlurmann, T. (2012), The potential co-use of aquaculture and offshore wind energy structures. In *Proceedings of the Sixth Chinese-German Joint Symposium on Hydraulic and Ocean Engineering (CGJOINT 2012)* (Vol. 6, pp. 597–603), Keelung: National Taiwan Ocean University.

Grabitz, E., Hilf, M., & Nettesheim, M. (2010–2015). *Das Recht der Europäischen Union.* Munich, Germany: Beck.

Grandori, A., & Soda, G. (1995). Inter-firm networks: Antecedents, mechanisms and forms. *Organisational Studies, 16,* 183–214.

Griffin, R., & Krause, G. (2010). *Economics of wind farm-mariculture integration.* Working Paper, Department of Environmental and Natural Resource Economics, University of Rhode Island.

Grote, B. (2016). Bioremediation of aquaculture wastewater: evaluating the prospects of the red alga *Palmaria palmata* (Rhodophyta) for nitrogen uptake. *Journal of Applied Phycology.* doi:10.1007/s10811-016-0848-x

Grote, B., & Buck, B. H. (2017). The IMTA-approach for nutrient balanced aquaculture: Evaluating the potential of turbot (*Scophthalmus maximus*) and dulse (*Palmaria palmata*) from onshore RAS to offshore wind farm environments. *Aquaculture.*

Gündling, L. (1983). *Die 200 Seemeilen-Wirtschaftszone: Entstehung eines neuen Regimes des Meeresvölkerrechts.* Berlin, Germany: Springer.

Guiry, M. D., & Blunden, G. (1991). *Seaweed resources in Europe: Uses and potential.* Chichester, UK: Wiley.

Hatch, U., & Thai, C. F. (1997). A survey of aquaculture production economics and management. *Aquaculture Economics & Management, 1,* 13–27.

Hesley, C. (1997). Open ocean aquaculture: Chartering the future of Ocean farming. In *Proceedings of an International Conference, April 23–25, 1997, Maui, Hawaii.* UNIHI-Seagrant-CP-98-08, University of Hawaii Sea Grant College Program: Maui.

Hickman, R. W. (1992). Mussel cultivation. In E. Gosling (Ed.), *The mussel Mytilus: Ecology, physiology, genetics and culture* (pp. 465–510). Amsterdam/London: Elsevier.

Howell, W. H., & Chambers, M. D. (2005). Growth performance and survival of Atlantic halibut (*Hippoglossus hippoglossus*) growth in submerged net pens. In *Proceedings of the Contributed Papers of the 21 super(st) Annual, AAC* (pp. 35–37), St. Andrews, NB (Canada), Aquaculture Canada super(OM) 2004.

Hundt, M., Goseberg, N., Wever, L., Ebeling, M., Schlurmann, T., & Dubois, J. (2011). Multiple Nutzung und co-management von Offshore Strukturen: Marine Aquakultur und Offshore Windparks, Tagungsband, 8. FZK Kolloquium. Maritimer Wasserbau und Küsteningenieurwesen. Forschungszentrum Küste (FZK). *Gemeinsame Einrichtung der Leibniz Universität Hannover und der technischen Universität Braunschweig, 8,* 127–140.

Indergaard, M., & Minsaas, J. (1991). Animal and human nutrition. In M. D. Guiry & G. Blunden (Eds.), *Seaweed resources in Europe: Uses and potential* (pp. 21–64). Chichester, New York, Brisbane: John Wiley & Sons.

Jensen, Ø., Dempster, T., Thorstad, E. B., Uglem, I., & Fredheim, A. (2010). Escapes of fishes from Norwegian sea-cage aquaculture: Causes, consequences and prevention. *Aquatic Environmental Interaction, 1,* 71–83.

Jobling, M. (1988). A review of the physiological and nutritional energetics of cod, *Gadus morhua* L., with particular reference to growth under farmed conditions. *Aquaculture, 70,* 1–19.

Johnson, S., & Houston, M. (2000). A reexamination of the motives and gains in joint ventures. *Journal of Finanyial and Quantitative Analysis, 35,* 67–85.

Joschko, T. J., Buck, B. H., Gutow, L., & Schröder, A. (2008). Colonisation of an artificial hard substrate by *Mytilus edulis* in the German Bight. *Marine Biology Research, 4,* 350–360.

Kain, J. M. (1979). A view of the genus *Laminaria. Oeeanography Marine Biology Annual Review, 17,* 101–161.

Kain, J. M. (1991). Cultivation of attached Seaweeds. In M. D. Guiry & G. Blunden (Eds.), *Seaweed resources in Europe: Uses and potential* (pp. 309–377). Chichester, UK: Wiley.

Kain, J. M., & Dawes, C. P. (1987). Useful European seaweeds: Past hopes and present cultivation. *Hydrobiologia, 151*(152), 173–181.

Kersandt, P. (2012). Rechtliche Rahmenbedingungen der Aquakultur im Meeres- und Küstenbereich von Nord- und Ostsee. In T. Bosecke, P. Kersandt & K. Täufer (Eds.), *Mereesnaturschutz. Erhaltung der Biodiversität und andere Herausforderungen im „Kaskadensystem" des Rechts* (pp. 181–206). Berlin: Springer.

Köller, J, Köppel, J., & Peters, W. (Eds.) (2006). *Offshore wind energy—Research on environmental impacts.* Springer.

Korsøen, O. J. (2011). Biological criteria for submergence of physostome (Atlantic salmon) and physoclist (Atlantic cod) fish in sea-cages. *Dissertation*, The University of Bergen, Bergen.

Krause, G., Buck, B. H., & Rosenthal, H. (2003). Multifunctional use and environmental regulations: potentials in the offshore aquaculture development in Germany. In A. Sjoestroem (Ed.), *Rights and Duties in the Coastal Zone, Proceedings of the Multidisciplinary Scientific Conference on Sustainable Coastal Zone Management Rights and Duties in the Coastal Zone, 12–14 June 2003*. Sweden: Stockholm.

Krause, G., Griffin, R. M., & Buck, B. H. (2011). Perceived concerns and advocated organisational structures of ownership supporting 'Offshore Wind Farm – Mariculture Integration'. In G. Krause (Ed.), *From turbine to wind farms: Technical requirements and spin off products* (pp. 203–218). Rijeka, Croatia: InTech, Open Access Publisher.

Krause, G., Brugere, C., Diedrich, A., Troell, M., Ebeling, M. W., Ferse, S. C., et al. (2015). A Revolution without people? Closing the people-policy gap in aquaculture development. *Aquaculture, 447*, 44–55.

Krone, R. (2012a). Artificial habitat in polyhedron for crustaceans in marine soft bottoms and application (Künstliches Habitat in Polyederform für Krebstiere auf marinen Weichböden und Anwendung). German Patent DE102010049049B3.

Krone, R. (2012b). Offshore wind power reef effects and reef fauna roles. *Dissertation*, University Bremen.

Krone, R., & Krämer, P. (2011). Device for colonizing and harvesting marine hardground animals, (Vorrichtung zur Ansiedelung und Ernte von marinen Hartbodentieren); German Patent DE102009058278B3; United States patent US020110139083A1.

Krone, R., & Krämer, P. (2012). Transportable device for colonizing and harvesting invertebrates and its use, (Transportierbare Vorrichtung zur Ansiedlung und Erntung von wirbellosen Tieren und Anwendung davon. German Patent DE102009049083B3; Canadian patent CA000002764735A1; international patent WO002011042003A1.

Krone, R., & Schröder, A. (2011). Wrecks as artificial lobster habitats in the German Bight. Helgoland marine research. doi: 10.1007/s10152-010-0195-2

Krone, R., Schmalenbach, I., & Franke, H. D. (2013). Lobster settlement at the offshore wind farm Riffgat, German Bight (North Sea). WinMon. BE Conference, Environmental impact of offshore wind farms, Royal Belgian Institute of Natural Sciences, Brussels, Belgium, 26 November 2013–28 November 2013.

Krone, R., Schröder, A., & Krämer, P. (2012). Device for developing habitats in the underwater area of an offshore construction, (Vorrichtung zur Habitaterschließung im Unterwasserbereich eines Offshore-Bauwerks). German Patent DE102010021606B4; international patent WO002011147400A3.

Kumar, M. V. (2007). Asymmetric wealth gains in joint ventures: Theory and evidence. *Finance Research Letters, 4*, 19–27.

Laing, I., Walker, P., & Areal, F. (2006). Return of the native—Is European oyster (*Ostrea edulis*) stock restoration in the UK feasible? *Aquatic Living Resources, 19*, 283–287.

Lanari, D., D'Agaro, E., & Ballestrazzi, R. (2002). Growth parameters in European sea bass (*Dicentrarchus labrax* L.): Effects of live weight and water temperature. *Italian Journal of Animal Science, 1*, 181–185.

Langan, R., & Horton, F. (2003). Design, operation and economics of submerged longline mussel culture in the open ocean. *Bulletin of the Aquaculture Association of Canada, 103*, 11–20.

Le Gall, L., Pien, S., & Rusig, A. M. (2004). Cultivation of *Palmaria palmata* (Palmariales, Rhodophyta) from isolated spores in semi-controlled conditions. *Aquaculture, 229*, 181–191.

Liu, Y., & Sumaila, R. U. (2007). Economic analysis of netcage versus sea-bag production systems for salmon aquaculture in British Columbia. *Aquaculture Economics & Management, 11*, 371–395.

Løvstad-Holdt, S., & Kraan, S. (2011). Bioactive compounds in seaweed: functional food applications and legislation. *Journal of Applied Phycology, 23*, 543–597.

Lutges, S., & Holzfuss, H. (2006). Integrates coastal zone management in Germany: Assessment and steps towards a national ICZM strategy. German Federal Ministry for the Environment, Nature Conservation and Nuclear Safety.

Lüning, K. (1979). Growth strategies of three *Laminaria* species (Phaeophyceae) inhabiting different depth zones in the sublittoral region of Helgoland (North Sea). *Marine Ecology Progress Series, 1,* 195–207.

Lüning, K. (1990). *Seaweeds: Their environment, biogeography, and ecophysiology.* Chichester, New York, Brisbane: Wiley.

Lüning, K. (2001). SEAPURA: Seaweeds purifying effluents from fish farms, an EU project coordinated by the Wattenmeerstation Sylt, Wadden Sea Newsletter, 20–21.

Lüning, K., & Pang, S. J. (2003). Mass cultivation of seaweeds: current aspects and approaches. *Journal of Applied Phycology, 15,* 115–119.

Lüning, K., Wagner, A., & Buchholz, C. (2000). Evidence for inhibitors of sporangium formation in *Laminaria digitata* (Phaeophyceae) during the season of rapid growth. *Journal of Phycology, 36,* 1129–1134.

Matos, J., Costa, S., Rodrigues, A., Pereira, R., & Sousa-Pinto, I. (2006). Experimental integrated aquaculture of fish and red seaweeds in Northern Portugal. *Aquaculture, 252,* 31–42.

Matthiessen, G. C. (2001). *Oyster culture.* London: Wiley-Blackwell.

Mc Hugh, D. J. (2003). A guide to the seaweed industry. FAO Fisheries Technical Papers T441.

Mee, L. (2006). *Complementary benefits of alternative energy: Suitability of offshore wind farms as aquaculture sites.* Report of Project 10517, commissioned by Seafish.

Michler-Cieluch, T. (2009). *Co-management processes in integrated coastal management—The case of integrating marine aquaculture in offshore wind farms.* Hamburg: University of Hamburg, Department of Integrative Geography.

Michler-Cieluch, T., & Krause, G. (2008). Perceived concerns and possible management strategies for governing wind farm-mariculture integration. *Marine Policy, 32,* 1013–1022.

Michler-Cieluch, T., & Kodeih, S. (2008). Mussel and seaweed cultivation in offshore wind farms: an opinion survey. *Coastal Management, 36,* 392–411.

Michler-Cieluch, T., Krause, G., & Buck, B. H. (2009a). Reflections on integrating operation and maintenance activities of offshore wind farms and mariculture. *Ocean and Coastal Management, 52,* 57–68.

Michler-Cieluch, T., Krause, G., & Buck, B. H. (2009b). Marine aquaculture within offshore wind farms: Social aspects of multiple use planning. *GAIA, 18,* 158–162.

Mittendorf, K., Nguyen, B., & Zielke, W. (2001). User manual waveloads—A computer program to calculate wave loading on vertical and inclined tubes. Institut für Strömungsmechanik und Elektron. Rechnen im Bauwesen, Universität Hannover. Retrieved October 12, 2016, from

Moksness, E., Kjorsvik, E., & Olsen, Y. (2004). *Culture of cold-water marine fish.* London: Wiley-Blackwell.

Molenaar, F. J., & Breeman, A. M. (1997). Latitudinal trends in the growth and reproductve seasonality of *Delesseria sanguinea, Membranoptera alata and Phycodrys rubens* (RHODOPHYTA). *Journal of Phycology, 33,* 330–343.

Morgan, K. C., & Simpson, F. J. (1981). The cultivation of Palmaria palmata. Effect of light intensity and temperature on growth and chemical composition. *Botanica Marina, 24,* 547–552.

Morgan, K. C., Wright, J. L., & Simpson, F. J. (1980a). Review of chemical constituents of the red alga *Palmaria palmata* (dulse). *Economic Botany, 34,* 27–50.

Morgan, K. C., Shacklock, P. F., & Simpson, F. J. (1980b). Some aspects of the culture of *Palmaria palmata* in greenhouse conditions. *Botanica Marina, 23,* 765–770.

Morison, J. R., O'Brien, M. P., & Schaaf, S. A. (1950). The force exerted by surface waves on piles. *Petroleum Transactions, AIME 2846, 189,* 149–157

Neori, A., Chopin, T., Troell, M., Buschmann, A. H., Kraemer, G. P., & Halling, C. (2004). Integrated aquaculture: Rationale, evolution and state of the art emphasizing seaweed biofiltration in modern mariculture. *Aquaculture, 231,* 361–391.

Newkirk, G. F., Muise, B., & Enright, C. T. (1995). Culture of the Belon oyster, *Ostrea edulis*, in Nova Scotia. In A. D. Boghen (Ed.), *Cold-water aquaculture in Atlantic Canada*. Moncton, Canada: Canadian Institute for Research on Regional Development.

Nielsen, B. (2010). Strategic fit, contractual, and procedural governance in alliances. *Journal of Business Research, 63*, 682–689.

NKüFischO. (2006). Niedersächsische Küstenfischereiordnung, Nds. GVBl. S. 108—VORIS 79300.

Noble, B. (2000). Institutional criteria for co-management. *Marine Policy, 24*(1), 69–77.

OFT. (2010). *Aquapod—A submergible fish cage*. Searsmont, USA: Ocean Farm Technologies.

Özbilgina, H., & Glass, C. W. (2004). Role of learning in mesh penetration behaviour of haddock (*Melanogrammus aeglefinus*). *ICES Journal of Marine Science: Journal du Conseil, 61*, 1190–1194.

Pang, S. J., & Lüning, K. (2004). Tank cultivation of the red alga *Palmaria palmata*: Effects of intermittent light on growth rate, yield and growth kinetics. *Journal of Applied Phycology, 16*, 93–99.

Pang, S. J., & Lüning, K. (2006). Tank cultivation of the red alga *Palmaria palmata*: Year-round induction of tetrasporangia, tetraspore release in darkness and mass cultivation of vegetative thalli. *Aquaculture, 252*, 20–30.

Pazos, A. J., Ruíz, C., García-Martín, O., Abad, M., & Sánchez, J. L. (1996). Seasonal variations of the lipid content and fatty acid composition of *Crassostrea gigas* cultured in El Grove, Galicia, N.W. Spain. *Comparative Biochemistry and Physiology, 114*, 171–179.

Pelz, P. (1974). Rationalisierung aus unternehmerischer Sicht: Vorbereitung und technischwirtschaftliche Beurteilung. In P. Pelz (Ed.), *Die Priorität der Rationalisierung im Unternehmen* (pp. 58). Düsseldorf/Berlin, Germany: Verein Deutscher Ingenieure, VDI-Bildungswerk.

Person-Le Ruyet, J., Buchet, V., Vincent, B., Le Delliou, H., & Quemener, L. (2006). Effects of temperature on the growth of pollack (*Pollachius pollachius*) juveniles. *Aquaculture, 251*, 340–345.

Perez, P., Kaas, R., Campello, F., Arbault, S., & Barbaroux, O. (1992). *La culture des algues marines dans le monde*. Nantes, France: IFREMER.

Petrell, R. J., & Alie, S. Y. (1996). Integrated cultivation of salmonids and seaweeds in open systems. *Hydrobiologia, 326*, 67–73.

Pogoda, B. (2012). Farming the High Seas: Biological performance of the offshore cultivated oysters *Ostrea edulis* and *Crassostrea gigas* in the North Sea. *Dissertation*, Marine Zoology of the Department of Biology & Chemistry, University of Bremen.

Pogoda, B., Buck, B. H., & Hagen, W. (2011). Growth performance and condition of oysters (Crassostrea gigas and Ostrea edulis) farmed in an offshore environment (North Sea, Germany). *Aquaculture, 319*, 484–492.

Pogoda, B., Grote, B., & Buck, B. H. (2015). Aquakultur-Site-Selection für die nachhaltige und multifunktionale Nutzung von marinen Gebieten in stark genutzten Meeren am Beispiel der Nordsee—Offshore Site-Selection. Gefördert durch das BMELV—Förderkennzeichen GZ: 511-06.01-28-1-73.009-10.

Pogoda, B., Buck, B. H., Saborowski, R., & Hagen, W. (2013). Biochemical and elemental composition of the offshore-cultivated oysters *Ostrea edulis* and *Crassostrea gigas*. *Aquaculture, 400*, 53–60.

Pogoda, B., Jungblut, S., Buck, B. H., & Hagen, W. (2012). Infestation of oysters and mussels by mytilicolid copepods: Differences between natural coastal habitats and two offshore cultivation sites in the German Bight. *Journal of Applied Ichthyology, 28*(5), 756–765.

Polk, M. (1996). Open Ocean Aquaculture. In *Proceedings of an International Conference, May 8–10, 1996, Portland, Maine*. UNHMP-CP-SG-96-9, Portland, New Hampshire/Maine Sea Grant College Program.

Pomeroy, R. S., Parks, J. E., & Balboa, C. M. (2006). Farming the reef: Is aquaculture a solution for reducing fishing pressure on coral reefs? *Marine Policy, 30*, 111–130.

Raman-Nair, W., & Colbourne, D. B. (2003). Dynamics of a mussel longline system article in aquacultural. *Engineering, 27,* 191–212.

Rhode Island Coastal Resources Management Council. (2010). Rhode Island Ocean Special Area Management Plan.

Rick, H. J., Rick, S., Tillmann, U., Brockmann, U., Gaertner, U., Duerselen, C., et al. (2006). Primary productivity in the German bight (1994–1996). *Estuaries and Coasts, 29,* 4–23.

Robinson, D. (2008). Strategic alliances and the boundaries of the firm. *The Review of Financial Studies, 21,* 649–681.

Rosenberg, R., & Loo, L. O. (1983). Energy flow in a *Mytilus edulis* culture in western Sweden. *Aquaculture, 35,* 151–167.

Rosenlund, G., & Skretting, M. (2006). Worldwide status and perspective on gadoid culture. *ICES Journal of Marine Science, 63,* 194–197.

Ruíz, C., Martinez, D., Mosquera, G., Abad, M., & Sánchez, J. L. (1992). Seasonal variations in condition, reproductive activity and biochemical composition of the flat oyster, *Ostrea edulis,* from San Cibran (Galicia, Spain). *Marine Biology, 112,* 67–74.

SAG. (1965). Seeaufgabengesetz of 24.5.1965, Federal Law Gazette 1965 II p. 833, last amended by Act of 19.10. 2013, Federal Law Gazette I p. 3836.

Sahr, J. (2006). *Wind und Meer: Küstenproduktion von morgen. Anregungen für Marikultur in Offshore Windparks.* Bremerhaven: CG Gloy.

Sanderson, J. C., Dring, M. J., Davidson, K., & Kelly, M. S. (2012). Culture, yield and bioremediation potential of *Palmaria palmata* (*Linnaeus*) Weber & Mohr *and Saccharina latissima* (*Linnaeus*) C.E. Lane, C. Mayes, Druehl & G.W. Saunders adjacent to fish farm cages in northwest Scotland. *Aquaculture, 354–355,* 128–135.

Scarratt, D. (1993). *A handbook of northern Mussel culture.* Montague: Island Press.

Schmalenbach, I., & Krone, R. (2011). Secondary use of offshore wind farms—Settlement of juvenile European lobsters (*Homarus gammarus*). IMARE-Conference "Marine Resources and Beyond", 5–7 September 2011, Bremerhaven, Germany.

Schmalenbach, I., Mehrtens, F., Janke, M., & Buchholz, F. (2011). A mark-recapture study of hatchery-reared juvenile European lobsters, *Homarus gammarus,* released at the rocky island of Helgoland (German Bight, North Sea) from 2000 to 2009. *Fisheries Research, 108,* 22–30.

SFA. (2008). Annual data for mussel harvest in Lower Saxony (Germany) in the years 1950–2007 provided by the State Fisheries Agency Bremerhaven (Germany).

Smaal, A. C., Kater, B. J., & Wijsman, J. (2009). Introduction, establishment and expansion of the Pacific oyster *Crassostrea gigas* in the Oosterschelde (SW Netherlands). *Helgoland Marine Research, 63,* 75–83.

Sørensen, H. C., Hansen, J., & Vølund, P. (2001). Experiences from the establishment of Middelgrunden 40 MW offshore wind farm. In *Proceedings of the European Wind Energy Conference,* 2–6 July, Copenhagen, Denmark: EWEA.

Stelzenmüller, V., Diekmann, R., Bastardie, F., Schulze, T., Berkenhagen, J., Kloppmann, et al. (2016). Co-location of passive gear fisheries in offshore wind farms in the German EEZ of the North Sea: A first socio-economic scoping. *Journal of Environmental Management, 183,* 794–805. doi dx.doi.org/10.1016/j.jenvman.2016.08.027.

Stickney, R. R. (1998). Joining forces with industry—open ocean aquaculture. In *Proceedings of the Third Annual International Conference,* May 10–15, Corpus Christi, Texas. TAMU-SG-99-103, Corpus Christi, Texas Sea Grant College Program.

Sturrock, H., Newton, R., Paffrath, S., Bostock, J., Muir, J., Young, J., et al. (2008). Prospective analysis of the aquaculture sector in the EU. Part 2: Characterisation of emerging aquaculture systems. Seville, Spain: Institute for Prospective Technological Studies.

Subandar, A., Petrell, R. J., & Harrison, P. J. (1993). Laminaria culture for reduction of dissolved inorganic nitrogen in salmon farm effluent. *Journal of Applied Phycology, 5,* 455–463. doi:10. 1007/BF02182738

Thieltges, D. W., Krakau, M., Andresen, H., Fottner, S., & Reise, K. (2006). Macroparasite community in molluscs of a tidal basin in the Wadden Sea. *Helgoland Marine Research, 60,* 307–316.

Tortell, P. (1976). A new rope for mussel farming. *Aquaculture, 8,* 383–388.

Treasurer, J. W., Sveier, H., Harvey, W., Allen, R., Cutts, C. J., Mazorra de Quero, C., et al. (2006). Growth, survival, diet, and on-growing husbandry of haddock *Melanogrammus aeglefinus* in tanks and netpens. *ICES Journal of Marine Science, 63,* 376–384.

Troell, M., Rönnbäck, P., Halling, C., Kautsky, N., & Buschmann, A. (1999). Ecological engineering in aquaculture: Use of seaweeds for removing nutrients from intensive mariculture. *Journal of Applied Phycology, 11,* 89–97.

VüAs. (1997). Verordnung über Anlagen seewärts der Begrenzung des deutschen Küstenmeeres of 23.01.1997, Federal Law Gazette I p. 57, last amended by Ordinance of 29.8.2013 Federal Gazette of 30.08.2013.

Vollstedt, B. (2011). The role of perceptions and networks in multiple marine resource use: Integrating aquaculture and offshore wind farms in the North Sea. Master thesis University Kiel.

Walne, P. R., & Mann, R. (1975). Growth and biochemical composition in *Ostrea edulis* and *Crassostrea gigas*. In H. Barnes (Ed.), *Proceedings of 9th EMBS Aberdeen* (pp. 587–607).

Walter, U., & Liebezeit, G. (2003). Efficiency of blue mussel (*Mytilus edulis*) spat collectors in highly dynamic tidal environments of the Lower Saxonian coast (southern North Sea). *Biomolecular Engineering, 20,* 407–411.

Walter, U., & Liebezeit, G. (2001). Abschlußbericht des Projektes: Nachhaltige Miesmuschel-Anzucht im niedersächsischen Wattenmeer durch die Besiedlung natürlicher und künstlicher Substrate. Forschungszentrum Terramare.

Walter, U., Buck, B. H., & Rosenthal, H. (2002). Marikultur im Nordseeraum: Status Quo, Probleme und Tendenzen. In J. L. Lozán, E. Rachor, & K. Reise (Eds.), *Warnsignale aus Nordsee & Wattenmeer: eine aktuelle Umweltbilanz* (pp. 122–131). Hamburg: Büro Wiss. Auswertungen.

Weiss, M., & Buck, B. H. (2017). Blue Mussel *(Mytilus edulis)* Meat as a Partial Fish Meal Replacement for the Diet in Turbot Aquaculture. *Journal of Applied Ichthyology.*

Weiss, M., Wittke, S., Greim Fish Consulting, WeserWind, Louis Schoppenhauer GmbH & Co. KG. (2012). NutriMat: Nutritional material from fouling organisms—Analyse der Nutzungsmöglichkeiten von biologischem Aufwuchs von künstlichen Hartsubstraten für die Gewinnung alternativer Protein- und Lipidressourcen. Bremerhaven, Germany: Imare.

Werner, A., & Dring, M. (2011). *Aquaculture explained no 27.* BIM—Irish Sea Fisheries Board: Cultivating Palmaria palmata.

Wever, L. (2011). *The emergence of multi-use ocean governance: Potentials and constraints for offshore windfarm-mariculture integration in Germany.* International Conference "Marine Resources and Beyond 2011", September 5-7 2011, Bremerhaven, Germany. Book of Abstracts. 46 p.

Wever, L., Buck, B. H., & Krause, G. (2014). Lessons from stakeholder dialogues on marine aquaculture in offshore wind farms: Perceived potentials, constraints and research gaps. *Marine Policy, 51,* 251–259.

Whitmarsh, D. J., Cook, E. J., & Black, K. D. (2006). Searching for sustainability in aquaculture: An investigation into the economic prospects for an integrated salmon-mussel production system. *Marine Policy, 30,* 293–298.

Whyte, J. N. C., Englar, J. R., & Carswell, B. L. (1990). Biochemical composition and energy reserves in *Crassostrea gigas* exposed to different levels of nutrition. *Aquaculture, 90,* 157–172.

Wildish, D. J., & Kristmanson, D. D. (1984). Importance to mussels of the benthic boundary layer. *Canadian Journal of Fisheries and Aquatic Sciences, 41,* 1618–1625.

Wildish, D. J., & Kristmanson, D. D. (1985). Control of suspension feeding bivalve production by current speed. *Helgoländer Wiss Meeresuntersers, 39,* 237–243.

Williamson, O. (1979). Transaction-cost economics: The governance of contractual relations. *Journal of Law and Economics, 22,* 233–261.

Williamson, O. (1981). The economics of organisation: The transaction cost approach. *American Journal of Sociology, 87,* 548–577.

Wiltshire, K. H., & Manly, B. F. J. (2004). The warming trend at Helgoland Roads, North Sea: Phytoplankton response. *Helgoland Marine Research, 58,* 269–273.

Chapter 12
The EU-Project "TROPOS"

Nikos Papandroulakis, Claudia Thomsen, Katja Mintenbeck,
Pedro Mayorga and José Joaquín Hernández-Brito

Abstract The global population is growing and the demand for food and energy is
steadily increasing. Coastal space all over the world becomes increasingly limited
and near-shore resources are often already heavily exploited. The use of offshore
regions may provide new opportunities, but also involves major challenges such as
the development of designs and technologies suitable for offshore condition. The
floating TROPOS 'Green & Blue' modular multi-use platform concept introduced
in this chapter is especially designed for offshore conditions and provides solutions
for the problems and obstacles involved in "moving offshore". The Green & Blue
platform concept integrates fish and algae aquaculture with a wind farm. The
floating multi-use approach allows for platform operation in deep waters and the
promotion of synergies such as joint logistics, shared infrastructure and services,
thereby making the use of offshore resources viable and profitable.

N. Papandroulakis (✉)
Hellenic Centre for Marine Research, Institute of Marine Biology Biotechnology
and Aquaculture, AquaLabs, Gournes, Heraklion GR 71500, Crete, Greece
e-mail: npap@hcmr.gr

C. Thomsen
Phytolutions GmbH, Campus Ring 1, Bremen 28759, Germany

K. Mintenbeck
Alfred Wegener Institute Helmholtz Centre for Polar and Marine Research (AWI), Integrative
Ecophysiology/Marine Aquaculture, Am Handelshafen 12, Bremerhaven 27570, Germany

P. Mayorga
EnerOcean S.L, Bulevar Louis Pasteur 5, Of.321, Málaga 29010, Spain

J.J. Hernández-Brito
PLOCAN (Consorcio Para La Construcción, Equipamiento Y Explotación de La Plataforma
Oceánica de Canarias), Crta. de Taliarte, s/n, Telde 35200, Spain

B.H. Buck and R. Langan (eds.), *Aquaculture Perspective of Multi-Use Sites
in the Open Ocean*, DOI 10.1007/978-3-319-51159-7_12

12.1 Introduction

12.1.1 Moving Offshore

Coastal space all over the world has become increasingly limited and near-shore resources are often already heavily exploited. As a consequence, there is a growing call to move towards offshore aquaculture technologies. The advantages of offshore aquaculture compared to coastal aquaculture include, e.g., fewer restrictions (environmental, geographical, political), less competition for space with other activities (Ryan 2004), and better environmental conditions which result in improved fish growth, more efficient feed conversion, and a lower risk for pathogen infection (Ryan et al. 2007). Additionally, moving from protected areas of restricted water circulation to more exposed sea conditions allows farms with higher production capacity without increasing environmental impacts (Lu et al. 2014). However, moving offshore involves some major challenges, among others, high costs for the required infrastructure and transportation, more difficult access to services, supply and markets, and the need to adapt methodologies and procedures to offshore conditions. The best approach to overcome these problems is the development of shared solutions with other offshore sectors, i.e. to find and make use of synergies for a common use of sea space, infrastructure, resources (energy, water, supplies), services (maintenance, monitoring, surveillance) and transportation.

Accordingly, the European Union has launched the "The Ocean of Tomorrow" call in the 7th Framework Programme (FP7) for Research and Development, with one topic being "Multi-use Offshore Platforms" (OCEAN 2011.1). In total, three projects were selected and received funding for the design of multi-use offshore platforms: H2OCEAN, MERMAID and TROPOS. The latter one, the TROPOS project, is presented in this chapter.

12.1.2 About TROPOS

The full title of the TROPOS project is "*Modular Multi-use Deep Water Offshore Platform Harnessing and Servicing Mediterranean, Subtropical and Tropical Marine and Maritime Resources*". TROPOS had a total project duration of 3 years (January 2012–January 2015). The project involved 20 partners from 9 different countries and was coordinated by PLOCAN (PLataforma Oceánica de las CANarias, Gran Canaria, Canary Islands, Spain).

The TROPOS project aimed at developing a floating modular multi-use platform system for use in deep waters, with an initial focus on Mediterranean and sub-tropical regions (Quevedo et al. 2012). Multi-use offshore platforms should allow for sustainable and eco-friendly uses and a synergistic exploitation of oceanic resources. A floating design facilitates access to deep sea areas and resources where deployment of conventional platform types is not possible. A modular multi-use

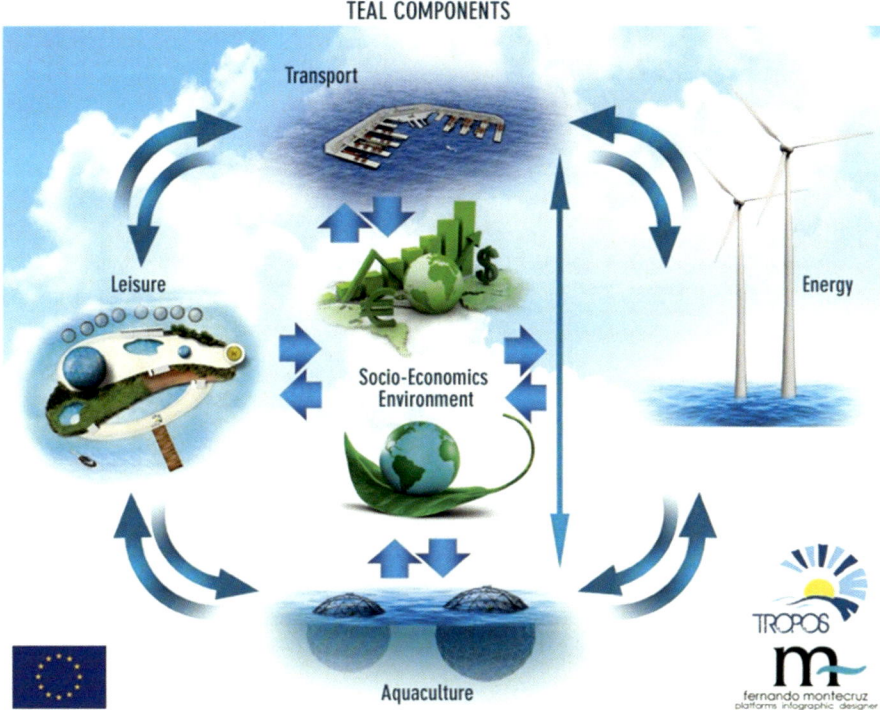

Fig. 12.1 The TROPOS TEAL (Transport, Energy, Aquaculture, Leisure) components (Fernando Montecruz for the TROPOS project)

approach allows for the integration of a range of functions from different sectors. In the case of TROPOS, functions of four different sectors were integrated, namely *Transport*, *Energy*, *Aquaculture*, and *Leisure* (in short: TEAL; Fig. 12.1). *Marine Transport* (T) provides critical services to the society ranging from building commercial and leisure ships, shipping of goods and fuel around the world, passenger transfer, to servicing offshore structures. The development of renewable *Energies* (E) is essential to address the dramatic depletion of fossil fuel reserves and to fight climate change, which has become one of the most critical global issues in recent years. Natural marine living resources are already heavily exploited, while the demand for these resources is steadily increasing. To reduce the fishing pressure on wild stocks, the demand needs to be increasingly met additionally by *Aquaculture* (A). The tourism industry represents the third largest socio-economic activity in the EU and space is needed for the development of new *Leisure* (L) activities. Not only in Europe, but all over the world there is an increasing demand for innovative, eco-friendly solutions in the tourism sector.

The development process involved successive iterations to find feasible solutions restricted by technical, environmental and socio-economic constrains. A multidisciplinary and cross sectorial team of specialists developed several tools to explore suitable sites, combinations of resources, synergies, structural solutions and their technical, environmental and socio-economic viabilities. The analyses also included logistics, security, installation, operational, decommissioning and maintenance requirements. The outcome of the development process included different theoretical platform concepts, designed and specified in great detail based on simulations and models which were validated by experimental tests of realistic small-scale models. The developed TROPOS concepts included the so-called "Sustainable Sevice Hub", "Leisure Island" and the "Green & Blue" concept. The latter one, the Green & Blue concept, integrates offshore aquaculture and renewable energies (Papandroulakis et al. 2014).

12.1.3 Aquaculture in TROPOS

Human population is growing and the demand for seafood is steadily increasing. Marine organisms are a source of high quality nutrients (proteins, essential fatty acids, vitamins, minerals, etc.) which are fundamental in human nutrition, but are also an important component in animal feeds and non-food sectors (e.g. cosmetic products). As the availability of natural resources is limited, an increasing amount of the demand has to be covered by aquaculture products. Aquaculture is the fastest growing economic area in the food industry sector and the proportion of the world's food fish coming from aquaculture has risen in 2006 to almost half (FAO 2014). With the increasing demand for seafood the fish-farming industry will have to continue this trend (Cressey 2009). Accordingly, the integration of aquaculture in the TROPOS multi-use concepts was of major interest.

The European aquaculture sector has access to dynamic and cutting-edge research and technologies, advanced equipment and fish feed. Production currently meets the requirements for environmental protection, and the final products are traceable and meet high quality standards. The European Aquaculture Technology and Innovation Platform (EATiP) has set several objectives for the sector to achieve by 2030. Particularly for the Mediterranean, the target is to increase production by 200%. This production scenario will increase the current production from 230,000 to 690,000 t. To meet this objective, EATiP has defined the development of efficient technologies (such as developed in the scope of TROPOS) to support continued growth as one of its strategic goals.

Considering the expected growth of the aquaculture industry and the market globalization, the targeted species, regardless of group (i.e. fish, shellfish, algae), have to shift to those with global distribution and market. The target species should be a fast growing and well adapted to the environmental conditions offshore and clear oceanic waters. Among finfish, typical target species are large, fast growing pelagic and deep demersal species, such as the greater amberjack and the wreckfish; both species recently became targets for developing innovative and efficient rearing technologies.

In TROPOS, the main focus was not only on finfish but also on microalgae aquaculture.

Microalgae are not only of high interest in the food and feed sectors, but also for the production of biofuel (Enzing et al. 2014). The market volume and value of microalgae range over a broad scale from 1,000–5,000 €/t for high volume biofuels to 10,000–35,000 €/t for high value food. The value of microalgae used for pharmaceuticals is even higher. In the aquaculture industry microalgae are important as feed for early developmental stages of several species (larvae of molluscs, echinoderms, crustaceans, fish), with a worldwide annual production of approximately 10,000 t/year and a market value of about 30–250 €/kg. Approximately $^1/_5$ of this biomass is used for larval stages of fish and shellfish reared in hatcheries (Muller-Feuga 2004). The most frequently used species are *Chlorella, Tetraselmis, Dunaliella, Isochrysis, Pavlova, Nannochloropsis, Phaeodactylum, Chaetoceros, Skeletonema* and *Thalassiosira* (Spolaore et al. 2006).

The greatest demand for microalgae lies in its potential to replace the current sources of PUFA (PolyUnsaturated Fatty Acids; Omega 3/6 fatty acids). The increasing public debate on the use of fishmeal for aquaculture puts microalgae in the focus as substitute for fishmeal, and Omega-3 industries increasingly produce concentrated EPA (eicosapentaenoic acid) and DHA (Docosahexaenoic acid) oils for human consumption.

Microalgae (and macroalgae) are also considered as possible substitute for fossil fuels. Despite a large body of research in the area of photosynthesis for carbon sequestration, little work has been done to develop practical algae production systems that do not ignore land availability limitations and can be installed at any location for the production of fuels, food and animal feed. Offshore algae production could fill that gap and allow the energy and cost efficient high volume conversion into biofuels.

12.2 The Conceptual Approach of TROPOS

12.2.1 General Platform Design

The key characteristics of the TROPOS platform concepts are (i) the floating design which enables platform operation in deep waters, (ii) the multi-use concept which supports the integration of different functions and services at one site and facilitates synergies, e.g. by joint logistic, and (iii) the modular approach which allows for a flexible combination of different types of modules adapted to requirements. In this way, the TROPOS multi-use platform system is able to integrate a full range of functions from the four TEAL sectors (Fig. 12.1) and can be perfectly adapted and used in a much greater number of global geographical locations than a non-floating, non-modular and/or a single-use platform design.

The conceptual design of the TROPOS platform involves a central unit associated with modules and satellites (TROPOS 2013b, 2015). This floating central unit can be moored to the seafloor and builds the core of the platform. The lower part of the central unit is equipped with a double hull to prevent oil spills. Modules with different functions can be directly attached to the central unit, and/or satellite units can be indirectly connected (via undersea cables), each according to requirements. Satellite units are fixed with their own mooring.

In Fig. 12.2, a schematic representation of platform elements and their functional connections is shown as an example. In this example the platform is composed of the central unit, two modules (substation and operation & maintenance base) and one type of satellite (renewable energy and fish aquaculture). Functional units are shown as black boxes, with the specific inputs and outputs relevant for their functionality (TROPOS 2014b).

The design and engineering specifications of the platform concept considers extreme weather conditions and unplanned events to be prepared for all kinds of emergencies. Safety and security of crew and visitors, and protection of the environment, are the top priority in the design of the TROPOS scenarios.

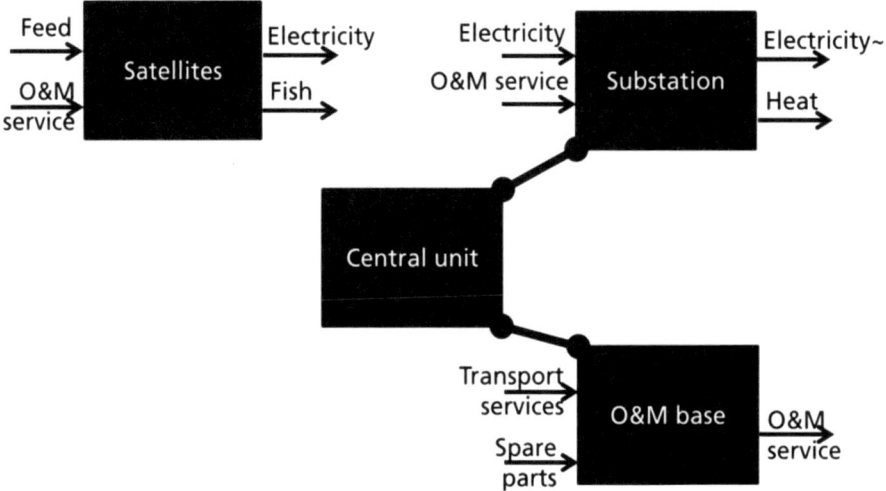

Fig. 12.2 Schematic example of TROPOS platform functional units and their connections. The example shows the main in and outputs of the satellite unit and two modules integrated into the central unit: a substation module representing the electrical connection between central unit and satellites, and an operation and maintenance (O&M) module (*Source* TROPOS 2014b)

Different concepts with different combinations of functional units from the TEAL sectors were developed in the project, among these, the so-called "*Green & Blue*" concept which is specified below.

12.2.2 The Green & Blue Concept

The Green & Blue concept is focusing on the use of physical and biological resources, intending to follow the strategies and actions of "Blue Growth[1]" as defined by the European Commission for the development of aquaculture and renewable energies in the EU. The Green & Blue concept integrates offshore aquaculture and renewable energies.

Depending on platform location and local conditions and dynamics, the renewable energy source to be harnessed might be wind, waves, solar or ocean thermal energy (Quevedo et al. 2013). Accordingly, the renewable energy component of the concept may be represented by a wind farm, wave energy converters, photovoltaic (PV) panels or an OTEC (Ocean Thermal Energy Conversion) plant.

The aquaculture component of the Green & Blue concept primarily includes fish and algae aquaculture. Generally, the fish culture unit is planned in the form of floating, submerged or drifting cages. Depending on the location, secondary cultures around the fish cages may be developed for bivalves and/or macroalgae. Additionally, if the location's depth allows such an activity, bottom cultures could be established. The algae culture unit can include several components, e.g. production of biomass (microalgae and macroalgae), products with high market value, and biofuel/gasification from macro and microalgae, and CO_2 fixation from flue gases. Wastewater e.g. domestic wastes from a central platform can be used to nourish algae, thereby combining algae biomass production with wastewater treatment. The source for the flue gas may be emissions from the satellite or the central platform or from on-shore emission sites as CO_2-stripping at a biogas plant.

The aquaculture units (both fish and algae) take advantage of the facilities that the platform offers, namely energy, protection, anchoring and logistics with regard to installation, maintenance and operation. An example for a Green & Blue concept integrating offshore fish and algae aquaculture and wind turbines is shown in Fig. 12.3.

One Green & Blue case scenario, developed in TROPOS and designed to be located in the Mediterranean Sea in deep water locations, is described in more detail below.

---

[1]"Blue Growth" refers to the European Commission's long term strategy to support sustainable growth in the marine and maritime sectors.

Fig. 12.3 Green & Blue concept integrating offshore fish and algae aquaculture and wind turbines (Fernando Montecruz for the TROPOS project)

12.3 The Green & Blue Concept in the Mediterranean

One Green & Blue platform was designed for a location north of Crete (southern Aegean Sea) at about 100 km distance from the shore in about 450 m water depth (Fig. 12.4). The site was chosen based on a GIS supported multi-criteria decision analysis, which also considered socio-economic and environmental aspects. Climatic, geographic and oceanographic characteristics and dynamics in this area were found to be appropriate for this platform concept (TROPOS 2013a).

In this particular scenario, fish and microalgae aquaculture are combined with wind turbines and some PV panels (see Fig. 12.3). The scenario includes six modules (integrated into the central unit) which aim to service, maintain and monitor the satellite units and to ensure smooth and optimal energy and biomass production flow. In all, there are 30 floating satellite units placed around the central unit, which comprise the wind turbines, some PV panels, as well as fish cages and algae floats.

12.3.1 Central Unit

In this concept, the central unit extends over several decks and hosts all functions and services that are needed for the daily operational requirements of the platform. This includes, among others, the engine, diesel oil tanks, waste and wastewater

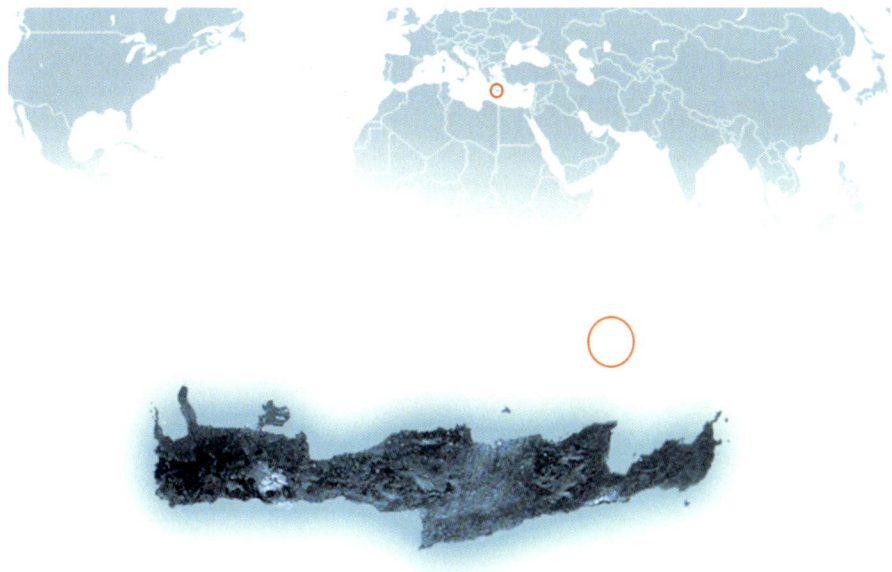

Fig. 12.4 Approximate Green & Blue platform location in the Mediterranean, north of Crete (Map created by Fernando Montecruz for the TROPOS project)

storage and treatment plants, transformers, a desalination unit for freshwater production, freshwater tanks, different kind of workshops, maintenance and repair areas, air conditioning and storage rooms. Moreover the central unit hosts fire-fighting system, control, surveillance & monitoring rooms, and also offices, common and leisure rooms, laundry, kitchen and hospital (TROPOS 2014b). The energy demand of the platform will be completely met by the locally produced renewable energy; the diesel engine mainly serves as a backup in case of emergency or deficiency of the natural energy resource. Waste and wastewater produced during the operational phase of the platform will be treated following best practice and stored on board until it can be transferred to shore. Both aquaculture and energy production require specific facilities for proper operation. These facilities are organized in different modules on the central unit.

12.3.2   Modules

The major objectives of the modules integrated into the central unit in the Green & Blue concept are for monitoring, service, maintenance and processing of the energy and aquaculture production on the satellite units (TROPOS 2014a). These modules include

- The fish processing plant
- The aquaculture workshop
- The aquaculture support
- The algae biorefinery
- The substation
- Accommodation facilities

The fish processing plant provides all facilities required for processing the fish production. Generally, fish biomass produced in a farm may be either being processed on site or transported to another processing facility according to the business plan that will be followed. While for land-based or nearshore aquaculture the processing units are close to markets and the flow of product is continuous. This is not the case for offshore farms where connection with land is neither easy nor daily. Hence, an on-site processing unit is a requisite in order to prolong product's shelf life from days to weeks depending on the processing level. Also, the disadvantages of the offshore farms due to distance may turn to benefits with the production of value added products. Existing alternatives to this approach represent the use of a "boat-factory" which, however, presents some disadvantages related to dependencies on external services and the minimization of flexibilities for marketing.

In the plant, the harvested fresh fish can be directly processed without any delay. The module includes the processing facilities leading to 3 groups of products: (i) fresh fish on ice, (ii) degutted fish, subsequently IQF (Individually Quick Frozen), and (iii) processed fish (fillets or steaks) either IQF or packed in MAP (Modified Atmosphere Packaging). Special care is given to byproducts (estimated at $\sim 20\%$ of the biomass) that are also marketed. There are cooling and freezing rooms, storages, warehouse for packaging material and an ice-production unit.

The aquaculture workshop includes a maintenance unit for repairing equipment, warehouses for storage of harvesting equipment, nets and spare parts.

The aquaculture support module includes a central communication room for control, surveillance and monitoring of the aquaculture production and environmental parameters, and for operation and control of the automated feeding system. The module also includes laboratories both for biological analysis and electronics.

In the algae biorefinery, the harvested microalgae may undergo further processing, such as extraction of proteins, lipids, carbohydrates or pigments, and conversion to biofuel.

The substation is connected to the wind farm and serves as infield cable node, voltage transformation and/or current conversion system and export cable connector between the satellite units, the central unit and the onshore grid. To establish such connection, it is necessary for the substation to have specific power transmission capabilities in combination with distinctive voltage transformation functionalities.

The accommodation module provides living space for crew and visitors.

12.3.3   Satellite Units

The Green & Blue scenario off Crete consists of 30 floating triangular satellite units. Each semi-submersible satellite unit is equipped with two 3.5 MW wind turbines (Fig. 12.5). Some units also include small 434 kW PV panels. The wind turbines have a rotor diameter of 112 m and total production of all 60 turbines (2 per satellite unit) was estimated with about 198 MW (TROPOS 2013c).

According to the design, 30 units will be installed allowing a maximum production capacity of 750 t fish per week. In each unit an open sea surface cage is integrated at the inert part of the triangular base (Figs. 12.4 and 12.6). All daily operational equipment including feed distributor, feed storage silos (66 t, 100 m^3) and monitoring equipment are incorporated to the base, and are remotely controlled from the central platform. For the Green & Blue platform in the Mediterranean, the production plan includes the rearing of 3 different species in two-year rearing cycles: (i) the European Sea Bass (*Dicentrarchus labrax*), (ii) the Meagre (*Argyrosomus regius*), and (iii) the Greater Amberjack (*Seriola dumerili*).

The cage is a typical cylindrical floating offshore cage of 33/36 m (internal/external) diameter. The volume contained in the net cage is about 25,000 m^3. In some of the units, the top of the fish cage is covered by solar panels for additional energy production (TROPOS 2013c).

Fig. 12.5 3D view of a floating wind satellite unit designed by EnerOcean S.L. for the TROPOS Green & Blue concept based on W2Power patented design (*Source* TROPOS 2013c)

Fig. 12.6 A model of the Green & Blue satellite unit triangular base structure with the fish net cage inside; the model was tested at the UCC-BEAUFORT (HMRC)—National Ocean Wave Basin in Cork (Ireland), model scale 1:100 (*Source* TROPOS 2014c)

The microalgae production unit or float containing the microalgae cultures is connected to the triangular construction, one float per satellite unit, making a total of 30 algae floats (Fig. 12.3). These photo-bioreactors are a closed system. The dimensions of a float are 50×200 m with an inserted phytoplankton unit of 400 m³. Two thrusters, located at the end of the plant, help to manoeuvre the submersible culture system (Fig. 12.7).

The satellite units are unmanned and monitored, operated and controlled from the central unit. The fish cages and the algae photo-bioreactors are equipped with sensors for online control and monitoring from the aquaculture modules.

Energy demands of the sensors as well as of the automated fish feeding system are met by the local production from wind and/or solar energy. Service and maintenance are provided by a crew based on the central unit.

In summary, the final layout of the satellite is comprised of:

- Two turbines (3.5 MW each/90 m height) installed on a triangular shaped platform (column distance (base) 90 m/column distance (triangle legs) 80 m) and their auxiliary component to enable an appropriate grid connection.
- A cage for aquaculture moored between the triangular legs (25,000 m³/diameter 33 m/depth 30 m).

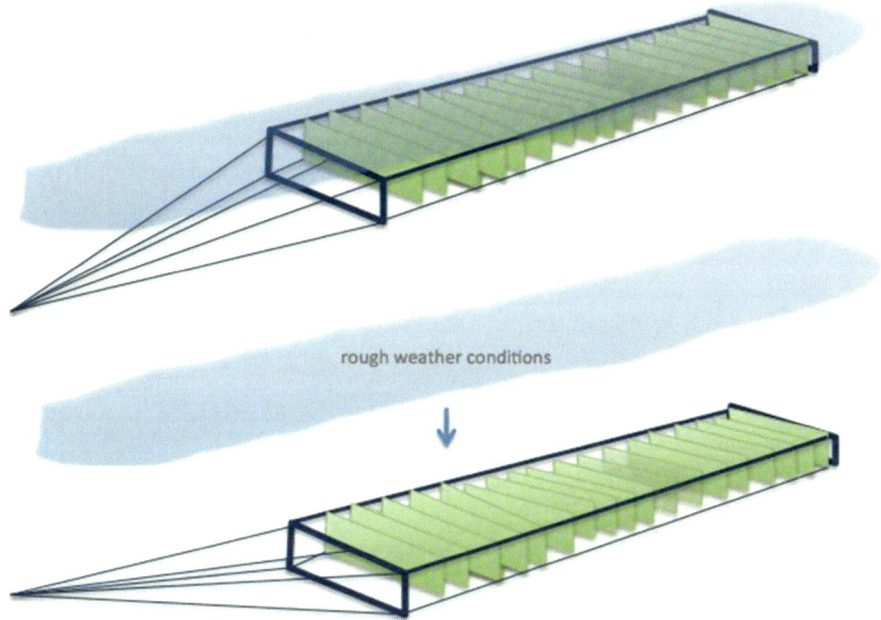

Fig. 12.7 Algae aquaculture part of the Green & Blue satellite unit; the submersible algae float is equipped with two thrusters (*Source* TROPOS 2014c)

- An Algae Production Unit connected to the wind satellite unit at a single point.
- An automatic feeding system (64 t in two silos with two different feeds and a distribution system located on the "vertical columns" of the triangle).
- An automatic feeding system for algae production (50 t CO_2 storage + nutrients).
- PV panels are installed on the top of the module for an additive 0.5 MW per satellite

12.3.4   Aquaculture Production Flow

Based on a theoretical production analysis the total aquaculture yield in the Green & Blue scenario in the Mediterranean is estimated to be 7,000 t of fish in a biannual production cycle and 2,000 t of algae in continuous production.

12.3.4.1   Finfish Aquaculture

The fish production plan developed for this Green & Blue scenario involves the rearing of three different species in a two-year rearing cycles. Taking advantage of the different natural spawning period of the selected species and in order to avoid an overlap in harvesting, stocking of cages is performed annually with half of the dedicated cages for each species as follows:

- European sea bass (*Dicentrarchus labrax*) a total of 8 cages; stocking 1 cage per month in February, March, April and May every year;
- Greater amberjack (*Seriola dumerili*) a total of 10 cages; stocking 5 cages in July every year;
- Meagre (*Argyrosomus regius*) a total of 10 cages; stocking five cages in October every year.

The 2 remaining fish cages serve as backup units. The stocking with fry will be followed by a rearing period of 2 years until market-size is reached (European sea bass 0.6 kg, greater amberjack 3 kg, meagre 2.5 kg). Fish will be fed with artificial feeds delivered by the automated feeding system. The amount of feed is based on the demand. Feeds will be transferred from cargo boats directly to the satellites and the silos will be loaded monthly.

Fish harvesting from the cages will be performed with a fish-pump mounted on a working barge and the harvested biomass will be transported to the central unit. The same fish-pump is used to upload the fish to the aquaculture module on the central platform (TROPOS 2013c).

In Table 12.1 the results of a production analysis performed on a monthly basis are shown. For all three fish species the amount of required feed, energy, ice for

Table 12.1 Production analysis (input/output table) of the complete Green & Blue fish aquaculture farm in the Mediterranean

	Species	Fish fry (1000 ind.)	Feeds (t)	Energy (kWh)	Ice (t)	Harvested fish (1000 ind.)	Harvested biomass (t)
January	European seabass	641	735	145,306	152	507	304
February		641	774	134,536	152	507	304
March		641	807	145,156	152	507	304
April		641	829	141,316	152	507	304
May	Greater amberjack		819	146,138	256	342	512
June		410	748	142,736	292	195	584
July		410	717	146,276	292	195	584
August		205	745	144,808	146	97	292
September	Meagre		763	142,598	256	427	512
October		513	706	146,372	304	243	608
November		513	657	142,832	304	243	608
December		256	678	144,856	152	122	304
Total		**4873**	**8980**	**1,722,930**	**2610**	**3892**	**5220**
Monthly average (t)		487	748	143,578	218	324	435

Source (TROPOS 2014b)

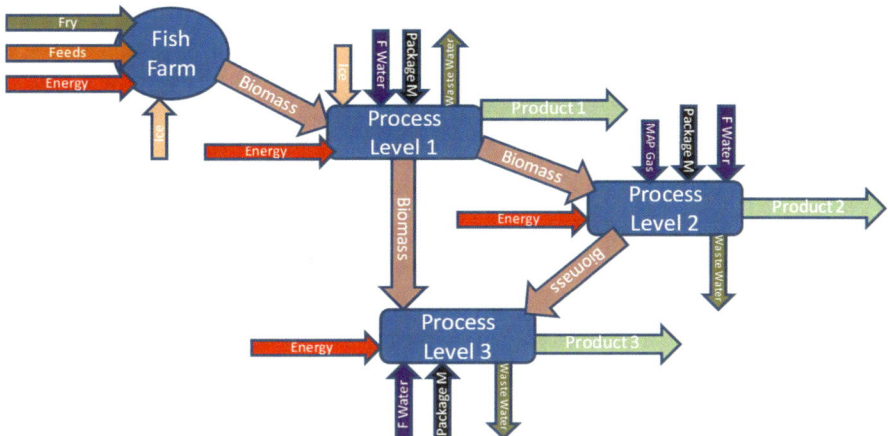

Fig. 12.8 Fish aquaculture and processing plant flow chart including all inputs and outputs in the Green & Blue concept (*Source* TROPOS 2014b)

conservation of the final product, and the number and total biomass of harvested fish was calculated.

After harvesting, the fish biomass processing in the processing module is organized in 3 levels (Fig. 12.8):

- Process Level 1, Basic Processing—in this step fish are processed after harvest either as fresh product (product 1) or pretreated for further processing in process levels 2 or 3.
- Process Level 2, Filet Processing and Modified Atmosphere Packaging—in this step fish are processed to produce filets and other products that require packaging in a modified atmosphere.
- Process Level 3, IQF (Individually Quick Frozen) Processing—in this step the final product is frozen biomass. Biomass inputs to this process level 3 are whole fish from process level 1 and filet from process level 2.

The outputs from levels 2 and 3, apart from the final product include a significant amount of byproducts, which are not considered waste but as a primary material for other industries (e.g. fish meal production). Therefore, all byproducts are refrigerated and exported (TROPOS 2014b).

In terms of biomass, the total products from the Green & Blue fish aquaculture are about 25% fresh fish, 55% transformed (Individually Quick Freezing/Modified Atmosphere Product), 20% byproducts.

All the energy required for the operation of the fish aquaculture in the Green & Blue scenario will be provided by the local renewable energy production.

12.3.4.2   Microalgae Aquaculture

The algae production starts with the seeding of a microalgae starter-culture in the closed photo-bioreactors on the algae floats. The key parameters for the algae production are irradiance, nutrients and size of installation. As the microalgae are in a closed system, the medium will be supplemented with some nutrients, such as nitrogen (9 t/year), phosphorus and micronutrients (1.5 t/year) (Table 12.2). CO_2 is also applied to enhance the production yield. This will be done by direct injection from a 50 t cryogenic CO_2 storage with assistance of an air blower.

The online control system allows monitoring and controlling of the production process inside the photo-bioreactors from board of the central unit. The unit is operating valves and pumps and measures, among other parameters, temperature, the pH value and the optical density in the system continuously. Environmental ambient parameters like irradiance and temperature are also recorded.

When the microalgae culture reaches the desired concentration the culture is harvested. The actual harvesting step of algae from the photo-bioreactors is carried out on the satellite and the biomass is then shuttled to the central platform for storage and further downstreaming. A two-step algae harvest is implemented with a first step consisting in a pre-concentration of the algae suspension followed by a second step of centrifugation. The dewatered algae biomass is then dried by spray-dryer or solar-dryer and either cold stored until transport to the shore or

Table 12.2 Details on the algae aquaculture production unit, including dimension of floats, inputs in terms of nutrients and output in terms of production as calculated based on a production analysis

Algae aquaculture				
Component	Subsystem		Number	Units
Units	Number of units		30	
	Dimension	Length	200	m
		Beam	50	m
		Area	10,000	m²
	Production per unit		65	t/a
	Total production (6–8 months, 30 units)		1950	t/a
Store nutrients	Nitrogen consumption per unit		9.0	t/a
	Phosphorus and micronutrients consumption per unit		1.5	t/a
	Volume per unit (Phytoplant)		400	m³
Production capacity	Max. production per week		750	t
	Production per day		300	t

Source (TROPOS 2013c)

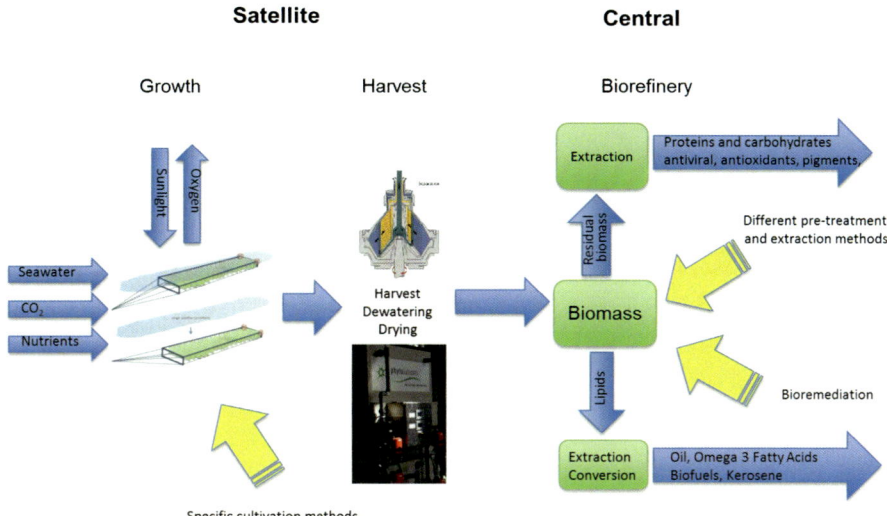

Fig. 12.9 Algae aquaculture production and processing; flow chart including all inputs and outputs in the Green & Blue concept (© by Phytolutions)

Table 12.3 Estimated microalgae production per month considering all 30 satellites units of the Green & Blue scenario	Months	Product (t)
	Jan	60
	Feb	80
	Mar	130
	Apr	150
	May	200
	Jun	280
	Jul	290
	Aug	250
	Sep	210
	Oct	160
	Nov	80
	Dec	60
	Estimate per year	1950

Source (TROPOS 2014b)

directly further processed in the algae biorefinery module on the central platform. A flow model of the entire algae aquaculture production and processing is shown in Fig. 12.9.

In total, considering all 30 satellite units of the Green & Blue scenario, the annual production capacity of the facility was estimated with 1950 t/year (dry weight) (Tables 12.2 and 12.3).

Analysis of the economic impact confirmed that the Green & Blue concept in the Mediterranean has a strong economic case for development (TROPOS 2014d) and a virtual environmental impact assessment showed that the concept has minimal negative environmental impact (TROPOS 2014e).

12.4 Summary

Moving offshore involves a lot of advantages such as reduced competition with other activities compared to the near shore zone, optimum culture conditions in term of water quality and absence of pathogens, and rapid dispersal and dilution of any effluents to the wide open ocean (i.e. less negative impact on the environment; Lu et al. 2014). Moving offshore, however, also goes along with a major challenge, namely the suitable technology. Weather and wave conditions are often harsh in the open ocean and deep waters make the fixation of aquaculture facilities a difficult task. Moreover, the distance to the coast significantly complicates logistic service and supply, and operation and maintenance. A solution to overcome these problems and obstacles is represented by the TROPOS multi-use platform approach. The floating structure was especially designed for offshore conditions and allows for use in deep waters. As the annual offshore wind energy capacity will experience an average net increase of 21.5% every year until 2020, the combination of marine aquaculture farming with offshore wind energy exploitation seems obvious. The Green & Blue concept intends to integrate an offshore wind farm with aquaculture facilities for fish and algae, so as to create synergies among the activities: (i) Since many components of the system need energy (e.g. workshop, laboratory, feeding unit, etc.), aquaculture can achieve high levels of autonomy when combined with the wind turbines triggering the development of the offshore aquaculture further; (ii) Placing the cages between the turbines will save space and mooring lines; (iii) Since maintenance tasks have to be provided in a frequent basis for both fish and wind farm, the service infrastructure can be shared; (iv) Based on this, having a common accommodation unit on the platform, downtime periods will be decreased for both industries (since there will be permanently located personnel providing continuous inspections and routine maintenance tasks, and able to respond fast in case needed). Apart from this, sharing of divers and diving infrastructure, logistics, maintenance tools, workshops, environmental monitoring, etc. will reduce the service/maintenance costs.

12.5 Outlook

Multi-purpose offshore platforms are designed to incorporate modules of several compatible activities, and thus, provides the opportunity for the aquaculture industry to operate viably and profitably in offshore sites. Since mooring is one of the major constrains for offshore aquaculture (the industry operates in depth <100 m), the synergy with the platform is vital for a feasible industry at deep offshore sites. The innovative synergies developed on modular multi-purpose platforms provide the opportunity for aquaculture to expand in offshore sites and meet future demands. Important constraints for success are the co-development and shared use of infrastructure, the selection of sites and technologies appropriate for environmental conditions, and the synergies in operational planning.

Until now, the aquaculture industry has moved offshore slowly and conservatively due to high costs and immaturity of innovative technologies. Some prototypes, consisting of submersible or semi-submersible cages, are currently available, allowing farms to be located farther from the coast than today. However, a significant research effort is needed on the development of new culture technologies as, although several prototypes have been proposed, few have been tested in commercial scale.

With its innovative design and technology the TROPOS concept represents an advanced solution for an integrated and synergistic offshore multi-use approach involving aquaculture and microalgae production. However, the next essential step is to move from theoretical approaches and modelled designs towards "real world" deployments. Even if financial support is possibly required at the beginning, pilot scale deployments are vital to proceed in the field of multi-use offshore installations.

References

Cressey, D. (2009). Aquaculture: Future fish. *Nature, 458,* 398–400.

Enzing, C., Ploeg, M., Barbosa, M., & Sijtsma, I. (2014). *Microalgae-based products for the food and feed sector: An outlook for Europe*. JRC Scientific and Policy Reports, Luxembourg: Publications Office of the European Union.

FAO. (2014). *The state of world fisheries and aquaculture*. Rome: Food and Agriculture Organization of the United Nations.

Lu, S. Y., Jason, C. S., Wesnigk, J., Delory, E., Quevedo, E., Hernández, J., et al. (2014). Environmental aspects of designing multi-purpose offshore platforms in the scope of the FP7 TROPOS Project. *Paper presented at the OCEANS 14 MTS/IEEE*, Taipei, Taiwan.

Muller-Feuga, A. (2004). Microalgae for aquaculture: The current global situation and future trends. In A. Richmond (Ed.), *Handbook of microalgal culture* (pp. 352–364). Oxford: Blackwell Science.

Papandroulakis, N., Anastasiadis, P., Thomsen, C., Koutandos, E., Mayorga, P., Quevedo, E. et al. (2014). Modular multipurpose offshore platforms: Innovative opportunities for aquaculture. In *Book of abstracts* (pp. 955–956) European Aquaculture Society Special Publication.

Quevedo, E., Delory, E., Castro, A., Llinás, O. & Hernández, J. (2012, October). Modular multi-purpose offshore platforms, the TROPOS project approach. In *Paper presented at the 4th International Conference on Ocean Energy*, Dublin.

Quevedo, E., Cartón, M., Delory, E., Castro, A., Hernández, J., Llinás, O., et al. (2013, June). Multi-use offshore platform configurations in the scope of the FP7 TROPOS Project. In *Paper presented at the OCEANS 13 MTS/IEEE*, Bergen.

Ryan, J. (2004). Farming the deep blue. In G. Mills & D. Maguire (Eds.), Dublin: Bord Iascaugh Mhara—Irish Sea Fisheries Board (BIM) & Irish Marine Institute. Retrieved from http://www. bim.ie/media/bim/content/downloads/FarmingtheDeepBlue.pdf

Ryan, J., Jackson, D., & Maguire, D. (contributing authors) (2007). Offshore aquaculture development in Ireland; next steps. In: L. Watson, A. Drumm (Eds.), *Technical report jointly commissioned by Bord Iascaugh Mhara—Irish Sea Fisheries Board (BIM) & Irish Marine Institute*, p. 35.

Spolaore, P., Joannis-Cassan, C., Duran, E., & Isambert, A. (2006). Commercial applications of microalgae. *Journal of Biosciences and Bioengineering, 101*, 86–96.

TROPOS (2013a). D2.3—Sample locations and setups for further design. Retrieved from www. troposplatform.eu

TROPOS (2013b). D3.2—Technical concept dossier for the central unit. Retrieved from www. troposplatform.eu

TROPOS (2013c). D3.4—Technical concept dossier for the satellite unit. Retrieved from www. troposplatform.eu

TROPOS (2014a). D3.3—Technical dossier for each of the established modules including key features. Retrieved from www.troposplatform.eu

TROPOS (2014b). D3.5—Integrated concept offshore platform system design. Retrieved from www.troposplatform.eu

TROPOS (2014c). D4.1—Report on hydrodynamic simulations of platform models and model validation. Retrieved from www.troposplatform.eu

TROPOS (2014d). D5.2—An assessment of the economic impact, on local and regional economics, of the large scale deployment. Retrieved from www.troposplatform.eu

TROPOS (2014e). D6.2—Report on environmental impact assessment and mitigation strategies. Retrieved from www.troposplatform.eu

TROPOS (2015). D4.3—Complete design specifications of 3 reference TROPOS systems. Retrieved from www.troposplatform.eu

Chapter 13
Offshore Platforms and Mariculture in the US

Jeffrey B. Kaiser and Michael D. Chambers

Abstract Global demand for seafood is increasing. Supply from wild caught sources has been essentially flat for twenty years and, depending on the specific fishery, in decline for many species that are considered fully exploited or over-exploited. As the fastest growing sector of world food production, aquaculture is increasingly playing a major role and currently accounts for nearly half of the total aquatic production worldwide. Marine cage culture in particular provides an opportunity to utilize vast amounts of the world's surface area to produce fish, shellfish, and plants, while avoiding land-use conflicts in crowded coastal regions. Currently in the US, very small volumes of marine fish are produced and very large volumes are imported. This trend shows no signs of slowing down with an ever increasing annual seafood trade deficit. In an effort to initiate more domestic production, private companies, research institutions, and government agencies have all been involved in various types of aquaculture production. Aquaculture can be generally categorized as land-based, near shore, or offshore. Offshore marine fish culture utilizing cages has been conducted on both the east and west coast of the US as well as in the Gulf of Mexico (GoM). Specifics on the projects in the GoM are described in the following sections.

13.1 Background

The world's population is estimated to be approaching 9 billion people by 2050. UN estimates project the total global food supply will need to increase by 70% to keep up with demand and the growth of a larger and more urban population

J.B. Kaiser (✉)
Marine Science Institute's Fisheries and Mariculture Laboratory,
The University of Texas at Austin, Port Aransas, TX, USA
e-mail: jeff.kaiser@utexas.edu

M.D. Chambers
School of Marine Science and Ocean Engineering,
University of New Hampshire, Morse Hall, Room 164, Durham, NH, USA

© The Author(s) 2017   375
B.H. Buck and R. Langan (eds.), *Aquaculture Perspective of Multi-Use Sites in the Open Ocean*, DOI 10.1007/978-3-319-51159-7_13

(FAO 2009). The world's oceans cover over 70% of the earth's surface and provide a large area to situate aquaculture operations that produce marine fish, shellfish, and plants. Currently, the ocean is underutilized as a location for protein production with 5% of the global protein supply (Oceanworld 2014) and a mere 2% of the total human calorie intake (UNEP 2012) acquired from this biome. Demand for food combined with a shortage of new arable land will require the global population to attempt to maximize production from the world's oceans. Potable freshwater and its distribution and use in the future is another critical reason to utilize the marine environment for aquaculture (Famiglietti 2014). All living organisms must have freshwater to live, including humans. The efficiency of rearing a million tons of protein in the marine environment when compared to producing the same amount of terrestrial protein becomes evident when the freshwater demand is considered. The current composition and location of water on the earth makes an obvious case for the use of the marine environment for aquaculture—that is where the water is! (Fig. 13.1).

The world is embracing aquaculture. Global output has increased from 7 million tons in 1980 to nearly 70 million tons currently and cultured product volume is projected to overtake capture fisheries within a few years. The US on the other hand, while being a major consumer of seafood products (ranked second behind

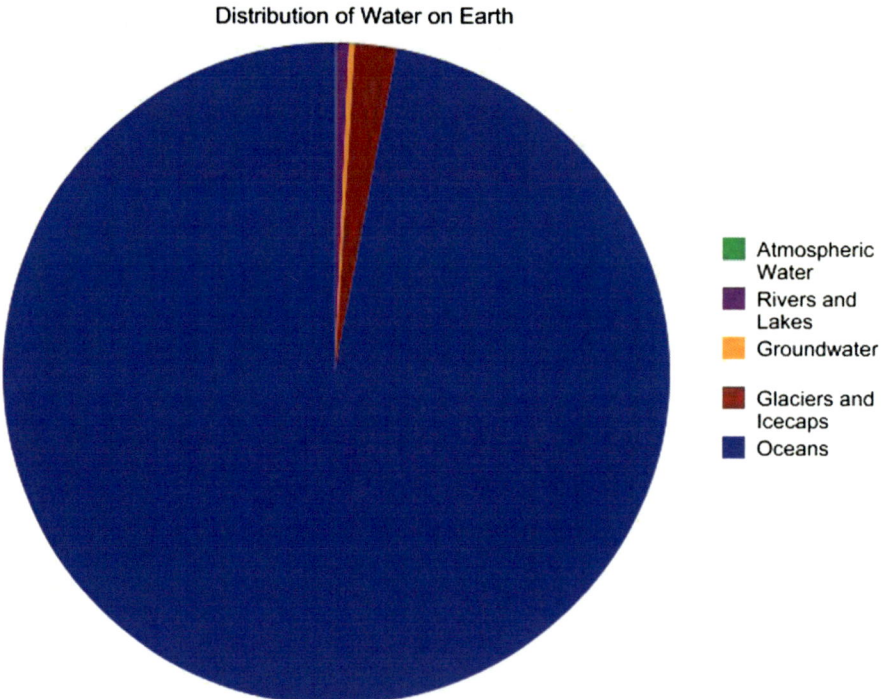

Distribution of Water on Earth

Atmospheric Water

Rivers and Lakes

Groundwater

Glaciers and Icecaps

Oceans

Fig. 13.1 Distribution of the earth's water (*Image* SanibelSeaschool.org)

China), is a minor producer of product and barely ranks in the top 15 producing countries (FAOFishStat 2016). The US currently imports around 90% of its edible seafood supply and has an annual trade deficit value of more than $10 billion—and that number is increasing each year. Ironically, for a country that has not effectively promoted domestic aquaculture expansion through easing of regulations and permitting processes, it is estimated that half of the product the US imports each year is produced by aquaculture (NOAA FishWatch.gov 2014).

Nearly half of the global population currently lives within 200 km of a coastline and that figure is expected to double by 2025 (PRB 2003). User conflicts and land demands for development will compete for space to conduct commercial scale marine aquaculture utilizing land-based pond or tank systems. Increasingly, both researchers and investors are considering the benefits of moving into the offshore environment where both water volume and water quality ameliorate many of the issues faced by traditional aquaculture production (Price and Morris 2013). While typically considered near shore, the explosion of the 1.8 million ton/yr salmon culture industry in the last 30 years is just one example of the potential rearing organisms in the marine environment offers. The vast majority of cage or net-pen operations currently in place are next to or within sight of land, which in most cases locates them within 10 miles of shore. Most, therefore, are situated with some sort of land mass nearby and are not exposed on all sides to prevailing wind and sea conditions.

With regards to the US coastline, several factors for siting of offshore aquaculture projects must be taken into consideration such as: distance from shore, water depth, logistics of maintaining personnel, feed, and equipment, transportation issues, environmental impact, culture species availability, product markets, among others. The GoM presents a unique situation because it is a sub-tropical saltwater environment which currently has nearly 3000 platforms in place associated with energy production. Certainly not all of these structures are in suitable locations for siting projects, but when the aforementioned factors are considered several of the existing platforms would be candidates for aquaculture. Supporting the energy industry are numerous vessels, terminal facilities, heavy equipment and experienced personnel, all of which would be valuable assets for the development of a commercial aquaculture venture in the offshore environment.

The mariculture projects utilizing offshore platforms described in the following sections were conducted in totally exposed locations and as such, presented the operators with unique challenges. Since 1990, there have only been three projects in GoM waters attempting to raise fish in cages either nearby or attached to oil and gas platforms. The platforms were used for monitoring the cage systems in addition to being a base of operation in two cases and strictly for observation purposes during the third project. It is important for the reader to note that the perspective offered by the co-authors is from direct personal experience working on two of these projects over several years—installing cage systems, stocking and feeding fish, inspecting and maintaining the cages, and living or working on the platforms involved. Results of these efforts will be discussed along with considerations for future use of oil and gas platforms for offshore aquaculture.

13.2   The Gulf of Mexico

Location is arguably the most important factor for any offshore cage culture project. Siting will dictate the expected sea conditions, type of cage system utilized, water parameters, and all transportation and logistical planning. These factors, among others, will ultimately determine the success or failure of a project. The GoM is characterized by short period wind driven waves for much of the year combined with tremendously powerful episodic tropical events during the June–November hurricane season. Offshore platforms, while engineered to withstand tremendous forces for decades, can be damaged or destroyed during hurricanes and tropical storms (Fig. 13.2) and an aquaculture project needs to have worst case contingency plans in place. Cage systems at totally exposed locations should be designed to survive a tropical event with nominal damage and above all, retain the fish within the system. Escaping the energy at the ocean surface is critical and submergible or submerged cages will likely be the preferred system in the GoM.

The distance the continental shelf extends into the GoM varies greatly depending on the specific location. It is the authors' assertion based on cage designs and diving requirements that for the purposes of siting, water depths of 30–60 + m would be considered adequate for offshore cage culture systems. This depth would allow for a cage to be submerged 10 m to escape the high energy surface forces and still have enough space for the cage volume and some allowance to maintain the system off of the bottom. The locations of the two projects off the coast of Texas, one 10 km

Fig. 13.2 Offshore platform damaged during a hurricane off of Louisiana

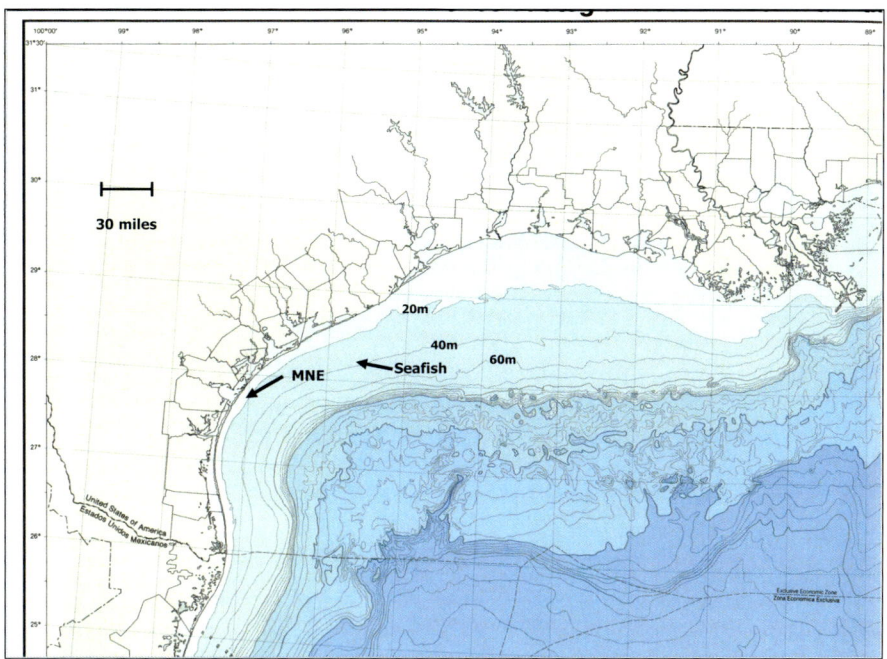

Fig. 13.3 Site locations for MNE Inc and Seafish aquaculture projects in the Gulf of Mexico

(21 m deep) and the other 42 km (40 m deep) offshore, are designated in Fig. 13.3. Conditions during these particular projects were seas calm to 7+ m, winds calm to above hurricane strength (78 knots and then the anemometer disappeared), and currents 0 to an estimated 130 cm/sec or 2.5 knots. During the course of several years, the weather pattern observed from worst to best conditions during seasons was periodic tropical events, followed by spring, winter, fall and summer. The period from February to May is particularly damaging to cage systems since the wind velocity and subsequent sea state often prevent site visits for maintenance from several days to as much as 2 weeks.

Wave action, currents, and fouling of net or cage materials will constantly wear and ultimately test all the materials of any offshore aquaculture system. Since weather and sea state determines when components can be safely inspected and repaired, the chosen design must assume extended maintenance intervals from days to even weeks in extreme conditions. Unfortunately, in several instances this inability to address a problem immediately, as can often be with onshore or nearshore systems, has proven to be the greatest challenge in the GoM. Personnel simply cannot safely operate in heavy seas and once a net, stanchion, mooring rope or chain, or any other part of the cage is compromised, waiting for the weather to subside is the only option. Frequently this means that what begins as a minor issue, combined with adverse conditions over several days, eventually causes major damage and there is absolutely nothing on-site personnel can do to prevent it. In some cases main component issues such as mooring failures or broken structural pieces will not result in a permanent setback to a project. Other situations, however, such as gaping holes in netting, cages breaking free and going adrift/sinking, or catastrophic failure of an entire cage system present an entirely different set of circumstances. The offshore environment is extremely challenging and it should be noted that every project in the GoM to date has had one or several of these events occur, often simultaneously.

In addition to the high energy environment of the GoM, another serious issue with netting and other containment materials is biofouling. The wave and current forces exerted on the cage systems stress the components and as barnacles, hydroids, and other organisms accumulate this stress increases. Antifouling paint and other coatings are available and were applied to cage materials however these are meant to reduce biofouling and are not capable of eliminating it altogether. Each year during the spring and early summer the GoM waters begin to warm and a significant amount of fouling occurred very quickly on the cage systems. The only options available are to either clean a cage in place or replace components when the accumulation becomes severe. Needless to say, the challenge of changing a 30–40 m diameter net that weighs several tons in the offshore GoM environment resulted in the cleaning option being employed. Figure 13.4 shows the preventative effect of copper based antifoulant paint on fiberglass panels on the left, with the top row being the treated samples. On the right the drag created by a virtual carpet of hydroids is seen on cage netting in a strong current. In addition to creating additional stress on all the system components, the deformation of much of the netting also effectively reduced the cage volume significantly. Any future offshore projects in the Gulf, however, will have to seriously consider biofouling and how to prevent and adequately deal with the inevitable growth of numerous marine organisms, particularly on the netting or containment materials.

Logistically speaking, operating at exposed locations offshore is significantly more expensive than onshore and near shore sites, largely due to transportation costs. Personnel, cage systems, maintenance equipment, fish, and feed are some of items that have to be transported by either air or sea to the project site whether the platform involved is manned or unmanned. Both platform situations have been tried during the GoM projects and each had its advantages as well as disadvantages. Certainly having a large, stable structure with multiple decks, living quarters, a power supply and crane in an otherwise uninhabitable offshore environment is an asset to an aquaculture venture. In the two projects off the coast of Texas, cages were either attached directly to the legs of a platform or mooring lines from the system were connected to the structure(s) at various points as shown in Fig. 13.5.

Fig. 13.4 *Top* Antifouling
effect of copper based paint
(*top row* are treated samples).
Bottom virtual carpet of
hydroids on cage netting

13.3 Past Projects

The history of mariculture utilizing offshore platforms in US waters begins in 1990 off the coast of Texas (Chambers 1998; Kaiser 2003). During the next five years MNE, a wholly owned subsidiary of Occidental Petroleum, embarked on a project to assess the feasibility of using oil and gas platforms as a base of operations for fish production. These included both manned and unmanned structures located 11–54 km offshore at depths of 20–80 m in both state and federal waters. Several cage designs were deployed, including both surface and submerged systems all of which were connected at some point to the platform itself. Various types of feed units were designed and placed on the platforms and company biologists would visit the projects sites every 1–2 weeks to inspect the cages and fish, re-supply the feeders, and perform general maintenance wherever required.

Red drum *Sciaenops ocellatus* was the primary species used during the project although on one occasion a smaller prototype cage was stocked with Florida pompano *Trachinotus carolinus*. Red drum from 30–85 g were acquired and transported to the project sites utilizing offshore crew boats in a fiberglass holding tank with supplemental pure oxygen. Stress and subsequent mortality was a

Fig. 13.5 Dunlop Tempest 4™ 18,000 m³ cage system in 40 m of water at the platform site 54 km offshore on the central coast of Texas

challenge and it was not uncommon to experience losses of 25% or more during the transport depending on a number of factors. Once in the cage and after commencing feeding however, the fish exhibited very good overall health and no incidence of disease or parasitic outbreaks were observed. Grow out of red drum for 6–12 months to a market size of 0.7–1.0 kg at final harvest densities of 7–40 kg/m³ was successful on three occasions totaling approximately 7300 kg (Chambers 1998). Though not a significant amount of production in terms of commercial volumes, this project stands alone as the only example of fish successfully stocked, fed, and harvested offshore in the Gulf of Mexico to date (Fig. 13.6).

The cages that survived the extreme GoM conditions during the course of the project were smaller, submerged designs whose cost per cubic meter would make them uneconomical for a large scale operation. Results of the MNE project demonstrated that the fish growth and acclimation to the cage system were very good and maintaining them from a biological standpoint was not particularly difficult. The maintenance of cage system components on the other hand, combined with the logistics of supplying feed and the production costs of simply operating offshore resulted in the shutdown of the project in 1995.

The next project along the Texas coast was a joint venture with Shell Oil company called Seafish Mariculture which took place from 1997–1999 (Kaiser 2003). The site chosen for the cages was an active platform complex 54 km offshore of the middle Texas coast in 40 m of water. This project differed from MNE in that it was the first commercial scale cage installation effort with biologists living

Fig. 13.6 A portion of the 1.0 kg red drum harvest during the MNE/Occidental project in 1995

on the platform full time to monitor and maintain the systems. Two person crews alternated during a 7 day on, 7 day off schedule which was coordinated with planned helicopter transport of platform personnel to minimize costs. Cage system supplies, fish food, fuel, water, and anything else needed for the project was delivered by crew boats out of Galveston, Texas and also coordinated with previously scheduled trips to increase efficiency. The platform was outfitted with permanent living quarters, a 7.3 m service vessel that could be craned off the platform, and a pneumatic Akva automated feed system with hoppers that held several tons of fish feed.

The cages that were tried during the project were both systems used in the salmon industry consisting of a circular or hexagonal floating collar portion, stanchions along the circumference, and a 30–40 m diameter cage with a weighted net portion approximately 15 m deep. The first system was similar to many currently used around the world and was constructed entirely of high density polyethylene (HDPE) and featured a double collar ring at the surface. The cage was constructed onshore at a terminal in Galveston and then towed to the site, moored, and had the main net installed along with an interior nursery cage to allow transport of smaller fish to the site (Fig. 13.7).

The system was situated between two platforms with half of the eight mooring lines connected to the legs of the structures and the other half anchored several hundred yards away with concrete blocks and anchors common to the aquaculture industry. Initially, the system performed adequately and was compliant enough to absorb the daily equipment stress as well as the periodic heavy seas that characterize the Gulf (Fig. 13.8).

Fig. 13.7 10,000 m^3 double ring HDPE 30 m diameter net pen moored between two platforms off of the central Texas coast

Fig. 13.8 Typical 4–5 m winter seas moving through the cage and platform site

While HDPE is a strong and versatile material used globally in numerous cage culture applications, the high energy surface action in the GoM exerted forces on the cage which eventually began to break various components, especially the stanchions and polysteel rope connections to the top of the net. This damage combined with a particularly active tropical season during 1998 proved to be too much and the system was totally destroyed by a storm and several thousand red drum were lost in the process.

The next system deployed was an 18,000 m^3 Dunlop Tempest™ cage system consisting of 0.4 m diameter 16 m long pressurized rubber hose sections connected by metal flanges at each juncture of the octagonal structure. This cage was located 100 yards from the platforms and attached to the legs of the structures in addition to mooring lines spread in different directions with concrete blocks and anchors (Fig. 13.9). Designed for use at very exposed offshore locations, the cage was extremely strong and the floating collar portion in particular was able to withstand tremendous forces during the project. As is the case in most cage culture operations however, the netting can become easily compromised for a variety of reasons and a significant amount of time was spent in the water patching holes as well as cleaning biofouling on the net. Staying ahead of the maintenance required on a large system proved problematic during the project and it was a constant challenge for personnel

Fig. 13.9 Dunlop cage deployed adjacent to the platform complex during the Seafish Mariculture project in 1998 during typical 1–2 m seas

to maintain the integrity of the cage. Damage would occur during a storm or heavy seas, repairs would be made quickly if time allowed between weather events, and then the next round of damage would occur. There were several thousand red drum stocked into the cage which grew well and the feed unit used on the project performed very well when the feed tube stayed attached as designed.

The project was ended abruptly and unexpectedly in 1999 when it was decided the cage needed to be moved as soon as possible to make room for pipeline construction in the area. Moving a large cage and moorings in the offshore GoM waters is a daunting task in perfect conditions and an impossible one in marginal or bad conditions. During this process the cage ended up entangled in the legs of a platform, the net was compromised, and for all practical purposes that was the end of the Seafish project. Several fish samples were taken along the way, but no significant amounts were ever brought to shore or harvested during the two year venture. A great deal was learned however in terms of what to expect and plan for during installation, feeding, and maintenance of large cage culture systems in the open GoM.

Though not directly utilizing a platform, there has also been one federally funded effort to conduct offshore aquaculture research in the GoM. This research took place from 2000–2003 and was initiated with the formation of the Offshore Aquaculture Consortium (OAC) which was a collaborative, Gulf-wide research and development program tasked with gathering primary scientific data regarding offshore aquaculture. The project site was in federal waters 40 km off the Mississippi coast at a depth of 26 m and within a few hundred meters of an active platform whose only involvement was for observation of the cage. The position of the OAC investigators was one of specifically not using a platform in an effort to have an autonomous cage system in the offshore Gulf waters. A commercially available Sea Station™ 600 m^3 system was installed with a single point mooring whose efficacy, ironically, was questioned at a workshop organized by the OAC in 2000. After being deployed for 50 days a component of the mooring failed and the cage was adrift for 40 days before being located and towed back to shore for repairs. The cage was subsequently redeployed with a three point mooring system and although fish were stocked into the system no significant amounts were harvested during course of the research. Funding was discontinued and unfortunately what began as an excellent opportunity for an offshore aquaculture demonstration project in the GoM ended abruptly in 2003. A detailed description of the entire OAC project can be found in the final report to NOAA (Bridger 2004).

13.4 The Future

Despite the challenging conditions, interest in offshore aquaculture production in the Gulf of Mexico still exists. From a technology standpoint, new and improved aquaculture systems are available that can survive and produce fish but the primary question is can this be done profitably offshore? Thus far, in the GoM the answer to that question has been no based on the experiences of the projects to date. A comprehensive economic feasibility study on using oil and gas structures in the GoM for mariculture concluded that at the present time it not likely to be a profitable venture (Kaiser et al. 2011a). Liability and decommissioning issues for the structure itself have long been identified as the major impediments to mariculture using platforms in the GoM and this situation remains true to the present (Dougall 1999; Kaiser et al. 2011a). The regulatory environment and permitting issues in offshore waters of the US have hindered development for two decades and investors have simply moved their offshore cage culture projects to other countries in many cases (Upton and Buck 2010). Internationally, there are only a couple of locations using dedicated platforms for mariculture production, one in the Mediterranean and the other in Japan. The platform in each case has multiple cages attached to it and acts as the operational center from which the growout systems are fed and maintained (Kaiser et al. 2011b). Considering the high energy environment, possibility for hurricanes, platform cost and liability issues, logistical challenges, and economic reality of offshore mariculture, it is not surprising there are no cage systems currently in the Gulf of Mexico.

That being said, it is the authors assertion that in the future a project with the right location, substantial capital investment, experienced personnel, and a good cage system will profitably grow fish in the Gulf of Mexico. The scenario will likely involve multiple submersible cages near a platform with a crane that houses a feeding unit capable of holding 200+ tons of feed which can be operated remotely if required. The structure may or may not be associated with an energy company and it may turn out that a platform put into place and designed specifically for a mariculture project, while certainly expensive, might be a better long-term option. The potential of refurbishing and reusing a decommissioned platform has also been suggested (Kaiser et al. 2011b) and may offer some advantages with regards to the liability issues that are associated with ownership.

An alternative to a platform proposed in the OAC final report is an aquaculture support vessel (ASV) which could be a large catamaran style vessel or more likely a lift boat type commonly used for nearshore maintenance on platforms (Bridger 2004). Several issues come to mind that make that option less attractive considering the high energy real world conditions in the GoM. Lift boats are very slow vessels (5–7 knots) that can only travel in decent seas—this complicates the logistics of anything in the 30+ km range. Because of depth requirements, these are the most likely areas for a cage project requiring tons of feed every day so faster crew boats would be required weekly to transfer the feed to a lift boat on a continual basis. In addition, once an operation is established with multiple cages and a substantial investment in fish stock, there can be no interruptions due to lift boat vessel maintenance, annual US Coast Guard inspections, and other mechanical issues. Because of this need for continual operation a second, backup ASV would be required for any large scale project that is conducted in GoM offshore waters. Assuming the cage site is in the preferred 30–50 m water depth, the cost of lift boats that can operate at these locations would be in the $5–20 + million dollars per boat range and would require the expense of two rotating teams of crew members as well. Lift boats cannot stay on site in all types of weather meaning that by pre-emptively heading in before severe conditions, not only are no personnel at the project site, but the feeders will not be getting filled on schedule, and fish will not be growing. Taken together, the cost of at least two ASV vessels, crews to man them, weekly supply boat runs, and the need for multiple feed buoys that must work flawlessly, the platform option starts to look better in comparison. Except for one hurricane evacuation during the Seafish project, the platform as a base of operations allowed personnel to remain on site continuously for two years regardless of the sea state and weather. This ability, combined with having a topside automated feed system instead of feed buoy(s) and virtually limitless feed storage capacity on multilevel decks begins to present a scenario where millions of kilograms of product could be cultured in the open ocean environment.

With regards to feeding, while a large buoy seems like an obvious alternative to using a platform there are several issues that warrant consideration for the GoM application of such a system. First, survivability of such a buoy is paramount since having several hundred thousand kilos of fish in culture is irrelevant if there is no way to reliably feed them. Designing a buoy that is able to maintain a specific position relative to the cages during normal choppy Gulf conditions, frequent 3–4 m seas, occasional 7+ m wave heights, and hurricanes is an engineering challenge at the very least. This technology is very expensive, but it has been demonstrated at an offshore site in New Hampshire (Fig. 13.10) and some of this research could be applied to future Gulf projects (Atlantic Marine Aquaculture Center 2007, http://ooa.unh.edu/). Feed barges up to 450 ton are used in protected waters (1–2 m) in Norway, Canada and Chile. Unfortunately, these pneumatic feeders would not survive in the GoM environment. Other possibilities for feeding a large scale farm include retrofitting a 40 m fishing vessel that can plug into a mooring and feed distribution system to each cage. If a severe storm is approaching, the vessel could detach and head back to port. Having a portable feeder would allow you to conduct maintenance, refueling, personnel exchange and fill the feed silos at a safe harbor instead of offshore. Offshore mariculture on a commercial scale involving millions of dollars of equipment and fish will likely need to be manned operations as well, a role that a platform and appropriate tender vessel could accomplish. Without thinking along this economy of scale involving systems that provide tens of thousands of cubic meters of water volume in which to grow marine fish offshore, the benefits of operating in the open GoM will not be realized.

Experiences gained after many years working on offshore aquaculture systems using platforms in the GoM can be distilled down into a partial list of some important points.

1. Eliminate modes of failure in all parts of the production system. Keep it strong and simple wherever possible as the offshore environment will reveal any component weaknesses, usually very quickly and often catastrophically.
2. Ocean conditions and weather determine everything offshore so plans must be made and adjusted accordingly. Nothing can be forced if conditions don't allow it as the ocean will always win in that situation.

Fig. 13.10 Aquamana 20 ton feed buoy deployed at the University of New Hampshire Open Ocean Aquaculture site located 13 km offshore. The hydraulic feeder fed four submerged cages in sea conditions seas up to 10 m

3. Redundancy and backup systems where applicable, especially with regards to moorings, which are arguably the most critical components of the cage system.
4. Prepare as much equipment while the system is onshore as possible. It becomes exponentially more difficult to work on nets, stanchions, cage ties, shackles, and anchor lines the moment it enters the water.

The future of mariculture worldwide is a bright one. Demand for marine plants and animals is increasing and, while wild caught fisheries are essentially flat or in decline, the culture of these products is attempting to meet that demand with increased supply each year. Where the US fits into this future is unclear given our past record of low marine aquaculture production, onerous permitting requirements, and vague regulatory picture, especially in offshore waters. In the US, the Gulf of Mexico is an area that may play a role in domestic offshore mariculture production, should it ever develop on a commercial scale. This "audacious hope" for the GoM, as described in Sims (2015), will hopefully become a reality in the near future when the right concept is put into action with adequate investment. This will not happen, however, until the liability and regulatory issues in US federal waters and economic challenges of production are resolved in order make it a more attractive venture to investors.

References

Atlantic Marine Aquaculture Center. (2007). UNH Atlantic Marine Aquaculture Center. UNH, Durham, NH. Accessed January 12, 2016, at http://amac.unh.edu/

Bridger, C. J. (Ed.) (2004). *Efforts to develop a responsible offshore aquaculture industry in the Gulf of Mexico: A compendium of offshore aquaculture consortium research*. Ocean Springs, MS: Mississippi-Alabama Sea Grant Consortium. MASGP-04-029.

Chambers, M. D. (1998). Potential offshore cage culture utilizing oil and gas platforms in the Gulf of Mexico. In: C. E. Helsley (Ed.), *Proceedings of an International Conference on Open Ocean Aquaculture '97, Charting the Future of Ocean Farming* (pp. 77–87). April 23–25, 1997. Maui, HI: University of Hawaii Sea Grant College Program #CP-98-08.

Dougall, D. (1999). Platforms and fish pens-an operator's perspective. In R. R. Stickney (Ed.), *Proceedings of the Third International Conference on Open Ocean Aquaculture on Joining forces with industry* (pp. 39–43). May 10–15, 1998. Corpus Christi, Texas. TAMU-SG-99-103.

Famiglietti, J. S. (2014). The global groundwater crisis. *Nature Climate Change, 4*, 945–948.

FAO. (2009). *How to feed the world in 2050*. Rome, Italy, Food and Agriculture Organization of the United Nations. Accessed November 1, 2014, at http://www.fao.org/fileadmin/templates/wsfs/docs/expert_paper/How_to_Feed_the_World_in_2050.pdf

FAOFishStat. (2016). Retrieved August 1, 2016, from http://www.nmfs.noaa.gov/aquaculture/docs/aquaculture_docs/world_prod_consumtion_value_aq.pdf

Kaiser, J. B. (2003). Offshore aquaculture in Texas: Past, present, and future. In C. J. Bridger & B. A. Costa-Pierce (Eds.), *Open ocean aquaculture: From research to commercial reality* (pp. 269–272). Baton Rouge, LA: The World Aquaculture Society.

Kaiser, M. J., Snyder, B., & Pulsipher, A. G. (2011a). *Assessment of opportunities for alternative uses of hydrocarbon infrastructure in the Gulf of Mexico*. U.S. Dept. of the Interior, Bureau of Ocean Energy Management, Regulation and Enforcement, Gulf of Mexico OCS Region, New Orleans, LA. OCS Study BOEMRE 2011-028.

Kaiser, M. J., Snyder, B., & Yu, Y. (2011b). A review of the feasibility, costs, and benefits of platform-based open ocean aquaculture in the Gulf of Mexico. *Ocean & Coastal Management, 54*, 721–730.

NOAA FishWatch.gov. (2014) Accessed November 1, 2014, athttp://www.fishwatch.gov/farmed_seafood/in_the_us.htm

Oceanworld. (2014). Accessed November 1, 2014, at http://oceanworld.tamu.edu/students/fisheries/fisheries1.htm

Population Reference Bureau. (2003). Accessed November 1, 2014, at http://www.prb.org/Publications/Reports/2003/RippleEffectsPopulationandCoastalRegions.aspx

Price, C. S., & Morris, J. A., Jr. (2013). *Marine cage culture and the environment: Twenty-first century science informing a sustainable industry*. NOAA Technical Memorandum NOS NCCOS 164.

Sims, N. A. (2015). An audacious hope: To be able to grow fish in the Gulf of Mexico! *Aquaculture Magazine*. Retrieved January 23, 2015, from http://aquaculturemag.com/magazine/december-january-2014/2014/12/23/an-audacious-hope-to-be-able-to-grow-fish-in-the-gulf-of-mexico

UNEP. (2012). *United Nations Environmental Programme, Grid-Adrenal.* Accessed March 13, 2012, at http://www.grida.no/publications/rr/food-crisis/page/3562.aspx

Upton, H. F., & Buck, E. H. (2010). Congressional research service. *Open Ocean Aquaculture.* RL32694. Retrieved November 12, 2014, from http://www.respecttheocean.org/linked/congressional_doc.pdf

Part V
Conclusion and Outlook

Chapter 14
Epilogue—Pathways Towards Sustainable Ocean Food Production

Bela H. Buck and Richard Langan

Abstract While there is a great deal of global interest in the development of combined uses of open ocean installations, for commercial scale multi-use platforms for food and energy production and other potential applications, the transition from concept to reality has yet to come to fruition. While much is known about the economics, environmental, political and societal effects of individual production sectors, there are many unknowns and challenges with regard to economics, engineering, liability and social aspects of multi-use. Mutually agreed upon principles, such as those articulated in the Bremerhaven Declaration, and EU directives and grant funding opportunities to advance research and development indicate that progress, although measured, is being made. The development of true commercial-scale multi-use offshore platforms will require investment in demonstration projects and multi-national cooperation and collaboration across public and private sectors.

14.1 Introduction

In putting together this volume, we have attempted to capture the various aspects and complexities of implementing open ocean aquaculture as an additional use of offshore platforms. As we have stated in our introduction, we see great potential for

B.H. Buck (✉)
Alfred Wegener Institute Helmholtz Centre for Polar and Marine Research (AWI),
Marine Aquaculture, Maritime Technologies and ICZM, ZMFE Zentrum für Maritime
Forschung und Entwicklung, Bussestrasse 27, 27570 Bremerhaven, Germany
e-mail: bela.h.buck@awi.de

B.H. Buck
Faculty of Applied Marine Biology, University of Applied Sciences Bremerhaven,
An der Karlstadt 8, 27568 Bremerhaven, Germany

R. Langan
School of Marine Science and Ocean Engineering, University of New Hampshire,
Judd Gregg Marine Research Complex, 29 Wentworth Road, Durham,
NH 03854, USA

B.H. Buck and R. Langan (eds.), *Aquaculture Perspective of Multi-Use Sites in the Open Ocean*, DOI 10.1007/978-3-319-51159-7_14

maximizing the productivity of ocean installations for seafood, energy and scientific pursuits, and there has been a considerable amount of thinking and discussion that has taken place globally in recent years to realize this goal. We are deeply grateful for the contributions of the co-authors of this volume, as their expertise and experience represents the state of knowledge on the topic of aquaculture as a potential multi-use of ocean platforms.

We have covered technical, operational, biological, economic, social and political considerations, as well as case studies, where aquaculture was implemented at pilot scale on offshore oil rigs and wind turbine platforms.

One important area that was not covered in this book was environmental and ecological risks associated with aquaculture. While we acknowledge that environment issues are extremely important, we felt that they have been covered exhaustively over the past three decades and that further discussion would not have added new information for this volume. A recent review paper by the US Department of Commerce's National Oceanic and Atmospheric administration (Price and Morris 2013) does an excellent job of synthesizing thirty years of literature on all environmental aspects of marine fish cage culture, so we recommend that report to readers with interest of this topic. While most environmental impacts have been associated with marine finfish culture, there are some concerns with shellfish and macroalgae culture though to a much lesser extent. These issues are also covered extensively in the literature (Fabia et al. 2009; Hatcher et al. 1994; Langan 2007, 2012; Lloyd 2003; Paul 1999; Plew et al. 2005; Price et al. 2015)

14.2   What We Have Learned

While we have extensive knowledge of the technical, social, political, environmental and biological aspects of aquaculture and offshore energy as separate entities, we have less knowledge and experience with these activities taking place within the same ocean footprint. Though there are several projects in planning stages. Particularly in the EU, most of the projects to date have been small scale and have essentially been opportunistic retrofits on existing structures or hypothetical scenarios that have yet to be realized. We do know that there are complexities in site assessment and selection as well as permitting and licensing for aquaculture and energy installations as stand-alone activities, however, we do not know if the process would become even more complex if the two activities were proposed for the same project location, though we suspect they might.

We also know that there is potential for mutual benefits and cost savings for things like maintenance and operations, though the greatest advantage would be for aquaculture due to the robust structure provided by energy platforms to serve as attachment points for e.g. fish cages and bivalve and/or seaweed longlines and potential inclusion of feeders and possibly even hatcheries and nurseries mounted on the platforms. Additionally, any power requirements of the fish, shellfish and macroalgae farms could be provided by wind turbines.

14.3 What Still Needs to Be Learned

14.3.1 Technologies and O&M

The installation of aquaculture devices offshore, either in close proximity to or direct attachment to the offshore wind farm or oil rig, needs as a first step a full complement of oceanographic, environmental and site-specific data set as well as a full understanding of the water motions on the farm structure as well as its associated candidates. To avoid the major stress on the installation induced by storm conditions the aquaculture installations should be submergible. That means that a significant part of the entire culture unit is located beneath the surface and/or in direct contact to the multi-use platform, while the bulk of the structure's mass is mounted below the surface to allow buoyancy and stabilization. This mooring component should be below the zone of turbulence and wave action (Starchild 1980). None of these technologies and system designs are available on the market, nor are they currently in development.

Depending on the species and culture designs, more insight into the operation and maintenance including deployment and harvest by taking local site-specific criteria into account. As none of these multi-use devices currently exist or are only on pilot scale, O&M strategies cannot be developed except theoretically (see Chap. 4). To get more insight into this emerging issue, large-scale multi-use platforms have to be installed (see Sect. 14.5.1 below).

For some uses such as aquaculture, floating platforms already exist. However, for other uses such as wind farms and service platforms the floating design is in development. The combination of floating platforms in a multi-use concept is a new challenge and still in its initial stages (e.g. wind farm and desalination = Stefanakou et al. 2016; wind farms and aquaculture = TROPOS 2016; MERMAID 2016).

14.3.2 Environmental Impacts

As already mentioned above, volumes of information on risks from any type of aquaculture on the local ecosystem in the nearshore and offshore realm are available. However, potential risks originating from a combination of uses following the co-use concept is not known, especially when it comes to potential cumulative effects from wind energy turbines or oil rigs in association with aquaculture. The compilation of Beiersdorf and Wollny-Goerke (2014) provides information on ecological impacts resulting from offshore wind energy installations on the seabed and its associated organisms (Gutow et al. 2014). In addition, information on potential impacts on the pelagic habitat including invertebrates, ichthyofauna (Krägefsky 2014) and mammals (Dähne et al. 2014; Skov et al. 2014) as well as on avifauna and bat fauna (Damian and Merck 2014; Hill et al. 2014; Mendel et al.

2014) is available (Krause 2014; Kühn and Schneehorst 2014). A similar collection of information regarding the impact of oil rigs on the environment exists (e.g. Kingston 1992; Olsgard and Gray 1995) as well as from other offshore structures. However, when decommissioning these structures after their expected lifetime, the impact on the ecosystem can be quite diverse. That includes the fact that restoration of habitats may lead to a severe impact of organisms associated with the reef structure (Claisse et al. 2015; see also "Rigs-to-Reefs" program in Chap. 1 of this volume; Reggio 1987). Therefore, not only the impact of structures on marine habitats alone but also in combination depending on their respective use should be taken into account in future projects and/or commercial realisation.

14.3.3 Ownership and Insurance

As already addressed in Chaps. 10 and 13 ownership is an emerging issue currently not solved. Krause et al. (2011) indicated the importance of the social dimension of the different mariculture-wind farm integration processes and how this develops with regard to the various forms of ownership and management such a venture might take. The ownership versions discussed in Chap. 13, (1) sole owner, (2) negotiated contract, and (3) legislated contract, are only three possibilities among even more. Finally, for all current offshore users the political allocation of ocean space is licensed for specific purposes only not in combination with multi-users. In the future, there should be a new version of assignments of ocean space to avoid a complex mix of ownership, associated commons and private property.

The issue of having a shared insurance or every stakeholder its own insurance has not been addressed. This is even more complicated if the multi-use installation is owned or operated by one legal entity and a problem arises with one use having an effect on the other use (e.g. aquaculture installations will be dislodged during storm conditions and get entangled in the ships propeller of the wind farm maintenance vessel). There is an urgent need to come up with potential solutions.

14.4 Future Challenges and Opportunities

The combination of several activities (e.g. renewable energy, aquaculture, maritime transport, and related services) in the same marine space, including in multi-use platforms, it is quite possible that the costs of offshore operations and the demand on the space needed can be reduced. The research on multi-use platforms funded under the EU-Initiative FP7 call "The Oceans of Tomorrow" has already provided promising designs, technological solutions and models for combining activities in terms of economic potential and environmental impact. However, before reaching a demonstration pilot stage, further research is needed, which will be funded under the umbrella of two larger calls for proposals described below.

One barrier to multi-use the offshore realm is that different environmental, safety and regulatory regimes and practices apply to different sectors and to different national jurisdictions. Furthermore, there is a lack of common understanding of the nature of operations within different sectors and the feasibility of combining these in a way that provides mutual benefits. The first proposal call, "Multi-use of the oceans' marine space, offshore and near-shore: compatibility, regulations, environmental and legal issues" (BG-03-2016, 2 million Euros per proposal) was started in November 2016 (e.g. Project No. 17 "MUSES" in Chap. 13). The challenge is to identify the real and perceived barriers to integration. Therefore, there is a need for a clear overview of compatibility, regulatory, environmental, safety, societal and legal issues within the context of the maritime spatial planning directive and how they impact on the combining of different marine and maritime activities (EU 2016).

In the second proposal call, it is postulated that technological research and innovations are needed to reduce risks for operators and investors. Therefore, the call "Multi-use of the oceans marine space, offshore and near-shore: Enabling technologies" (BG-04-2017, 8 million Euros per proposal) was initiated in October 2016. The scope of the proposals should cover aspects with regard to combinations of innovative, cost-effective technologies and methods including automation and remote monitoring technologies, flexible structures and facilities in order to test concepts of multi-use platforms leading to pilot demonstration phases. Tests of the sustainable operability of co-located maritime activities around coastal or deep sea environments as well as health and safety issues associated with multi-use marine platforms should be addressed as well as environmental and economic viability and societal acceptance (EU 2016). However, as this call more or less covers a vast collection of emerging issues (offshore to nearshore, technology, economy, environmental issues, social issues, and many suggested multi-use combinations, etc.) we recommend to focus on the most promising multi-uses resulting from previous nation and international projects.

14.5 Recommendations to Accelerate Development in the Future

14.5.1 International Test Station

Scientists working in the field of offshore platforms, either for marine bio-resources or renewable energy and oil, agree that it is a very cost-intensive plan to install a larger aquaculture multi-use installation, which needs to be scaled up from scientific pilot scale to a size that simulates commercial production. It is therefore necessary to pool national and international expertise and resources to co-finance an offshore multi-use platform to allow the next step of research and development. This is even more important for other stakeholders interested in moving offshore, such as off-shore (container) terminals for commercial shipping as well as offshore energy suppliers (used by energy consumers off the coast to avoid long sea passages to coastal harbours, such as fisheries, fish processing platforms, deep-sea mining, etc.), and other related services.

14.5.2   Bremerhaven Declaration

The initial concept for preparing the "Bremerhaven Declaration" was conceived at the "Marine Resources and Beyond Conference" conducted in 2011 and finalized in 2012 during the "International Workshop on Open Ocean Aquaculture", both held in Bremerhaven, Germany. Recognizing that the gap between demand and supply is increasing on a global scale, it is not surprising that the pressure to pursue offshore aquaculture development is growing. This is of prime importance in countries which are not able to install aquaculture operations nearshore. Some of these countries foster the concept of co-using offshore space. To better organize the wide range of scientific work in social, technological, economic and natural resources disciplines with focus on the co-use internationally, the Bremerhaven Declaration gathers issues on open ocean farming systems in co-management with the strong participation of other stakeholders interested in offshore natural resource uses, such as wind farms, oil rigs or other offshore installations. One major goal was to support inter- and transdisciplinary cooperation on local, national and international level and to avoid cost-intensive overlap of research. The participants believed that this would have the best output this agreement offers ample opportunities to bring aquaculture production to new levels.

A number of pertinent issues related to open ocean aquaculture and multi-use with other offshore installations was discussed, by

- **recognizing** that global food security, human health and overall human welfare are in serious jeopardy since the production of living marine resources for vital human foods cannot be sustained by natural fisheries production, even if these resources are properly managed at levels of optimum sustainable yields;
- **realizing** that the gap between seafood supply and demand is increasing at an alarming rate as these are nutrient-dense foods are considered extremely important for human health and well-being. On the other hand, the development of aquaculture has been remarkable and today provides more than half of all seafood destined for human consumption;
- **confirming** that conventional land-based and coastal aquaculture will continue to grow, thereby playing in the future a growing role in quality food supply. However, this much needed development will only delay the widening of the gap in seafood supply and new and modern technologies such as offshore farming systems are required to significantly assisting in closing this gap;

- **noting** that the world is too dependent, however, on aquaculture development and its exports, as aquaculture is threatened by coastal urbanization, industrialization, and water pollution. Weighing these trends we believe that it is urgent that the world develop offshore aquaculture, while complying with the FAO Code of Conduct for Responsible Fisheries and Aquaculture as well as with other environmental regulatory frameworks in support of sustainable aquaculture development;
- **finding** that Offshore aquaculture will require much higher inputs of capital but also needs a new level of cooperation from a wide range of social, technological, economic, and natural resource users;
- **discovering** that over the past decade major advances and new concepts have evolved, and several of them have been successfully tested at the pilot scale level, while others have failed.
- **learning** that these experiments and scale-up trials have led us to believe that offshore aquaculture does have substantial potential to bring global aquaculture production to new levels to meet future human needs;
- **believing** firmly that strategies need to be developed with strong participation of all affected stakeholders interested in the social-ecological design and engineering of innovative offshore aquaculture food systems;
- **recognizing** that the integration of offshore food and energy systems (e.g. aquaculture systems and windfarms; oil and gas) appear to be especially promising, but will require a high level of innovative technology, the use of marine spatial planning, and transparent, adaptive management for spatial efficiency and conflict resolution;
- **concluding** also that open ocean aquaculture if intelligently designed can be incorporate into overall cooperative fisheries restoration and management strategies.

The participants, which included a core group of the global expertise on the subject, formulated a series of specific recommendations and called upon national, international, intergovernmental agencies, as well as the industries, potential investors, scientists, regulators and NGOs of the respective countries to strongly support these recommendations with the aim to provide a healthy and environmentally sustainable bio-resource system that can substantially contribute to meet the future demands of our societies. All recommendations and justifications are summarised in Rosenthal et al. (2012a, b).

14.6   Conclusion

When we began this project several years ago, we were motivated by our firm conviction that a sustainable seafood supply for future generations is dependent on expansion of marine aquaculture and that open ocean environments are likely the best option for increasing production. We also recognized that maximizing the output of goods and services, in this case energy and food, from a human developed footprint of seafloor is key to wise ecological use of our oceans. While we concede that this is an enormous challenge that requires unprecedented commitment from a wide range of public and private sectors, we believe that the beneficial outcomes will far exceed the effort and investment. We applaud all efforts to move the multi-use concept forward and were pleased to see a recent article in an international aquaculture online publication encouraging its development (Holmyard 2016).

Our intent with producing this volume was to inform governments, research institutions and private industry about the potential for the multi-use concept, and to inspire these entities to continue the pursuit of transforming concept into reality. We hope that our efforts will contribute to the quest for creative solutions to food and energy production to the benefit of human health and well-being, and to ecologically sound and sustainable use of our oceans.

It is apparent that the orchestration of a multi-use concept, such as an integration of marine aquaculture with wind energy or oil production in the offshore realm, is still in its infancy. The major issue extracted from this volume clearly indicates that practical multifunctional use of offshore areas requires technical and economic feasibility as a basic prerequisite to assure that all operators will support a multi-use concept. Consequently, more information on the economic and technical viability of this joint venture is the key factor. If these issues will get more insight and result in best solutions, this will be a practical approach towards rationalizing marine stewardship in the offshore setting. The current initiatives on EU level are a perfect pre-condition to achieve this goal.

References

Beiersdorf, A., & Wollny-Goerke, K. (2014). *Ecological research at the offshore windfarm alpha ventus: Challenges, results and perspectives.* Wiesbaden: Springer.

Claisse, J. T., Pondella, D. J., Love, M., Zahn, L. A., Williams, C. M., & Bull, A. S. (2015). Impacts from partial removal of decommissioned oil and gas platforms on fish biomass and production on the remaining platform structure and surrounding shell mounds. *PLoS ONE, 10.* doi:10.1371/journal.pone.0135812

Dähne, M., Peschko, V., Gilles, A., Lucke, K., Adler, S., Ronnenberg, K., et al. (2014). Marine mammals and windfarms: Effects of *alpha ventus* on harbour porpoises. In A. Beiersdorf & K. Wollny-Goerke (Eds.), *Ecological research at the offshore windfarm alpha ventus: Challenges, results and perspectives.* Wiesbaden: Springer.

Damian, H. P., & Merck, T. (2014). Cumulative impacts of offshore windfarms. In A. Beiersdorf & K. Wollny-Goerke (Eds.), *Ecological research at the offshore windfarm alpha ventus: Challenges, results and perspectives.* Wiesbaden: Springer.

EU (2016) Accessed October 20, 2016, at https://ec.europa.eu/research/participants/portal/desktop/en/opportunities/h2020/calls/h2020-bg-2016-2017.html#c,topics=callIdentifier/t/H2020-BG-2016-2017/1/1/1/default-group&callStatus/t/Forthcoming/1/1/0/default-group&callStatus/t/Open/1/1/0/default-group&callStatus/t/Closed/1/1/0/default-group&+identifier/desc

Fabia, G., Manoukian, S., & Spagnoloa, A. (2009). Impact of an open-sea suspended mussel culture on macrobenthic community (Western Adriatic Sea). *Aquaculture, 289,* 54–63.

Gutow, L., Teschke, T., Schmidt, A., Dannheim, J., Krone, R., & Gusky, M. (2014). Rapid increase of benthic structural and functional diversity at the *alpha ventus* offshore test site. In A. Beiersdorf & K. Wollny-Goerke (Eds.), *Ecological research at the offshore windfarm alpha ventus: Challenges, results and perspectives*. Wiesbaden: Springer.

Hatcher, A., Grant, J., & Schofield, B. (1994). Effects of suspended mussel culture (*Mytilus spp.*) on sedimentation, benthic respiration and sediment nutrient dynamics in a coastal bay. *Marine Ecology Progress Series, 115*, 219–235.

Hill, R., Hill, K., Aumüller, R., Schulz, A., Dittmann, T., Kulemeyer, C., et al. (2014). Of birds, blades and barriers: Detecting and analysing mass migration events at *alpha ventus*. In A. Beiersdorf & K. Wollny-Goerke (Eds.), *Ecological research at the offshore windfarm alpha ventus: Challenges, results and perspectives*. Wiesbaden: Springer.

Holmyard, N. (2016). Can aquaculture flourish in a more symbiotic sea? Accessed October 19, 2016, at http://advocate.gaalliance.org/can-aquaculture-flourish-in-a-more-symbiotic-sea/ October 19, 2016.

Kingston, P. F. (1992). Impact of offshore oil production installations on the benthos of the North Sea. *ICES Journal of Marine Science, 49*, 45–53.

Krägefsky, S. (2014). Effects of the *alpha ventus* offshore test site on pelagic fish. In A. Beiersdorf & K. Wollny-Goerke (Eds.), *Ecological research at the offshore windfarm alpha ventus: Challenges, results and perspectives* (pp. 83–94). Wiesbaden: Springer.

Krause, G., Griffin, R. M., & Buck, B. H. (2011). Perceived concerns and advocated organisational structures of ownership supporting "offshore wind farm–mariculture integration". In G. Krause (Ed.), *From turbine to wind farms–technical requirements and spin-off products*. Rijeka, Croatia: InTech, Open Access Publisher.

Krause, J. (2014). Conservation features: Sensitive species and habitats in the German exclusive economic zone. In A. Beiersdorf & K. Wollny-Goerke (Eds.), *Ecological research at the offshore windfarm alpha ventus: Challenges, results and perspectives* (pp. 39–44). Wiesbaden: Springer.

Kühn, B., & Schneehorst, A. (2014). Oceanographic and geological research at *alpha ventus*: Instruments for predicting environmental conditions and interactions. In A. Beiersdorf & K. Wollny-Goerke (Eds.), *Ecological research at the offshore windfarm alpha ventus: Challenges, results and perspectives* (pp. 53–65). Wiesbaden: Springer.

Langan, R. (2007). Results of environmental monitoring at an experimental offshore farm in the Gulf of Maine: Environmental conditions after seven years of multi-species farming. In C. S. Lee & P. J. O'Bryen (Eds.), *Open ocean aquaculture—Moving forward* (pp. 57–60). Waimanalo, Hawaii: Oceanic Institute.

Langan, R. (2012). Mussel culture: Innovations in deep water. In: B. A. Costa-Pierce & G. G. Page (Eds.), *Sustainability science in aquaculture. Encyclopedia of sustainability science and technology*. doi10.1007/978-1-4419-0851-3

Lloyd, B. D. (2003). *Potential effects of mussel farming on New Zealand's marine mammals and seabirds: A discussion paper*. Wellington, NZ: Department of Conservation.

Mendel, B., Kotzerka, J., Sommerfeld, J., Schwemmer, H., Sonntag, N., & Garthe, S. (2014). Effects of the *alpha ventus* offshore test site on distribution patterns, behaviour and flight heights of seabirds. In A. Beiersdorf & K. Wollny-Goerke (Eds.), *Ecological research at the offshore windfarm alpha ventus: Challenges, results and perspectives* (pp. 95–110). Wiesbaden: Springer.

MERMAID (2016) Accessed October 27, 2016, at http://www.vliz.be/projects/mermaidproject/

Olsgard, F., & Gray, J. S. (1995). A comprehensive analysis of the effects of offshore oil and gas exploration and production on the benthic communities of the Norwegian continental shelf. *Marine Ecology Progress Series, 122*, 277–306.

Paul, W. (1999). Reducing the risk of open ocean aquaculture facilities to protected species. In NOAA (Ed.), *National strategic initiative project, summaries 1999* (pp. 1–11). Aquaculture Information Center: DOC/NOAA.

Plew, D. R., Stevens, C. L., Spigel, R. H., & Hartstein, N. D. (2005). Hydrodynamic implications of large offshore mussel farms. *IEEE Journal of Oceanic Engineering, 30*, 95–108.

Price, C. S., & Morris, J. A., Jr. (2013). *Marine cage culture and the environment: Twenty-first century science informing a sustainable industry* (Vol. 164). NOAA Technical Memorandum NOS NCCOS.

Price, C. S., Morris, J. A., Keane, E., Morin, D., Vaccaro, C. & Bean, D. (2015). *Protected Species and Mussel Longline Interactions.* NOAA Technical Memorandum NOS/NCCO.

Reggio, V. C. (1987). *Rigs-to-Reefs: The use of obsolete petroleum structures as artificial reefs.* New Orleans: U.S. Department of the Interior, Minerals Management Service, Gulf of Mexico OCS Region, (OCS Report MMS 87-0015).

Rosenthal, H., Costa-Pierce, B. A., Krause, G., & Buck, B. H. (2012a). Bremerhaven declaration on the future of global open ocean aquaculture, part I: Preamble and recommendations. Aquaculture forum on open ocean aquaculture development—From visions to reality: The future of offshore farming. Funded by: Investment in sustainable fisheries co-financed by the European Union (European Fisheries Fund—EFF), Ministry of Economics, Labour and Ports (Free Hanseatic City of Bremen), The Bremerhaven Economic Development Company Ltd.

Rosenthal, H., Costa-Pierce, B. A., Krause, G., & Buck, B.H. (2012b). Bremerhaven declaration on the future of global open ocean aquaculture—Part II: Recommendations on subject areas and justifications. Aquaculture forum on open ocean aquaculture development—From visions to reality: the future of offshore farming. Funded by: Investment in sustainable fisheries co-financed by the European Union (European Fisheries Fund—EFF), Ministry of Economics, Labour and Ports (Free Hanseatic City of Bremen), The Bremerhaven Economic Development Company Ltd.

Skov, H., Heinänen, S., Arreborg-Hansen, D., Ladage, F., Schlenz, B., Zydelis, R., et al. (2014). Marine habitat modelling for harbour porpoises in the German Bight. In A. Beiersdorf & K. Wollny-Goerke (Eds.), *Ecological research at the offshore windfarm alpha ventus: Challenges, results and perspectives* (pp. 151–169). Wiesbaden: Springer.

Starchild, A. (1980). Seaward Ho!—The ocean frontier. In Office of Technology Assessment (Ed.), *Energy from open ocean kelp farms* (pp. 83–98). University Press of the Pacific.

Stefanakou, A., Dagkinis, I., Lilas, T., Maglara A., & Vatistas, A. (2016). Development of a floating wind-desalination multi-use platform (MUP) in the context of optimal use of maritime space. In *Offshore energy and storage symposium (OSES) and industry connector event.* University of Malta, Valletta, Malta 13–15 July 2016.

TROPOS (2016) Accessed October 27, 2016, at http://www.troposplatform.eu/